Mathematical Engineering

Today, the development of high-tech systems is unthinkable without mathematical modeling and analysis of system behavior. As such, many fields in the modern engineering sciences (e.g. control engineering, communications engineering, mechanical engineering, and robotics) call for sophisticated mathematical methods in order to solve the tasks at hand.

The series Mathematical Engineering presents new or heretofore little-known methods to support engineers in finding suitable answers to their questions, presenting those methods in such manner as to make them ideally comprehensible and applicable in practice.

Therefore, the primary focus is—without neglecting mathematical accuracy—on comprehensibility and real-world applicability.

To submit a proposal or request further information, please use the PDF Proposal Form or contact directly: Dr. Thomas Ditzinger (thomas.ditzinger@springer.com)

Indexed by SCOPUS, zbMATH, SCImago.

More information about this series at http://www.springer.com/series/8445

Yuri B. Zudin

Non-equilibrium Evaporation and Condensation Processes

Analytical Solutions

Third Edition

Springer

Yuri B. Zudin
National Research Center
Kurchatov Institute
Moscow, Russia

ISSN 2192-4732 ISSN 2192-4740 (electronic)
Mathematical Engineering
ISBN 978-3-030-67552-3 ISBN 978-3-030-67553-0 (eBook)
https://doi.org/10.1007/978-3-030-67553-0

This Springer imprint is published by the registered company Springer Nature Switzerland AG
The registered company address is: Gewerbestrasse 11, 6330 Cham, Switzerland

To my beloved wife Tatiana,

who has always been my source of inspiration,

to my children Maxim and Natalya,

and to my grandchildren Alexey and Darya

Preface

The present third edition substantially augments the second edition of the book (Non-equilibrium Evaporation and Condensation Processes. Analytical Solutions, Springer, 2019) by the author. Non-equilibrium evaporation and condensation processes play an important role in a number of fundamental and applied problems. When using laser methods for the processing of materials, it is important to know the laws of both evaporation (for thermal laser ablation from the target surface) and condensation (for the interaction with the target of an expanding vapor cloud). Some accident situations in the energetic industry develop from a sudden contact of bulks of cold liquid and hot vapor. Shock interaction of two phases produces a pulse rarefaction wave in vapor accompanied by an abrupt variation of pressure in vapor and intense condensation. Spacecraft thermal protection design calls for the modeling of the depressurization of the protection cover of nuclear propulsion units. To this end, one should be capable of calculating the parameter of intense evaporation of the heat transfer medium as it discharges into vacuum. Solar radiation on a comet surface causes evaporation of its ice core with the formation of the atmosphere. Depending on the distance to the Sun, the intensity of evaporation varies widely and can be immense. The process of evaporation, which varies abruptly in time, has a substantial effect on the density of the comet atmosphere and the character of its motion.

The specific feature of intense phase transitions is the formation of the non-equilibrium Knudsen layer near the surface. In this setting, the standard gas-dynamic description within the Knudsen layer becomes illegitimate: the phenomenological parameters of the gas, as determined by statistical averaging rules, cease to have their macroscopic sense. Under non-equilibrium conditions, the joining conditions of the condensed and gaseous phases turn out to be much more involved than those adopted in the equilibrium approximation. From the consistent consideration of molecular-kinetic effects on the phase boundary, one can get important nontrivial information about the thermodynamic state of vapor under phase transitions.

An important problem of safety assurance in nuclear power plants is the calculation of the process of discharge of the heat transfer medium through pipeline ruptures. This can be accompanied by the explosive boiling of superheated liquid resulting in the substantial restructuring of the flow structure. The explosive boiling regime is manifested most vividly when the liquid attains the limit thermodynamic temperature (the spinodal temperature). This is accompanied by homogeneous nucleation (fluctuation generation of vapor bubbles in the mother phase).

Despite the fluctuation character of nucleation and the short lifetime of vapor bubbles, the phenomenon of gaseous (vapor) bubbles in liquid has many manifestations: underwater acoustics, sonoluminescence, ultrasonic diagnosis, reduction of friction by surface nanobubbles, and bubble boiling. In applications pertaining to the physics of boiling, it is required to know the dependence of the growth rate of a vapor bubble on a number of parameters: thermophysical properties of liquid and vapor, capillary, viscous, and inertial forces, and molecular-kinetic laws on the phase boundary.

Modern progress in microelectronics and nanotechnologies calls for further analysis of the behavior of the phase boundary in microscopic objects, and in particular, the behavior of the liquid–gas boundary. Here, of great value is the study of the joint action of intermolecular and surface forces, which control the motion of evaporating microscopically thin films.

The cooling of heated surfaces by droplet jets is widely spread in various engineering applications: energy industry, metallurgy, cryogenics, space engineering, and firefighting. The progress in this area is hindered by the insufficient comprehension of all the phenomena accompanying the impingement of a jet on a surface. The key problem here is the study of the interaction of liquid droplets with a rigid surface.

The problem of vapor condensation in a transverse flow past a horizontal cylinder (in contrast to the "classical" case of stationary steam on a vertical plate) is much more involved, which requires a comprehensive analysis, including the correct consideration of all forces acting on a falling condensate film. With film condensation of vapor, even a small portion of noncondensable gas can significantly reduce the intensity of heat- and mass transfers. Due to the loss of vapor due to condensation, the velocity of the steam–gas flow decreases downstream, while the concentration of the noncondensable gas increases. Therefore, the negative effect of noncondensable gas can be quite substantial, even though its initial content could be small. In contrast to the case of pure vapor condensation, it is necessary to calculate the local characteristics of the heat- and mass transfer.

Bubble boiling, which is an integral part of various techniques, is an extremely effective method for cooling solid surfaces exposed to high-intensity thermal effects (structural elements of thermonuclear fusion plants, high-power lasers, physical targets, etc.). The very strong dependence of the heat flux density on the wall overheating makes it possible to divert energy flows of huge density at a relatively small temperature difference. The unique combination of the resistance to "strong" and the instability to "weak" external influences provides a fundamental obstacle to creating a consistent theory of bubble boiling. Special attention should be paid to the debatable issue of the influence of thermophysical characteristics of a heat-transmitting wall on the characteristics of bubble boiling.

Exotic non-equilibrium effects accompany the boiling of liquid helium in the state of superfluidity, which is a macroscopic quantum state. Of fundamental interest here is the analysis of thermodynamic principles of superfluid helium from two alternative positions: the macroscopic approach, which is based on the two-fluid model, and the microscopic analysis, which depends on the quantum-mechanical model of quasi-particles.

Of special interest is also the physical concept of pseudoboiling, which describes the laws of heat exchange in the range of supercritical pressures of a single-phase liquid. The model of pseudoboiling enables one to calculate the heat exchange with turbulent flow in a channel of medium with highly variable thermophysical properties.

Gas bubbles that rise in a stationary liquid under the action of the gravity force can have shapes of a sphere, an ellipsoid of rotation flattened in the direction of the pop-up, and a spherical cap. Depending on the volume of the bubble, the trajectory of its motion can be straight, zigzag, or helical. The unusual shape and trajectory of gas bubbles have been the subject of a huge number of studies starting with Leonardo da Vinci. Experimental studies of rising bubbles cover the last 120 years. The only exact solution to the problem of rising bubbles in a liquid volume was obtained by Hadamard and Rybczynski in the case of very small bubbles. At present, the dependence of the rise velocity on the size of moderate and large gas bubbles is described by semi-empirical and empirical relations.

Helium at supercritical pressures is used as a coolant for cryogenic superconductivity-based objects. Pulsations of flow rate and temperature may develop in a flow of supercritical helium under certain combinations of regime parameters. Due to the strong dependence of the thermophysical properties of supercritical helium on the temperature during its flow in the channels, a thermohydraulic instability of the "density wave" type can occur. The theoretical analysis of this specific instability requires a physical model which is fundamentally different from the standard method of small oscillations.

In the case of a turbulent flow in channels of two-phase flows, the effect of dispersed inclusions (bubbles, droplets, etc.) on the thermal hydrodynamics of the continuous phase plays an important role. To correctly account for this effect, it is necessary to conduct a theoretical study of the process of bubble and droplet breakup as they interact with the turbulent continuous medium.

The present book is solely concerned with analytical approaches to the statement and solution of problems of this kind. The analytical approach is capable of providing a solution to the mathematical model of a physical problem in the form of compact formulas, expansions into series, and integrals over a complete family of eigenfunctions of a certain operator. The study involves the application of the available methods and the discovery of new methods of solutions of a given mathematical model of a real process, given as a differential or integral equation or a system of differential or integro-differential equations. The resulting analytical relation provides an adequate description (even for a simplified model) of the essence of a physical phenomenon. From analytical solutions, one is capable to understand and represent in a transparent form the principal laws, especially in the study of a new phenomenon or a process. This is why analytical methods are always employed in the first stage of mathematical modeling. Analytical solutions are also used as test models for the validation of results of numerical solutions.

In Chap. 1, the molecular-kinetic theory is looked upon as a link between the microscopic and macroscopic levels of the description of the structure of the material. Historical aspects of the creation by Ludwig Eduard Boltzmann of his seminal equation are discussed; we also dwell upon the discussions following this discovery. We give a precise solution to the Boltzmann equation in the case of space-homogeneous relaxation. Applied problems of intense phase transition are discussed. The problem of specifying boundary conditions on the phase interface of the condensed and gaseous phases is discussed. Methods of kinetic analysis of evaporation and condensation processes are discussed.

Chapter 2 is concerned with the non-equilibrium effects on the phase interface. We give the conservation equations of molecular flows of mass, momentum, and energy, and describe the classical problem of evaporation into a vacuum. Actual and extrapolated boundary conditions are analyzed for the gas-dynamic equations in the external domain. It is shown that in the non-equilibrium Knudsen layer (adjacent to the phase boundary), the velocity distribution function of molecules can be conventionally split into two parts. We also discuss the problem of determination of the accommodation coefficients of mass, momentum, and energy. We present the fundamentals of the linear kinetic theory. Approximate kinetic models of the strong evaporation problem are described.

Chapter 3 is devoted to the approximate kinetic analysis of strong evaporation. On basis of the mixing model, we give analytical solutions for temperatures, pressures, and mass velocities of vapor and match them with the available numerical and analytical solutions. The mechanism of reflection of molecules from the condensed-phase surface is analyzed. The effect of the condensation coefficient on the conservation equations of molecular flows of mass, momentum, and energy, and also on the thermodynamic state of the resulting vapor is studied. "Thermal conductivity in target–intensive evaporation" conjugate problem is calculated. The asymptotic behavior of the solutions in terms of the key parameters of the systems is obtained and analyzed from the physical viewpoint. The conjugate problem for the hyperbolic heat conduction equation was considered. The integral Laplace transform was applied to find an analytical solution of the hyperbolic heat conduction equation in the general case when the temperature and the heat flux on the body surface are arbitrary time functions. A general solution to the problem is constructed using the concept of the relative HTC. A two-zone approximation of the solution was given, using which the following characteristic parameters of the conjugate problem were identified: the delay time, the height of the hyperbolic shelf, and the hyperbolic and parabolic zones of the evaporation process.

Chapter 4 proposes a semi-empirical model of strong evaporation based on the linear kinetic theory. Extrapolated jumps of density and temperature on the condensed-phase surface are obtained by summing the linear and quadratic components. The expressions for the linear jumps are taken from the linear kinetic theory of evaporation. The nonlinear terms are calculated from the relations for a rarefaction shock wave with due account of the corrections for the acceleration of the egressing flow of gas. Analytical dependences of the vapor parameters in the gas-dynamic region on the Mach number, the condensation coefficient, and the number of degrees of freedom of gas molecules are put forward. For a kinetic shock evaporation wave, its model is proposed based on the composition of classical results of the linear kinetic evaporation theory and the theory of gas-dynamic shock rarefaction wave. A kinetic shock wave is shown to be stable based on the second law of thermodynamics. It is proved that the maximum entropy principle, which follows from Prigogine's theorem, is satisfied during transonic evaporation. A highly unstable solution of the inverse problem of intense evaporation is obtained. It is shown that the solution of the intense evaporation problem is conservative with respect to the method of approximation of the distribution function.

In Chap. 5, the approximate kinetic analysis of strong condensation is developed. The "mixing model" is used to calculate regimes of subsonic and supersonic condensation. Peculiarities of supersonic condensation with increased Mach number are studied: inversion of the solution, bifurcation of the solution, transition to two-valued solutions, and the limit Mach number, for which a solution exists. The effect of the condensation coefficient on the conservation equations for mass, the normal component of the momentum, and the energy of molecular flows is studied. The "condensation lock" phenomenon due to the reduced permeability of the condensed-phase surface is examined.

In Chap. 6, the mixing model is used for the analysis of linear kinetic problems of phase transition. The asymmetry of evaporation and condensation, which occurs for intensive processes, remains even for the case of linear approximation. The expressions for pressure and temperature jumps are obtained for the evaporation problem: these results almost coincide with those of the classical linear theory. The dependence of the vapor pressure on its temperature is shown as having a minimum near the margin between the anomalous and normal regimes of condensation. The results are extended to the case of diffusion reflection of molecules from the phase boundary.

Chapter 7 is concerned with the spherically symmetric growth of a vapor bubble in an infinite volume of a uniformly superheated liquid. We consider the influence of each effect within the framework of the limiting schemes. A detailed analysis of the energetic thermal scheme of a bubble is carried out. As the next step, we come to the "binary" schemes of growth that describe the simultaneous effect of two factors on the growth of a bubble. The evaporation–condensation coefficient was estimated by comparing the theoretical solution with experimental data on the growth of a vapor bubble under reduced gravity conditions. The growth mechanism of bubbles formed as a result of the homogeneous bubble nucleation is studied.

Chapter 8 is concerned with the study of the growth of a vapor bubble in the case when the superheating enthalpy exceeds the phase transition heat is considered. The Plesset–Zwick formula was extended to the region of strong superheating. It was that when the Stefan number exceeds 1 there arises a feature of the mechanism of heat input from the liquid to the vapor leading to the effect of pressure blocking in the vapor phase. To calculate the Stefan number in the metastable region, we use the scaling law of change in the isobar heat capacity. The problem of the effect of the experimental conditions on the effervescence of the butane drop is solved. An algorithm was proposed for constructing an approximate analytical solution for the range of Stefan numbers greater than unity.

Chapter 9 provides an evaporating meniscus on the interface of three phases. An approximate solution method is presented capable of finding the influence of the kinetic molecular effects on the geometric parameter of the meniscus and on the heat transfer intensity. The method depends substantially on the change of the boundary-value problem for the fourth-order differential equation (describing the thermo-hydrodynamics of the meniscus) by the Cauchy problem for a second-order equation. Analytical expressions for the evaporating meniscus parameters are obtained from the analysis of the interaction of the intermolecular, capillary, and viscous forces, and the study of the kinetic molecular effects. The latter effects are shown to depend substantially on the evaporation–condensation coefficient.

Chapter 10 is concerned with the kinetic effects for a spheroidal state. The question on the influence of the kinetic molecular effects on the drop equilibrium conditions is considered for the first time. Results of the linear kinetic theory of evaporation are used to evaluate the kinetic pressure difference due to the non-equilibrium conditions of the evaporation process. It is shown that, depending on the value of the evaporation/condensation coefficient, the kinetic pressure with respect to a drop may have either repulsing or attracting character. The analytical dependence for the thickness of the vapor film for a wide range of evaporation/condensation coefficients is found.

Chapter 11 provides vapor condensation upon the transversal flow around a cylinder. The analytical solutions for the limiting heat-exchange laws, which correspond to the effect of only one factor, are obtained under the assumption that there is no effect of the remaining factors. The results of the solution are presented as relative (with respect to the case of steady-state vapor) heat-exchange laws. The qualitative analysis of the effect of mode parameters on heat transfer upon condensation is carried out. The analysis of the limiting heat-exchange laws demonstrates their mutual interdependence, which impedes the isolation of simple asymptotics of the problem under consideration with respect to individual parameters. The mechanistic model of condensation from a vapor–gas mixture is considered. We propose an iteration-free procedure of the solution of the main equation of the mechanistic model, from which parametrical analysis with arbitrary different mass contents of the inert gas can be carried out. For high gas contents, the mechanistic model is shown to be unphysical, which is manifested in the inversion of the diffusion component of the heat flow (the diffusion paradox) and vanishing of the total heat flow (the condensation lock paradox). We develop a modified mechanistic model of condensation from a vapor–gas mixture based on the introduction of the effective heat of phase transition.

Chapter 12 describes the principal constituents of the general problem of the boiling phenomenon: conditions for the inception of boiling, formation of nucleation sites, and boiling regimes. Growth laws of a vapor bubble in a bulk of liquid and on a rigid surface are described. A microlayer of liquid under a vapor bubble, a macrolayer under vapor conglomerates, and dry spots on the heat surfaces are studied. A brief description of heat-exchange models for nucleate boiling is given; these models are based on the bubble dynamics and integral characteristics of the process. Special attention is given to a debating problem on the effect of thermophysical characteristics of a heat-transmitting wall. An approximate model for the periodic conjugate heat-exchange problem for boiling is given. The calculation results of the conjugation factor for boiling and transition boiling regimes are given.

Chapter 13 describes the superfluidity phenomenon due to the formation of "particle condensate" in one quantum state. Here, we consider specific peculiarities of heat exchange with film boiling of superfluid helium (He-II) related to molecular-kinetic effects on the phase boundary. The analysis of thermodynamic principles of He-II in the framework of the two-fluid model is carried out. A method of construction of thermodynamics from the first principles is considered. The use of the quantum-mechanical conception of quasi-particles enables us to prove the equivalence of the macroscopic and microscopic levels of He-II thermodynamics analysis.

Chapter 14 describes the heat transfer problem under a turbulent flow in a coolant channel in the zone of supercritical pressures. The modified surface renewal model was developed capable of calculating the effect of variable thermophysical properties on the friction and heat exchange. The approximation solution is shown as being legitimate in describing the general case of variation of thermophysical properties. The model was validated on problems with available solutions: flow in a turbulent boundary layer of a viscous compressible gas, a permeable wall past incompressible fluid, etc. The law of heat transfer for the turbulent flow in the channel in the zone of supercritical pressures was calculated.

In Chap. 15, the derivation of the problem of gas bubble rise in a liquid at rest under the action of the gravity force is given. We present a model of bubble rise for the region free from the effect of viscosity (moderately large and large bubbles). In the model based on the "base underpressure" concept, a bubble is assumed to be composed of two parts. The analysis of the "traditional" and "capillary" asymptotics of the dependence of the rise velocity on the equivalent radius is given. A general analytical solution of the problem is put forward, which holds for the "inviscid range" of the bubble rise velocity. The problem of the rise of the Taylor bubble in a round tube is considered. An analytical solution of the problem is obtained based on the collocation method and the asymptotical analysis of the solution to the Laplace equation. The method employed was validated on an example of the solution of the corresponding flat problem. As a result, a correct approximate solution of the rise problem of a Taylor bubble in a round tube is presented.

Chapter 16 describes the problem of the dynamics of gas bubbles in a liquid under various conditions. Based on the analysis of the Laplace equation for the velocity potential in an ideal fluid, a generalized Rayleigh equation is proposed for the dynamics of a bubble in a circular pipe. The spherical and cylindrical asymptotics are analyzed. An exact analytical solution of the bubble collapse problem in a tube is obtained. A critical analysis of the problem of homogeneous nucleation of vapor bubbles in an unlimited volume of liquid is given. A quantum-mechanical model of homogeneous nucleation is proposed. The problem of bubble size in a turbulent fluid flow is considered. A description of the Kolmogorov–Hinze model of bubble breakup, which is based on the balance of capillary and inertial forces, is given. The modified Kolmogorov–Hinze model eliminating the above paradox is presented.

The main idea of the modified model is the consideration of the balance of the forces acting on the bubble of liquid pulsations caused by the gravitational of bubble rise.

In Chap. 17, we study how the heat-exchange characteristics depend on the droplets dispersed in a turbulent steam flow in a channel. The process of droplet breakup is investigated on the basis of a semi-empirical resonance model. A final relation for the maximum stable droplet diameter is obtained. The classical Prandtl turbulence model was used to study the effect of droplets on the heat exchange. An analytical solution of the energy equation with thermal effluents caused by the droplet evaporation is obtained. The relative law of heat transfer is calculated, allowing one to determine the quantitative measure of the effect of the droplets on the heat exchange with a dispersed flow. The asymptotic variants of the analytical solution thus obtained are analyzed.

In Chap. 18, the derivation of the thermal–hydraulic "density wave"-type instability occurring in flows of supercritical helium in channels is given. It is shown that this instability is described by three dimensionless quantities: the extension parameter, the pressure parameter, and the homochronicity parameter. We consider two problems described by two parameters and distinguished by the type of pressure losses over the channel length. The Maple computer algebra software system was used to derive the exact analytical solutions determining the stability boundary, the frequency of developed perturbations, and the characteristics of growth increment and damping decrement. The asymptotical behavior of the analytical solution is studied and its approximations are constructed. It is shown that with increasing extension parameter the system crosses in succession new stability boundaries, as a result, high-frequency exponentially increasing perturbations of more and more increasing frequency come into play.

Chapter 19 presents the results of an experimental investigation of heat transfer in a pebble bed for flows of single-phase boiling liquid. The experiments involved the measurements of the temperature of a heated wall, as well as of the temperature distribution over the channel cross section at the outlet from the pebble bed. Use was made of a method of processing of experimental data, which enables one to determine the turbulent thermal conductivity without differentiation of the experimentally obtained temperature profile. The solution of unsteady-state heat equation, obtained for the conditions of the initial thermal segment, is used for this purpose. The experimental data for single-phase flow are described using the mathematical model of the process with two free parameters (the temperature on the boundary of the flow core and the turbulent thermal conductivity) and then processed by numerical optimization methods.

Appendix A considers the problem of heat transfer under film boiling. We obtain analytical solutions capable of taking into account the effects of vapor superheat in a film and the influence of the convection on the effective values of thermal conductivity and heat of phase transition of superheated vapor. Universal calculation formulas are presented describing the dependence of these values on the Stefan number for the cases of linear and parabolic distributions of velocities in the vapor film.

Appendix B analyzes the physical relation of the hydrodynamic model of pool boiling crisis and the development of instability for evaporation from the interphase boundary (the Landau instability). It is shown that in both cases, the resulting dependences for the limit heat flow are identical in both cases and differ only by a numerical coefficient. An analogy with the development of the Kelvin–Helmholtz instability is given. Calculations of the critical heat flow at liquid boiling in a pipe under conditions of a turbulent flow in a channel are performed. It is found that the use of the centrifugal acceleration in Kutateladze's formula provides a satisfactory agreement with the experimental data.

I would like to deeply thank the Director of ITLR, Series Editor Mathematical Engineering of Springer-Verlag, Prof. Dr.-Ing. habil. B. Weigand, for his strong support for my aspiration to successfully accomplish this work, as well as for his valuable advice and numerous fruitful discussions concerning all aspects of the analytical solution methods. Prof. B. Weigand repeatedly invited me to visit Institute of Aerospace Thermodynamics to perform joint research. Our collaboration was of great help to me in the preparation of this book. I am deeply indebted to Dr. T. Ditzinger, Editor of Springer-Verlag, for his interest in the publication and very good cooperation during the preparation of this manuscript.

The work on this book would be impossible without the long-term financial support of my activity at German universities (Uni Stuttgart, TU München, Uni Paderborn, HSU/UniBw Hamburg) from the German Academic Exchange Service (DAAD), from which for quarter of century I was awarded eight (!) grants. I also wish to express my sincere thanks to Dr. P. Hiller, Dr. W. Trenn, Dr. H. Finken, Dr. T. Prahl, Dr. G. Berghorn, Dr. M. Krispin, Dr. A. Hoeschen, M. Linden-Schneider, and also to all other DAAD employees both, in Bonn and in Moscow.

I would like to thank my dear wife Tatiana for her invaluable moral support for my work, especially in these tough and challenging times.

I am also thankful to Dr. A. Alimov (Moscow State University) for his very useful comments, which contributed much toward the considerable improvement of the English translation of this book.

In conclusion, I cannot but stress the most crucial role played in my career by the prominent Russian scientist Prof. Labuntsov who was my scientific advisor. I would consider my task accomplished if in this book, I was able to develop some of Prof. Labuntsov's ideas that could lead to some new modest results.

Stuttgart, Germany
November 2020

Yuri B. Zudin

Contents

Chapter 1
Introduction to the Problem

Abbreviations

BC Boundary condition
KL Knudsen layer

Symbols

f Distribution function
I Collision integral
$\boldsymbol{n}$ Unit vector
t Time
r Distance between particles
u Argument
$\mathbf{v}$ Molecular velocity
w Volume element
x Spatial coordinate

Greek Letter Symbols

μ Argument
ϕ Fourier transform of a distribution function

1.1 Kinetic Molecular Theory

Statistical mechanics (at present, the statistical physics), which is considered as a new trend in theoretical physics and is based on the description of involved systems with an infinite number of molecules, was created by Maxwell, Boltzmann, and Gibbs. An important constituent of statistical mechanics is the kinetic molecular theory, which resides on the Boltzmann integral-differential equation. In 1872, Ludwig Boltzmann published his epoch-making paper [1], in which, on the basis of his Boltzmann equation, he described the statistical distribution of the molecules of gas. The equilibrium

Y. B. Zudin, *Non-equilibrium Evaporation and Condensation Processes*,
Mathematical Engineering, https://doi.org/10.1007/978-3-030-67553-0_1

distribution function of molecules with respect to velocities, as derived by Maxwell in 1860, is a particular solution to the Boltzmann equation in the case of statistical equilibrium in the absence of external forces. The famous H-theorem, which theoretically justifies that the gas growth irreversibly in time, was formulated in [1].

Metaphysically, the kinetic molecular theory promoted the decisive choice between two alternative methods of describing the structure of matter: the continual and discrete ones. The continual approach operates with continuous medium and by no means is concerned with the detailed inner structure of matter. The system of Navier–Stokes equations is considered as its specific tool in the application to liquids. The discrete approach traditionally originates from the antique atomistic structure of matter. By the end of the nineteenth century, it was already generally adopted in chemistry. However, in the time of Boltzmann no final decision in theoretical physics was made. It may be said that Boltzmann's theory played a crucial role in the solution of this central problem: the description of the structure and properties of a substance should be based on the discrete kinetic approach.

The time period at the end of the nineteenth century is noticeable in European science by notorious philosophical discussions between the leading natural scientist. Wilhelm Ostwald, the author of "energy theory" in the natural philosophy considered energy as the only reality, while matter is only a form of its manifestation. Being skeptical about the atomic–molecular view, Ostwald interpreted all natural phenomena as various forms of energy transformation and thus brought the laws of thermodynamics to the level of philosophic generalizations. Ernst Mach, a positivist philosopher and the founder of the theory of shock waves in gas dynamics, was a great opponent of atomism. Since at his times, atoms were unobservable, Mach considered the "atomistic theory" of matter as a working hypothesis for explaining physical and chemical phenomena.

Disagreeing with the "energists" (Ostwald) and "phenomenologists" (Mach), Boltzmann, nevertheless tried to find in their approaches a positive component and sometimes spoke almost in the spirit of Max's positivism. In his paper [2], he wrote: "I felt that the controversy about whether matter or energy was the truly existent constituted a relapse into the old metaphysics which people thought had been overcome, an offence against the insight that all theoretical concepts are mental pictures".

Irrespective of the fact that Boltzmann's theory depends on the simple kinetic molecular model (which now seems quite transparent), it looked fairly challenging for many physicists 150 years ago. The principal moment of the theory is the following postulate: all phenomena in gases can be completely described in terms of interactions of elementary particles: atoms and molecules. Consideration of the motion and interaction of such particles had enabled to put forward a general conception combining the first and the second laws of thermodynamics. The crux of Boltzmann's perceptions can be expressed in a somewhat simplified form as follows [3]: atoms and molecules do really exist as elements in the outside world, and hence there is no need to artificially "generate" them from hypothetical equations. The study of the interaction of molecules on the basis of the kinetic molecular theory provides comprehensive information about gas behavior.

It is also worth pointing out that until the mid-1950s, theoretical physics contained the "caloric theory", which looked quite good from the application point of view. This theory was capable of adequately describing a number of facts, but was incapable of correctly describing transitions of various forms of energy into each other. It was the kinetic molecular theory that made it possible to ultimately and correctly solve the problem of the description of the heat phenomenon. So, from the metaphysical point of view, the kinetic molecular theory is an antithesis to both the "energetic" and the "phenomenological" approaches.

Boltzmann introduced into science the concept of the "statistical entropy", which later played a crucial role in the development of quantum theory [4]. When Planck was deriving his well-known formula on the spectral density of radiation, he first wrote it down from empirical considerations. Later, Planck obtained this formula by theoretical considerations with the help of the statistical concept of entropy. In extending this concept for the radiation of a black body, he required the conjecture of discrete portions of energy. As a result, Planck had arrived at the definition of an elementary quantum of energy with a fixed frequency. This being so, quantum theory in its modern form could not in principle be formulated without an appeal to statistical entropy [5]. A few years after Einstein, Planck introduced the concept of a quantum of light. The Bose–Einstein statistics and the Fermi–Dirac statistics both have their roots in Boltzmann's statistical method. Finally, the second law of thermodynamics (an increase of the entropy for a closed system) is obtained as an equivalent of the H-theorem.

The Boltzmann equation, which was obtained, strictly speaking, for rarefied gases, proved applicable also to the problem of the description of a dense medium. Succeeding generations of scientists investigated in this way plasmas and mixtures of gases (simple and polyatomic ones), and molecules were being considered as small solid balls. It is worth observing here that the kinetic molecular theory was a link between the microscopic and macroscopic levels of the description of matter. The solution to the Boltzmann equation by Chapman–Enskog's method of successive approximations (expansion in terms of a small parameter near the equilibrium) had enabled one to directly calculate the heat conduction and the viscosity coefficients of gases.

For many years, due to its very involved structure, the Boltzmann equation had been looked upon as a mathematical abstraction. It suffices here to mention that the Boltzmann equation involves a fivefold integral collision integral and that in it the distribution function varies in the seven-dimensional space: time, three coordinates, and tree velocities. From the applied point of view, the need for solving the Boltzmann equation was at first unclear. Various continual-based approximations proved quite successful for near-equilibrium situations.

However, in the 1950s, with the appearance of high-altitude aviation and the launch of the first artificial satellite, it became eventually clear that the description of motion in the upper atmosphere is only possible in the framework of the kinetic molecular theory. The Boltzmann equation also proved to be indispensable in vacuum-engineering applications and in the study of the motion of gases under low-pressure conditions. Later it seemed opportune to develop methods of the kinetic molecular theory in far-from-equilibrium situations (that is, for processes of high intensity).

It appeared later that the Boltzmann equation can give much more than it was expected 100 years ago. The Boltzmann equation proved capable of describing involved nonlinear far-from-equilibrium new type phenomena. It is worth noting that such phenomena were formulated originally from pure theoretical considerations as a result of the solution of some problems for the Boltzmann equation.

1.2 Discussing the Boltzmann Equation

The kinetic molecular theory depends chiefly on Boltzmann's H-theorem, which underlies the thermodynamics of irreversible processes. According to this theorem, the mean logarithm of the distribution function (the H-function) for an isolated system decreases monotonically in time. By relating the H-function to the statistical weight, Boltzmann showed that the state of heat equilibrium in a system will be the most probable. Considering as an example a perfect monatomic gas, he showed the H-function as being proportional to the entropy and derived a formula relating the entropy to the probability of a macroscopic state (Boltzmann's formula).

Boltzmann's formula directly yields the statistical interpretation of the second law of thermodynamics based on the generalized definition of the entropy. This relation unites in fact classical Carnot–Clausius thermodynamics and the kinetic molecular theory of matter. It is the probabilistic interpretation of the second law of thermodynamics that manages to reconcile the property reversibility of mechanical phenomena with the irreversible character of thermal processes. However, at first this most important location provision of statistical thermodynamics was vigorously opposed by fundamentalist scientists.

The first objections against the new Boltzmann's theory had appeared already in 1872 right after the appearance of the paper [1]. With some simplification, these objections can be phrased as follows [3].

- why do the reversible laws of mechanics (the Liouville equation) allow the irreversible evolution of a system (Boltzmann's H-theorem)?
- does the Boltzmann equation contradict classical dynamics?
- why does the symmetry of the Boltzmann equation not agree with that of the Liouville equation?

The Liouville equation, which is of primary importance for classical dynamics, features the fundamental symmetry property: the reversion of velocity leads to the same result as that for time. In contrast to this, the Boltzmann equation, which describes the evolution of the distribution function, does not have the symmetry property. The reason for this is the invariance of the collision integral in the Boltzmann equation with respect to the reversion of velocity: Boltzmann's theory does not distinguish between the collisions reversed in the positive or negative directions of time (that is, "in the past or in the future").

This remarkable property of the Boltzmann equation had led Poincaré to the conclusion that the trend in the entropy growth contradicts the fundamental laws of classical mechanics. Indeed, according to the well-known Poincaré recurrence theorem (1890) [3], after some finite time interval, any system should return to a state which is arbitrarily close to the initial one. This means that to each possible increase of the entropy (when leaving the initial state), there should correspond a decrease of the entropy (when returning to the initial state).

In 1896, Zermelo, a pupil of Planck, derived the following corollary to the Poincaré recurrence theorem: no single-valued continuous and differentiable state function (in particular, the entropy) may increase monotonically in time. It turns out that irreversible processes in classical dynamics are impossible in principle when excluding the singular initial states. Boltzmann, when raising objections to Zermelo, pointed out the statistical basis of the kinetic molecular theory, which operates with probabilistic quantities. For a statistical system, which is composed of a huge number of molecules, the recurrence time should be astronomically large and hence has negligible probability.

So, the Poincaré recurrence theorem remains valid, but in the context of a gas system it acquires the abstract sense: in reality, only irreversible processes with finite probability are realized. In 1918 Caratheodory claimed that the proof of the Poincaré recurrence theorem is insufficient, for it does not make use of Lebesgue's (1902) concept of a "measure of a set point". In reply to Zermelo's criticism, Boltzmann wrote: "Already Clausius, Maxwell and others have shown that the laws of gases have statistical character. Very frequently and with the best possible clarity I have been emphasizing that Maxwell's law of distribution of velocities of gas molecules is not the law of conventional mechanics, but rather a probabilistic law. In this connection, I also pointed out that from the viewpoint of molecular theory the second law is only a probability law…". In 1895, in reply to Kelvin's strong criticism, Boltzmann wrote: "My theorem on the minimum (or the H-theorem) and the second law of thermodynamics are only probabilistic assertions".

The discussion on the H-theorem was concluded by Boltzmann in his last lifetime publication [6]: "Even though these objections are very potent in explaining theorems of kinetic theory of gases, they by no means disprove the simple theorems of probability …The state of thermal equilibrium differs only in that that to it there correspond the most frequent distributions of vis viva between mechanical elements, whereas other states are rare, exceptional. Only by this reason, an isolated gas quantum which is in a state different from thermal equilibrium will go over into thermal equilibrium and will permanently stay there…".

In 1876 Loschmidt put forward the following fundamental objection to the kinetic molecular theory: the time-symmetric dynamic equations exclude in principle any irreversible process. Indeed, reverse collisions of molecules "mitigate" the consequences of direct collisions, and hence in theory the system should return to the initial state. Hence, following its decrease, the H-function (or the inverse entropy) must again increase from a finite value to the initial value. Correspondingly, following its growth the must again decrease. Boltzmann in his polemics with Loschmidt pointed out the conjecture of "molecular chaos", underlying his statistical approach.

According s to this conjecture, in a real situation there is no correlation of any pair of molecules prior to their collision. In a simplified form, the line of Boltzmann's reasoning is as follows.

Loschmidt's idea of intermolecular interaction postulates the existence of some "storage of information" for gas molecules in which they "store" their previous collisions. In the framework of classical dynamics, the role of such storage should be played by correlations between molecules. Let us now trace the consequences of a "time-backward" evolution of a system which is accepted by the Liouville equation. It turns out that certain molecules (however far they were at the time of velocities reversion) are "doomed" to meet at a predetermined time instant and be subject to a predetermined transformation of velocities. But this immediately implies that the reversion of velocities in time generates a highly organized system, which is antipodal to the state of molecular chaos.

This being so, Boltzmann's elegant physical considerations formally disprove Loschmidt's rigorous observation. As a result, the kinetic molecular theory had enabled to justify a passage from classical dynamics to statistical thermodynamics or, figuratively speaking, "from order to chaos ". Such a passage is most natural in rarefied gases, which determined the main domain of applicability of the Boltzmann equation.

Boltzmann's legacy is extremely broad and very deep in its contents. The philosophical idea of the atomic structure matter weaves through his work in a striking manner. He uncompromisingly defended this idea from Mach and Ostwald as representatives of the phenomenological (or "pure") description of natural phenomena. In his polemics with Ostwald, who stated that any attempts of the mechanistic interpretation of energetic laws should be rejected, Boltzmann wrote: "From the fact that the differential equations of mechanics are left unchanged by reversing the sign of time without changing anything else, Ostwald concludes that the mechanical view of the world cannot explain why natural processes already run preferentially in a definite direction. But such a view appears to me to overlook that mechanical events are determined not only by differential equations, but also by initial conditions ". In his numerous speeches and popular talks, Boltzmann always pointed out the real existence of atoms and molecules: "Thus he, who believes he can free himself from atomism by differential equations, does not see the wood for the trees... We cannot doubt that the scheme of the world, that is assumed with it, is in essence and structure atomistic".

One should also mention the original Boltzmann's idea pertaining to the time nature, which he did not succeed in bringing in the scientific form. A year before his tragic death, he wrote to the philosopher von Brentano: "I am just now occupied with determining the number which plays the same role for time as the Loschmidt number for matter, the number of time-atoms = discrete moments of time, which make up a second of time".

The synthesis between classical dynamics and the kinetic molecular theory was achieved in the 1930s. Bogolyubov [7] gave an elegant derivation of the Boltzmann equation from the Liouville equation. This derivation, which depends on the "hierarchy of characteristic times", takes into account binary collisions of molecules.

Later, Bogolyubov in collaboration with other researchers developed systematic methods capable of producing more general equations (which take into account triple and multiple collisions). These methods were subsequently used as a basis for the derivation of equations describing dense gases. According to Ruel [8]: "... La vie de Boltzmann a quelque chose de romantique. Il s'est donné la mort parce qu'il était, dans un certain sens, un raté. Et pourtant nous le considérons maintenant comme un des grands savants de son époque, bien plus grand que ceux qui furent ses opposants scientifiques. Il a vu clair avant les autres, et il a eu raison trop tôt...".

1.3 Precise Solution to the Boltzmann Equation

Numerous studies show that considerable mathematical difficulties are encountered trying to solve precisely the Boltzmann equation. Bobylev [9] seems to be the first to obtain the only known particular precise solution to the Boltzmann equation. Below we shall briefly enlarge on the results of the pioneering work [9]. In the classical kinetic theory of monatomic gases, the gas state at time $t \geq 0$ is characterized by one-particle distribution function of molecules over spatial coordinates $\boldsymbol{x}$ and velocities $\mathbf{v}$ in the three-dimensional Euclidean space: $f(\boldsymbol{x}, \mathbf{v}, t)$. With some simplification, this function can be looked upon as the number of particles (molecules) per unit volume of the velocity-configuration phase space at a time t. Its space–time evolution is described by the Boltzmann equation

$$\frac{\partial f}{\partial t} + \mathbf{v}\frac{\partial f}{\partial \boldsymbol{x}} = I[f, f]. \tag{1.1}$$

The right-hand side of (1.1), the collision integral, this is the nonlinear integral operator, which can be represented as

$$I[f, f] = \int d\boldsymbol{w} d\boldsymbol{n} g\left(u, \frac{\boldsymbol{un}}{u}\right)\left\{f(\mathbf{v}')f(\boldsymbol{w}') - f(\mathbf{v})f(\boldsymbol{w})\right\}. \tag{1.2}$$

Here, $\boldsymbol{w}$ is the volume element, $\boldsymbol{n}$ is the unit vector, and $|\boldsymbol{n}| = 1$, $d\boldsymbol{n}$ is the unit sphere surface element; the integration is taken over the entire five-dimensional space of molecular velocities. In (1.2), we used the following notation

$$\begin{aligned} &\boldsymbol{u} = \mathbf{v} - \boldsymbol{w}, u = |\boldsymbol{u}|, g(u, \boldsymbol{\mu}) = u\sigma(u, \boldsymbol{\mu}), \\ &\boldsymbol{v}' = 1/2(\mathbf{v} + \boldsymbol{w} + u\boldsymbol{n}), \boldsymbol{w}' = 1/2(\mathbf{v} + \boldsymbol{w} - u\boldsymbol{n}). \end{aligned} \tag{1.3}$$

We shall assume that the collision of molecules follow the laws of the classical mechanics of particles, which interact with the pair potential $U(r)$ where r is the distance between particles. The function $\sigma(u, \boldsymbol{\mu})$ in (1.3) is the differential scattering cross section for the angle $0 < \theta < \pi$ in the center-of-mass system of colliding molecules, where $u > 0$, $\boldsymbol{\mu} = \cos\theta$ are the arguments. The quantity

$g(u, \mu) > 0$ in (1.2) is considered as a given function, which depends on the chosen model of molecules. For the model of molecules under consideration (rigid balls of radius r_0), we have $g(u, \mu) = ur_0^2$. A more involved expression appears for the model of molecules, in which they are considered as point particles with power–law interactions

$$U(r) = \alpha / r^n (\alpha > 0, n \geq 2) g(u, \mu) = u^{1-4/n} g_n(\mu),$$

where $g_n(\mu)(1-\mu)^{3/2}$ is a bounded function.

The principal mathematical difficulties in solving the Boltzmann equation are related to the nonlinearity and involved structure of the collision integral (1.2). The very first had shown that the boundary-value problem for the Boltzmann equation is much more challenging than the initial-value problem. The problem of relaxation (an approximation to the equilibrium) can be stated in the most simple way as follows

$$\frac{\partial f}{\partial t} = I[f, f], \; f|_{t=0} = f_0(\mathbf{v}). \tag{1.4}$$

Equation (1.4) describes the space-homogeneous Cauchy problem of independent interest. Problems of existence and unique solvability of the Boltzmann equation (both for the Cauchy, and for boundary-value problems) were studied extensively. Gilbert, Chapman–Enskog, and Grad were the first to study approximate solutions by their classical methods. Various extensions of such approaches are also available. Maxwell molecules are particles interacting with the repelling potential $U(r) = \alpha / r^4$. For this model, the scattering cross section $\sigma(u, \mu)$ is inversely proportional to the absolute value of the velocity u.

Hence, the function $g(u, \mu)$ from (1.2) is independent of u, which substantially simplifies the evaluation of the collision integral. This remarkable advantage of Maxwell molecules, which was known already to Boltzmann, was for researchers. Bobylev [9] was the first to show that the nonlinear operator (1.2) can be substantially simplified by using the Fourier transform with respect to the velocity. Setting

$$\phi(\boldsymbol{x}, \boldsymbol{k}, t) = \int d\mathbf{v} \exp(-i\boldsymbol{k}\mathbf{v}) f(\boldsymbol{x}, \mathbf{v}, t) \tag{1.5}$$

and changing (1.1) to the Fourier representation, we arrive at the following equation for $\phi(\boldsymbol{x}, \boldsymbol{k}, t)$

$$\frac{\partial \phi}{\partial t} + i \frac{\partial^2 \phi}{\partial \boldsymbol{k} \partial \boldsymbol{x}} = J[\phi, \phi] = \int d\mathbf{v} \exp(-i\boldsymbol{k}\mathbf{v}) I[f, f]. \tag{1.6}$$

For any function $g(u, \mu)$ in (1.2) which is independent of u, the operator $J[\phi, \phi]$ has a much simpler form versus the operator $I[f, f]$. It is easily shown that this property is satisfied only by Maxwell molecules among all available models of molecules. This leads to a substantial simplification of the transformed Eq. (1.6). However, the appearance of the mixed derivative on the left of (1.6) does not allow one to efficiently solve the spatial-inhomogeneous problems. This impediment disappears in examining the relaxation problem (1.4), which has the form in the Fourier representation

$$\frac{\partial \phi}{\partial t} = J[\phi, \phi]. \tag{1.7}$$

Let us consider the Cauchy problem for the spatial-homogeneous Boltzmann equation

$$f_t = I[f, f] = \int d\boldsymbol{w} d\boldsymbol{n} g\left(u, \frac{\boldsymbol{un}}{u}\right)\{f(\mathbf{v}')f(\boldsymbol{w}') - f(\mathbf{v})f(\boldsymbol{w})\} \tag{1.8}$$

as written in the notation (1.3). Here, the subscript means the derivative in t. The initial condition for (1.8) reads as

$$f|_{t=0} = f_0(\mathbf{v}) : \int d\mathbf{v} f_0(\mathbf{v}) = 1, \int d\mathbf{v}\mathbf{v} f_0(\mathbf{v}) = 0, \int d\mathbf{v}\mathbf{v}^2 f_0(\mathbf{v}) = 3. \tag{1.9}$$

By the laws of conservation of the number of particles, moment and energy, the solution $f(\boldsymbol{v}, t)$ to the problems (1.8)–(1.9) satisfies the same requirements for all $t > 0$

$$\int d\mathbf{v} f(\mathbf{v}) = 1, \int d\mathbf{v}\mathbf{v} f(\mathbf{v}) = 0. \tag{1.10}$$

The corresponding Maxwellian distribution reads as

$$f_{\mathrm{M}}(\mathbf{v}) = (2\pi)^{-1/2} \exp(-\vartheta^2). \tag{1.11}$$

An approach to the solution of the above problem can be written as the following formal scheme.

- Changing to the Fourier representation

$$\phi(\boldsymbol{k}, t) = \int d\mathbf{v} f(\mathbf{v}, t) \exp(-i\boldsymbol{k}\mathbf{v}), \tag{1.12}$$

gives us, instead of (1.8), the following more simple equation

$$\phi_t = J[\phi, \phi] = \int d\boldsymbol{n} g\left(\frac{\boldsymbol{kn}}{k}\right)\left\{\phi\left(\frac{k + \boldsymbol{kn}}{2}\right)\phi\left(\frac{k - \boldsymbol{kn}}{2}\right) - \phi(0)\phi(\boldsymbol{k})\right\}. \tag{1.13}$$

- The following initial condition for (1.13) is set as

$$\phi|_{t=0} = \phi_0(\boldsymbol{k}) = \int d\mathbf{v} f_0(\mathbf{v}) \exp(-i\boldsymbol{k}\mathbf{v}),\ \phi_0|_{\boldsymbol{k}=0} = 1,\ \left.\frac{\partial \phi_0}{\partial \boldsymbol{k}}\right|_{\boldsymbol{k}=0} = 0,\ \left.\frac{\partial^2 \phi_0}{\partial \boldsymbol{k}^2}\right|_{\boldsymbol{k}=0} = -3. \tag{1.14}$$

- The solution $\phi(\boldsymbol{k}, t)$ to the problems (1.13), (1.14) is studied.
- Using the inversion formula

$$f(\mathbf{v}, t) = (2\pi)^{-3} \int d\mathbf{v}\phi(\boldsymbol{k}, t) \exp(i\boldsymbol{k}\mathbf{v}), \tag{1.15}$$

we formulate the final results for the distribution function $f(\mathbf{v}, t)$. Here, we assume that the integral (1.15) is convergent.

The Fourier analogues of formulas (1.10), (1.11) read as

$$\phi(0, t) = 1, \left.\frac{\partial\phi(\boldsymbol{k}, t)}{\partial\boldsymbol{k}}\right|_{k=0} = 0, \left.\frac{\partial^2\phi(\boldsymbol{k}, t)}{\partial\boldsymbol{k}^2}\right|_{k=0} = -3, f_{\mathrm{M}}(\boldsymbol{k}) = \exp\left(-\frac{k^2}{2}\right). \tag{1.16}$$

From the above equation, it follows that the precise solutions to the Boltzmann equation can be obtained only in very rare special cases.

1.4 Intensive Phase Change

At present, processes of intensive phase change find more and more practical applications. This involves the physics of air-dispersed systems, air dynamics, microelectronics, ecology, etc. The study of intensive phase change is relevant for the purposes of practical design of heat-exchange equipment, systems of integrated thermal protection of aircraft, and vacuum engineering. We indicate some important applications related to the intensive phase change

- simulation of the evaporation of a coolant into the vacuum under the theoretical loss of leak integrity of the protective cover of a nuclear reactor of a space vehicle,
- organization of materials–laser interaction [10] (intensive evaporation from heated segments and intensive condensation in the cooling area),
- simulation of Space Shuttle airflow during their re-entry [11].

Intensive phase change plays a governing role in engineering processes accompanying laser ablation [10]. Materials–laser interaction involves a number of mutually related physical processes: radiation transfer and absorption in a target from the condensate phase, heat transfer in a target, evaporation and condensation on the target surface, and gas dynamics of the surrounding medium.

Anisimov [12] seems to be the first to give a theoretical description of laser ablation in vacuum. Studying the non-equilibrium Knudsen Layer (KL), the author of [12] found a relation between the target temperature and the parameters of egressing vapor. Extending the approach of [12], Ytrehus [13] proposed the model of intensive evaporation. The heat model of ablation in exterior atmosphere relating the gas parameters with the radiation intensity was considered by Knight [14, 15], who examined the system of gas dynamics equations conjugated with the heat transfer equation in the target. Under this approach, the Boundary Conditions (BC) were specified from the solution of the kinetic problem of intensive evaporation. The further development in the heat model of laser ablation was related to the numerical study of radiation pulses of arbitrary form and with the study of the phase change (melting/consolidation) in a target [16, 17].

An important application of the intensive phase change is the problem of simulation of comet atmosphere [18–21]. According to modern theory, the comet core is chiefly composed of aquatic ice with an admixture of mineral particles [18]. Subject to radiation, the ice begins to evaporate, forming the inner gas–dust atmosphere. Depending on the distance from the Sun, the intensity of ice evaporation and the density of the near-core comet atmosphere vary substantially. At large distances from the Sun, in the atmosphere is small, the flow regime being free-molecular. In the Earth's orbit, the flow regime in dense regions of the atmosphere on the illuminated (day) side is described by the solid medium laws. The gas density decreases away from the comet core, the continual flow regime changing first by the transient regime, and then by the free-molecular regime. The conjugation of the gas-dynamic region with the comet surface leads to a very involved mathematical problem, for which some particular solutions are known [19–21].

However, in the general case (relaxing gas, arbitrary surface geometry, and time-variable evaporation intensity solution), the above problem has no solution. Various approximate approaches were found to be useful in setting the BC for gas-dynamic equations. The system of Navier–Stokes equations in a local plane-parallel approximation was considered in [19]. The boundary conditions on the comet core surface were set as on the rarefaction expansion shock. In [20, 21], various integrated calculation schemes were used involving the Navier–Stokes equations in the gas-dynamic region with the specification of boundary conditions in the dense flow region.

The new direction of kinetic analysis related to turbulence modeling [22, 23] seems to be quite intriguing. In this case, the solution to the Boltzmann equation is sought by expanding the distribution function into a series in Knudsen numbers, which play the role of the rarefaction parameter (the Chapman–Enskog expansion). A decrease in the Knudsen number results in a transition from stable to unstable flows, which corresponds to a transition from a laminar to a turbulent flow region.

In the subcritical (laminar) regime, the solution to the Boltzmann equation for macroscopic parameters is known to be close to the solution to the Navier–Stokes equations. In the supercritical (turbulent) region, the solution becomes both unstable and non-equilibrium. Besides, the distribution function becomes rapidly changing in time, with the viscous stress and heat transfer rates increasing discontinuously. To the increasing values of the dissipating quantities, one may correspond some values of the turbulent viscosity and the turbulent heat conduction. This being so, the Boltzmann equation is capable of giving a closed model for the description of turbulence, without requiring closing conjectures (as in the classical Reynolds equations). It is worth noting, however, that this direction of the kinetic is in an early stage of development.

The simulation of an intensive phase change depends primarily on setting the boundary conditions on the interfacial surface between the condensed and gaseous phases. From the kinetic analysis, it is known that the distribution functions of the molecules that emit from the interface, and of the molecules approaching it from the vapor, are substantially different. This results in a heavy non-equilibrium condition in the KL, which is adjacent to the interface surface and whose thickness is of the order of the mean free path of molecules. The one-dimensional problem of evaporation/condensation in a half-space for the Boltzmann equation can be obtained using the Hilbert expansion in the powers of Knudsen numbers [24]. Its solution gives the boundary conditions for the Navier–Stokes equations in the outer (with respect to the interface surface) gas volume.

Landau and Lifshitz [25] proposed an elegant way of determining the number of boundary conditions from the linear analysis of one-dimensional Euler equations. Any small gas-dynamic perturbation is split in the general case into two acoustic waves (which propagate with or against the stream) and the perturbation of the entropy (propagating with the stream). A small disturbance of the phase interface may also be resolved into the components corresponding to the above three types of linear waves. Note that the flow in the near-interface gaseous region may depend only on the waves that propagate from the interface to the gas. In this case, the number of BC will be equal to the number of the components of the velocity of the outgoing wave.

In the subsonic evaporation, there are two linear waves propagating in the vapor: one of the acoustic waves and the perturbation of the entropy. This calls for two BCs. Besides, it directly implies the physical impossibility of supersonic evaporation, in which there are no perturbations propagating from the gaseous region toward the interface [25]. For subsonic condensation, only one acoustic wave penetrates the vapor from the interface side. Hence, only one BC suffices here. At present, there is no general agreement about the implementation of supersonic condensation. Numerical studies show that for some gas parameters, no supersonic condensation is possible. The above physical considerations clearly demonstrate the asymmetry of the two alternative processes of intensive phase change: evaporation (two BCs) and condensation (one BC).

Historically, the first kinetic analysis of phase change was made on the basis of the linearized Boltzmann equation. This resulted in approximate analytical solutions underlying further theoretical studies. However, the linear analysis is only capable of providing the asymptotics for the hypothetical general solution for small departures from the equilibrium. Hence, it does not seem possible to precisely assess its range of applicability. The Boltzmann equation is known to be a very involved integro-differential equation, which is conventionally solved by employing numerical methods, which provide a powerful and continuously developing means for evaluating the parameters of intensive phase change. However, the efficiency of numerical methods is resisted by the duration of calculations and the accuracy of the solution may decrease due to the statistical noise.

Of late, new perspectives of the efficient solution of the Boltzmann equation have become available based on the parallelization of data processing. Nevertheless, so far the Boltzmann equation is conventionally replaced by its simplified analogue, and in particular, by the Krook model relaxation equation [24]. This equation, which secures important properties of the collision integral (the conservation laws, H-theorem), proved useful for describing a wide class of kinetic molecular processes in various media. The relative simplicity of the Krook equation enables one, in particular, to perform a detailed investigation of the problem of inhomogeneous gas relaxation.

In the majority of realizations of streams, in parallel with the regions described by the kinetic equation (the boundary layer, the absorbing or evaporating surface, etc.), there appear zones that are subject to the laws of the continuous medium (the principal stream for a flow in a channel, the jet nucleus). This calls for the design of hybrid schemes of numerical calculation, which combine the kinetic and gas-dynamic parts. The problem here is in constructing a general algorithm for the calculation of such composed flows.

A certain part of such flows is far from the thermodynamical equilibrium and is described by the Boltzmann equation. Another part, which is close to the equilibrium state, is described by the Navier–Stokes equations. The hybrid approximation paves the way for the investigation of a number of important problems, which are not amenable to the solution in the frameworks of the only Boltzmann equation due to numerical difficulties (of which the principal one is the considerable amount of computer time).

The natural desire to employ numerically efficient gas-dynamical models (based on the Navier–Stokes equations) for the simulation of intensive phase change leads to the solution of the following two problems

- ascertaining the application range of the gas-dynamical approach,
- statement of BC for the gas dynamics equations.

The gas-dynamical approach is incorrect in flow regions in which the continuity condition of the medium is violated. Physically this means that the length of the free path of molecules becomes comparable with the characteristic flow size. The phenomenological properties of a continuous medium become invalid also in the thin KL, which is adjacent to the evaporation surface. In this layer, the distribution function of molecules over velocities, which describes the evaporation process, changes strongly from the local equilibrium. Under ordinary circumstances, the thickness of the KL is quite small and hence can be neglected in the gas-dynamical approximation.

The difficulty here is in the statement of BC, which needs to be set on the outer boundary of the non-equilibrium KL. The approximate analytical approach to the solution of the problems of intensive phase change started to develop from the papers [12–15, 26, 27]. This approach depends on the conservation equations for molecular fluxes of mass, momentum, and energy within the KL, as well as additional physical considerations. As distinct from numerical methods, the approximate approach is capable of providing analytical solutions in the wide range of variations of the Mach number.

There are a lot of studies dealing with the numerical analysis of intensive phase changes, in which remarkable results were obtained, important both in theoretical and applied aspects. For example, the book [28] considers in detail the methods of direct numerical solution of the Boltzmann equation, describes the results of numerical simulation of classical flows (structure of a shock wave, heat transfer) and of two- and three-dimensional flows. A new class of nongradient non-equilibrium flows was found. The present book is mostly focused on the exposition of the author's approximate analytical methods for the solution of the problems of intensive phase change.

References

1. Boltzmann L (1872) Weitere Studien über das Wärmegleichgewicht unter Gasmolekülen. Sitzungsberichte der Kaiserlichen Akademie der Wissenschaften. Wien Math. Naturwiss. Classe 66:275–370. English translation: Boltzmann L (2003) Further studies on the thermal equilibrium of gas molecules. The kinetic theory of gases. History Modern Phys Sci 1:262–349
2. Boltzmann L (1900) Über die Entwicklung der Methoden der theoretischen Physik in neuerer Zeit. Jahresbericht der Deutschen Mathematiker-Vereinigung 8:71–95
3. Cercignani C (2006) Ludwig Boltzmann: the man who trusted atoms. Oxford
4. Haug H (2006) Statistische Physik - Gleichgewichtstheorie und Kinetik. 2. Auflage. Springer
5. Müller-Kirsten HJW (2013) Basics of statistical physics, 2nd edn. World Scientific
6. Boltzmann L, Nabl J (1907) Kinetische Theorie der Materie. Enzyklopiidie Math. Wissenschaften 5(1):493–557. Leipzig, Teubner
7. Bogolyubov NN (1946) Kinetic equations. J Exper Theoret Phys (in Russian) 16(8):691–702
8. Ruel D (1991) Hasard et Chaos. Princeton University Press
9. Bobylev AV (1984) Exact solutions of the nonlinear Boltzmann equation and of its models. Fluid Mech Soviet Res 13(4):105–110
10. Gusarov AV, Smurov I (2002) Gas-dynamic boundary conditions of evaporation and condensation: numerical analysis of the Knudsen layer. Phys Fluids 14(12):4242–4255
11. Micol JM (1995) Hypersonic aerodynamic/aerothermodynamic testing capabilities at langley research center: aerodynamic facilities complex. AIAA Paper 95-2107
12. Anisimov SI (1968) Vaporization of metal absorbing laser radiation. Sov Phys JETP 27(1):182–183
13. Ytrehus T (1977) Theory and experiments on gas kinetics in evaporation. In Potter JL (ed) Rarefied gas dynamics n.Y., vol 51, no 2, pp 1197–1212
14. Khight CJ (1979) Theoretical modeling of rapid surface vaporization with back pressure. AIAA J 17(5):519–523
15. Khight CJ (1982) Transient vaporization from a surface into vacuum. AIAA J 20(7):950–955
16. Ho JR, Grigoropoulos CP, Humphrey JAC (1995) Computational study of heat transfer and gas dynamics in the pulsed laser evaporation of metals. J Appl Phys 78(6):4696–4709
17. Gusarov AV, Gnedovets AG, Smurov I (2000) Gas dynamics of laser ablation: Influence of ambient atmosphere. J Appl Phys 88:4352–4364
18. Crifo JF (1994) Elements of cometary aeronomy. Curr Sci 66(7–8):583–602
19. Crifo JF, Rodionov AV (1997) The dependence of the circumnuclear come structure on the properties of the nucleus. I. Comparison between an homogeneous and an inhomogeneous spherical nucleus with application to P/Wirtanen. Icarus 127:319–353
20. Crifo JF, Rodionov AV (2000) The dependence of the circumnuclear come structure on the properties of the nucleus. IV. Structure of the night-side gas coma of a strongly sublimating nucleus. Icarus 148:464–478
21. Rodionov AV, Crifo JF, Szegö K, Lagerros J, Fulle M (2002) An advanced physical model of cometary activity. Planet Space Sci 50:983–9102
22. Aristov V (1999) Study of unstable numerical solutions of the Boltzmann equation and description of turbulence. Proc 21st Intern Symp Raref Gas Dynam Cepadues Editions 2:189–196
23. Aristov V, Ilyin O (2010) Kinetic model of the spatio-temporal turbulence. Phys Let A 374(43):4381–4438

24. Cercignani C (1990) Mathematical methods in kinetic theory. Springer
25. Landau LD, Lifshits EM (1987) Fluid mechanics. Butterworth-Heinemann
26. Labuntsov DA, Krykov AP (1979) An analysis of intensive evaporation and condensation. Int J Heat Mass Transf 22:989–1002
27. Rose JW (2000) Accurate approximate equations for intensive sub-sonic evaporation. Int J Heat Mass Transfer 43:3869–3875
28. Aristov VV (2001) Direct methods for solving the Boltzmann equation and study of non-equilibrium flows. Kluwer Academic Publishers, Dordrecht

Chapter 2
Non-equilibrium Effects on the Phase Interface

Abbreviations

BC	Boundary condition
CPS	Condensed-phase surface
DF	Distribution function
KL	Knudsen layer

Symbols

$\mathbf{c}$	Molecular velocity
c_p	Isobaric heat capacity
c_v	Isochoric heat capacity
c_f	Friction coefficient
f	Distribution function
H	Enthalpy
J	Molecular flow
j	Flow in the Navier–Stokes region
Kn	Knudsen number
k	Thermal conductivity
M	Mach number
m	Molecular mass
n	Molecular number density
p	Pressure
q	Heat flux
s	Speed ratio
St	Stanton number
T	Temperature
$\mathbf{u}_\infty$	Hydrodynamic velocity vector
u_∞	Hydrodynamic velocity
v	Thermal velocity

Y. B. Zudin, *Non-equilibrium Evaporation and Condensation Processes*,
Mathematical Engineering, https://doi.org/10.1007/978-3-030-67553-0_2

Greek Letter Symbols

α	Accommodation coefficient
β	Evaporation–Condensation coefficient
ν	Kinematic viscosity
μ	Dynamic viscosity
ρ	Density

Subscripts

e	Emitted
l	Liquid
m	State at mixing surface
r	Reflected
s	Saturation state
v	Vapor
w	State at condensed-phase surface
∞	State at infinity

2.1 Conservation Equations of Molecular Flows

The description of intense phase changes calls for the solution to the flow problem in the ambient spaces of evaporating (condensing) matter, as described by gas-dynamic equations. The specific feature of an intense phase change lies in the formation near the Condensed-Phase Surface (CPS) of the Knudsen Layer (KL) of a thickness of the order of the mean free path of molecules. The existence of the KL depends on the non-equilibrium character of evaporation (condensation) resulting in the anisotropy of the velocity Distribution Function (DF) near CPS.

In this setting, the gas-dynamic description becomes unjustified—the phenomenological gas parameters (temperature, pressure, density, and velocity), as defined according to the conventional rules of statistical averaging, lose their macroscopic sense. Such a situation can be described at a simplified level with the help of an "imaginary experiment". Assume that the KL has hypothetical micrometers of pressure and temperature. Then their readings will not agree with the statistically averaged values, but will rather depend on the structure of a micrometer. Such anomalies disappear beyond the KL, in the outer region, where the Navier–Stokes equations hold. The outer region is also called the "Navier–Stokes region".

Under a rigorous approach, one needs to specify the Boundary Condition (BC) for the equations of gas dynamics in the Navier–Stokes region on the outer boundary of the KL. To this aim, one needs to know the DF on this interface. This, in turn, leads to the problem of the solution of the Boltzmann equation in the KL. Only in this case, one may evaluate the corresponding gas-dynamic parameters as the corresponding moments of the DF. It should be also pointed out that the outer boundary of the KL is defined only up to several mean free paths of molecules. So, the setting of the

BC for the equations of gas dynamics is a highly nontrivial macroscopic problem. Its relation with the problem of microscopic parameters for the Boltzmann equation in the KL governs the specifics and difficulty of the kinetic analysis of the intense phase changes [1].

The spectrum of molecules in the KL is formed by the two oppositely directed molecular flows, the one emitted by the CPS and the one incident on it from the Navier–Stokes region. In a somewhat simplified form, the physical scheme of emission of molecules can be described as follows. Some parts of liquid molecules, which are near the CPS and are in the state of chaotic thermal motion, temporarily acquire the kinetic energy, which exceeds the bond energy of molecules. As a consequence, "fast molecules" escape from the surface into the gaseous domain. This suggests that the intensity of the surface emission of molecules is uniquely determined by the temperature T_w of the CPS. This leads to the physically plausible conjecture: the spectrum of the emitted molecules are described by the equilibrium Maxwell distribution [2, 3]

$$f_w^+ = \frac{n_w}{\pi^{3/2} \mathrm{v}_w^3} \exp\left(-\left(\frac{\mathbf{c}}{\mathrm{v}_w} \right)^2 \right). \tag{2.1}$$

Here, $n_w = p_w / k_B T_w$ is the molecular gas density, $\mathbf{c}$, $\mathbf{u}_\infty$ are, respectively, the vectors of molecular and hydrodynamic velocity, c_z is the normal component of molecular velocity,

$\mathrm{v}_\infty = \sqrt{2R_g T_\infty}$, $\mathrm{v}_w = \sqrt{2R_g T_w}$ is the thermal velocity, the index "w" denotes conditions on the CPS, the index "∞" denotes the conditions at infinity, $R_g = k_B/m$ is the individual gas constant, is the Boltzmann constant, m is the mass of the molecule.

Maxwell considered gas as an ensemble of perfectly elastic balls moving chaotically in a closed volume and colliding with each other. Ball molecules can be subdivided into three groups in terms of their velocities, in a stationary state the number of molecules in each group being constant, even though they may change their velocities after collisions. This setting suggests that in equilibrium, particles have different velocities, their velocities distributed according to the Gauss curve (the Maxwell distribution).

Using the so-obtained DF, Maxwell calculated a number of quantities of great value in transport phenomena: the number of particles in a definite range of velocities, the average velocity, and the average squared velocity. The complete DF was calculated as the product of DF for each of the coordinates.[1] The flow of molecules flying toward CPS is formed as a result of their collisions away from the CPS over the entire KL. Hence, its DF should reflect a certain averaged state of vapor in the surface region. As a result, the total DF on the CPS can be conventionally split into two parts, which "genetically" differ from each other

---

[1] This implied their independence, which at this time was unclear to many researchers and required justification (which was done later).

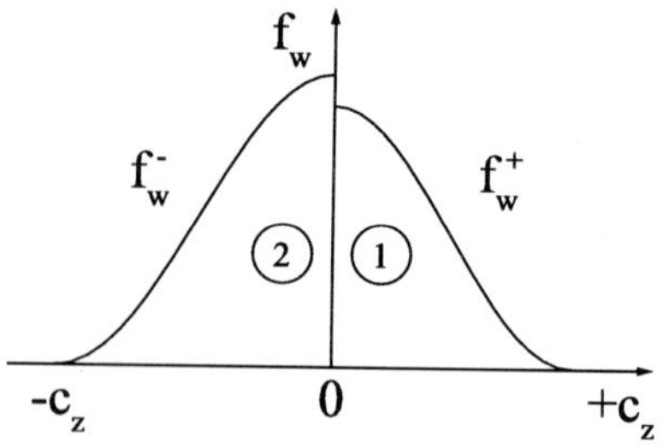

Fig. 2.1 Distribution function of molecules on the condensed-phase surface

$$\left.\begin{aligned} 0 \le c_z < \infty : f_w = f_w^+, \\ -\infty < c_z \le 0 : f_w = f_w^-. \end{aligned}\right\} \tag{2.2}$$

Thus, the distribution of molecules in terms of velocities on the CPS (Fig. 2.1) should be discontinuous. Away from the phase interface, the discontinuity in the DF is smoothed due to intermolecular collisions, the principal reconstruction occurring within the KL. Let us consider the one-dimensional problem of evaporation/condensation in the half-space for a vapor (of monatomic perfect gas) at rest. In this case, the vector of hydrodynamic velocity $\mathbf{u}_\infty$ is degenerated into the scalar velocity u_∞ in the direction of the evaporation flow. It is assumed that on the phase interface, the constant temperature T_w is maintained by virtue of an external heat surface. On the phase interface, there are coexisting molecular flows: the emitting ones J_i^+ and the incident ones J_i^-. In the equilibrium case, we have $J_i^+ = J_i^-$. For $J_i^+ \neq J_i^-$ in the Navier–Stokes region, there are flows ($j_i \neq 0$), respectively, of mass ($i = 1$), momentum ($i = 2$), and energy ($i = 3$)

$$J_1^+ - J_1^- = j_1, \tag{2.3}$$

$$J_2^+ - J_2^- = j_2, \tag{2.4}$$

$$J_3^+ - J_3^- = j_3. \tag{2.5}$$

So, the flows of evaporation/condensation in the Navier–Stokes region appear as the difference between the one-way gas flows for CPS. These net transfers are written in terms of the macroscopic parameters in the Navier–Stokes region

$$j_1 = \rho_\infty u_\infty,$$

$$j_2 = \rho_\infty u_\infty^2 + p_\infty,$$

$$j_3 = \frac{1}{2}\rho_\infty u_\infty^3 + \frac{5}{2}p_\infty u_\infty.$$

Here, ρ_∞ is the density, p_∞ is the pressure, u_∞ is the hydrodynamic velocity, and the index "∞" denotes the conditions at infinity, $J_i^+ > J_i^-$, $j_i > 0$ during evaporation, $J_i^+ < J_i^-$, $j_i < 0$ during condensation.

Equations (2.3)–(2.5) can be looked upon as conservation equations of mass, momentum, and energy for the KL.[2] According to the theory [1], the flows of quantities are calculated by integrating the DF with the corresponding over the three-dimensional field of molecular velocities c_x, c_y, c_z. Accordingly, equations for the flows are as follows: for the emitting ones J_i^+ we have

$$\left.\begin{aligned} J_1^+ &= m \int_{-\infty}^{\infty} dc_x \int_{-\infty}^{\infty} dc_y \int_{0}^{\infty} dc_z \left(c_z f_w^+\right), \\ J_2^+ &= m \int_{-\infty}^{\infty} dc_x \int_{-\infty}^{\infty} dc_y \int_{0}^{\infty} dc_z \left(c_z^2 f_w^+\right), \\ J_3^+ &= \tfrac{1}{2} m \int_{-\infty}^{\infty} dc_x \int_{-\infty}^{\infty} dc_y \int_{0}^{\infty} dc_z \left(\mathbf{c}^2 c_z f_w^+\right), \end{aligned}\right\} \tag{2.6}$$

and for the incident ones J_i^-

$$\left.\begin{aligned} J_1^- &= m \int_{-\infty}^{\infty} dc_x \int_{-\infty}^{\infty} dc_y \int_{-\infty}^{0} dc_z \left(c_z f_w^-\right), \\ J_2^- &= m \int_{-\infty}^{\infty} dc_x \int_{-\infty}^{\infty} dc_y \int_{-\infty}^{0} dc_z \left(c_z^2 f_w^-\right), \\ J_3^- &= \tfrac{1}{2} m \int_{-\infty}^{\infty} dc_x \int_{-\infty}^{\infty} dc_y \int_{-\infty}^{0} dc_z \left(\mathbf{c}^2 c_z f_w^-\right). \end{aligned}\right\} \tag{2.7}$$

Substituting the positive part f_w^+ of the DF from (2.1) into Eq. (2.6) and integrating, we arrive at the following expressions for the emitting flows

$$J_1^+ = \frac{1}{2\sqrt{\pi}} \rho_w \mathrm{v}_w, \tag{2.8}$$

$$J_2^+ = \frac{1}{4} \rho_w \mathrm{v}_w^2, \tag{2.9}$$

$$J_3^+ = \frac{1}{2\sqrt{\pi}} \rho_w \mathrm{v}_w^3. \tag{2.10}$$

The quantity J_1^+ is also known as the "one-way Maxwell flow". In order to evaluate the incident flows from integrals (2.7), one needs to specify the negative part f_w^- of the DF, which is unknown a priori. Theoretically, it can be found from the solution of the Boltzmann equation in the KL. It should be pointed out that on the hypothetical precise solution to the Boltzmann equation, the system of conservation equations becomes the system of identities by definition.

---

[2] Here, it is assumed that the mass, momentum, and energy flows in the stationary state are equal through any plane parallel to the CPS.

However, at present strong solutions of this highly involved integro-differential equation are available only in some special cases. If f_w^- is given from some or other model view, then the system of equations (2.3)–(2.5) is superdefinite. Hence, to close the problem this system must be augmented with a free parameter, which gives the semi-empirical character to the solution. It is worth noting that departures from the local thermodynamic equilibrium are manifested exclusively in the gas phase. In most cases, the non-equilibrium effects in the condensed phase can be neglected: if they nevertheless occur, this happens only with anom intensities of transport processes.

2.2 Evaporation into Vacuum

The kinetic molecular theory was founded by Maxwell [2, 3], who in 1860 obtained his famous formula (2.1) for the velocity Distribution Function (DF) for gas molecules in thermal equilibrium. In 1872, Boltzmann put forward an equation describing the statistical distribution of gas molecules (the Boltzmann equation) [4]. Function (2.1) is a particular solution to the Boltzmann equation for the case of statistical equilibrium in the absence of external forces. Being highly involved, for a long time the Boltzmann equation was considered as a mathematical abstraction. It seems that only in the 1960s it was understood that problems related to low gas density, high velocities of its motions, and with noticeable departures from the thermodynamic equilibrium can be investigated solely on the basis of the Boltzmann equation.

Historically, the first applied kinetic molecular problem was the problem of "evaporation into vacuum". In 1882, Hertz published his classical paper [5] on the evaporation of mercury at low pressure. Analyzing the results of his experiments, Hertz arrived at the following fundamental conclusion: for any substance there exists the maximal evaporation flux, which depends only on the temperature of the surface and the specific properties of a given substance. The maximal evaporation flux cannot be higher than the number of molecules of vapor that hit in unit time the surface of the condensate in the state of equilibrium. Hence, the upper limit during the evaporation is the achievement for the one-way Maxwell mass flow, as defined by relation (2.8).

In 1913, Langmuir [6] employed formula (2.8) to evaluate the vapor pressure of tungsten during its evaporation in a vacuum tube. In 1915, Knudsen [7] performed new experiments on mercury evaporation. He found that the maximal velocity of evaporation is in line with relation (2.8). However, this pertained only to highly purified mercury: the velocity of evaporation of impure mercury was found to be lower by almost three orders. To interpret these experimental data, Knudsen introduced into formula (2.8) the "evaporation coefficient" β as a cofactor

$$j_1 \equiv J_1^+ = \frac{1}{2\sqrt{\pi}}\beta\rho_w \mathrm{v}_w. \tag{2.11}$$

The evaporation coefficient shows that among all vapor molecules that hit the CPS only the part β is absorbed by it, the remaining part $1 - \beta$ of molecules is reflected from the interface and goes off into the vapor. Knudsen [7] also introduced the "condensation coefficient". In the majority of cases, one adopts the assumption that the coefficients of evaporation and condensation are equal. In the present chapter, we shall adopt this hypothesis and use the "evaporation–condensation coefficient". Taking into account the ideal gas law

$$p = \rho R_g T = \frac{1}{2}\rho \mathrm{v}^2, \tag{2.12}$$

relation (2.11) can be rewritten as

$$j_1 = \beta \frac{p_w}{\sqrt{2\pi R_g T_w}}. \tag{2.13}$$

It is very interesting that both Langmuir and Knudsen came to use formulas of (2.13) from very different positions. Langmuir was led to inquire about the evaporation flux by examining the reactions of tungsten with oxygen. Knudsen's interest in the problem of evaporation appeared in connection with the study of the dynamics of rarefied gases and in connection with the development of the effusion method for finding the vapor pressure. In the papers by Hertz, Langmuir, and Knudsen, relations (2.11) and (2.13) were interpreted as the maximal intensity of evaporation into vacuum (see the survey [8]).

It was clear for the classics of the kinetic molecular theory that in the framework of the one-dimensional problem, the stationary process of evaporation into vacuum is in reality impossible, and hence, the mass flux, as defined by formulas (2.11) and (2.13), cannot be achieved. In the actual fact, the presence of a "cloud of molecules" in the CPS (a typical expression from the early period of investigations) with density ρ_∞ and pressure p_∞ will result in a decrease in the velocity of evaporation. This decelerating effect can be taken into account by introducing into relation (2.13) the corresponding difference of pressures

$$j_1 = \beta \frac{p_w - p_\infty}{\sqrt{2\pi R_g T_w}}. \tag{2.14}$$

Relation (2.14), which is known in the literature as the Hertz–Knudsen equation, is widely useful in calculations of processes of evaporation/condensation (in particular, for experimental evaluation of β), to the present day. Relation (2.14) says that the net transfer is proportional to the difference of two one-way Maxwell flows. This introduces the following two assumptions: near a macroscopic surface, the vapor is at rest. The vapor state can be described by the local equilibrium of the Maxwell distribution.

In 1933, Risch [9] proposed a modification of the Hertz–Knudsen equation, by assuming that the flow incident to the CPS has an equilibrium Maxwell spectrum with density ρ_∞ and pressure p_∞

$$j_1 = \beta\left(\frac{p_w}{\sqrt{2\pi R_g T_w}} - \frac{p_\infty}{\sqrt{2\pi R_g T_\infty}}\right). \tag{2.15}$$

The Hertz–Knudsen equation (2.14) is a consequence of (2.15) with $T_w \approx T_\infty$. In 1956, Schrage [10] modified the empirical setup of [9] by taking into account the effect of the flow of evaporation/condensation on the molecular flow incident to the CPS

$$J_1^- = \frac{p_\infty \Gamma}{\sqrt{2\pi R_g T_\infty}}. \tag{2.16}$$

Here, $\Gamma(s) = \exp(-s^2) - \sqrt{\pi} s \operatorname{erfc}(s)$, $s = u_\infty/\mathrm{v}_\infty$ is the speed ratio, u_∞ is the hydrodynamic velocity, $\mathrm{v}_\infty = \sqrt{2R_g T_\infty}$ is the thermal velocity, (all the quantities are taken on the infinity). For $u_\infty = 0$ ($\Gamma = 1$), the relations (2.15) and (2.16) are identical. The backflow will become slower during evaporation by the motion of vapor: $u_\infty > 0 \Rightarrow \Gamma < 1$. On the other hand, during condensation the hydrodynamic velocity will be summed with the velocity of the backflow, thereby accelerating it: $u_\infty < 0 \Rightarrow \Gamma > 1$. The Hertz–Knudsen equation modified in this way reads as

$$j_1 = \beta\left(\frac{p_w}{\sqrt{2\pi R_g T_w}} - \frac{p_\infty \Gamma}{\sqrt{2\pi R_g T_\infty}}\right). \tag{2.17}$$

A detailed survey of the early period of the kinetic molecular analysis may be found in the old survey [8], which is still relevant. Here, it is worth mentioning that the aforementioned attempts to modify the Hertz–Knudsen equation on the same basis had led to unsatisfactory results. The thing is that the momentum conservation equation (2.4) and the energy conservation equation (2.5) must be satisfied in addition to the mass conservation equation (2.3) for a correct description of a phase change. An attempt to satisfy all three conservation equations within the same rigid scheme results in the superdefiniteness of mathematical description and physical absurdity. The above incorrectness of Eqs. (2.14) and (2.15), (2.17) results, in particular, in the uncertainty of the temperature T_∞. These difficulties demonstrate the necessity of a stringent kinetic molecular formulation of the problem of phase change, which would reside in the actual picture of a non-equilibrium gas state near the CPS.

An important step in the kinetic molecular study of evaporation was made in 1960 by Kucherov and Rikenglaz [11]. As distinct from the empirical approach of Schrage, Kucherov and Rikenglaz correctly took into account the actual motion of vapor in the normal direction to the surface with the velocity u_∞ and wrote the DF of the molecular backflow in the form of the displaced Maxwell distribution

$$f_w^- = \frac{n_w}{\pi^{3/2} \mathrm{v}_w^3} \exp\left(-\left(\frac{\mathbf{c} - u_\infty}{\mathrm{v}_w}\right)^2\right). \tag{2.18}$$

The function f_w^- is also called the "volume DF". This function and various modifications thereof were successfully used in the majority of theoretical studies on the processes of evaporation and condensation from the positions of the kinetic molecular theory. Parameters of strong evaporation were numerically calculated in [12–14]. In [12], the method of molecular dynamics was employed to demonstrate the legitimacy of using the equilibrium Maxwell spectrum for the emitted flow (relation (2.1)). In [13, 14], the Lennard-Jones potential of intermolecular interaction was used to show that the molecular flow of mass exceeds 3.6 times the quantity calculated from the Hertz–Knudsen equation. The conclusion [12] on the equilibrium Maxwell spectrum of the emitted flow was also justified. In a number of papers, the inadequacy of the real physical process of collisions was numerically demonstrated, which occurs due to the fact that the frequency of collisions does not depend on the velocities of collided particles.

2.3 Extrapolated Boundary Conditions

The degree of rarefaction of gas near the CPS is characterized by the Knudsen number $\mathrm{Kn} = l_{\mathrm{mol}}/l_0$ [1]. Here, l_0 is the linear scale of the region in which transport processes in the gas phase occur (the thickness of the Prandtl boundary layer and the transverse section of the channel), and l_{mol} is the mean free path of molecules. In the limit case, $\mathrm{Kn} \gg 1$ (actually, already for $\mathrm{Kn} \geq 1$), the gas flow can be calculated without the consideration of collisions of molecules between each other, taking into account only impacts of molecules on the CPS. Such a flow regime, which is also called the free molecule regime, is manifested in practice already when $\mathrm{Kn} \sim 1$. In the limit case, $\mathrm{Kn} \ll 1$ (actually, already for $\mathrm{Kn} \leq 0.1$), the phenomenological prescriptions of continuum mechanics hold in the gas region. In such a continuum regime, the thickness of the KL is immaterial in comparison with the macroscopic geometrical scales: $l_{\mathrm{mol}} \ll l_0$. Here, the flow can be calculated on the basis of the Navier–Stokes equations. However, the BCs for them are obtained by gluing together the KL with the Navier–Stokes region.

The continual method is approximately valid and is used in practice already for $\mathrm{Kn} \leq 10^{-3}$. In the range $10^{-3} \leq \mathrm{Kn} < 1$, various flow regimes of the rarefied gas, which lie between the free molecule regime and the continuum regime, are realized. Outside the KL, the phenomenological laws of heat transfer (Fourier's law) and momentum transfer (Newton's friction law) now apply. The scheme of the region near the CPS (see Fig. 2.2) consists of the KL (I) with the adjacent Navier–Stokes region (II).

In comparison with the outer linear scale $l_{\mathrm{mol}} \ll l_0$, the KL in the majority of cases has vanishingly small thickness. Hence, the detailed description of the fields

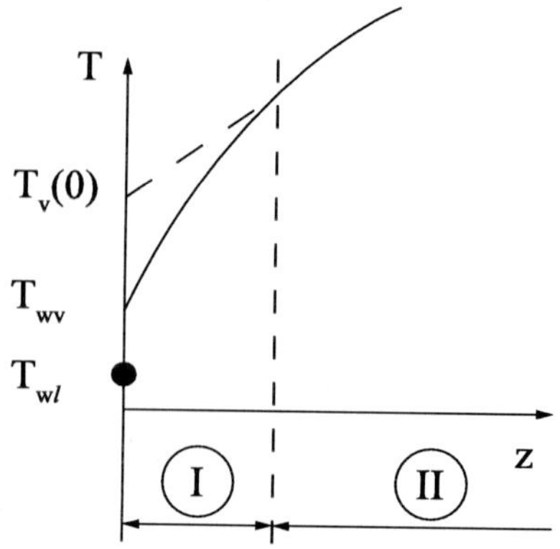

Fig. 2.2 Temperature distribution in the region near the condensed-phase surface. I. The Knudsen layer and II. the Navier–Stokes region

of quantities (velocities, temperatures, pressures, and densities) in region I is purely theoretical. If these parameters are imagined to be extrapolated from the transverse coordinate up to the CPS, ignoring the KL, we obtain some conditional values of the quantities. The difference between these extrapolated values from the true parameters results in the appearance of "kinetic jumps" on the CPS (which, of course, have only conditional character). It is worth remarking that the ultimate purpose of the applied kinetic molecular analysis is the evaluation of the extrapolated BC.

As an example, we shall consider the heat transfer through an impermeable surface. Figure 2.2 schematically shows the actual distribution of the gas temperature near the CPS, including the KL. The extrapolation of the temperature profile from the Navier–Stokes region is shown by dotted lines. As a result, on the CPS we have two temperature jumps: the actual $\Delta T_w = T_{wl} - T_{wv}$ and the extrapolated $\Delta T = T_{wl} - T_v(0)$ ones. Here, T_{wv}, T_{wl} is the actual gas temperature on the surface, respectively, from the gas and liquid side, $T_v(0)$ is the extrapolated value of the gas temperature on the surface (the conditional quantity). Thus, we have two alternative settings of the problem.

- The minimum program. To set the BC on the CPS for the equations of gas dynamics in the Navier–Stokes region, it suffices to specify the extrapolated temperature jump. Hence, giving the available conditional gas temperature, one may construct the distribution of temperatures over the entire volume of gas (except the KL), which is important in practice.
- The maximum program. To find the distribution of temperatures within the KL, one needs to solve the Boltzmann equation for the DF. Next, using it in the corresponding integral as a weight function, one may in theory obtain the precise profile of temperatures in the entire gas space. Since the Boltzmann equation is highly involved, on a certain value of the transverse coordinate one has to stop the process of solution (which can be only numerical) and "sew" the KL with the Navier–Stokes region.

It is worth pointing out that it frequently happens that in the literature on applied problems, no mention of the true parameters (or on actual jumps) is given. In the present chapter, we shall be concerned only with analytical solutions, as a part of the minimum program. This means that the purpose of the solution is to find conditional temperature jumps, whereas by parameters of the gas on the CPS we shall mean the corresponding extrapolated quantities.

2.4 Accommodation Coefficients

The concept of the accommodation coefficient was first introduced by Maxwell [2, 3], who considered two limit variants: all the molecules incident on the CPS are completely absorbed by it. All the molecules incident on the CPS are completely reflected by it. In accordance with the conservation equations of mass, momentum, and energy, one may define three corresponding accommodation coefficients. Knudsen [7] gave a concrete physical form to Maxwell's perceptions. In particular, he wrote DB in the following form

$$f^+ = f_e + (1 - \beta) f_r. \tag{2.19}$$

From relation (2.19), it follows that in the general case, among all molecules incident on the CPS, only their part defined by the quantity β is absorbed by the surface, while the other part $(1 - \beta)$ of the molecules is reflected from it. Hence, the function f^+ of the molecules flying out from the CPS can be represented in two parts. The first part (f_e) describes the evaporated molecules, while the second part (f_r) describes the molecules reflected from the interface. Knudsen calls β the coefficient of evaporation (as the vapor moves away from the CPS) or the coefficient of condensation (as the vapor moves away toward the CPS). Note that the quantity β in formula (2.19) is the accommodation coefficient of mass.

In a similar way, below we shall use the concepts of the momentum coefficients and the thermal accommodation coefficient. Knudsen [7] introduced the following definitions: the condensation coefficient is the ratio of the number of molecules absorbed by the surface to the total number of the molecules incident on it. The evaporation coefficient is the ratio of the flow molecules emitted by the interface to the number of flow molecules generated by the CPS in the reference case: the equilibrium Maxwell distribution, the vapor density corresponds to the CPS temperature on the saturation line.

Knudsen's scheme of evaporation is known as the "diffusion scheme". At present, there are very different approaches to the theoretical definition of the coefficients of evaporation and condensation. For a survey of theoretical and experimental studies of the evaporation–condensation coefficient of water, we refer the reader to the paper [15]. In [16], the quantity β was defined on the basis of the transition state theory, which in turn resides on the barrier of potential energy between vapor and liquid molecules. Nagayama and Tsuruta [16] combined the potential energy of molecules and the activation energy barrier. In order to transfer a molecule from one phase into a different one, it has to gain (for evaporation) or lose (for condensation) the activation energy.

In a number of papers [10, 16–19], the condensation coefficient was modeled by the molecular dynamics method. Schrage [10] points out the simplified character of the calculation of β from the quantity of vapor molecules reflected from the CPS. Instead, they calculated this quantity by analyzing the energy exchange between the gas and liquid molecules. Tsuruta et al. [17] used the energy criterion as a condition for the surface to capture an incident molecule: the kinetic energy of a gas molecule must decrease discontinuously to the energy of heat motion of liquid molecules. In [13], the method of molecular dynamics was employed to model various variants of processes of evaporation/condensation: a pure liquid in equilibrium and non-equilibrium conditions, liquid mixture.

Four main types of behaviors of gas molecules near the CPS were considered: (a) evaporation, (b) reflection, (c) condensation, and (d) molecular exchange. The quantity β was shown by calculations to markedly depart from the temperature. Matsumoto [19] expressed skepticism about the standard definition of the evaporation–condensation coefficient for the case of intense phase change. In 1916, Langmuir [20] was the first to perform a theoretical analysis of condensation with due account of the energy exchange of gas and liquid molecules. He assumed that the time of energy exchange on the CPS is equal in order to the period of oscillations of liquid molecules near the equilibrium. Since this period is extremely small, the energy exchange takes place practically instantaneously, which implies that $\beta \approx 1$. It is worth noting that a similar result was obtained by modern modeling of the process of condensation by the molecular dynamics method [21].

A considerable number of papers were concerned with the experimental evaluation of the evaporation and condensation coefficients. The experimental results from the survey [15] give the range $\beta \approx 6*10^{-3}$–1.0 for the condensation coefficient. A slightly more narrow range of experimental data is given in [21]: $\beta \approx 10^{-2}$–1.0. It may be assumed that such a wide scatter in the experimental results is indicative of the dependence of the result on the method of measurement. The generally accepted method of experimental evaluation of the condensation coefficient is based on the kinetic molecular model. Besides, this method postulates that there exists a clear geometric boundary between gas and liquid. In reality, instead of modeling an "infinitely thin" CPS, there exists a thin (of the scale comparable to the thickness of the KL) transient layer in which the medium density changes monotonically from the liquid state to the gas state. According to [18], the quantity of "blurriness" of the phase interface amounts to several distances between molecules in liquid.

At first sight, it seems attractive to analyze the characteristics of the phase interface in the framework of continuum mechanics. In this case, it is absolutely correct to consider the CPS as a geometrical line that has no thickness. Indeed, here the scales of blurriness of the vapor–liquid boundary are always negligible in comparison with the characteristic linear scale in the Navier–Stokes equations. However, in this case we come to a different contradiction: the standard definition of β becomes meaningless.

Indeed, the molecules incident on CPS may be decelerated by numerous interactions with the "vapor cloud" long before they reach the CPS. In this case, the

CPS cannot be considered as the only source of reflected molecules. Besides, in a certain limit situation, a molecular flow flying away from the CPS due to evaporation completely reflects the molecular flow flying toward the interface. In this hypothetical variant none of the gas molecules will reach the CPS, hence, the experiment should yield $\beta \to 0$. Conversely, if all the molecules flying toward CPS "adhere" to it during the time of kinetic relaxation, then the opposite limit variant should be implemented: $\beta \to 1$.

In the survey [22], on evaporation into vacuum it is noted that the Hertz–Knudsen equation is frequently used up until now to evaluate the evaporation–condensation coefficient. Under this approach, the departure between the calculated values may be as high as three orders. Julin et al. [21] analyzed the possible causes for such a wide scatter. The analysis of a great number of theoretical and experimental papers and studies by the molecular dynamics method has shown that the Hertz–Knudsen equation is unreliable.

Indeed, this equation reflects only one of the three conservation equations—the conservation law of mass flux—and it does not take into account the conservation laws of momentum and energy. Julin et al. [21] also put forward a modified Hertz–Knudsen equation, the results of 127 experiments on the evaporation of water and ethanol being used to justify this equation. A survey of various methods for measuring evaporation–condensation coefficients may be found in [23].

2.5 Linear Kinetic Theory

The quantitative measure of the intensity of a phase change is the speed ratio s, which is the ratio of the absolute value of the velocity of vapor motion and the most probable thermal velocity of molecules

$$s = \frac{u_\infty}{\sqrt{2R_g T}}. \tag{2.20}$$

This quantity is close to the Mach number

$$\mathrm{M} = \frac{u_\infty}{\sqrt{(c_p/c_\mathrm{v})R_g T_\infty}}$$

and is related to it as follows

$$s = \sqrt{\frac{c_p}{2c_\mathrm{v}}}\mathrm{M}. \tag{2.21}$$

Here, c_p, c_v are, respectively, the isochoric and isobaric heat capacities of gas.

For a number of applications, the intensity of transfer processes is quite small in comparison with that of molecular mixing. Hence, for the kinetic molecular analysis it is admissible to use only the first powers of the departure of parameters from equilibrium and drop higher powers. Such a method is known as linearization; the linear kinetic theory is the non-equilibrium theory based on this method.

The linear kinetic theory of evaporation/condensation was first developed by Labuntsov [24] and Muratova and Labuntsov [25]. The authors [25] were concerned with the solution of the Boltzmann equation (by the method of moments) and of the Krook model relaxation equation. The main results of [25], as obtained in several variants by numerical solution of equations, determine the fields of actual and extrapolated parameters in the KL. Moreover, the difference between the solutions on the basis of the Boltzmann and Krook equations was found to be immaterial. The linear kinetic theory was developed later, in particular, in the papers [26–29]. Below, we shall consider some results obtained in the pioneering paper [25].

In the absence of phase change, we have an impermeable phase interface. This can be either liquid or a hard surface. Here there is no mass transport, while the heat is transported through the interface according to the mechanism of heat conductivity. In this case, the linear kinetic theory gives the conclusion that the gas temperature on the surface $T_v(0)$ does not agree with that of the condensed phase on the boundary: $T_{wl} \neq T_l(0)$ (Fig. 2.2). The temperature jump $T_l(0) - T_v(0)$ is found to be proportional to the near-surface heat flux of the gas phase: $q = -k(\partial T/\partial x)_{x=0}$. The quantity $T_l(0) - T_v(0)$ also depends on the thermal accommodation coefficient α, which reflects the efficiency of the energy exchange when gas molecules interact with CPS. The concluding relation of linear kinetic theory reads as

$$\theta(0) = \sqrt{\pi}\frac{1 - 0.41\alpha}{\alpha}\tilde{q}. \tag{2.22}$$

Here, $\theta(0) = \frac{T_l(0) - T_v(0)}{T}$ is the dimensionless temperature jump on the surface, and $\tilde{q} = \frac{q}{q_*} = \frac{q}{p_v \text{v}}$ is the dimensionless heat flux on the surface.

In the frameworks of linear analysis, as a characteristic temperature T is involved in the dimensionless parameters one may take any of the temperatures phases, that is, $T \approx T_l(0) \approx T_v(0)$. The heat flux $\tilde{q}$ in relation (2.22) is considered positive if the heat is transferred from the interface toward the gas. With $\alpha = 1$, relation (2.22) implies that

$$\theta(0) = 1.05\tilde{q}. \tag{2.23}$$

The scaled quantity for the heat flux q_* is proportional to the one-way flow of energy transported through the unit reference surface due to the heat motion of gas molecules

$$q_* = p_v \text{v} = R_g \rho_v T_v.$$

Hence, the relation $\tilde{q} = q/q_*$ can be looked upon as the non-equilibrium parameter in the process of heat transport in gas. Relation (2.23) is the consistency condition on the CPS, which refines the approximate equilibrium relation: $\theta(0) = 0$. It is quite natural that when the equilibrium approximation is more justified, the less is the value of the non-equilibrium parameter $\tilde{q}$. The decrease of the gas pressure with q = const will result in an increase in the temperature jump $T_l(0) - T_v(0)$ thanks to a decrease in q_*.

Let us now consider an external flow of a surface by a high-velocity gas flow. Note that the relation for the heat flux can be expressed in terms of the Stanton number St

$$q = \rho_v u_{v\infty}(H_{vc} - H_{v\infty})\mathrm{St}, \tag{2.24}$$

where $H_{vc} - H_{v\infty}$ is the difference in the total gas enthalpies. We have the following estimate relations

$$\frac{u_{v\infty}}{v_\infty} \equiv s \approx \mathrm{M},\ R_g \approx c_{pv}. \tag{2.25}$$

Using Eq. (2.25), this gives

$$\frac{c_{pv}(T_l(0) - T_v(0))}{H_{vc} - H_{v\infty}} \approx \mathrm{St}\,\mathrm{M}. \tag{2.26}$$

Thus, an increase in the Mach number leads to an increase in the temperature jump. The flow of viscous gas along an impenetrable interface results in the transport through it of the tangential component of the momentum, which is responsible for the appearance of the friction stress. According to the kinetic molecular description, in the actual fact the gas velocity on the interface (in the frame where it is fixed) is not zero, as is adopted in the equilibrium scheme (Fig. 2.3).

The linear theory shows that the gas velocity on the interface $u_v(0)$ (which is called the "slip velocity") is proportional to the tangential stress on the surface τ. Collision and reflection of molecules with the interface results in a loss of the longitudinal component of the momentum. In this case, we have

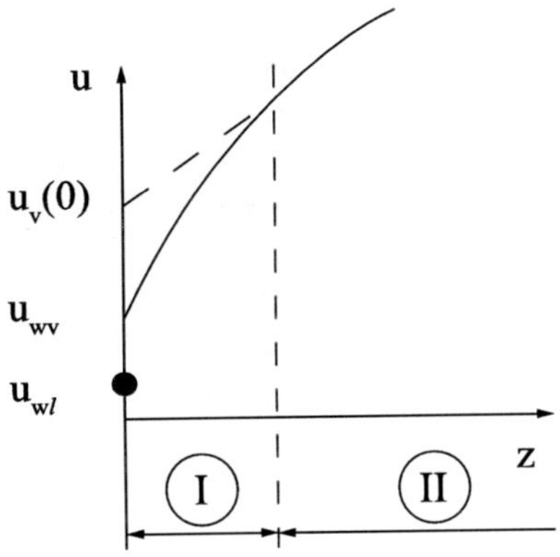

Fig. 2.3 Velocity distribution in the region near the condensed-phase surface. I. The Knudsen layer and II. the Navier–Stokes region

$$\tilde{u}_{\rm v}(0) = \tilde{\tau}. \tag{2.27}$$

Here, $\tilde{u}_{\rm v}(0) = u_{\rm v}(0)/{\rm v}$ is the dimensionless slip velocity, $\tilde{\tau} = \tau/p_{\rm v} = 2\tau/\rho_{\rm v}{\rm v}^2$ is the dimensionless tangential stress. Formula (2.27) shows that for small values of the non-equilibrium parameter ($\tilde{\tau} \equiv \tau/p_{\rm v} \ll 1$), the state will be close to equilibrium ($u_{\rm v}(0) = 0$). A decrease in the pressure in the system with $\tilde{\tau}$ = const increases the slip velocity. Let us now express the tangential stress in terms of the friction coefficient c_f

$$\tau = \frac{c_f}{2}\rho_{\rm v}u_{{\rm v}\infty}^2. \tag{2.28}$$

Using relation (2.28) in formula (2.27) and taking into account the approximate estimate $u_{{\rm v}\infty}/{\rm v} \approx {\rm M}$, we have

$$\frac{u_{\rm v}(0)}{u_{{\rm v}\infty}} \approx c_f{\rm M}. \tag{2.29}$$

So, $u_{\rm v}(0)/u_{{\rm v}\infty}$ increases with the Mach number. It immediately follows that the slip phenomena are considerable during flights of high-speed planes and space vehicles. In this case, due to the high rarity of the atmosphere, the kinematical viscosity $\nu_{\rm v} = \mu_{\rm v}/\rho_{\rm v}$ will be anomalously high. Hence, the flow pattern of an aircraft surface may prove to be laminar even for very high motion speeds. Since for a laminar flow we have $c_f \sim 1/\sqrt{\rho_{\rm v}}$, the friction coefficient will increase as the gas density decreases.

The BCs during evaporation and condensation on a phase interface are found to be much more involved that those assumed in the equilibrium approximation. In order to consider the results of the kinetic molecular description, it is appropriate to introduce the following quantities.

- $T_l(0)$ is the surface temperature of a condensed phase.
- $T_l(0)$ is the saturated pressure corresponding to the surface temperature, that is, $p_{ws} = p_{ws}(T_l(0))$.
- $p_{{\rm v}\infty}$ is the actual vapor pressure near the surface (beyond the KL).
- $T_{\rm v}(0)$ is the vapor temperature on the CPS (the extrapolated value).
- j is the flow of substance crossing a unit area on the CPS.
- q is the heat flux crossing a unit area on the CPS (positive values of j and q correspond to flows delivered in the vapor phase).
- β is the evaporation–condensation coefficient.
- It is worth pointing out that p_{ws} is a purely theoretical value, which may be different from the actual pressure in the system. The results of the linear theory [25] may be briefly summarized as follows.
- The pressure within the KL is constant and equal to $p_{v\infty}$, so that the condensed phase is under the same pressure as the vapor (without consideration of the surface tension on the curved boundary).

- Let T_s be the theoretical saturation temperature with the actual pressure in the vapor phase $p_{v\infty}$. Then both $T_l(0)$ and T are different from T_s.
- On the interface surface there is a temperature jump, which is proportional to the flows of mass j and heat q.
- The process is characterized by the difference $p_{ws} - p_{v\infty}$, which is the difference between the actual pressure $p_{v\infty}$ in the system and the calculated saturation pressure p_{ws}, as defined from the temperature $T_l(0)$ of the CPS.

Quantitative relations of the linear theory can be conveniently written down using the following dimensionless quantities:

- the heat flux $\tilde{q} = \frac{q}{q_*} = \frac{q}{p_{v\infty} v}$,
- the mass flow $\tilde{j} = \frac{j}{\rho_v v} = \frac{u_v}{v}$,
- the temperature $\theta(0) = \frac{T_l(0) - T_v(0)}{T}$,
- the pressure difference $\Delta\tilde{p} = \frac{p_{ws} - p_{v\infty}}{p}$.

Using the above notation we may write down the special BC, which take into account the non-equilibrium effects on the CPS

$$\theta(0) = 0.45\tilde{j} + 1.05\tilde{q}, \tag{2.30}$$

$$\Delta\tilde{p} = 2\sqrt{\pi}\frac{1 - 0.4\beta}{\beta}\tilde{j} + 0.44\tilde{q}. \tag{2.31}$$

Relations (2.30) and (2.31) are necessary refinements of the equilibrium consistency condition. They contain very interesting information about the specifics of non-equilibrium phenomena during phase transitions. Assume that there is no mass flux through the CPS $\left(\tilde{j} = 0\right)$. Then relations (2.30) and (2.31) describe the temperature and pressure jumps on an impenetrable CPS

$$\theta(0) = 1.05\tilde{q}, \tag{2.32}$$

$$\Delta\tilde{p} = 0.44\tilde{q}. \tag{2.33}$$

Under standard conditions of a phase transition of finite intensity, when there are no considerable overheats of vapor away from the CPS, we have the condition: $\tilde{q} \ll \tilde{j}$. Besides, relations (2.30) and (2.31) assume the form

$$\theta(0) = 0.45\tilde{j}, \tag{2.34}$$

$$\Delta\tilde{p} = 2\sqrt{\pi}\frac{1 - 0.399\beta}{\beta}\tilde{j}. \tag{2.35}$$

Using the Clausius–Clapeyron relation, we express the quantity $p_{ws} - p_{v\infty}$ in terms of the corresponding difference of temperatures $T_l(0) - T$

$$\left(\frac{dp}{dT}\right)_s = \frac{L\rho_v\rho_l}{(\rho_l - \rho_v)T} \approx \frac{L\rho_v}{T} = \frac{Lp_v}{R_g T^2}, \tag{2.36}$$

$$p_{sc} - p_{v\infty} = \left(\frac{dp}{dT}\right)_s (T_l(0) - T_s) = \frac{L}{R_g T}\frac{T_l(0) - T_s}{T} p_v. \tag{2.37}$$

Here, L is the heat of phase transition. Hence, relation (2.35) assumes the form

$$\frac{T_l(0) - T_s}{T} = 2\sqrt{\pi}\frac{1 - 0.4\beta}{\beta}\frac{R_g T}{L}\tilde{j}. \tag{2.38}$$

Using relations (2.32) and (2.38), we have

$$\frac{T_s - T_v(0)}{T} = \left(0.45 - 2\sqrt{\pi}\frac{1 - 0.4\beta}{\beta}\frac{R_g T}{L}\right)\tilde{j}. \tag{2.39}$$

From formula (2.38) it is seen that for evaporation $\left(\tilde{j} > 0\right)$ we have $T_l(0) > T_s$, while for condensation $\left(\tilde{j} < 0\right)$ we have $T_l(0) < T_s$. So, we have proved that the temperature on the CPS with evaporation is larger (for condensation, is lower) than the saturation temperature with the actual pressure in the system. This is a physically natural conclusion. From (2.39) it is seen that, in the same processes, the sign of the difference $T_s - T_v(0)$, and hence, of the vapor temperature $T_v(0)$ on the CPS, depends on the sign of the bracketed expression on the right of (2.39). This quantity can be estimated using the well-known Trouton's rule: $L/R_g T \approx 10$ (under normal conditions). It turns out that the resulting signs might be different depending on the value of the evaporation–condensation coefficient. For $\beta = 1$ we have

$$0.45 - 2\sqrt{\pi}\frac{1 - 0.4\beta}{\beta}\frac{R_g T}{L} \approx 0.237 > 0. \tag{2.40}$$

Hence, for $\beta = 1$, evaporation $\left(\tilde{j} > 0\right)$ results in a subcooled vapor: $T_v(0) < T_s$. During condensation, the vapor on the CPS is superheated: $T_v(0) > T_s$. The difference in the temperatures decreases with β: it is zero for $\beta \approx 0.6$ and then changes the sign. So, with $\beta \approx 0.3$ in the case of evaporation we have $T_v(0) > T_s$, for condensation, $T_v(0) < T_s$.

Figure 2.4 shows the relative position of the temperatures $T_l(0)$, $T_v(0)$, and T_s during evaporation and condensation with various values of β, by the results of the above analysis. Relations (2.37) and (2.38) show that the quantity

$$\tilde{j} = \frac{j}{\rho_v v} \tag{2.41}$$

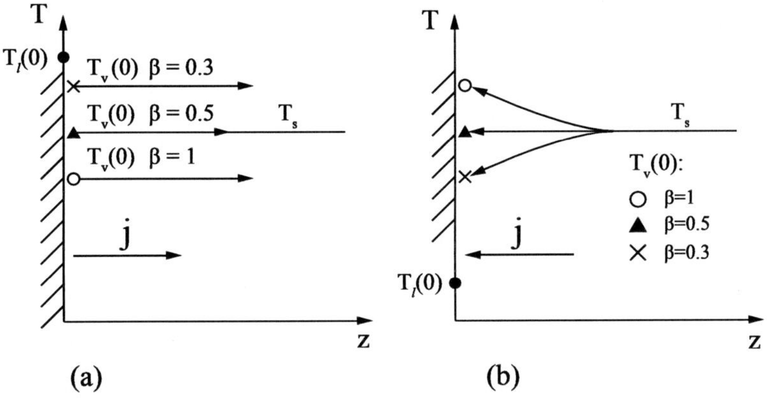

Fig. 2.4 Relative position of the temperatures $T_l(0)$, $T_v(0)$, and T_s during evaporation (**a**) and condensation (**b**) with various values of β

is the non-equilibrium parameter during phase transitions. The smallness of this parameter $\left(\tilde{j} \to 0\right)$ justifies the approximation of the local thermodynamic equilibrium. So, the relations (2.38) and (2.39) from the pioneering paper [25] contain very interesting subtle information about the actual parameters of vapor on the CPS during evaporation/condensation.

2.6 Introduction to the Problem of Strong Evaporation

2.6.1 Conservation Equations

Of special importance in the analysis of the strong evaporation is establishing its limit possible intensity. Landau and Lifshitz [30] gave a detailed analysis of the development of small disturbance of the phase interface. In the general case, it splits into two acoustic waves (propagating upstream and downstream the gas flow), the perturbation of entropy (upstream and downstream the gas flow), and the propagation of entropy (propagating together with the gas flow). If the velocity of evaporation reaches the sound velocity, then the acoustic wave propagating upstream the gas flow "stands" on the interface.

It is a common belief nowadays that supersonic evaporation is impossible. Indeed, assume that the gas flow in the outside region is described by the equations for a perfect gas. Then the analysis of the characteristic properties of the system of Euler equations shows that in the supersonic flow any perturbation is moved away from the interface. Thus, even if the domain of supersonic flow would exist ab initio, it must eventually separate from the interface. It immediately implies the physical impossibility of supersonic evaporation for which no perturbation propagates from the gas domain toward the interface.

Hence, the attainment of the sonic evaporation state should be looked upon as the limit possible case. Let us estimate the nonlinear effects of the evaporation problem. Assume that the spectrum of incident molecules is described by the volume DF (2.18), which takes into account the effect of evaporation flow with velocity u_∞. Substituting f_w^- into the integrals (2.7) and transforming, we obtain a system of equations consisting of the mass, momentum, and energy conservation laws

$$\frac{\sqrt{\tilde{T}_\infty}}{\tilde{p}_\infty} + \sqrt{\pi} s \operatorname{erfc}(s) - \exp(-s^2) = 2\sqrt{\pi} s, \tag{2.42}$$

$$\frac{\sqrt{\pi}}{\tilde{p}_\infty} + \sqrt{\pi}(1 + 2s^2)\operatorname{erfc}(s) - 2s \exp(-s^2) = 2\sqrt{\pi}(1 + 2s^2), \tag{2.43}$$

$$\frac{4}{\tilde{p}_\infty \sqrt{\tilde{T}_\infty}} + \sqrt{\pi}(5 + 2s^2)\operatorname{erfc}(s) - 2\sqrt{\pi}(2 + s^2)\exp(-s^2) = 2\sqrt{\pi}(5s + 2s^3). \tag{2.44}$$

Here, $\tilde{p}_\infty = p_\infty / p_w$, $\tilde{T}_\infty = T_\infty / T_w$ are the dimensionless values, respectively, of pressure, temperature, and s is the speed ratio given by formula (2.20).

It should be noted that this system of equations was obtained for the limit case of complete absorption of molecules incident on the CPS: $\beta = 1$. The evaporation problem can be stated as follows in the form natural for applications. Assume that we know the temperature T_w of the CPS, and hence, the density of saturated vapor with this temperature: $\rho_w = \rho_s(T_w)$. Assume further that we are given some parameter of vapor in the Navier–Stokes region (for example, the pressure p_∞). It is required to find two unknowns: the temperature T_∞ and the mass flux $j_1 = \rho_\infty u_\infty$. The form of the system of equations (2.42)–(2.44) suggest the following more formal statement of the same problem: find the dependences $\tilde{T}_\infty(\mathrm{M})$, $\tilde{p}_\infty(\mathrm{M})$. Here, M is the Mach number related to the speed ratio by formula (2.21).

This system of three equations is superdefinite. This is a direct corollary to the stringent setting of the negative part of the DF as relation (2.18). With this point of view, we consider the above early analytical solutions to the problem of evaporation into vacuum.

- the solution (2.13) by Langmuir and Knudsen was obtained from the mass conservation law (2.42) using the DF (2.18) with $u_\infty = 0$.
- the Hertz–Knudsen equation (2.14) was obtained by heuristic introduction of the difference of pressures into relation (2.13), which takes into account the presence of the "cloud of molecules" for the CPS.
- the "improved" relations (2.15) and (2.17) were obtained by semi-empirical modifications of the Hertz–Knudsen equation.

- the authors of [11, 29] used, in the linear approximation, the mass and energy conservation laws (respectively, (2.42) and (2.44)) and ignored the momentum conservation law (2.43).

Let us now consider the results of solutions [11, 29] in the linear approximation

$$\theta(0) = 0.443s, \tag{2.45}$$

$$\Delta\tilde{p} = 1.995s. \tag{2.46}$$

We write the relations of the linear theory (2.34) and (2.35) with $\beta = 1$

$$\theta(0) = 0.45s, \tag{2.47}$$

$$\Delta\tilde{p} = 2.13s. \tag{2.48}$$

So, strange as it seems, in the linear approximation the results of [11, 29] differ from the precise ones by 2.5% (the temperature jump) and 6.5% (the pressure jump). The solution to the system of equations (2.42) and (2.44) is written as very cumbersome analytical relations. The dimensionless mass flux

$$\tilde{J} = 2\sqrt{\pi}\frac{\rho_\infty u_\infty}{\rho_w \mathrm{v}_w} \tag{2.49}$$

is the additional parameter of the solution. The quantity $\tilde{J}$ is the ratio of the mass flux and the one-way Maxwell flow of molecules emitted by the CPS and defined by relation (2.8). It can be evaluated using the relation following from the Clausius–Clapeyron equation, as written for the CPS conditions and the Navier–Stokes region

$$\tilde{p}_\infty = \tilde{\rho}_\infty \tilde{T}_\infty. \tag{2.50}$$

Here, $\tilde{\rho}_\infty = \rho_\infty/\rho_w$ is the dimensionless value density.

The numerical results of [31, 32] were used as reference results for the verification of the solution obtained from Eqs. (2.42) and (2.44). In these papers, the processes in vapor were described by the spatially one-dimensional Boltzmann equation with the Bhatnagar–Gross–Krook collision term [1].

From Figs. 2.5, 2.6 and 2.7 it is seen that in the nonlinear approximation, the solutions of [11, 29] agree with the numerical solutions of [31, 32] with marked error, which attains its maximum value with sonic evaporation (M = 1): 10% for $\tilde{T}_\infty$, 20% for $\tilde{p}_\infty$, and 30% for $\tilde{J}$. The method of papers [11, 29] based on ignoring one of the three conservation equations does not provide any proof, and in addition, has quantitative errors. With this proviso, the two remaining combinations of the conservation equations have the same "right to life": the "mass + momentum" and the "momentum + energy" equations. However, calculations with these pairs of

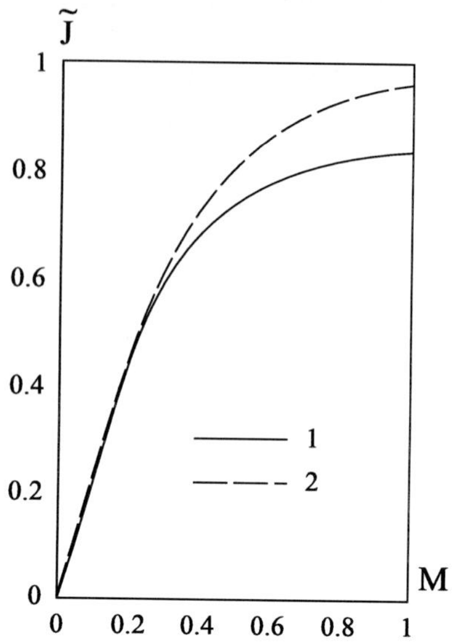

Fig. 2.5 Dependence of the dimensionless mass flux in the Navier–Stokes region on the Mach number. 1. Numerical solutions of [32] and 2. solutions of [11]

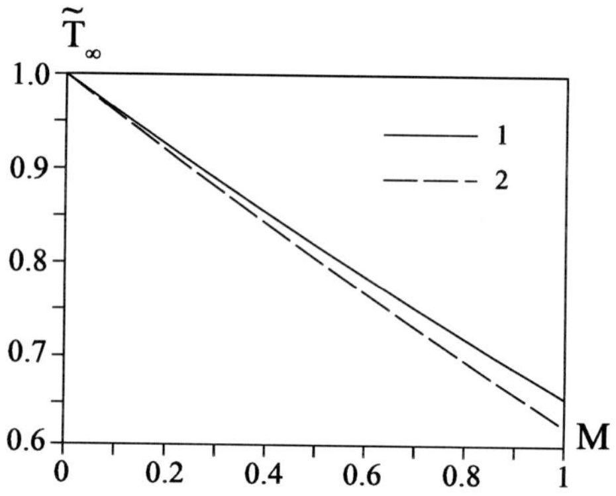

Fig. 2.6 Dependence of the dimensionless temperature in the Navier–Stokes region on the Mach number. 1. Numerical solutions of [32] and 2. solutions of [11]

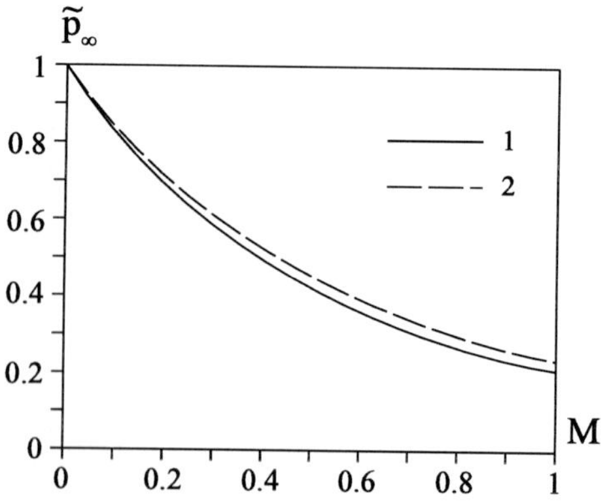

Fig. 2.7 Dependence of the dimensionless pressure in the Navier–Stokes region on the Mach number. 1. Numerical solutions of [32] and 2. solutions of [11]

equations lead to anomalous results. Thus, the above example clearly suggests the necessity of having a correct analytical solution to the problem of evaporation.

2.6.2 The Model of Crout

In 1936, Crout [33] proposed the first correct model of strong evaporation. In this pioneering paper, the description of the process was built on the physical analysis of the evolution of the DF of emitted molecules between CPS (the section "w") and the conditional section "e" inside the KL. Crout [33] used the following two main assumptions.

(1) Initially, the equilibrium spectrum (2.1) of the emitted molecular flow under the effect of intermolecular collisions in the KL is "blurred" and in the section "e" acquires the "ellipsoidal character"

$$f_e^+ = \frac{n_e}{\pi^{3/2} \mathrm{v}_r^2 \mathrm{v}_z} \exp\left(-\left(\frac{c_x^2 + c_y^2}{\mathrm{v}_r^2}\right) - \left(\frac{(c_z - u_z)^2}{\mathrm{v}_z^2}\right)\right). \tag{2.51}$$

Here, $\mathrm{v}_r = \sqrt{2R_g T_r}$, $\mathrm{v}_z = \sqrt{2R_g T_z}$, $\mathrm{v}_\infty = \sqrt{2R_g T_\infty}$, $\mathrm{v}_w = \sqrt{2R_g T_w}$, T_r, T_z are, respectively, the longitudinal and transverse temperatures, c_z is the normal component of the molecular velocity, c_x, c_y are the molecular velocities parallel to the CPS, the index "w" denotes the conditions on the CPS, and the index "∞" denotes the conditions at infinity. Since the flow is one-dimensional, we have $c_x = c_y = c_r$, where c_r is the transverse molecular velocity. The ellipsoidal distribution function f_e^+ differs from the Maxwell distribution function f_w^+ by the presence of different measures of the mean velocity of motion of molecules in the longitudinal and transverse directions (the longitudinal and transverse temperatures). Besides, relation (2.51) takes into account the shift of u_z over the axis of longitudinal velocities c_z.

(2) Between the sections "w" and "e", the molecular flows of mass

$$\int_{-\infty}^{\infty} dc_x \int_{-\infty}^{\infty} dc_y \int_{0}^{\infty} dc_z \left(c_z \left(f_e^+ - f_w^+\right)\right) = 0 \tag{2.52}$$

is conserved, as well as the molecular flows of the momentum

$$\int_{-\infty}^{\infty} dc_x \int_{-\infty}^{\infty} dc_y \int_{0}^{\infty} dc_z \left(c_z^2 \left(f_e^+ - f_w^+\right)\right) = 0 \tag{2.53}$$

and the energy

$$\int_{-\infty}^{\infty} dc_x \int_{-\infty}^{\infty} dc_y \int_{0}^{\infty} dc_z \big(c_z c^2 \big(f_e^+ - f_w^+\big)\big) = 0. \tag{2.54}$$

The function f_e^+ involves four unknowns: the hydrodynamic velocity u_z, the density n_e, and two thermal velocities: the longitudinal v_z and the transverse v_r ones. The same unknowns come over to Eqs. (2.52)–(2.54). In turn, Eqs. (2.42)–(2.44) involve two unknowns: the temperature T_∞ and the pressure p_∞. Thus, the system of equations (2.42)–(2.44), (2.52)–(2.54) involves four unknowns and is closed. As a result, Crout obtained a complete and qualitatively correct solution of the evaporation problem of arbitrary intensity. A flaw of [33] is that the adopted approximation of the distribution function on the surface is adapted to the BC on the surface evaporation only in the mean (in the terminology of the book [1]). Besides, in the domain of small intensity of the process, this solution is inaccurate; it quantitatively poorly agrees with relations (2.34) and (2.35) of the linear theory.

2.6.3 *The Model of Anisimov*

In 1968, Anisimov [34] proposed an original idea of the closure of the system of equations (2.42)–(2.44). He assumed that the DF of the molecules incident on the CPS is proportional to the volume DF

$$f_w^- = A f_{w0}^-. \tag{2.55}$$

Here,

$$f_{w0}^- = \frac{n_w}{\pi^{3/2}\mathrm{v}_w^3} \exp\left(-\frac{2c_r^2}{\mathrm{v}_r^2} - \left(\frac{c_z - u_\infty}{\mathrm{v}_w}\right)^2\right), \tag{2.56}$$

A is the free parameter. Substituting f_w^- from Eqs. (2.55) and (2.56) into integrals (2.7), we obtain the following system of equations

$$\frac{\sqrt{\tilde{T}_\infty}}{\tilde{p}_\infty} + A\big[\sqrt{\pi} s\,\mathrm{erfc}(s) - \exp(-s^2)\big] = 2\sqrt{\pi} s, \tag{2.57}$$

$$\frac{\sqrt{\pi}}{\tilde{p}_\infty} + A\big[\sqrt{\pi}(1+2s^2)\mathrm{erfc}(s) - 2s\exp(-s^2)\big] = 2\sqrt{\pi}(1+2s^2), \tag{2.58}$$

$$\frac{4}{\tilde{p}_\infty\sqrt{\tilde{T}_\infty}} + A\big[\sqrt{\pi}(5+2s^2)\mathrm{erfc}(s) - 2\sqrt{\pi}(2+s^2)\exp(-s^2)\big] = 2\sqrt{\pi}(5s+2s^3). \tag{2.59}$$

With a given velocity, the factor s of the system of equations (2.57)–(2.59) involves three unknowns: $\tilde{p}_\infty$, $\tilde{T}_\infty$, and A. For the case of monatomic gas, we have $\mathrm{M} = u_\infty/\sqrt{5/3RT_\infty}$and $s = \sqrt{5/6}\mathrm{M}$. In [34], the limit case of sonic evaporation $\left(\mathrm{M}_\infty = 1, s = \sqrt{5/6}\right)$ was considered and the following limit parameters were calculated

$$\mathrm{M} = 1 : \tilde{T}_\infty = 0.6691, \tilde{p}_\infty = 0.2062, \tilde{J}_\infty = 0.8157. \tag{2.60}$$

Relations (2.60) show that acoustic evaporation generates vapor with the parameters: $T_\infty \approx 2/3T_w$, $p_\infty \approx 1/5p_w$, and$J_\infty \approx 4/5J_1^+$. Here, J_1^+ is the one-way Maxwell flow emitted by the CPS; it can be found by formula (2.8). Physically this means that, with the maximal velocity of evaporation, approximately 1/5 of the emitted molecules are decelerated by the incident flow. Accordingly, they come back to the interface and are condensed on it. Like many pioneers, Anisimov wrote a small, very informal note [34] (only three pages!). However, the brilliant idea of [34] opened a line of research on the strong evaporation on the basis of mass, momentum, and energy conservation laws. In this connection, it suffices to mention the papers [35–43].

In 1977, Labuntsov and Kryukov [35] and independently Ytrehus [36] applied the method of [34] to the entire region of variation of the Mach number: $0 \le \mathrm{M} \le 1$. The resulting analytical solution [35, 36] reads as

$$\sqrt{\tilde{T}_\infty} = \sqrt{1 + B^2} - B, \tag{2.61}$$

$$\tilde{p}_\infty = \frac{1}{2}\exp\left(s^2\right)\left(C + D\sqrt{\tilde{T}_\infty}\right), \tag{2.62}$$

$$\tilde{J}_\infty = 2\sqrt{\pi}\frac{s\tilde{p}_\infty}{\sqrt{\tilde{T}_\infty}}. \tag{2.63}$$

Here we used the following notation

$$B = \sqrt{\pi}/8s, C = \exp\left(-s^2\right) - \sqrt{\pi}s \ \mathrm{erfc}(s),$$
$$D = \left(1 + 2s^2\right)s \ \mathrm{erfc}(s) - 2/\sqrt{\pi}s \ \exp\left(-s^2\right).$$

In 1979, Labuntsov and Kryukov [37] in more detail their method of [35]. The results of [37] accord well with the numerical results of [31, 32], the departure of the dependence $\tilde{T}_\infty(\mathrm{M}_\infty)$ being the greatest ($\approx 5\%$). Later, Knight [38, 39] used the system of equations (2.61)–(2.63) to construct the thermal model of laser ablation in the outer atmosphere. This model relates the gas parameters with the intensity of evaporation. The author of [38, 39] deals with the dual problem, which involves the system of equations of gas dynamics and heat transfer equation in a radiated target.

2.6.4 *The Model of Rose*

In 2000, Rose [40] proposed a model of strong evaporation, which is a modification of Schrage's old model [10]. Schrage [10] defined the negative part of DF as follows

$$f_w^- = (1 + Ac_z) f_{w0}^+, \, c_z < 0. \tag{2.64}$$

Here, f_{w0}^+ is the equilibrium Maxwell distribution, as defined by formula (2.1), A is the free parameter. The form of Eq. (2.64) was based on the work on diffusion in binary systems by the author of [44]. Rose [40] replaced in formula (2.64) f_{w0}^+ by the function f_{w0}^-, as given by relation (2.56). The results of [40] in the form of the dependences $\tilde{p}_\infty$, $\tilde{T}_\infty$, $\tilde{J}_\infty$ on the Mach number agree quite well with the numerical results [31, 32], the dependence $\tilde{T}_\infty$(M) being even better than that of Ytrehus [36] and Labuntsov and Kryukov [37].

2.6.5 *The Mixing Model*

Inside the KL we introduce the conditional mixing surface "m" and write for it the mass, momentum, and energy conservation laws (2.3)–(2.5)

$$\rho_w \mathrm{v}_w - \rho_m \mathrm{v}_m I_1 = 2\sqrt{\pi}\rho_\infty u_\infty, \tag{2.65}$$

$$\rho_w \mathrm{v}_w^2 - \rho_m \mathrm{v}_m^2 I_2 = 4\rho_\infty u_\infty^2 + 2\rho_\infty \mathrm{v}_\infty^2, \tag{2.66}$$

$$\rho_w \mathrm{v}_w^3 - \rho_m \mathrm{v}_m^3 I_3 = \sqrt{\pi}\rho_\infty u_\infty^3 + \frac{5}{2}\sqrt{\pi}\rho_\infty \mathrm{v}_\infty^2 u_\infty. \tag{2.67}$$

Equations (2.65)–(2.67) take into account the state equation for a perfect gas $p = \rho v^2/2$; the following notations were used

$$I_1 = \exp(-s_m^2) - \sqrt{\pi} s_m \operatorname{erfc}(s_m), \tag{2.68}$$

$$I_2 = \frac{2}{\sqrt{\pi}} s_m \exp\left(-s_m^2\right) - \left(1 + 2s_m^2\right) \operatorname{erfc}(s_m), \tag{2.69}$$

$$I_3 = \left(1 + \frac{s_m^2}{2}\right) \exp\left(-s_m^2\right) - \left(\frac{5\sqrt{\pi}}{4} s_m + \frac{\sqrt{\pi}}{2} s_m^3\right) \operatorname{erfc}(s_m). \tag{2.70}$$

Here, $s_m = u_m/\mathrm{v}_m$ is the speed ratio, u_m is the hydrodynamic velocity, and $\mathrm{v}_m = \sqrt{2R_g T_m}$ is the thermal velocity (all the quantities are taken on the mixing surface).

It is assumed that due to the molecular flow, the mixing parameters in the section will differ from those in the Navier–Stokes region. We shall assume that between the sections "∞" and "m", the molecular mass flows are conserved (the condition mixing)

$$\rho_\infty u_\infty = \rho_m u_m. \tag{2.71}$$

Let us consider the solution of the system of equations (2.68)–(2.71). Assume that we are given the following quantities: the density ρ_w and the thermal velocity v_w on the CPS and the hydrodynamic velocity u_∞ in the Navier–Stokes region. Then the system of equations (2.68)–(2.71) contains 5 unknowns: the density ρ_m, the thermal velocity v_m, and the hydrodynamic velocity u_m on the mixing surface, as well as the density ρ_∞ and the thermal velocity v_∞ in the Navier–Stokes region. For the closure of the system of equations, we adopt the following assumption $v_m = v_\infty$, which physically means the equality of temperatures: $T_m = T_\infty$.

The mixing model was developed in the papers of the author of the present book [41–43]. If we assume that $u_m = u_\infty$ and exclude the mixing condition (2.71), we arrive at Anisimov's model. Thus, the mixing model is a further development of the last one. We note that the introduction inside the KL of some conditional surface correlates in a certain sense with Crout's model, even though there are principal differences here: in Crout's model, one modifies the positive part of the DF, whereas in the mixing model, the negative part.

References

1. Kogan MN (1995) Rarefied gas dynamics. Springer
2. Maxwell JC (1860) Illustrations of the dynamical theory of gases: Part I. On the motions and collisions of perfectly elastic spheres. Philos Mag 19:19–32
3. Maxwell JC (1860) Illustrations of the dynamical theory of gases: Part II. On the process of diffusion of two or more kinds of moving particles among one another. Philos Mag 20:21–37
4. Boltzmann L (1872) Weitere Studien über das Wärmegleichgewicht unter Gasmolekülen. Sitzungsberichte der Kaiserlichen Akademie der Wissenschaften. Wien Math. Naturwiss. Classe 66:275–370. English translation: Boltzmann L (2003) Further studies on the thermal equilibrium of gas molecules. The kinetic theory of gases. Hist Mod Phys Sci 1:262–349
5. Hertz H (1882) Über die Verdünstung der Flüssigkeiten, inbesondere des Quecksilbers, im luftleeren Räume. Ann Phys Chem 17:177–200
6. Langmuir I (1913) Chemical reactions at very low pressures. II. The chemical cleanup of nitrogen in a tungsten lamp. J Am Chem Soc 35:931–945
7. Knudsen M (1934) The kinetic theory of gases. Methuen, London
8. Knacke O, Stranski I (1956) The mechanism of evaporation. Prog Metal Phys 6:181–235
9. Risch R (1933) Über die Kondensation von Quecksilber an einer vertikalen Wand. Helv Phys Acta 6(2):127–138
10. Schrage RW (1953) A theoretical study of interphase mass transfer. Columbia University Press, New York
11. Kucherov RY, Rikenglaz LE (1960) On hydrodynamic boundary conditions for evaporation and condensation. Soviet Phys JETP 10(1):88–89

12. Zhakhovsky VV, Anisimov SI (1997) Molecular-dynamics simulation of evaporation of a liquid. J Exp Theor Phys 84(4):734–745
13. Hołyst R, Litniewski M (2009) Evaporation into vacuum: mass flux from momentum flux and the Hertz-Knudsen relation revisited. J Chem Phys. 130(7):074707. https://doi.org/10.1063/1.30770-7206
14. Hołyst R, Litniewski M, Jakubczyk D (2015) A molecular dynamics test of the Hertz-Knudsen equation for evaporating liquids. Soft Matter 11(36):7201-6. https://doi.org/10.1039/c5sm01508a
15. Marek R, Straub J (2001) Analysis of the evaporation coefficient and the condensation coefficient of water. Int J Heat Mass Transf 44:39–53
16. Nagayama G, Tsuruta T (2003) A general expression for the condensation coefficient based on the transition state theory and molecular dynamics simulation. J Chem Phys 118(3):1392–1399
17. Tsuruta T, Tanaka H, Masuoka T (1999) Condensation/evaporation coefficient and velocity distributions at liquid–vapor interface. Int J Heat Mass Transf 42:4107–4116
18. Matsumoto M (1996) Molecular dynamics simulation of interphase transport at liquid surfaces. Fluid Phase Equilib 125:195–203
19. Matsumoto M (1998) Molecular dynamics of fluid phase change. Fluid Phase Equilib 144:307–314
20. Langmuir I (1916) The evaporation, condensation and reflection of molecules and the mechanism of adsorption. Phys Rev 8:149–176
21. Julin J, Shiraiwa M, Miles RE, Reid JP, Pöschl U, Riipinen I (2013) Mass accommodation of water: bridging the gap between molecular dynamics simulations and kinetic condensation models. J Phys Chem A 117(2):410–420
22. Persad AH, Ward CA (2016) Expressions for the evaporation and condensation coefficients in the Hertz-Knudsen relation. Chem Rev 116(14):7727–7767
23. Davis EJ (2006) A history and state-of-the-art of accommodation coefficients. Atmos Res 82:561–578
24. Labuntsov DA (1967) An analysis of the processes of evaporation and condensation. High Temp 5(4):579–647
25. Muratova TM, Labuntsov DA (1969) Kinetic analysis of the processes of evaporation and condensation. High Temp 7(5):959–967
26. Loyalka SK (1990) Slip and jump coefficients for rarefied gas flows: variational results for Lennard–Jones and n(r)–6 potentials. Phys A 163:813–821
27. Siewert E (2003) Heat transfer and evaporation/condensation problems based on the linearized Boltzmann equation. Eur J Mech B: Fluids 22:391–408
28. Latyshev AV, Uvarova LA (2001) Mathematical modeling. Problems, methods, applications. Kluwer Academic/Plenum Publishers, New York, Moscow
29. Bond M, Struchtrup H (2004) Mean evaporation and condensation coefficient based on energy dependent condensation probability. Phys Rev E 70:061605
30. Landau LD, Lifshits EM (1987) Fluid mechanics. Butterworth-Heinemann
31. Gusarov AV, Smurov I (2002) Gas-dynamic boundary conditions of evaporation and condensation: numerical analysis of the Knudsen layer. Phys Fluids 14(12):4242–4255
32. Frezzotti A (2007) A numerical investigation of the steady evaporation of a polyatomic gas. Eur J Mech B Fluids 26:93–104
33. Crout PD (1936) An application of kinetic theory to the problems of evaporation and sublimation of monatomic gases. J Math Phys 15:1–54
34. Anisimov SI (1968) Vaporization of metal absorbing laser radiation. Sov Phys JETP 27(1):182–183
35. Labuntsov DA, Kryukov AP (1977) Intense evaporation processes. Therm Eng 4:8–11
36. Ytrehus T (1977) Theory and experiments on gas kinetics in evaporation. In: Potter JL (ed) Rarefied gas dynamics: technical papers selected from the 10th international symposium on rarefied gas dynamics. Snowmass-at-Aspen, CO, July 1976. In: Progress in astronautics and aeronautics. Am Inst Aeronaut Astronaut 51:1197–1212

37. Labuntsov DA, Krykov AP (1979) An analysis of intensive evaporation and condensation. Int J Heat Mass Transf 22:989–1002
38. Khight CJ (1979) Theoretical modeling of rapid surface vaporization with back pressure. AIAA J 17(5):519–523
39. Khight CJ (1982) Transient vaporization from a surface into vacuum. AIAA J 20(7):950–955
40. Rose JW (2000) Accurate approximate equations for intensive sub-sonic evaporation. Int J Heat Mass Transf 43:3869–3875
41. Zudin YB (2015) Approximate kinetic analysis of intense evaporation. J Eng Phys Thermophys 88(4):1015–1022
42. Zudin YB (2015) The approximate kinetic analysis of strong condensation. Thermophys Aeromech 22(1):73–84
43. Zudin YB (2016) Linear kinetic analysis of evaporation and condensations. Thermophys Aeromech 23(3):437–449
44. Furry WH (1948) On the elementary explanation of diffusion phenomena in gases. Am J Phys 16:63–78

Chapter 3
Approximate Kinetic Analysis of Strong Evaporation

Abbreviations

BC	Boundary conditions
CPS	Condensed-phase surface
DF	Distribution function
HTC	Heat transfer coefficient
HHCE	Hyperbolic heat conduction equation
KL	Knudsen layer
PHCE	Parabolic heat conduction equation

Symbols

B	Kinetic parameter
$\mathbf{c}$	Molecular velocity vector
c_p	Isobaric heat capacity
f	Distribution function
g	Permeability coefficient
h	Heat transfer coefficient (HTC)
h_{fg}	Heat of phase transition
J	Molecular flux
j	Mass flux
k	Thermal conduction
m	Molecular mass
M_∞	Mach number
p	Pressure
$\tilde{p}$	Pressure ratio
q	Heat flux
t_r	Relaxation time
T	Temperature
$\tilde{T}$	Temperature ratio
u_∞	Hydrodynamic velocity

Y. B. Zudin, *Non-equilibrium Evaporation and Condensation Processes*,
Mathematical Engineering, https://doi.org/10.1007/978-3-030-67553-0_3

$\tilde{u}_\infty$	Speed ratio
V	Thermal velocity
x	Coordinate

Greek Letter Symbols

α	Thermal diffusivity
β	Condensation coefficient
ρ	Density
ν	Kinematic viscosity
ψ	Evaporation parameter
Ψ	Relative heat transfer coefficient

Subscripts

m	State at mixing surface
s	Saturation state
w	State at condensed-phase surface
∞	State at infinity
$*$	Reference case

3.1 Introduction

The knowledge of the laws governing intense evaporation is important for vacuum technologies, exposure of materials to laser radiation, outflow of a coolant on the loss of sealing in the protective envelope of an atomic power plant, and for other applications. The problem of evaporation from a condensed-phase surface into a half-space filled with vapor represents a boundary-value problem for the gas dynamics equations. Its distinctive feature is that near the surface there exists a Knudsen Layer (KL) in which the microscopic description becomes inapplicable due to the anisotropy of the velocity Distribution Function (DF) of gas molecules. Inside this non-equilibrium layer, the thickness of which is of the order of the molecular mean free path, any flow obeys the microscopic laws described by the Boltzmann equation [1].

The specifics of the kinetic analysis resides in the necessity of solving a complex conjugate problem—a macroscopic boundary-value problem for the gas dynamics equations in the region of continuous medium flow (also called the Navier–Stokes region) and a microscopic problem for the Boltzmann equation in the KL. Moreover, the Boundary Conditions (BC) for the first problem are to be determined from the solution of the second problem. In the extrapolation of the gas temperature and pressure distributions from the Navier–Stokes region to the condensed-phase surface, there arise kinetic jumps, i.e., the BC for a continuous medium do not coincide with their actual values.

If the evaporation flux is much less than the most probable velocity of the thermal motion of molecules, it is allowable to use the linearized Boltzmann equation in the kinetic analysis. The linear kinetic theory of evaporation and condensation in its final form was presented for the first time in [2] and also described in [3]. The outflow of vapor with a normal velocity component comparable with the velocity of sound is called intense evaporation. In this case, kinetic jumps of the parameters are compared with the absolute pressure and temperature values in the Navier–Stokes region [4].

The kinetic Boltzmann equation represents a nonlinear integro-differential equation for a three-dimensional DF, the accurate solution of which is possible only in special cases [5]. The numerical solution of this equation encounters difficulties due to its high dimensionality and to the complex structure of the collision integral entering into this equation [6]. However, for the majority of applications, information on the DF in a thin Knudsen layer is insignificant. In solving applied problems, it is necessary to only correctly specify fictitious BC for the equations of gas dynamics in the Navier–Stokes region. Therefore, up to the present time, the area of the kinetic analysis associated with the approximate determination of gas-dynamical BC without solving the Boltzmann equation [4, 7–10] remains highly topical.

A very complex analysis of the linear problem of evaporation that obviated the need for the Boltzmann equation was applied in [7]. The mathematical procedure involved the transformation of the linearized Boltzmann equation into the Wiener–Hopf integro-differential equation with the transformation of the latter into a matrix form with subsequent factoring and investigation of the matrix equation on the basis of the Gohberg–Krein theorem on self-conjugated matrices. It is noteworthy that in his subsequent work [8], Pao refuted the results obtained by him in [7] because of the mathematical error committed by him.

An essentially new method of determining gas-dynamical BC for the case of sonic evaporation (the vapor velocity is equal to the velocity of sound) was suggested in [9]. The next important event was the publication of the papers [4, 10] in which the method of [9] was generalized to the case of vapor flow with arbitrary velocities. The idea behind the approach of [4, 9, 10] was the approximation of the DF from reasonable physical considerations and the subsequent solution of conservation equations for molecular fluxes in the KL.

3.2 Conservation Equations

We consider a one-dimensional stationary problem of evaporation from a condensed-phase surface into a half-space filled with vapor (with a monatomic ideal gas). The equations for the conservation equations of mass, momentum, and energy flows, respectively, are as follows

$$J_1^+ - J_1^- = \rho_\infty u_\infty, \tag{3.1}$$

$$J_2^+ - J_2^- = \rho_\infty u_\infty^2 + p_\infty, \tag{3.2}$$

$$J_3^+ - J_3^- = \frac{1}{2}\rho_\infty u_\infty^3 + \frac{5}{2}p_\infty u_\infty. \tag{3.3}$$

Here, ρ_∞ is the density, p_∞ is the pressure, u_∞ is the hydrodynamic velocity, the index "∞" denotes the conditions at infinity, J_i^+ and J_i^- are the molecular flows of the enumerated quantities in the KL that are emitted by the surface and are incident on the surface from the gas space ($i = 1, 2, 3$).

The values of J_i^+ and J_i^- are calculated by a familiar method as integrals of the DF f with respect to the three-dimensional field of molecular velocities [1]. The unbalance of molecular flows in the KL $\left(J_i^+ > J_i^-\right)$ leads to the appearance of macroscopic flows of evaporation in the Navier–Stokes region on the right-hand sides of Eqs. (3.1)–(3.3), i.e., of mass ($i = 1$), momentum ($i = 2$), and energy ($i = 3$) flows. The standard assumption of the kinetic molecular theory is the one that neglects the reflection of molecules from the surface and their secondary emission. It is assumed that the spectrum of the molecules emitted by the surface is independent of the distribution of the molecules collided with it and is entirely determined by the surface temperature

$$f_w^+ = \frac{p_w}{k_B T_w}\left(\frac{m}{2\pi k_B T_w}\right)^{3/2} \exp\left(-\frac{mc_w^2}{2\pi k_B T_w}\right). \tag{3.4}$$

Here, k_B is the Boltzmann constant, m is the molecular mass, T_w is the temperature on the surface.

Relation (3.4) specifies the Maxwell equilibrium distribution (half-Maxwellian) that corresponds to the temperature T_w and to the vapor saturation pressure at this temperature $p_w(T_w)$. Note that the physically plausible relation (3.4) has no rigorous theoretical substantiation. Thus, it is acknowledged in [11] that "we are not aware of any serious derivation of such boundary condition". To determine the spectrum of the molecules emitted by the surface, Zhakhovskii and Anisimov [11] carried out a numerical simulation of evaporation into vacuum by the method of molecular dynamics and, using the results of the investigation, they concluded that "thus, in the case of low vapor density, the use of the half-Maxwellian distribution as a boundary condition in solving gas-dynamical problems seems to be a reasonable approximation". The velocity distribution of the molecules flying to the surface f_∞^- is prescribed in the form of a half-Maxwellian (3.4) in the coordinate system related to the evaporation flux at infinity u_x. Thus, in the coordinate system fixed on the condensed-phase surface, the function f_∞^- will be shifted by the value along the velocity component normal to the surface

$$f_\infty^- = \frac{p_\infty}{k_B T_\infty}\left(\frac{m}{2\pi k_B T_\infty}\right)^{3/2} \exp\left(-\frac{m(\mathbf{c}_\infty - u_\infty)^2}{2k_B T_\infty}\right). \tag{3.5}$$

Within the framework of the rigorous kinetic approach, the non-equilibrium DF in the KL is determined from the solution of the boundary-layer problem for the Boltzmann equation with BC (3.4) and (3.5). As the distance from the surface increases, the DF approaches an equilibrium function, and, starting from a certain distance, it goes over into the local Maxwellian distribution (3.5). This distance is taken as a conventional external boundary of the KL beyond which the gas motion obeys the equations of gas dynamics.

As is known, the conservation Eqs. (3.1)–(3.3) are the first three momentum equations following from the Boltzmann equation [1]. Therefore, when substituting an exact DF (in terms of its integral expressions for J_i^+ and J_i^- into Eqs. (3.1)–(3.3), the latter must transform into identities. There is a fundamentally different situation within the framework of the approach used in [9, 10]. Here, one solves a system of conservation Eqs. (3.1)–(3.3) with a given DF having a discontinuity on the condensed-phase surface. Here, the positive half-Maxwellian f^+ is already known from relation (3.4). Its use leads to the following values of the integrals J_i^+ for the emitted molecular flows

$$\left.\begin{aligned} J_1^+ &= \tfrac{1}{2\sqrt{\pi}}\rho_w \mathrm{v}_w, \\ J_2^+ &= \tfrac{1}{4}\rho_w \mathrm{v}_w^2, \\ J_3^+ &= \tfrac{1}{2\sqrt{\pi}}\rho_w \mathrm{v}_w^3. \end{aligned}\right\} \tag{3.6}$$

Here, $\mathrm{v}_w = \sqrt{2k_B T_w/m}$ is the thermal velocity on the surface, ρ_w is the density on the surface.

The parameters of the flows of molecules flying to the surface remain unknown. In order to determine them, one has to assign a negative half-Maxwellian f^-. The macroscopic theory of intense evaporation at $M_\infty = 1$ (the vapor flow velocity in the Navier–Stokes region is equal to the velocity of sound) is presented in [9] where the author proceeded from the hypothesis that the function f_w^- is proportional to the rear half of the equilibrium DF in the Navier–Stokes region

$$f_w^- = Af_\infty^- \equiv A\, f_\infty|_{c<0}. \tag{3.7}$$

In [4, 10], the method of [9] was generalized to the entire range of variation of the Mach number, $0 < M_\infty \leq 1$. The use of (3.7) makes it possible to calculate the integrals J_i^- for the flows of molecules flying to the surface. The system of Eqs. (3.1)–(3.3), as subject to Eqs. (3.6) and (3.7), after some transformations can be presented in the form

$$\frac{\sqrt{\tilde{T}}}{\tilde{p}} - AI_1^- = 2\sqrt{\pi}\tilde{u}_\infty, \tag{3.8}$$

$$\frac{1}{\tilde{p}} - AI_2^- = 2 + 4\tilde{u}_\infty^2, \tag{3.9}$$

$$\frac{1}{\sqrt{\tilde{T}\,\tilde{p}}} - AI_3^- = \frac{5\sqrt{\pi}}{2}\tilde{u}_\infty + \sqrt{\pi}\tilde{u}_\infty^3. \tag{3.10}$$

Here, $\tilde{u}_\infty \equiv u_\infty / \mathrm{v}_\infty$ is the speed ratio related to the Mach number at infinity $\mathrm{M}_\infty \equiv u_\infty(5k_BT_\infty/3m)^{-1/2}$ as $\tilde{u}_\infty = \sqrt{5/6}M_\infty$, and $\mathrm{v}_\infty = \sqrt{2k_BT_\infty/m}$ is the thermal velocity in the Navier–Stokes region. The nondimensional molecular flows incident on the condensed-phase surface are written in the form

$$\left.\begin{aligned} I_1^- &= \exp\left(-\tilde{u}_\infty^2\right) - \sqrt{\pi}\tilde{u}_\infty \operatorname{erfc}\left(\tilde{u}_\infty\right), \\ I_2^- &= \tfrac{2}{\sqrt{\pi}}\tilde{u}_\infty \exp\left(-\tilde{u}_\infty^2\right) - \left(1 + 2\tilde{u}_\infty^2\right)\operatorname{erfc}\left(\tilde{u}_\infty\right), \\ I_3^- &= \left(1 + \tfrac{1}{2}\tilde{u}_\infty^2\right)\exp\left(-\tilde{u}_\infty^2\right) - \tfrac{\sqrt{\pi}}{2}\left(\tfrac{5}{2} + \tilde{u}_\infty^2\right)\operatorname{erfc}\left(\tilde{u}_\infty\right), \end{aligned}\right\} \tag{3.11}$$

where $\operatorname{erfc}\left(\tilde{u}_\infty\right)$ is the additional integral of probabilities.

The system of Eqs. (3.8)–(3.11) determines the dependences of the temperature ratio $\tilde{T} = T_\infty/T_w$, the pressure ratio $\tilde{p} = p_\infty/p_w$, and the free parameter A on the speed ratio $\tilde{u}_\infty$ (and thereby on M_∞). As a result, a refined analytical solution was obtained in [4, 10] that was far ahead of the later numerical investigations of intense evaporation [12, 13]. It is important to note that the solutions in [4, 10] for $M_\infty \to 0$ have a correct limiting transition to the results of the linear kinetic theory [2].

3.3 Mixing Surface

The aim of the present chapter is to further develop the approach of [9, 10] with the aid of the approximation of the function f_w^- which is more flexible than (3.7). As is known [1], the expressions for the molecular flows J_i^+ and J_i^- result from the integration of various combinations of molecular velocities with the weight function which is a three-dimensional velocity DF of molecules f. As is seen from (3.6), here the density ρ acts as a preintegral factor. The existence of the linear relationship $f \sim \rho$ allows one to reformulate condition (3.7) in the following form

$$\rho_m = A\rho_\infty, \tag{3.12}$$

where ρ_m is the density on the hypothetical mixing surface located inside the KL. We assume that the following relation is valid

$$\rho_m u_m = \rho_\infty u_\infty \tag{3.13}$$

subject to the isothermicity condition

$$T_m = T_\infty \tag{3.14}$$

With an account for (3.13) and (3.14), the system of Eqs. (3.11) can be rewritten as

$$\left.\begin{aligned} I_1^- &= \exp\left(-\tilde{u}_m^2\right) - \sqrt{\pi}\tilde{u}_m \operatorname{erfc}\left(\tilde{u}_m\right), \\ I_2^- &= \tfrac{2}{\sqrt{\pi}}\tilde{u}_m \exp\left(-\tilde{u}_m^2\right) - \left(1 + 2\tilde{u}_m^2\right)\operatorname{erfc}\left(\tilde{u}_m\right), \\ I_3^- &= \left(1 + \tfrac{1}{2}\tilde{u}_m^2\right)\exp\left(-\tilde{u}_m^2\right) - \tfrac{\sqrt{\pi}}{2}\left(\tfrac{5}{2} + \tilde{u}_m^2\right)\operatorname{erfc}\left(\tilde{u}_m\right). \end{aligned}\right\} \tag{3.15}$$

Here, $\tilde{u}_m$ is the speed ratio on the mixing surface. It is determined from equalities (3.13) and (3.14)

$$\tilde{u}_m = \frac{\tilde{u}_\infty}{A}. \tag{3.16}$$

If instead of (3.16) we write the equality $\tilde{u}_m = \tilde{u}_\infty$ and retain (3.14), we arrive at the base model of [4, 10]. Thus, the difference of the present model from that of [4, 10] is reduced to the introduction of the additional correlation (3.16) into the already formulated system of conservation equations. Thus, in the half-Maxwellian f_∞^- used for closing relation (3.7), the replacement $u_\infty \Rightarrow u_m \equiv Au_\infty$ was made.

Figure 3.1 presents the calculated dependences of T_∞/T_w and p_∞/p_w on the Mach number in the Navier–Stokes region, as obtained by solving the system of Eqs. (3.1)–(3.4) subject to Eqs. (3.6), (3.7), (3.15), and (3.16). As is seen from Fig. 3.1a, b, the calculated curves practically match the results of the analytical solution of [10], as well as the results of the later numerical investigation [12]. It is noteworthy that the numerical investigation encounters difficulties of attaining the regime of sonic evaporation in which $M_\infty = 1$. Thus in [12], the last calculated point was obtained for $M_\infty \approx 0.955$, whereas in [13] for $M_\infty \approx 0.87$. At the same time, the analytical approach of [4, 10] is free of the indicated limitation. Mention should also be made

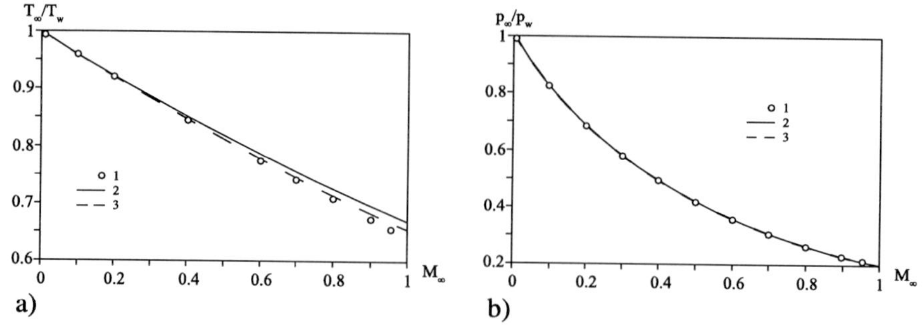

Fig. 3.1 Parameters in the Navier–Stokes region versus the Mach number. 1 Numerical solution of [12], 2 analytical solution of [10] and 3 the solution obtained by solving the system of Eqs. (3.8)–(3.10). **a** the dimensionless temperature and **b** the dimensionless pressure

of the analytical solution obtained in [14] in which the spectrum of the molecules flying to the surface was sought in the form

$$f_w^- = (1 + Ac_z)f_\infty^-. \tag{3.17}$$

Rose [14] did not justify physically the use of the weight integrand (3.17) that includes the molecular velocity c_z in the direction of the evaporation flux. Nevertheless, he obtained virtually the same calculated curves for T_∞/T_w and p_∞/p_w as those presented in Fig. 3.1. Thus, there is ample proof of the conservativeness of the problem of intense evaporation relative to the means of the introduction of the free parameter into the conservation Eqs. (3.1)–(3.3). It is worth mentioning here the conclusion drawn in [13] that "even a rough approximation of the velocity DF in the KL is capable of ensuring a satisfactory analytical description of the gas-dynamical conditions in evaporation".

3.4 Limiting Mass Flux

In the classical works [15, 16] that initiated the kinetic analysis of evaporation, the problem of evaporation into vacuum was posed. Based on physical considerations, it was predicted in [15, 16] that the molecular flow J_1^+ must decrease when moving through the Knudsen layer. The reason for the decrease in the molecular flow emitted by the surface was called the presence of "surface cloud of molecules" of density ρ_∞ (the typical term in earlier kinetic works). The question of the limiting possible mass flow of evaporation was also investigated in [17, 18]. In [19], as a result of the numerical solution of the problem of intense evaporation, it was established that the limiting mass flux J_1^+ (one-sided Maxwellian flow) is not reached in a stationary process.

An interesting numerical experiment was carried out in [20], where a nonstationary problem of vapor expansion on evaporation into vacuum was solved. It was established that after a certain period of relaxation the incident flow of molecules J_1^- increases from zero to a certain maximum value. As a result, the mass flux of evaporation $\rho_\infty u_\infty$ decreases from the maximum value J_1^+ by about 20%. Physically this means that approximately the fifth part of the molecules emitted after collisions in the Knudsen layer returns to the surface and condenses on it. In this connection, a question arises on the possibility of exceeding the velocity of sound by the flow $u_\infty^* = \sqrt{5k_B T_\infty/3m}$ in the Navier–Stokes region. On the one hand, the system of conservation Eqs. (3.1)–(3.3) does not impose limitations on the velocity u_∞. At the same time, it is shown in [10, 13, 17, 18] that, depending on the selected approximate method of calculation, the maximum evaporation mass flux lies in the range of Mach numbers at infinity: $M_\infty \equiv u_\infty(5k_B T_\infty/3m)^{-1/2} \approx 0.866 - 0.994$.

As is known [21], three types of disturbances can propagate in a gas: the density disturbance propagating with the flow velocity u, and two sound waves. The first sound wave moves against the flow with the velocity u_∞, and the second moves with the flow with the velocity $u_\infty - u_\infty^*$. During evaporation at a subsonic velocity $u_\infty < u_\infty^*$, the density disturbance propagates to the side of the gas and the single sound wave propagates with the velocity $u_\infty + u_\infty^*$.

Let the evaporation flux reach the velocity of sound. Then, in the Eulerian coordinate system connected with the surface, the sound wave propagating against the flow with the velocity $u_\infty - u_\infty^*$ will be at rest. From this the choking condition follows, that is, gas-dynamical disturbances cannot propagate against the evaporation flow and the surface conditions cease to influence the vapor flow. The given physical considerations point to the impossibility of supersonic evaporation. Now, within the framework of the presented model, we determine the degree of retardation of one-sided Maxwellian flow J_1^+. For this purpose, from Eqs. (3.1) and (3.6) we find the evaporation mass flux coefficient which is the ratio of the mass flux in the Navier–Stokes region to the emitted molecular flow

$$K = \frac{\rho_\infty u_\infty}{J_1^+}. \tag{3.18}$$

As is seen form Fig. 3.2, the evaporation mass flux coefficient increases with the Mach number from zero to a certain limiting value $K_{\max}$. The formal extension of calculation to the nonphysical branch of supersonic evaporation yields a distinct maximum of the function $K(M_\infty)$ at $M_\infty = 0.9325$.

Figure 3.3 compares the calculated curve $K(M_\infty)$ with the results of the analytical solution of [10] as well as with the results of the numerical investigation carried out in [12].

Figure 3.4 demonstrates the change in the Mach number on the mixing surface on an increase in the evaporation intensity. It follows from this figure that in the motion of vapor from the mixing surface up to the boundary between the KL and the Navier–Stokes region the velocity of its flow increases, with the qualitative character of the function $M_m(M_\infty)$ remaining the same as of the function $K(M_\infty)$ in Fig. 3.2.

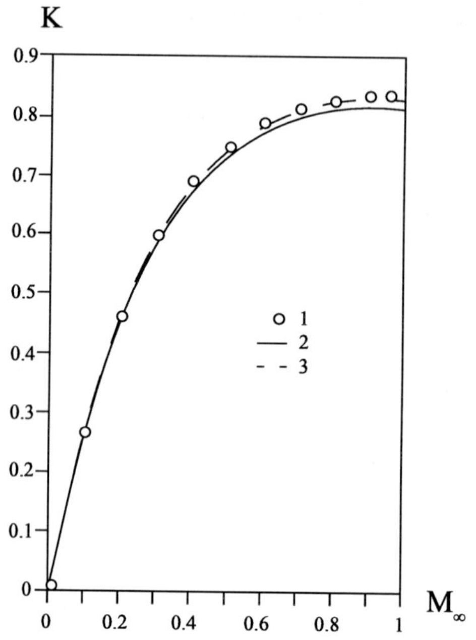

Fig. 3.2 Evaporation mass flux coefficient versus the Mach number in the Navier–Stokes region. 1 Numerical solution of [12], 2 analytical solution of [10] and 3 the solution obtained by solving the system of Eqs. (3.8)–(3.10)

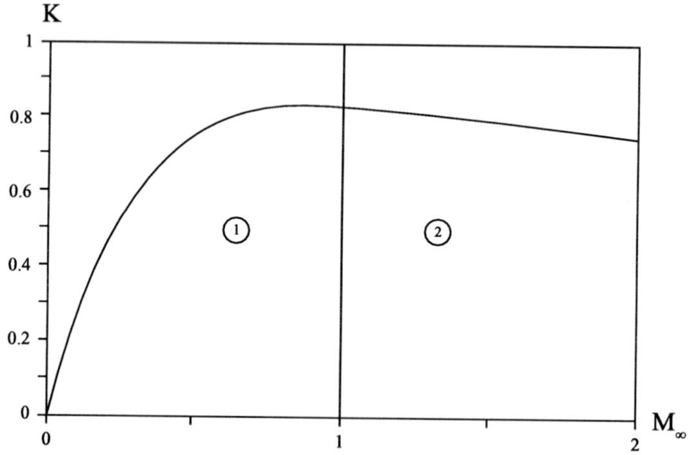

Fig. 3.3 Evaporation mass flux coefficient versus the Mach number in the Navier–Stokes region, obtained by solving the system of Eqs. (3.8)–(3.10). 1 Subsonic evaporation and 2 supersonic evaporation

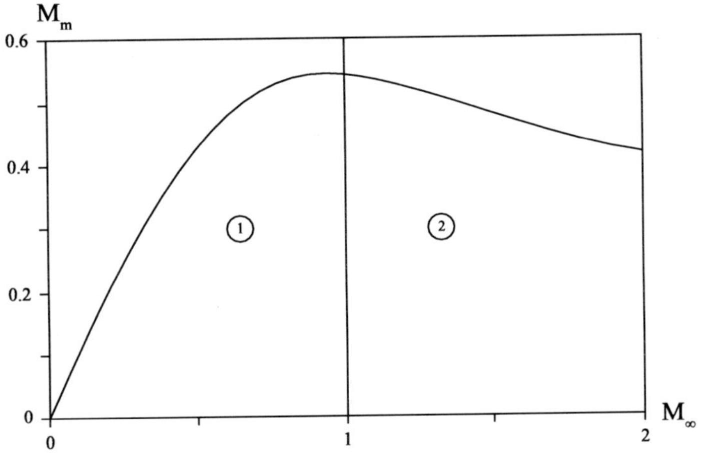

Fig. 3.4 Dependence of the Mach number on the mixing surface on the Mach number in the Navier–Stokes region. 1 Subsonic evaporation and 2 supersonic evaporation

The mixing model was developed in the papers of the author of the present book [22–24].

3.5 Reflection of Molecules from the Surface

3.5.1 Condensation Coefficient

The system of Eqs. (3.1)–(3.3) deals with an absolutely permeable CPS, which adsorbs all molecules that fall on it from the Navier–Stokes region. In the general case, only a part of the molecular flow is transmitted through the surface. This part is determined by the condensation coefficient β. The quantity β, which reflects the state of the surface and the physical nature of the condensed phase, may in general vary in the range $0 < \beta \leq 1$.

The concept of the "condensation coefficient" has a long history beginning with Maxwell [25, 26]. Maxwell defined the "accommodation coefficient" and considered two limit cases: the surface either completely absorbs or completely reflects all the molecules incident on it. In this case, in accordance with the equations for conservation of mass, momentum, and energy one can find the three corresponding accommodation coefficients.

Knudsen [16] gave a concrete physical interpretation of Maxwell's ideas. He assumed that in the general case only a part of all the molecules incident on the CPS is absorbed by the surface. This part is controlled by the quantity β. Correspondingly, the remaining part $(1 - \beta)$ is reflected from it. So, according to Knudsen, the condensation coefficient is the ratio of the number of molecules absorbed by the surface to the general number of molecules incident on it. Knudsen also introduced the "evaporation coefficient" as the ratio of two flows of molecules emitted from the CPS. The first flow pertains to the real case, while the second one to the reference case of an absolute permeable surface. Here, the reference state is defined as the state in which the vapor density corresponds to the CPS temperature on the saturation line and the emitted molecules have an equilibrium Maxwell distribution.

Theoretical analysis of condensation with due account of energy exchange between gaseous and liquid molecules was first made by Langmuir [27]. Langmuir assumed that the time for energy exchange on the CPS is equal in the order of magnitude to the period of oscillations of molecules of the liquid near the equilibrium state. This period is very small, and hence according to Langmuir, the energy exchange is "instantaneous", which implies that $\beta \approx 1$. It is interesting to note that similar results were obtained in modern models of the condensation process by the method of molecular dynamics [28].

By now, a fairly large body of theoretical and experimental data has been accumulated on the evaporation and condensation coefficients. In particular, from the survey [29] (for water) it follows that the approaches involved in the definition of the above coefficients and the results obtained on their basis can be substantially different. A detailed analysis of the problems of definition of the evaporation and condensation coefficients can be found in [30], which also indicates a great discrepancy between experimental and theoretical results and the absence of a unique viewpoint on the processes of accommodation of gas by the surface.

Below we shall assume that the coefficients of evaporation and condensation are approximately equal, and call them generally the "condensation coefficient". Let us examine the effect of β on the conservation equations of molecular flows. We denote by J_i^+ the flows of mass ($i = 1$), momentum ($i = 2$), and energy ($i = 3$), which are emitted by the CPS in the case it is absolute permeable. Assume now that not all the flow incoming from the liquid to the gaseous phase is passed through the surface, but only a part of it $J_{i\beta}^+ = \beta J_i^+$. In turn, assume that only the part β of the flow of the molecules J_i^- incident from the Navier–Stokes region on the CPS is captured by it (the remaining part $(1-\beta)J_i^-$ is reflected from it). Then the total molecular flow outgoing from the surface is

$$J_{i\beta}^+ = \beta J_i^+ + (1-\beta)J_i^-, \tag{3.19}$$

$i = 1, 2, 3.$

From (3.19) it is possible to find the macroscopic flows $J_{i\infty}$ in the Navier–Stokes region, which are defined as the difference of the emitted and incident molecular flows

$$J_{i\infty} = J_{i\beta}^+ - J_i^- = \beta\left(J_i^+ - J_i^-\right). \tag{3.20}$$

Let us introduce the "permeability coefficients" as the ratio of the flow outgoing from the CPS (the case $\beta < 1$) and the flow emitted by an absolutely permeable CPS (the case $\beta = 1$)

$$g_i = \frac{J_{i\beta}^+}{J_i^+}. \tag{3.21}$$

Hence, using (3.19), (3.20) in (3.21) gives

$$g_i = 1 - 2\sqrt{\pi}\frac{1-\beta}{\beta}\frac{J_{i\infty}}{J_i^+}. \tag{3.22}$$

The quantities J_i^+ are given by (3.6). The macroscopic flows in the Navier–Stokes region are found from Eqs. (3.8)–(3.10)

$$J_{1\infty} = \rho_\infty u_\infty, \tag{3.23}$$

$$J_{2\infty} = \rho_\infty u_\infty^2 + p_\infty, \tag{3.24}$$

$$J_{3\infty} = \frac{1}{2}\rho_\infty u_\infty^3 + \frac{5}{2}p_\infty u_\infty. \tag{3.25}$$

From Eqs. (3.21)–(3.25), we obtain expressions for the permeability coefficients for each of the three conservation equations

$$g_1 = 1 - 2\sqrt{\pi}\frac{1-\beta}{\beta}\frac{\tilde{p}\,\tilde{u}_\infty}{\sqrt{\tilde{T}}}, \tag{3.26}$$

$$g_2 = 1 - 2\frac{1-\beta}{\beta}\,\tilde{p}\left(1 + 2\tilde{u}_\infty^2\right), \tag{3.27}$$

$$g_3 = 1 - \frac{5\sqrt{\pi}}{2}\frac{1-\beta}{\beta}\,\tilde{p}\sqrt{\tilde{T}}\tilde{u}_\infty\left(1 + \frac{2}{5}\tilde{u}_\infty^2\right). \tag{3.28}$$

It follows that in the general case $\beta \leq 1$, the conservation equations of molecular flows (7)–(9) assume the form

$$g_1 J_1^+ - J_1^- = J_{1\infty}, \tag{3.29}$$

$$g_2 J_2^+ - J_2^- = J_{2\infty}, \tag{3.30}$$

$$\psi_3 J_3^+ - J_3^- = J_{3\infty}. \tag{3.31}$$

3.5.2 Diffusion Scheme for Reflection of Molecules

In the standard kinetic analysis of intensive evaporation, it is assumed that all three permeability coefficients are equal: $g_1 = g_2 = g_3$. Such a simplified scheme of reflection of molecules, which dates back to Knudsen, is known as the "diffusion scheme". This scheme was first applied in [2] in linear approximation and then extended in [31] to the case of intensive evaporation. Later, attempts were made to use more involved schemes of the interaction of molecules of an incident flow with the CPS [32]. The surface molecular flow emitted by the surface has an equilibrium Maxwell distribution given by (3.4). In the frameworks of the diffusion scheme, it is assumed that after the interaction with the CPS the reflected flow "forgets" its original spectrum and also acquires a Maxwell distribution. However, this scheme is self-contradictory.

- In the frameworks of the diffusion scheme, one finds only the permeability coefficient of the flow of mass g_1. Physically, this means that only the balance of the mass flow ($i = 1$) is achieved, whereas the flows of the momentum normal component ($i = 2$) and the energy ($i = 3$) through CPS remain unbalanced. So, the diffusion scheme involves relation (3.26) and ignores relations (3.27) and (3.28).
- The diffusion scheme contains the mechanism of "filtration" of molecules approaching the CPS from the liquid phase. However, the classical kinetic analysis is concerned with the processes which take place only in the gaseous phase. The role of the condensed phase is reduced to setting BC on the surface. Hence, its involvement in the interaction chain of molecular flows is in direct contradiction to the original statement of the problem.

In recent times, studies based on the modeling of evaporation/condensation processes by the molecular dynamics method become more and more popular [33]. This method, which appeared in the mid-twentieth century, operates with relatively small systems of particles. The main idea of the molecular dynamics method is to study various properties of substances by modeling the motion and interactions of molecules. This aim is achieved by using the hard-sphere potential, the Lennard–Jones potential, continuous potentials, etc. Studies of intensive evaporation by the molecular dynamics method showed the important role of fluctuations of the bonding energy in the surface layer of liquids. It was shown that a considerable contribution to the molecular flow comes from the molecules whose kinetic energy has the same order as the energy of heat motion. The structure of the transient layer between the gaseous and liquid phases was studied, and the fluctuations of the potential energy and pair correlation functions were calculated. However, the molecular dynamics method develops only at the "microscopic level" and in principle is incapable of producing applied correlation relations between averaged parameters. Based on the above arguments, we shall use below the traditional diffusion scheme of reflection of molecules.

The book [34] developed a semi-empirical model of intensive evaporation, which enabled one to significantly simplify the mathematical description of the problem. The model is based on the comparative analysis of the available approximate methods for the calculation of intensive evaporation with the use of various approximations of DFs (which sometimes substantially differ from each other). In the frameworks of the semi-empirical model, it is assumed that the physical mechanism of molecular-kinetic phenomena for evaporation is adequately described by the linear kinetic theory [2]. Under this approach, the transition to intensive evaporation is achieved by augmenting the linear kinetic terms describing the surface of the shock wave with a rarefaction wave (Rankine–Hugoniot's relation [35]). Let us now proceed with the calculation of the permeability coefficient g_1 in the case of intensive evaporation. According to (3.1)

$$J_1^+ = \frac{1}{2\sqrt{\pi}}\rho_w\sqrt{\frac{2kT_w}{m}}. \tag{3.32}$$

Assuming that the CPS temperature remains the same ($T_w = \text{idem}$) when changing from the general case $\beta \le 1$ to the limit case $\beta = 1$, we get from (3.21), (3.23)

$$g_1 = \frac{\rho_{w\beta}}{\rho_w}. \tag{3.33}$$

So, numerically the diffusion scheme of reflection of molecules is reduced to changing the real vapor saturation density at the surface temperature (ρ_w) by its modified value $(\rho_{w\beta})$ with unchanged CPS temperature ($T_w = \text{const}$)

$$\rho_w \to \rho_{w\beta} = \rho_w\left(1 - 2\sqrt{\pi}\frac{1-\beta}{\beta}\frac{\tilde{p}\,\tilde{u}_\infty}{\sqrt{\tilde{T}}}\right). \tag{3.34}$$

Let us write the state equation of a perfect gas first in the reference case of absolute permeability, $\beta = 1$, and then in the case $\beta \le 1$

$$\left.\begin{aligned} p_w &= \frac{\rho_w kT_w}{m}, \\ p_{w\beta} &= \frac{\rho_{w\beta} kT_w}{m}. \end{aligned}\right\} \tag{3.35}$$

Hence, using (3.34) and (3.35)

$$\tilde{p}^{-1} = \tilde{p}_*^{-1} + 2\sqrt{\pi}\frac{1-\beta}{\beta}\frac{\tilde{u}_\infty}{\sqrt{\tilde{T}}}. \tag{3.36}$$

Here, $\tilde{p}= p_\infty/p_w$, $\tilde{p}_* = p_\infty/p_{w\beta}$ is the ratio of pressures, respectively, for the reference and general cases. Relation (3.36) is fundamental for the calculation of the effect of β on the process of evaporation.

3.6 Thermodynamic State of Vapor

The use of (3.34)–(3.36) enables one to examine the effect of the condensation coefficient on the thermodynamic state of the resulting vapor. The initial parameters for the evaporation problem are the temperature T_w of the surface and the pressure p_∞ of gas and the CPS. Using the saturation line equation, from these basic quantities one can evaluate the saturation pressure at the surface temperature $p_w \equiv p_s(T_w)$ and the saturation temperature at the pressure in the system $T_s \equiv T_s(p_\infty)$. The theoretical quantity p_w is contained in the denominator of the pressure ratios $\tilde{p}= p_\infty/p_w$ and is also used in the state Eq. (3.35). Let us first consider the case when the evaporation intensity is negligible in comparison with that of molecular mixing. The situation in which we can content ourselves with linear departures of the parameters from equilibrium is the subject of study of the kinetic theory of evaporation/condensation. The findings of this theory, as obtained in [2], can be written as follows

$$\frac{T_w - T_\infty}{T_w} = \frac{\sqrt{\pi}}{4}\tilde{u}_w, \tag{3.37}$$

$$\frac{p_{w\beta} - p_\infty}{p_w} = \frac{6\sqrt{\pi}}{5}\tilde{u}_w. \tag{3.38}$$

Using in (3.38) Eq. (3.36), we get, in the linear approximation

$$\frac{p_w - p_\infty}{p_w} = 2\sqrt{\pi}\frac{1 - 0.4\beta}{\beta}\tilde{u}_w. \tag{3.39}$$

Here, $\tilde{u}_w = u_w/\sqrt{2R_g T_w}$ is the velocity ratio constructed from the parameters on the CPS. Clausius–Clapeyron's relation for the saturation line reads as

$$\left(\frac{dp}{dT}\right)_s = \frac{h_{fg}\rho_v\rho_l}{(\rho_l - \rho_v)T_w} \approx \frac{\rho_v\rho_l}{T_w}. \tag{3.40}$$

Here, h_{fg} is the heat of phase transition.

For the range of pressures which are far from the critical point, we have $\rho_l \gg \rho_v$. Taking this into account and employing the equation state for a perfect gas (3.35), we rewrite (3.40) as

$$\left(\frac{dp}{dT}\right)_s \approx \frac{h_{fg} p_w}{R_g T_w^2}. \tag{3.41}$$

Integrating (3.41), we get, in the linear approximation

$$p_w - p_\infty = \left(\frac{dp}{dT}\right)_s (T_w - T_s) \frac{h_{fg}}{R_g T_w} p_w. \tag{3.42}$$

Now an appeal to from (3.37), (3.42) shows that

$$\frac{T_w - T_s}{T_w} = 2\sqrt{\pi}\gamma \frac{1 - 0.4\beta}{\beta} \tilde{u}_w, \tag{3.43}$$

where $\gamma = R_g T_w / h_{fg} \sim 0.1$ is the Trouton parameter [3].

Equations (3.37), (3.43) link the temperature T_∞ of the vapor formed on the CPS and the saturation temperature T_s at the pressure p_∞

$$\frac{T_\infty - T_s}{T_w} = \left(2\sqrt{\pi}\gamma \frac{1 - 0.4\beta}{\beta} - \frac{\sqrt{\pi}}{4}\right) \tilde{u}_w. \tag{3.44}$$

A direct consequence of (3.44) is that in the general case, the thermodynamic state of the vapor emanating from CPS is characterized by the non-equilibrium depending on the condensation coefficient (Fig. 3.5).

- in the interval $0 < \beta < \beta_*$, a vapor is formed on the CPS superheated with respect to the saturation temperature: $T_\infty > T_s$.
- in the interval $\beta_* < \beta \leq 1$, the vapor emanating from the surface is subcooled (super saturated): $T_\infty < T_s$.

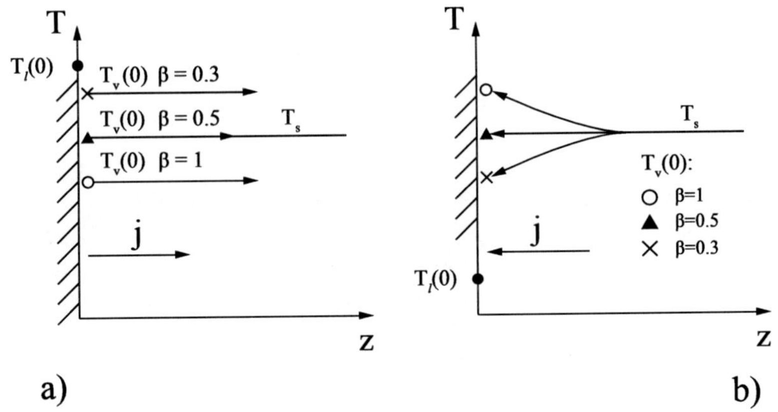

Fig. 3.5 Relative position of the temperatures $T_l(0)$, $T_v(0)$, and T_s during evaporation **(a)** and condensation **(b)** with various values of β

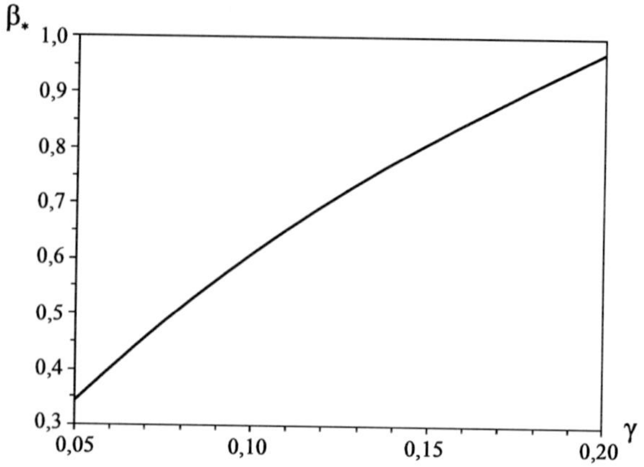

Fig. 3.6 Dependence of the equilibrium value of the condensation coefficient on the Trouton parameter at atmospheric pressure

- the equilibrium case defined by the formula

$$\beta_* = \frac{8\gamma}{1 + 3.2\gamma} \tag{3.45}$$

is an exception.

Figure 3.6 gives the dependence of the equilibrium value of the condensation coefficient on the Trouton parameter at atmospheric pressure. The figure shows that β_* can vary in the range from $\beta_* = 0.389$ (dodecane, $\gamma = 0.0857$) to $\beta_* = 0.932$ (helium $\gamma = 0.186$). The above dependences $\tilde{T}$ (M_∞), $\tilde{p}$ (M) for the reference case $\beta = 1$ can be approximated by the following formulas

$$\tilde{T}^{-1} = 1 + k_{T1}\mathrm{M}_\infty + k_{T2}\mathrm{M}_\infty^2 + k_{T3}\mathrm{M}_\infty^3, \tag{3.46}$$

$$\tilde{p}^{-1} = 1 + k_{p1}\mathrm{M}_\infty + k_{p2}\mathrm{M}_\infty^2 + k_{p3}\mathrm{M}_\infty^3. \tag{3.47}$$

The values of the numerical coefficients in (3.46), (3.47) are given in Table 3.1. As an important practical application of the intensive evaporation problem, we mention the combustion process on the interfacial surface. Here, the role of the condensed phase is played by the volume of liquid fuel and the process of evaporation is reduced to the emanation of gaseous exhaust products. As a characteristic example, let us consider evaporation at atmospheric pressure of dodecane, which is the principal component of organic fuels. We shall use the saturation line in the form

$$T(\mathrm{K}) = \frac{C_1}{C_2 - \ln(p(\mathrm{kPa}))} + C_3, \tag{3.48}$$

Table 3.1 Numerical coefficients in (3.46), (3.47)

	$i = 3$	$i = 5$	$i = 6$
k_{T1}	0.405	0.371	0.362
k_{T2}	−0.144	−0.217	−0.235
k_{T3}	0.226	0.182	0.170
k_{p1}	1.94	1.78	1.74
k_{p2}	1.38	1.14	1.09
k_{p3}	0.45	0.314	0.273

where $C_1 = 3744$, $C_2 = 14.06$, and $C_3 = 92.83$.

Now from Eqs. (3.35), (3.36), (3.46), and (3.47), one can find the "non-equilibrium degree" $\kappa = T_\infty / T_s$. The results of the calculations are depicted in Fig. 3.7. The figure shows that for $M_\infty = 0$, the vapor is in equilibrium due to the absence of evaporation

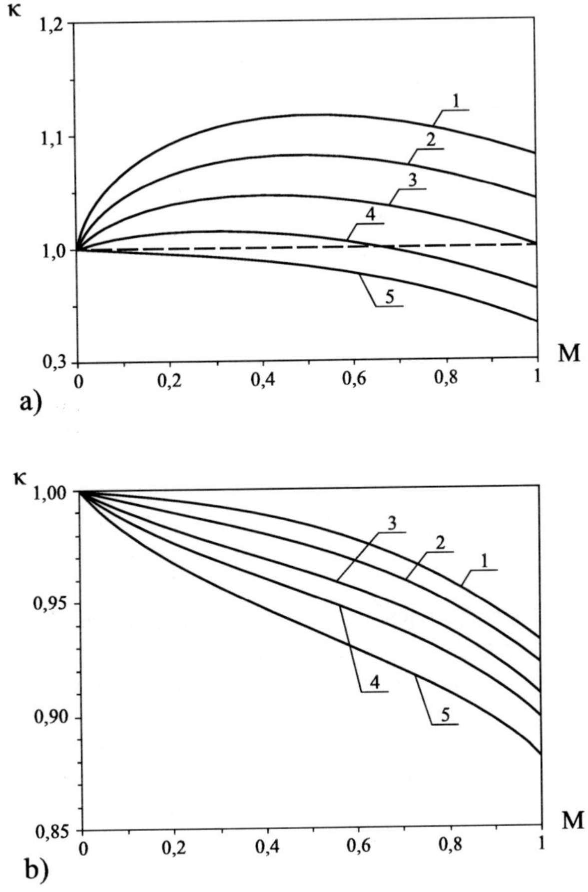

Fig. 3.7 Dependence of the non-equilibrium degree on the Mach number. **a** 1 $\beta = 0.15$, 2 $\beta = 0.2$, 3 $\beta = 0.279$, 4 $\beta = 0.4$, and 5 $\beta = 0.538$. **b** 1 $\beta = 0.538$, 2 $\beta = 0.6$, 3 $\beta = 0.7$, 4 $\beta = 0.8$, and 5 $\beta = 1$

($T_\infty = T_s : \kappa = 1$). In the range $0 < \beta \leq 0.279$, the vapor is superheated, $\kappa > 1$, for the entire range of variation of the Mach number. For $\beta = 0.279$, the vapor is again saturated at the point of "acoustic evaporation" ($M_\infty = 1 : \kappa = 1$). The area with subcooled (supersaturated) vapor, $\kappa < 1$, appears in the range of large Mach numbers as the condensation coefficient is increasing. The curve $\beta = 0.538$ is "boundary": for it $\kappa < 1$ for the entire range of variation of the Mach number. Finally, in the range $0.538 < \beta \leq 1$, the resulting vapor is always supersaturated for any Mach number.

The information derived above on the actual thermodynamic state of vapor for evaporation on the CPS depending on the condensation coefficient augments the relations of [2], which were obtained in the frameworks of the linear kinetic theory. For a more involved combustion in a different gas, one needs to have data on the temperature and density of exhaust products and atmospheric gas to calculate convection over the combustion surface. Note that the "boundary" curve $\kappa(M_\infty)$ has as $M_\infty \to 0$ the asymptotics $\gamma = 0.0538$, which is markedly different from that obtained in the linear approximation ($\gamma = 0.0857$). The cause of this disagreement is the empirical method of construction of the saturation line (3.48). Indeed, the above dependence (3.48) replicates, with minimal deviation, the saturation line for dodecane in a wide range of pressure variations. This dependence, however, is not based on the Clausius–Clapeyron relation (3.40), and hence, by definition, cannot provide the passage to the limit to small departures of pressure and temperature along the saturation line.

3.7 Laser Irradiation of Surface

Laser irradiation of the condensed-phase surface, which is known as a "target" in the literature, provides an efficient method of production of materials with special properties. The development of physical backgrounds for laser technology calls for a detailed theoretical investigation of the effect of high-power radiation on condensed substances. An important phenomenon related to laser irradiation is "laser ablation", which is the collection of physico-chemical processes removing the substance from the interfacial surface. The laser ablation results in intensive processes of evaporation and condensation, and as a corollary, in the formation of a substance flow through the CPS.

One of the first theoretical descriptions of laser ablation was proposed in [9], where a link between the target temperature and the vapor with "acoustic evaporation" was found. Later, models [36–38] appeared extending the theory of [9] to the entire range of Mach numbers of the vapor flow. Laser pulses are supplied to the target with periods of the order of a nanosecond and are characterized by a huge density of heat flow. Hence, even under the vacuum conditions, the vapor escaping under the conditions of strong evaporation has no time scatter. Hence, a "vapor cloud" is formed near the CPS with pressure exceeding that of the ambient medium. Since the free path length of vapor molecules is much smaller than the size of the vapor cloud, the regime of continuous gas-dynamic flow is realized at the initial state of its expansion.

Hence, the ablation regime is most important for applications related, for example, to pulse laser-induced evaporation. Note that gas-dynamic discontinuities were experimentally observed, for example, in the form of shock waves in the surrounding atmosphere and the contact boundary between the originating vapor and the atmospheric gas [39, 40]. The gas-dynamic conditions of evaporation on the CPS/vapor boundary depend substantially on its temperature, and hence laser ablation should be calculated when the gas-dynamic problem is conjugated with the heat transfer problem in the target. Note that involved processes on the CPS have not been taken into account in most early studies on laser ablation. Calculations were made using highly simplified approaches, where the one-dimensional [41] or two-dimensional [42] gas dynamics was studied in the nonconjugate setting. Besides, the evaporation rate and the vapor parameters near the CPS were estimated from experimental data or accepted from some model considerations.

The modern form of the "heat model" of laser ablation, which relates the parameters of vapor and the outside gas with the intensity of irradiation, was proposed in [37, 38]. In the target, the heat conduction equation related to the gas-dynamic equations by the BC of intensive evaporation was considered. In [43], to determine the evaporation rate it was assumed that the high-temperature condensed phase behaves like a dense gas. In [44], it was shown that once the laser pulse is complete, the CPS is mostly cooled by means of heat removal inside the target. As a result, for saturated vapor the pressure corresponding to the CPS temperature becomes smaller than the actual vapor pressure, which results in its retrograde condensation on the target. It is worth pointing out that all the studies mentioned above were in essence numerical. It is believed that an approximate analytical model of intensive evaporation is needed for a correct extension of the available results.

3.8 Integral Heat Balance Method

3.8.1 Analytical Solutions

It is well known that the majority of hydrodynamic and heat transfer problems are described by partial differential equations. So, the Navier–Stokes and energy equations are quasi-linear partial differential equations, which in the majority of cases can be solved only numerically. This may suggest the "natural" conclusion about the absolute priority of numerical solutions in this field. In the actual fact, analytical solutions to hydrodynamic and heat transfer problems play a significant role even in our computerized age. They have the following important advantages over numerical methods

- the merit of the analytical approach lies in the possibility of a closed qualitative description of a process, detection of the complete list of dimensionless parameters, and their hierarchical classifications in the order of importance,
- analytical solutions have the required generality, and parametric studies can be carried out by varying the boundary-value and initial conditions in analytical solutions,
- for the purpose of testing of numerical solutions of complete equations, one has to have basis analytical solutions of simplified equations (which are obtained after the assessment and rejection of some terms in the original equations),
- globally, analytical solutions can be used for direct validity checks of statements of numerical investigations of each specific problem.

In various applications, the thermal conduction problem is conjugated to some external problem: an inverse boundary-value thermoelasticity problem, an inverse heat conduction problem, an optimal control, etc. The subject of our study is the conjugate "heat conduction in the target–intensive evaporation from CPS" problem. One considers either the evaporation problem itself (in the case of a liquid target) or the sublimation problem (in the case of a hard target). It is known that mathematical modeling includes as components the solution of a complex of problems. The most important problem is that of the construction of a model adequately describing a given physical process. Ideally, the base model should be sufficiently simple and be adequate for the process under study. Each problem can be mathematically modeled both using numerical and analytical methods. In turn, the latter can be either exact (classical) or approximate to some extent.

- In spite of the vigorous development of modern computing hardware, direct methods of solution of conjugate problems remain poorly efficient in a number of cases. The problem under consideration has multiparameter characters, and hence any numerical solution necessarily gives as a f product a particular exact solution describing the given concrete conditions.
- The classical analytical methods include the methods of separation of variables, sources, heat potentials, as well as integral transformations (with finite or infinite limits). It should be recalled that exact analytical methods can be used only for the solution of linear differential equations. However, even in this case, the application of classical methods results in solutions in the form of series that poorly converge for small values of time. In separate special cases, the convergence of an exact analytical solution (the problem for an infinite plate with first-kind BC) can be observed only when using up to five hundred thousand (!) terms of the functional series.
- Approximate methods include variational methods, weighted residual methods, and integral methods. Besides being less accurate in calculations, approximate methods are superior to numerical methods due to their universality. Nevertheless, by no means all approximate methods are efficient. In particular, they are unfit for the solution of heat conduction problems for small values of time. First, to find eigenvalues of boundary-value problems, one should be capable of solving high-degree algebraic equations. Second, the satisfaction of initial conditions leads to

large systems of algebraic linear equations involving ill-conditioned matrices of coefficients [45–48]. Despite the availability of standard computer programs, the solution of such system involves as a rule considerable computational difficulties.

3.8.2 Heat Perturbation Front

An efficient approximate approach to the solution of boundary-value heat conduction problems is based on the concept of the "thermal perturbation front", which monotonically travels from the surface inside the body [45]. In turn, this approach is based on the integral method of heat balance or the method of averaging of functional corrections. In the frameworks of the integral method, the heat conduction equation is replaced by the heat balance integral. The field of temperatures is approximated by a polynomial, whose coefficients are found from some additional BC specified on the domain boundaries and on the front of thermal perturbation. The fulfillment of these conditions in the sought-for solution should be equivalent to their fulfillment at boundary points of the original differential equation.

Integral methods are capable of delivering simple approximate analytical solutions which exactly satisfy the initial and BC. However, such solutions feature the following "genetic disadvantage": they satisfy the original differential equation not exactly, but only "in the mean". Consequently, this leads to the necessity of an a priori choice of the temperatures field (for example, as a quadratic or a cubic parabola), which guarantees that the resulting solution cannot be exact.

The introduction of the concept of the thermal perturbation front significantly simplifies the boundary-value problem under consideration. Instead of solving the original partial differential equation, one integrates the ordinary differential equation with respect to the additional sought-for function. As a result, a series of approximate solutions starts to converge much faster, because the requirement that the initial solution of each separate problem is satisfied in the entire range of the spatial variables is replaced by the requirement that it should be satisfied only at one boundary point [49, 50].

Let us dwell upon one peculiarity characteristic of all integral methods. It is known that the parabolic heat conduction equation is derived on the basis of the fundamental hypothesis that heat propagates with an infinite rate in the Navier–Stokes region. At the same time, the integral method of the solution of the heat conduction equation involves the assumption about the finite propagation velocity of the "thermal perturbation front".

However, this "paradox" is only apparent. In the actual fact, a thermal layer of finite thickness propagating inside the body is an approximate image of a one-dimensional time-dependent field of temperatures in the original partial differential equation. As a result, the approximation of the thermal layer resembles Karman–Pohlhausen's method [51] in the theory of boundary layer. In the frameworks of this method, the system of Prandtl partial differential equations (the continuity equation and the equation of conservation of the momentum longitudinal component) is replaced by an ordinary differential equation for a layer of finite thickness propagating along a streamlined plate.

3.9 Heat Conduction Equation in the Target

To describe the interaction of the radiation flux with the target, one has to solve the conjugate problem involving the system of conservation equations of molecular flows in gas and the heat conduction equation in the heated body (liquid or hard). Let us consider the time-dependent one-dimensional heat conduction equation in a semi-infinite body $0 \leq x < \infty$

$$\frac{\partial \vartheta}{\partial t} = \alpha \frac{\partial^2 \vartheta}{\partial x^2}. \tag{3.49}$$

Here, t is the time elapsed from the beginning of irradiation, y is the coordinate measured from the CPS inside the target, $\vartheta = T - T_0$ is the difference in temperatures, and T_0 is the temperature of the body at infinity. Initially, the distribution of temperatures in the body is homogeneous

$$t = 0(0 \leq x < \infty) : \vartheta = 0. \tag{3.50}$$

Assume that for $t > 0$ the surface temperature T_w varies in time according to some law $\vartheta_w = \vartheta_w(t)$

$$x = 0(t > 0) : \vartheta = \vartheta_w, \tag{3.51}$$

where $\vartheta_w = T_w - T_0$.

In the case of simple homogeneous BC of the type $T_w = \text{const}$, $q_w = \text{const}$, one can introduce the similarity variable $\sim x/\sqrt{t}$ into Eqs. (3.49)–(3.51). As a result, (3.49) becomes an ordinary differential equation, which can be solved by using the Duhamel integral [52]. This solution can be represented in the error functions. To solve the conjugate problem, we shall use the approximate integral method of a layer of finite thickness δ. This method is based on the concept of the "thermal perturbation front". In accordance with this approach, we introduce the generalized variable $\eta = x/\delta$. The adiabatic condition is specified on the outer surface of the thermal layer which propagates inside the body. Assume that the temperature drop through the thermal layer can be written as $\vartheta = \vartheta_w f$, where $\vartheta_w = \vartheta_w(t)$, $f = f(\eta)$. Now the partial derivatives in (3.49) assume the form

$$\left.\begin{aligned} \frac{\partial \vartheta}{\partial x} &= \frac{\vartheta_w}{\delta}\frac{df}{d\eta}, \\ \frac{\partial^2 \vartheta}{\partial x^2} &= \frac{\vartheta_w}{\delta^2}\frac{d^2\vartheta}{d\eta^2}, \\ \frac{\partial \vartheta}{\partial t} &= \vartheta'_w f - \frac{\delta' \vartheta_w \eta}{\delta}\frac{df}{d\eta}. \end{aligned}\right\} \tag{3.52}$$

Here, prime denote time derivative: $\delta' = \frac{d\delta}{dt}$, $\vartheta'_w = \frac{d\vartheta_w}{dt}$. According to (3.50), (3.51), the BC for the function $f(\eta)$ read as

$$\left.\begin{array}{c}\eta = 0 : f = 1, \\ \eta = 1 : f = 0, \frac{df}{d\eta} = 0.\end{array}\right\} \quad (3.53)$$

The quadratic parabola

$$f = (1 - \eta)^2 \quad (3.54)$$

is the simplest function satisfying the BC (3.53).

According to the integral method, function (3.54) should be substituted into the original Eq. (3.49) and then the resulting equation should be integrated with respect to η from $\eta = 0$ to $\eta = 1$. As a result, we get the ordinary differential equation, which can be conveniently written in the following compact form

$$\frac{\delta'}{\delta} + \frac{\vartheta_w'}{\vartheta_w} = 6\frac{\alpha}{\delta^2}. \quad (3.55)$$

Equation (3.55) is a first-order linear inhomogeneous differential equation with respect to $\delta(t)$. Its general solution is trivial and has a quadrature (see [44])

$$\delta^2 = 12\frac{\alpha}{\vartheta_w^2}\int_0^t \vartheta_w^2 dt. \quad (3.56)$$

According to (3.54), the heat flow q and the temperature drop ϑ_w on the heated surface are related as follows

$$q = 2\frac{k\vartheta_w}{\delta}. \quad (3.57)$$

The law of motion of the front inside the target will be sought in the self-similar form, which follows from the dimensional analysis

$$\delta = m\sqrt{\alpha t}, \quad (3.58)$$

where m is the "growth modulus". Using (3.58) in (3.56), we get the following expression for the growth modulus

$$m^2 = \frac{12}{\vartheta_w^2}\frac{1}{t}\int_0^t \vartheta_w^2 dt. \quad (3.59)$$

By (3.57), (3.58) the parameters are related as

$$\vartheta_w = \frac{1}{2} mq \sqrt{\frac{t}{\rho k c_p}}. \tag{3.60}$$

Using dependence (3.60) in Eq. (3.59), we get the second expression for the growth modulus

$$m^2 = \frac{12}{q^2} \frac{1}{t^2} \int_0^t q^2 t dt. \tag{3.61}$$

From the equivalent expressions (3.59), (3.61), one can obtain the limit variants of the solution (3.57) satisfying the self-similar law (3.58) of motion of the thermal perturbation front. So, for $\vartheta_w = \text{const}$ (first kind BC), it follows from (3.59) that

$$\delta = \sqrt{12\alpha t} (m = 12). \tag{3.62}$$

In turn, for $q = \text{const}$ (second kind BC), we have from (3.61)

$$\delta = \sqrt{6\alpha t} (m = 6). \tag{3.63}$$

The surface conductance is found from formulas (3.57), (3.58) as

$$h \equiv \frac{q}{\vartheta_w} = 2\frac{k}{\delta} = \varsigma \sqrt{\frac{\rho k c_p}{t}}, \tag{3.64}$$

where $\varsigma = 2/m$.

Now using (3.62)–(3.64), $\varsigma = 1/\sqrt{3}$ for the BC $\vartheta_w = \text{const}$ and $\varsigma = \sqrt{2/3}$ for the BC $q = \text{const}$. For comparison, the exact solutions of these limit cases [45] read as $\varsigma = 1/\sqrt{\pi}$ for the BC $\vartheta_w = \text{const}$ and $\varsigma = \sqrt{\pi}/2$ for the BC $q = \text{const}$. So, relation (3.64) is capable of qualitatively correctly describing the heat transfer law of a semi-infinite body for two classical BCs. The relative errors Δ of approximate solutions can be obtained from the comparison with exact results: $\Delta \approx 2\%$ for the BC $\vartheta_w = \text{const}$ and $\Delta \approx 8\%$ for the BC $q = \text{const}$. Thus, the error of calculation in the first case is quite acceptable, and in the second case is quite high. This fact will be taken into account in the calculations that follow.

3.10 The "Thermal Conductivity–Evaporation" Conjugate Problem

Let us consider the problem of irradiation of a liquid target with a homogeneous initial distribution of temperatures $T_0 < T_s$ and a time-constant heat flow $q = \text{const}$ (Fig. 3.8). The temperature drop on the CPS grows in time according to the law

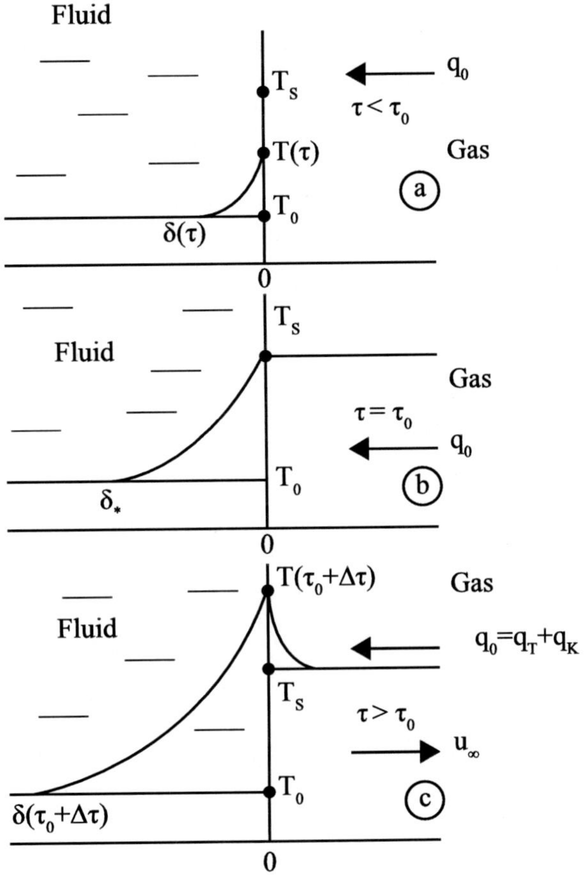

Fig. 3.8 "Thermal conductivity–evaporation" conjugate problem

$$\vartheta_w \equiv T_w - T_s = \sqrt{\frac{3}{2}}q\sqrt{\frac{t}{\rho k c_p}}. \tag{3.65}$$

Assume that at some time $t = t_0$, the surface temperature $T_w(t)$ attains the saturation temperature with pressure $T_s \equiv T_s(p_\infty)$ in the system. This triggers the process of liquid evaporation from the CPS. From (3.63), we find the thickness of the thermal layer for $t = t_0$ as

$$\delta_0 = \sqrt{6\alpha t_0}. \tag{3.66}$$

The time to reach the evaporation state is determined from relation (3.60) with $m = 6$ as

$$t_0 = \frac{2}{3}\frac{(q\vartheta_0)^2}{\rho k c_p}, \tag{3.67}$$

where $\vartheta_0 = T_s - T_0$ and $\alpha = k/\rho c_p$—thermal diffusivity.

After "triggering" the evaporation process, the heat flow q = const incident on the CPS will contain two components of heat balance

$$q = q_1 + q_2. \tag{3.68}$$

Here, q_1 is the heat flow spent for target warming, which can be found from formula (3.57), and q_2 is the heat flow spent for evaporation, which can be found from the kinetic analysis as follows

$$q_2 = h_{fg}\rho_v u_\infty. \tag{3.69}$$

For the later analysis, we introduce the following quantities: the dimensionless thickness of the thermal layer

$$\tilde{\delta} = \frac{\delta}{\delta_0}, \tag{3.70}$$

the dimensionless drop of temperatures measured from the saturation temperature

$$\theta = \frac{T_w - T_s}{\vartheta_0}, \tag{3.71}$$

the dimensionless drop of temperatures measured from the initial target temperature

$$\tilde{\vartheta} \equiv 1 + \theta = \frac{T_w - T_0}{\vartheta_0}, \tag{3.72}$$

the fraction of the heat flow spent for evaporation (the evaporation parameter)

$$\psi = \frac{q_2}{q}, \tag{3.73}$$

and the dimensionless time measured from the beginning of evaporation

$$\tau = \frac{t - t_0}{t_0}. \tag{3.74}$$

Using the above notation, we write in the dimensionless form the differential Eq. (3.55), which specifies the motion of the thermal perturbation front

$$\frac{d\left(\tilde{\delta}^2\right)}{d\tau} + 2\frac{d\ln\left(\tilde{\vartheta}\right)}{d\tau}\tilde{\delta}^2 = 1 \tag{3.75}$$

and the heat balance Eq. (3.68), which relates two components of the heat flow

$$\frac{\tilde{\vartheta}}{\tilde{\delta}} + \psi = 1. \tag{3.76}$$

Equations (3.75), (3.76) are fundamental for later calculations. In the case when the function $\tilde{\vartheta}\ (\tau)$ is given in an explicit form, the solution of Eq. (3.75) can be written as a quadrature, which can be found from (3.56)

$$\tilde{\delta}^2 = 1 + \frac{2}{\tilde{\vartheta}^2} \int_0^{\tau} \tilde{\vartheta}^2 \, d\tau. \tag{3.77}$$

Since in the actual fact the quantities $\tilde{\vartheta}$ and $\tilde{\delta}$ are related by condition (3.76), there appears the bonding quantity—the evaporation parameter ψ. This quantity can be found in the frameworks of the kinetic theory. The following system of equations is "triggered".

1. The input parameters are the pressure p_∞ in gas and the temperature T_w of the surface. From (3.46), (3.47) in view of the first Eq. (3.35), one finds the conditional gas pressure $\mathrm{p}_{w\beta}$ on the surface for the reference case $\beta = 1$.
2. From Eq. (3.36) using the saturation line Eq. (3.48), one gets the pressure $p_w \equiv p_s(T_w)$ as the saturation pressure at the surface temperature.
3. From the system of conservation equations of molecular flows (7)–(9), one finds the relation between the dimensionless temperature drop $\tilde{\vartheta}$ and the evaporation parameter ψ.
4. From Eq. (3.76), one finds the dimensionless thickness of the thermal layer $\tilde{\delta}$.
5. The resulting dependence $\tilde{\vartheta}\ (\tau)$ and $\tilde{\delta}\ (\tau)$ are substituted into the differential Eq. (3.75). The resulting dependence $\psi(\tau)$ is found by solving (3.75).

This chain of calculations shows that the overall procedure of the solution of the fundamental Eqs. (3.75), (3.76) is quite cumbersome. Estimates show that possible inaccuracies at any of the intermediate stages can result in considerable errors in the determination of the function $\psi(\tau)$.

3.11 Linear Evaporation Problem

We first consider the case of evaporation of small intensity, when one can consider only linear departures of the kinetic parameters from equilibrium: $\tilde{u}_w \equiv u_w/\sqrt{2R_g T_w} \ll 1$. In the frameworks of the linear problem, the chain of equations assumes a much simpler form.

- from Eqs. (3.30), (3.38), one finds the jump of temperatures $\frac{T_w - T_\infty}{T_w}$ and the conditional jump of pressures $\frac{p_{w\beta} - p_\infty}{p_w}$,
- from Eq. (3.39), one finds the actual jump of pressures $\frac{p_w - p_\infty}{p_w}$,
- from Eq. (3.43) and using the saturation line Eq. (3.42), one determines the function

$$\tilde{\vartheta} = 1 + B\psi. \tag{3.78}$$

Here,

$$B = \frac{\gamma F q T_s}{h_{fg} \rho_w v_w \vartheta_0} \tag{3.79}$$

is the kinetic parameter, $F(\beta) = 2\sqrt{\pi}\frac{1-0.4\beta}{\beta}$ and is the function of the condensation coefficient

- from Eq. (3.76), it follows that

$$\tilde{\delta} = \frac{1 + B\psi}{1 - \psi}, \tag{3.80}$$

- from Eq. (3.75), we get the dependence $\psi(\tau)$ in the implicit form

$$\tau = \frac{1}{2}(1 + B)^2 \frac{\psi(2 - \psi)}{(1 - \psi)^2} + B^2 \ln(1 - \psi) - \ln(1 + B\psi), \tag{3.81}$$

- from (3.79), it follows that the parameter B is complex.

It is responsible for the process of evaporation of the heat flow q, the condensation coefficient β, the temperatures of CPS and vapor, and the thermophysical properties. Figure 3.9 shows the dependences $\psi(\tau)$ for various parameters B.

It is seen that an increase of the kinetic parameter results in delaying the transition from the warming regime to the evaporation regime. The asymptotics of solution (3.81) can be written down explicitly as

$$\tau \to 0 : \psi = \frac{\tau}{1 + B}, \tag{3.82}$$

$$\tau \to \infty : \psi = 1 - \frac{1 + B}{\sqrt{2\tau}}, \tag{3.83}$$

$$B \to 0 : \psi = 1 - \frac{1}{\sqrt{1 + 2\tau}}, \tag{3.84}$$

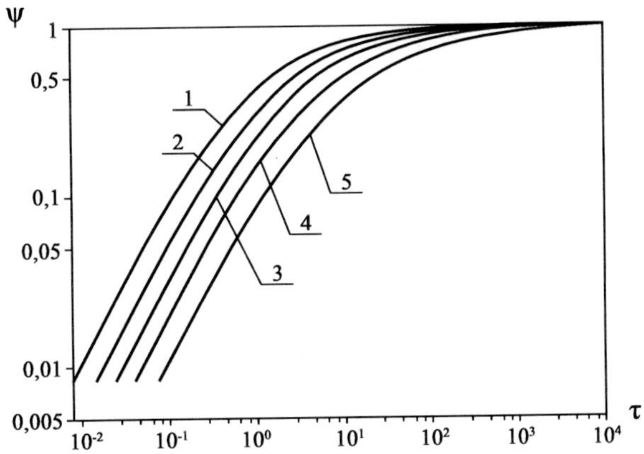

Fig. 3.9 Dependences $\psi(\tau)$ for various parameters B. 1 B $= 0$, 2 $B = 0.4$, 3 $B = 1$, 4 $B = 2$, and 5 $B = 4$

$$B \to \infty : \psi = \frac{\tau}{B^2}. \tag{3.85}$$

From (3.82) it follows that, in the initial period, the fraction of the heat flow spent on evaporation is negligible. Asymptotics (3.83) show that for sufficiently large exposure times, the volume has time to get completely warm, and all the supplied heat due to irradiation is spent for evaporation. Physically, these two limit cases are quite natural.

Asymptotics (3.84) is much different. From this asymptotics, it follows that even for the zero value of the kinetic parameter the problem remains to have the conjugate character. In other words, the redistribution of the heat balance components from the volume warm-up regime as $\tau \to 0$ to the complete evaporation regime as $\tau \to \infty$ still takes place. This means that the limit case of monotone warming of a body by a time-constant heat flow $q =$ const, as described by Eq. (3.65), is not realized for any τ.

The asymptotics $B \to 0$ can be physically explained as follows. From (3.79), it follows that the convergence to the zero of the kinetic parameter is equivalent to that of the heat flow incident on the target, $B \sim q \to 0$. In the actual fact, this means the degeneration of both components of the heat balance (3.68): $q_1 \sim q_2 \sim 0$. We note, however, that by (3.65) a similar situation also takes place with constant heating of the target by a constant heat flow without evaporation

$$q \to 0 : \vartheta_w \to T_w - T_s = \text{const}.$$

However, on the qualitative level, the above problems are totally different. The "nonconjugate case" $q =$ const does not take into account the possible stage of evaporation. On the other hand, asymptotics (3.84) keeps the "memory" about the statement of the conjugate "thermal conductivity plus evaporation" problem. So, we have found out that the heat flow balance Eq. (3.76) plays a fundamental role in the conjugate problem under consideration. The addition of this equation transforms the original system of equations into a qualitatively new state. Mathematically, the "physical paradox" described above is equivalent to the bifurcation phenomena. The discovery of such a special regime gives another evidence to support the advantage of approximate analytical modeling.

Finally, asymptotics (3.85) means that with an unboundedly large value of B, the evaporation regime cannot practically be triggered due to the inequality $\tau \ll B$. In this case, the problem remains formally conjugate, but for all values of time it "hangs" in the domain $\tau \to 0$. The above characteristic features of the conjugate problem can play a significant role in applications involving the combustion of a semi-bounded volume of fuel subject to an external heat flow. Below, it will be shown that the kinetic distinguishing feature of the problem is most sharply manifested in the case of small values of the condensation coefficient.

3.12 Nonlinear Evaporation Problem

In posing the general (nonlinear) problem, conditions 1–4 of the kinetic chain can be verified analytically, but this may make some mathematical calculations inordinately heavy. The fifth condition is reduced to a quadrature, which was calculated numerically using the computer algebra system Maple. Here, in contrast to the linear problem, instead of the universal dimensionless kinetic parameter (3.79) one should use dimensional parameters, and in particular, specify concrete values of the heat flow. The results of the calculation of the evaporation parameter are given in Figs. 3.10–3.13.

The solid line shows the results of the calculation for the general case, and the dashed line, for the linear approximation. The figures show that if the condensation coefficient decreases in the range $0.1 < \beta \leq 1$, then both the values of β and the heat flow have practically no effect on the dependence $\psi(\tau)$. Here, asymptotics (3.84)

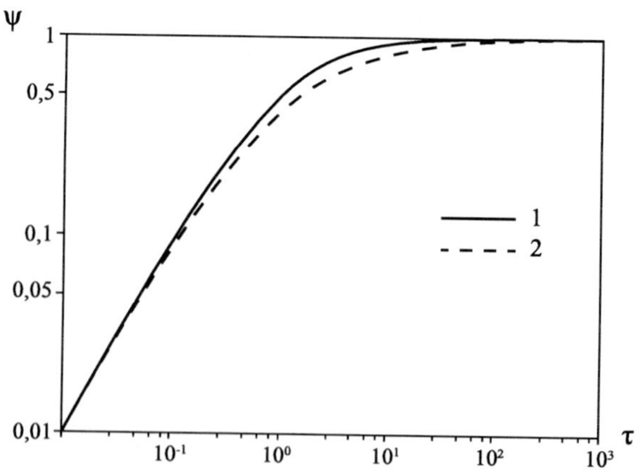

Fig. 3.10 Evaporation parameter versus the dimensionless time for $\beta = 1$, $q = 10^6$ W/m^2. 1 General case and 2 linear approximation

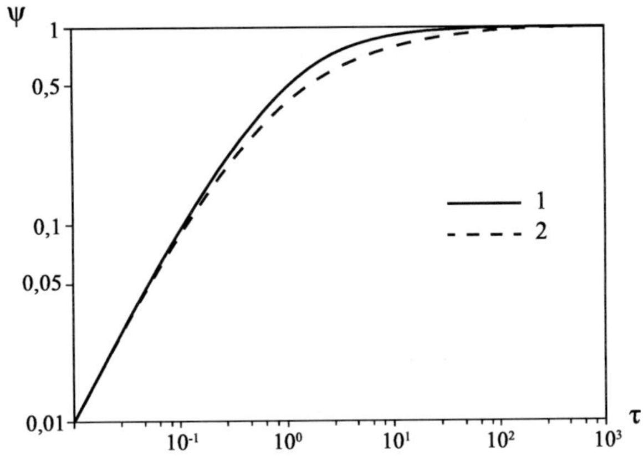

Fig. 3.11 Evaporation parameter versus the dimensionless time for $\beta = 10^{-1}$, $q = 10^6$ W/m^2. 1 General case and 2 linear approximation

of the linear problem is in fact realized. Besides, the departure of the dashed curve from the solid one is explained by the involvement of the saturation line Eq. (3.48) for dodecane from the Clausius–Clapeyron relation (3.40). The strong effect of β and q on the character of the curve $\psi(\tau)$ is observed in the range $10^{-3} < \beta \le 10^{-2}$. Figures 3.10–3.13 show that the numerical dependence of the evaporation parameter on time becomes more and more gradual for each fixed value of β with increasing q. The analysis of Fig. 3.10 shows that the effect of both these parameters increases monotonically as the condensation coefficient decreases.

It would be interesting to study the thermodynamic state of the outgoing vapor from the CPS in nonlinear approximation. To this end, a limit case was chosen in which the "acoustic evaporation" regime, $M_\infty = 1$, is realized when changing to the complete evaporation stage. This corresponds to the heat flow $q \approx 2.07*10^8$ W/m^2. Figure 3.14 shows the theoretical dependences of the non-equilibrium degree of the resulting vapor versus the dimensionless time for various values of the condensation coefficient.

For $\tau = 0$, due to the absence of evaporation, the vapor is in equilibrium $(M = 0 : \kappa = 1)$. The absence of a thermodynamic equilibrium becomes manifest with increasing time. In the range $0.15 \le \beta \le 0.279$, the resulting vapor for any time was found to be superheated with respect to the saturation temperature ($\kappa > 1$). It is worth noting that in this case the dependence $\kappa(\tau)$ passes through its maximum. For the "boundary" case $\beta = 0.279$, the vapor again returns to the equilibrium state ($\kappa = 1$) as $\tau \to \infty$. If the condensation coefficient is further increased, then the vapor becomes superheated up to a certain time ($\kappa > 1$), and then it becomes subcooled (supersaturated), $\kappa < 1$. With increasing β, the length of the interval in which the superheated vapor exists becomes smaller (and eventually it vanishes with $\beta = 0.538$). Finally, for $\beta > 0.538$, the vapor is always supersaturated at any time.

To conclude, let us consider the applicability of formula (3.84) in calculations. Physically, this formula corresponds to "weak evaporation". According to estimates, the use of asymptotics (3.84) in the range $0 < B < 10^{-2}$ gives an error at most

Fig. 3.12 Evaporation parameter versus the dimensionless time for $\beta = 10^{-2}$. 1 General case and 2 linear approximation. **a** $q = 10^6$ W/m^2, **b** $q = 5 * 10^6$ W/m^2, and **c** $q = 10^7$ W/m^2

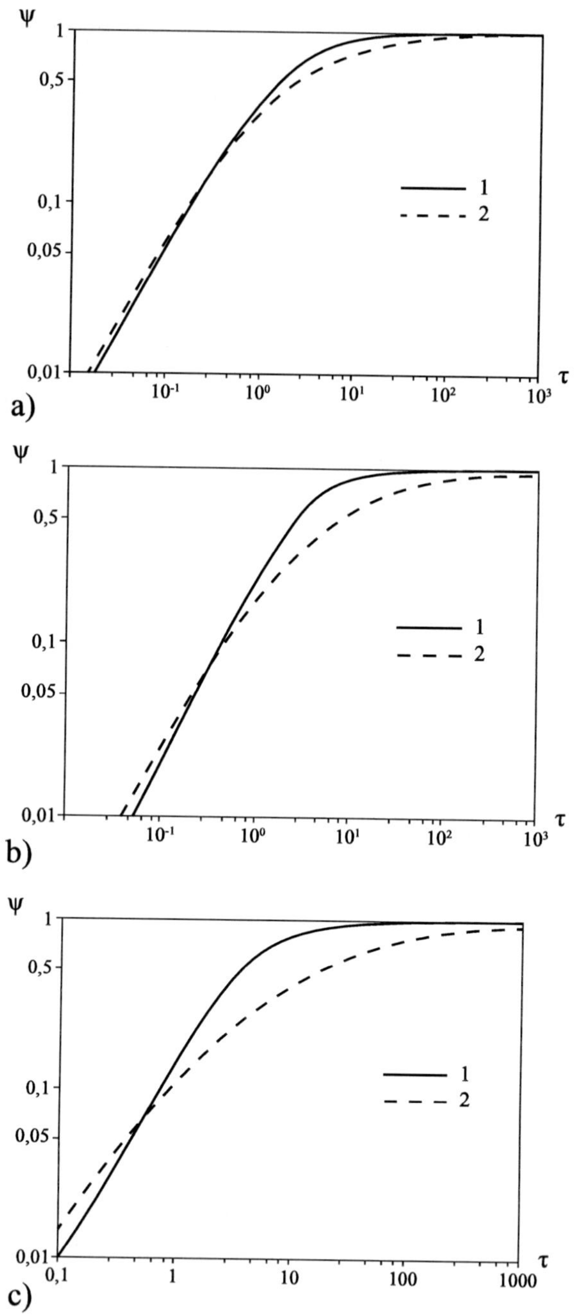

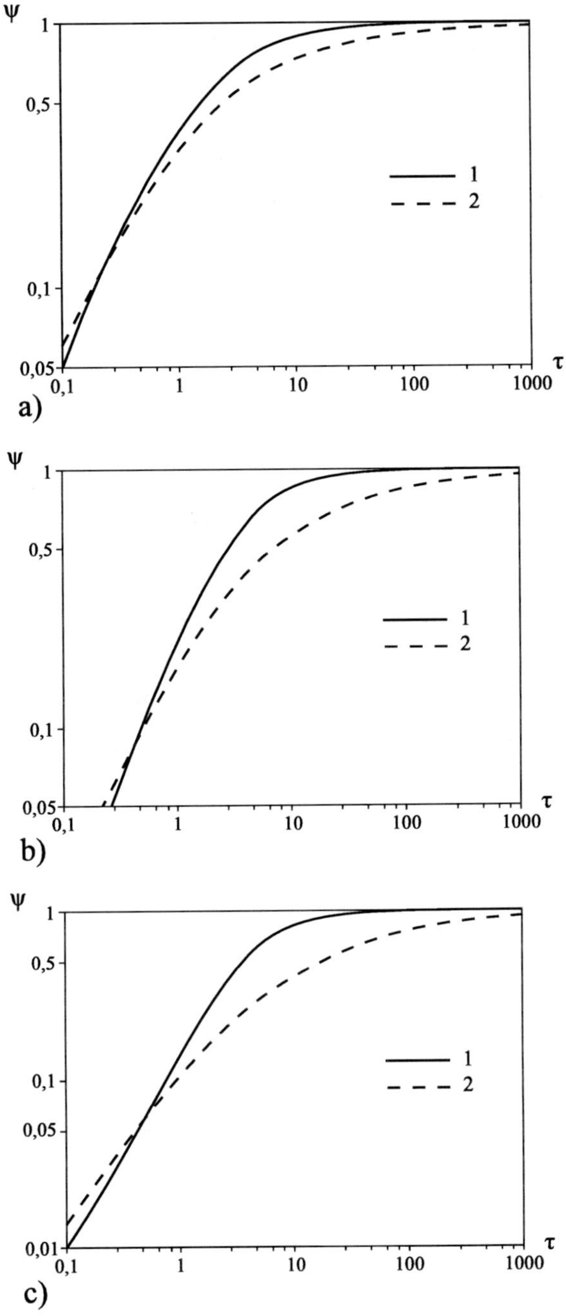

Fig. 3.13 Evaporation parameter versus the dimensionless time for $\beta = 10^{-3}$.1 General case, 2 linear approximation. **a** $q = 10^5$ W/m^2, **b** $q = 5 * 10^5$ W/m^2, and **c** $q = 10^6$ W/m^2

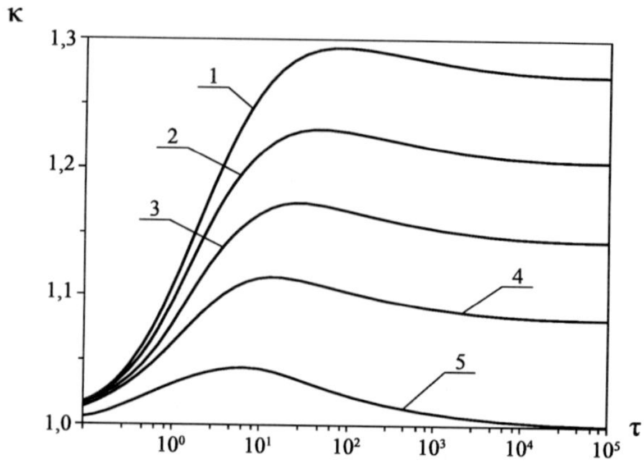

Fig. 3.14 Non-equilibrium degree of the resulting vapor versus the dimensionless time for $q = 2.07{*}10^8$. 1 $\beta = 0.15$, 2 $\beta = 0.2$, 3 $\beta = 0.279$, 4 $\beta = 0.4$, and 5 $\beta = 0.538$

2%. From the above conditions of evaporation of dodecane at atmospheric pressure, one can get the dependence of the maximally possible heat flow on the condensation coefficient

$$q_{\max} = \frac{33.5\beta}{1 - 0.4\beta} \frac{\text{kW}}{\text{m}^2}. \tag{3.86}$$

The dependence $q_{\max}(\beta)$ obtained from formula (3.86) is depicted in Fig. 3.15.

The results derived above show that the thermodynamic state of vapor behaves nontrivially as a function of the condensation coefficient. This problem is relevant in the study of vapor evaporation into a gaseous atmosphere. In this case, the evaporated and atmospheric gases mix with each other, which has an effect on the temperature and pressure in the Navier–Stokes region.

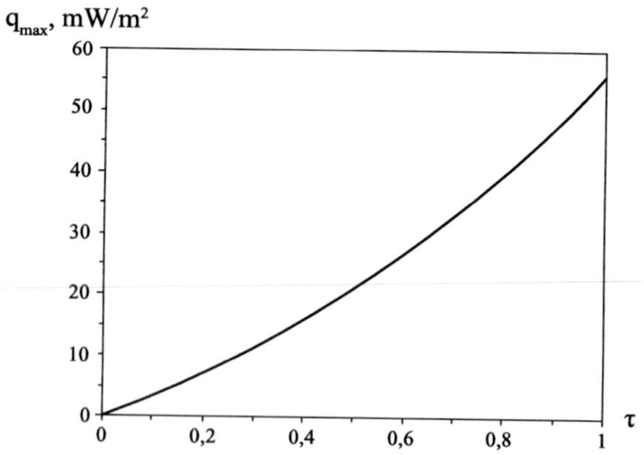

Fig. 3.15 The maximal heat flow versus the condensation coefficient

3.13 Irradiation of Saturated Fluid

The previous analysis was concerned with the general irradiation of subcooled liquid (Fig. 1). The elegant analytical solution (3.81) was obtained via the machinery of the linear kinetic theory of evaporation [2]. Our calculations show that this solution properly describes the process of sufficiently intense evaporation (see (3.86) and Fig. 3.15). Hence, below we shall use the linear approximation when dealing with more involved problems.

We first consider a particular case of solution (3.81) when the heat flux reaches the surface of a semi-infinite body uniformly heated to the saturated temperature: $T_0 = T_s$. This means that the heating time of the body surface to saturation and the initial subcooling of the liquid are both zero: $t_0 = \vartheta_0 = 0$. From (3.79), it follows that the kinetic parameter becomes infinite: $B \to \infty$. It should be noted that already for $t \geq 0$ the conjugation stage is realized, without going through the heat conduction stage. To clarify the physics of the process of irradiation of saturated fluid, consider the case of weak evaporation. Here, instead of t_0, as the natural time scale one should take the quantity $t_* = t_0 B^2$, which can be found from (3.67), (3.79)

$$t_* = \frac{2\pi}{3}\left(\frac{1-0.4\beta}{\beta}\right)^2\left(\frac{c_p}{R_g}\right)^2\left(\frac{R_g T_s}{h_{fg}}\right)^4\frac{\alpha_w}{v_w^2}. \tag{3.87}$$

Now the one-parameter solution (3.81) assumes the more simple universal form (Fig. 3.16)

$$\tau = \frac{\psi(2-\psi)}{2(1-\psi)^2} + \ln(1-\psi), \tag{3.88}$$

where $\tau = t/t_*$ is the dimensionless time.

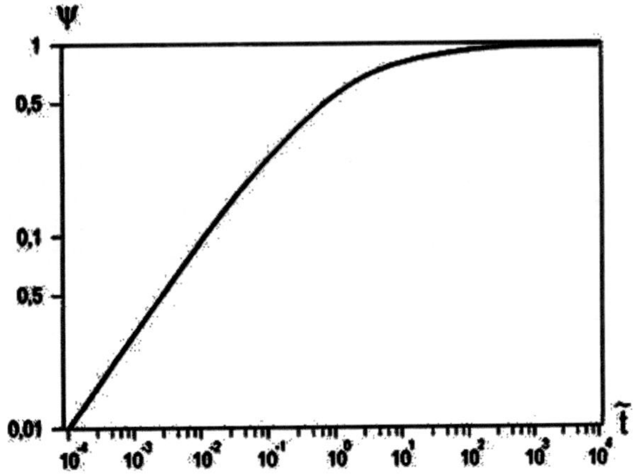

Fig. 3.16 Evaporation parameter versus dimensionless time for irradiation of saturated fluid

The implicit solution (3.88) can be approximated, with an error at most 5%, by the simple relation

$$\psi = \frac{\sqrt{\tau}}{1+\sqrt{\tau}}. \tag{3.89}$$

From (3.89), it follows that in the case of irradiation of saturated fluid the evaporation parameter ψ is independent of the heat flux incident on the target, but depends on the vapor temperature, its thermophysical properties, and on the condensation coefficient. Let us consider the asymptotics of solution (3.88). As in the general case of the irradiation of subcooled liquid, in the initial period the fraction of the heat flux spent for evaporation is negligibly small

$$\tau \to 0 : \psi = \sqrt{\tau}. \tag{3.90}$$

For sufficiently large irradiation time, the entire volume has time to get completely warm reaching the maximum possible temperature, and so all the supplied heat due to irradiation is spent for evaporation

$$\tilde{t} \to \infty : \psi = 1 - \frac{1}{\sqrt{2\tau}}. \tag{3.91}$$

According to the well-known physical–chemical Trouton's rule [3], $R_g T_s / h_{fg} \approx 10^{-1}$. Taking into account the relation $c_p / R_g = 5/2$ for the isobaric heat capacity for a monoatomic perfect gas, we get, for the conjugation period

$$t_c = 164 \left(\frac{1 - 0.4\beta}{\beta} \right)^2 \frac{\alpha_w}{\mathrm{v}_w^2}. \tag{3.92}$$

It would be interesting to compare this quantity with the characteristic time of intermolecular collisions [17]

$$t_{\mathrm{mol}} = \sqrt{2\pi} \frac{\nu_w}{\mathrm{v}_w^2}. \tag{3.93}$$

From (3.92), (3.93), we have

$$\chi \equiv \frac{t_c}{t_{\mathrm{mol}}} \approx \frac{0.41}{\mathrm{Pr}} \left(\frac{1 - 0.4\beta}{\beta} \right)^2,$$

where $\mathrm{Pr} = \nu_w / \alpha_w$ is the Prandtl number. Assuming $\mathrm{Pr} \approx 1$ for gas, we get

$$\chi \equiv \frac{t_c}{t_{\mathrm{mol}}} \approx 0.41 \left(\frac{1 - 0.4\beta}{\beta} \right)^2. \tag{3.94}$$

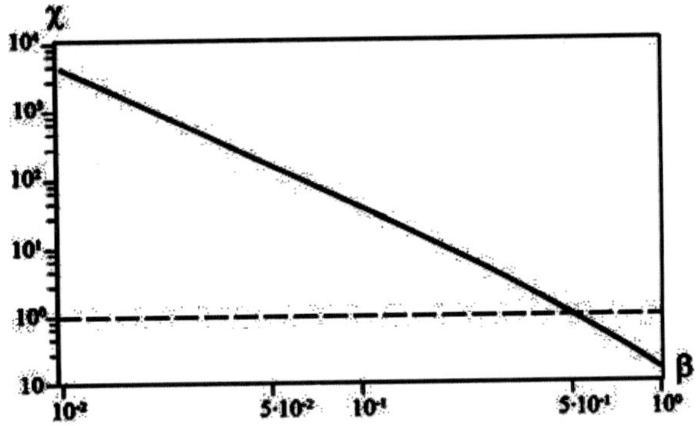

Fig. 3.17 Parameter χ versus the condensation coefficient

Figure 3.17 shows the dependence of the parameter χ on the condensation coefficient. It shows that in the reference case $\beta = 1$, the conjugation period is smaller by almost one order than the time of intermolecular collisions ($\chi = 0.148$). This means that the beginning of irradiation is accompanied by the practically instantaneous transition to the complete evaporation state. As the condensation coefficient decreases, the characteristic times of both processes become close to each other and for $\beta = 0.51$ they become equal ($\chi = 1$). Moreover, for $\beta = 10^{-2}$ we get $\chi \approx 4{*}10^3$. So, a decrease in the condensation coefficient increases the delay of the conjugation stage (this phenomenon is manifested most vividly for $\beta < 10^{-1}$).

3.14 Conjugate Problem for Hyperbolic Heat Conduction Equation

3.14.1 Generalized Form for the Solution

We write the heat flux on the body surface as $q_k = h\vartheta$, where h is the Heat Transfer Coefficient (HTC). According to the theory of heat conduction [49], the HTC for a semi-infinite body varies in the range $h_\vartheta \le h \le h_q$. Here, the subscripts denote the corresponding thermal BC: "ϑ", for the first kind BC ($\vartheta = \text{const}$) and "q", for the second kind BC ($q = \text{const}$). In further analysis, we shall use the following dimensionless HTC

$$\tilde{h} = h\sqrt{\frac{t_*}{kc_p\rho}},\ \tilde{h}_\vartheta = h_\vartheta\sqrt{\frac{t_*}{kc_p\rho}},\ \tilde{h}_q = h_q\sqrt{\frac{t_*}{kc_p\rho}}. \tag{3.95}$$

We now have the following double inequality $1/\sqrt{3\tau} \le \tilde{h} \le \sqrt{3/4\tau}$. In view of (3.95), formula (3.88) for the evaporation parameter can be written as

$$\psi = \left(1 + \sqrt{3}\,\tilde{h}\right)^{-1}.$$

The asymptotical analysis of solution (3.88) shows that in the process of irradiation of a body, the thermal state of its surface evolves between two limit variants: from $q = \text{const}$ to $\vartheta = \text{const}$. In these cases, we have by (3.95)

$$\left.\begin{array}{l} \tau \to 0: \ \tilde{h} = \sqrt{\frac{8}{9}}\tilde{h}_q = \sqrt{\frac{3}{4}\tau}, \\ \tau \to \infty: \ \tilde{h} = \tilde{h}_\vartheta = \frac{1}{\sqrt{3\tau}}. \end{array}\right\}$$

Hence, solution (3.88) can be written with good approximation as

$$\psi = \left\{1 + \left[\sqrt{3}\tilde{h}_\vartheta + \left(\sqrt{\frac{8}{3}}\tilde{h}_q - \sqrt{3}\tilde{h}_\vartheta\right)\varphi\right]\right\}^{-1}. \tag{3.96}$$

Relation (3.96) is an explicit form of solution (3.88) for the evaporation parameter. Here,

$$\varphi = \left(1 + \tau^{1/2}\right)^{-1/2}.$$

is the transition function describing the evolution of the heat flux components with time

$$\left.\begin{array}{l} \tau \to 0: \varphi = 1, \ \psi = \left(\sqrt{\frac{8}{3}}\tilde{h}_q\right)^{-1}, \\ \tau \to \infty: \ \varphi = 0, \ \psi = \left(\sqrt{3}\tilde{h}_\vartheta\right)^{-1}. \end{array}\right\}$$

In view of (3.96), the exact solution (3.88) can be approximately written in the following generalized form

$$\psi = 1 + \tau^{-1/2}\left[1 + \left(\sqrt{2} - 1\right)\left(1 + \tau^{1/2}\right)^{-1/2}\right]. \tag{3.97}$$

3.14.2 Fourier's Heat Conduction Hypothesis

The analytical theory of heat conduction originates from Fourier's report "On the Propagation of Heat in Solid Bodies" given in Paris in 1807. In its final form, Fourier presented his theory in 1822 in the paper "Théorie analytique de la chaleur", which was called the "great mathematical poem" by Lord Kelvin. As a result, the basic Fourier's conduction law (or Fourier's heat conduction hypothesis) claims that the heat flux q is proportional to the gradient of the temperature[1]

$$q(x,t) = -k\frac{\partial\vartheta(x,t)}{\partial x}. \tag{3.98}$$

For a complete description of heat propagation in a stationary medium, one should also take into account the energy conservation law

$$\frac{\partial\vartheta}{\partial t} = -\frac{1}{c_p\rho}\frac{\partial\vartheta}{\partial x}.$$

This implies Eq. (3.49), which is also known as the Parabolic Heat Conduction Equation (PHCE). Its specific features are illustrated by Lord Kelvin's point heat source solution (Thomson W, The Dynamical Theory of Heat, 1851, Edinburgh). Assume that at time $t = 0$ at point $x = 0$, the heat quantity Q is instantaneously generated. The solution of (3.49) with the initial conditions $\vartheta(x,0) = 0$ for $-\infty < x < \infty$ is given by

$$\vartheta(x,t) = \frac{1}{2\sqrt{\pi t}}Q\exp\left(-\frac{x^2}{4\alpha t}\right).$$

So, the solution obtained by Lord Kelvin describes the time decaying heat diffusion, moreover, for $t > 0$ we have $\vartheta(x,t) > 0$ on the entire interval $-\infty < x < \infty$. This leads to the paradox of infinite rate of heat propagation: a thermal disturbance at some point of the body results in an instantaneous variation of the temperature at all points of the body. Note that a similar paradox also takes place in the theory of Brownian motion.

This property of PHCE, which follows from Fourier's heat conduction hypothesis (3.98), contradicts the fundamental physical laws restricting infinite signal propagation velocities. This shows that Fourier's hypothesis is incomplete, because it does not involve parameters restricting the heat propagation rate. The necessity of the introduction of such parameters is due to the heat transfer mechanism by elementary particles of matter: electrons, molecules, and ionic lattices. This is why the result of heat transfer—the formation of the temperature field in the rigid body—should also be governed by processes related to periodic oscillations of these particles. As a result, the heat propagation law should be of wave character depending on the motion velocities of particles, the length and time of their free path, and their interaction at collisions.

Riemann was the first to cast some doubt on Fourier's hypothesis in the context of heat propagation in anisotropic media (Riemann B, Weber H, Lewy H. The collected works of Bernhard Riemann. Dover Publications. New York. 1953). He reformulated the problem as follows: "Determine the thermal state of an arbitrary solid body so that a system of isothermal curves given at some time should remain a system of

[1]Here and below we consider the one-dimensional heat conduction problem.

isothermal curves at all times, i.e., the temperature of a point should be expressed as a function of time and two auxiliary variables". Based on the study of Fourier's hypothesis, Riemann found a class of isothermal surfaces inside a rigid body. He proved that with each isothermal surface of a given set, one can uniquely associate a certain differential operator of heat conduction. Here, the linear operator of heat conduction of parabolic type forms only one subset of a more general set. It follows that Fourier's heat conduction hypothesis is a particular case of a general heat motion law in a rigid body.

Unfortunately, the original Riemann's work remained practically unnoticed and the theory of heat conduction was further developed along the way of searching the solutions of the parabolic operator with various ICs and BCs. This direction appeared from the practical need in the evaluation of temperature distributions in bounded bodies of various shapes. However, this approach contradicts Riemann's generalized law of propagation of heat in a rigid body. This corollary can be formulated in a simplified way as follows: "To the parabolic operator of heat conduction there corresponds a definite narrow class of isothermal surfaces. Moreover, it is impossible to go beyond this class by simple enumeration of initial and boundary conditions".

So, in some cases, the traditional methods of solution of boundary-value problems for PHCEs are reduced to the artificial "imposition" of various ICs and BCs to the parabolic heat conduction operator with the aim of obtaining unnatural temperature fields. The author thinks that it was the violation of Riemann's heat conduction law that resulted in the appearance of several paradoxes and ill-posed problems in the theory of heat conduction.

The paradox of infinite heat propagation rate evoked a real discussion in the 1950s. Several hypotheses on heat flow relaxations were introduced to eliminate this paradox. In this way, the studies arrived at the wave nature of heat conduction with a finite propagation rate of thermal excitations in a rigid body. A dualism of the theory of heat conduction with optical phenomena was identified.

- On the one hand, the mechanism of heat conduction is effected by the flow of interacting particles (atoms, molecules, etc.). The process of heat propagation is described by the parabolic operator of heat conduction. This operator is characterized by macroscopic parameters (heat capacity, heat conductivity, etc.), which are interpreted as thermophysical constants of the medium. The heat propagation rate is infinite.
- On the other hand, heat conductivity is a wave process with a finite propagation rate of a thermal wave and its dispersion as its quantitative characteristics.

It is generally accepted that, for the majority of heat conduction problems, the effect of the finite heat propagation rate is negligibly small. Hence, PHCEs are used practically always for the description of the heat transfer mechanism in a rigid body. The finiteness of the propagation rate of thermal perturbations is taken into account only when dealing with processes with extremal thermal effects and with super-fast variation of thermal fields. As a typical example of such phenomena, one can mention the effect on the target surface from powerful pulse heat fluxes with laser treatment of materials (they run in nano- or femtoseconds). Other examples include heating processes following friction with anomalously high velocity, heat stroke physical mechanisms, and process of local heating with the dynamical propagation of a fraction under transonic conditions.

3.14.3 Cattaneo–Vernotte's Heat Conduction Hypothesis

In 1958, Cattaneo [53] and Vernotte [54] independently proposed a generalization of the classical law of heat conduction by taking into account the time of heat relaxation t_r

$$q(x, t + t_r) = -k\frac{\partial \vartheta(x, t)}{\partial x} \tag{3.99}$$

in relation (3.98). Expanding the left-hand side of (3.99) in a Taylor series in t_r, we get in the zero approximation Fourier's hypothesis (3.98), and in the first approximation, Cattaneo–Vernotte's hypothesis

$$q(x, t) + t_r\frac{\partial q(x, t)}{\partial t} = -k\frac{\partial \vartheta(x, t)}{\partial x}. \tag{3.100}$$

The use of formula (3.100) in the energy conservation equation results in the Hyperbolic Heat Conduction Equation (HHCE)

$$\frac{\partial \vartheta}{\partial t} + t_r\frac{\partial^2 \vartheta}{\partial t^2} = \alpha\frac{\partial^2 \vartheta}{\partial x^2}. \tag{3.101}$$

Equation (3.101) takes into account the delay of the reaction of the non-equilibrium system followed by a thermal perturbation. This effect is due to the inertia of the heat flux, which responds to a variation of the temperature gradient not instantaneously (as in an equilibrium system), but after a finite time t_r. Equation (3.101) becomes the classical PHCE (3.49) as $t_r \to 0$.

A number of practically important problems of transient heat conduction for hyperbolic transfer model was considered in the book [55], which develops the analytical approach resulting in new integral relations convenient for numerical calculations.

3.14.4 Spatially Inhomogeneous Structures

In the majority of studies, it is traditionally assumed that the heat transfer rate has the same order as the sound velocity w. This means that the relaxation time is a characteristic of the matter, which can be estimated as $t_r \approx \alpha/w^2$. For rigid bodies, liquids, and gases, this gives the range $t_r \approx \left(10^{-12} - 10^{-8}\right)$ s. This leads to an unambiguous statement on the negligibly small effect of heat relaxation in the majority of practical problems. However, the results of experiments [56] on the measuring of nonstationary distribution of temperatures in pebble beds (wet sand, baking soda ($NaHCO_3$), glass balls, etc.) in a continuous air environment are in complete contradiction with these estimates. The analysis of experimental findings of [56] gave huge (in comparison with the commonly accepted ones) values of the relaxation time in the range 10–54 s. Close values ($t_r \approx 15$ s) were also obtained in experiments of [57] on measuring a nonstationary temperature field at the contact of two samples of biological tissue. In turn, a criticism of findings of [56, 57] was given in [58, 59]. For example, in [58, 59] with sand and biological tissue pebble beds, no effect of heat flow relaxation was registered at all. The experimental findings of [58, 59] are in good agreement with the solution of the classical PHCE.

In order to explain these contradictory findings, a detailed analysis of the experimental and data processing methods of [56–59] was carried out in [60]. The following conclusions were made in [60]. When determining the coefficient a_e of efficient thermometric conductivity of the granular medium, Kaminski [56] used the solution of the PHCE (3.49). This value was further used for the evaluation of the relaxation time involved in the HHCE, but this is incorrect. Moreover, in the description of experimental data in [56], there is a notational confusion with the use of the thermal conductivity and the thermal diffusivity, which makes problematic the interpretation of the results.

- In [57], the temperature field was measured with high accuracy in a specimen, and calibration experiments on heating of an aluminum plate were carried out. The drawbacks of the method of [57] include the measurement of each of the thermophysical parameters of the medium (the heat conduction k_e, the heat capacity c_e, and the density ρ_e) in separate independent experiments. This could result in substantial errors in the evaluation of the efficient thermal diffusivity ($\alpha_e = k_e/c_e\rho_e$) and, as a result, of the relaxation time.
- In experiments of [58] with a wet sand pebble bed, the method for finding α_e is debatable (as in [56]). Moreover, here the specimen was electrically heated, which corresponds to second kind BCs. In addition, the surface temperature was determined as the mean temperature over the heating period, but this seriously degrades the physicality of the process.
- In experiments of [59] (as well as in [56, 58]), a debatable method was employed for finding α_e. In addition, the first kind BCs cannot be reproduced due to specimen heating by hot water circulation. What is more, the measurement of the nonstationary temperature field in the specimen with one-second sampling of one thermocouple does not allow the correct estimation of the relaxation time.

Roetzel et al. [60] conducted a series of experiments in which the efficient thermometric conductivity a_e of the medium and the relaxation time t_r were measured in each experiment. Various kinds of pebble beds were used: artificial sand, powder (sodium hydroxide and aluminum oxide), copper-coated lead spheres, etc. Biological tissue was also used to replicate the conditions of the experiments in [57–59].

Table 3.2 Comparison of the values of thermometric conductivity a_e and the relaxation time t_r, from experiments in [56–60]

Material	$\alpha_e * 10^6 (m^2/s)$	t_r(s)	Source
Sand	0.226	2.26	[60]
Sand	0.408	20	[56]
Sand	0.218	0	[58]
Sand	0.169	0	[59]
Baking soda ($NaHCO_3$)	0.185	0.66	[60]
Baking soda ($NaHCO_3$)	0.310	28.7	[56]
Biological tissue	0.132	1.77	[60]
Biological tissue	0.140	15.5	[57]
Biological tissue	0.130	0	[59]

A cylinder of diameter 40 mm and height 80 mm with a pebble bed was used as a working body. Air, helium, nitrogen, and argon were used as gaseous filling agents. Experiments with various pressure and temperature of the gaseous filling agent were carried out. Harmonic pulsations of the temperature were generated on the specimen surface by using a Peltier element. The temperature distribution over the axis was measured by six thermocouples spaced at distance 10 mm. The phase shift of pulsations and the damping decrement of their amplitude measured in the experiments were used jointly with the results of analytical solutions of the HHCE for a semi-infinite body. The values of α_e, t_r thus obtained were compared with the results of [56–59] (see the table).

As a result, the values of t_r from [60] were found to be approximately one-order smaller than those obtained in [56, 57]. It was also found that t_r depends strongly on the characteristic size of particles in a pebble bed and on the medium temperature. Table 3.2 compares the values of the efficient thermometric conductivity a_e and the relaxation time t_r, as measured in the experiments in [56–60].

3.14.5 Estimate of the Relaxation Time

From the representative results obtained in the experiments of [60], one can theoretically estimate the thermal relaxation time in materials with spatially inhomogeneous structure. A conjectural cause of the anomalously high (as compared with the generally accepted ones) values of t_r from [56–60] is the specific cell-like character of heat transfer in the pebble bed. Here, the mechanism of heat conductivity is suppressed due to the point contact of touching particles. Hence, the heat transfer is effected via free convection in the gaseous filler. Hence, the efficient heat conductivity of the specimen should depend on the particle size, their packing type in the pebble bed, and on the kind of the gas and its temperature (as was demonstrated in experiments of [60]).

In the experiments in [60], for a pebble bed of lead spheres of diameter 4.4 mm with gaseous argon–nitrogen filler, the following values were found: $\alpha_e \approx 8 * 10^{-6}$ m^2/s and $t_r \approx 14$ s. For estimates, we shall use the analytical solution [61] of the temperature shock problem on the ball surface. Using the integral heat-balance method, it was obtained in [61] that

$$\Delta = \exp\left(-27.9\alpha t/d_0^2\right).$$

Here, $\Delta = (\vartheta_s - \vartheta_0)/\vartheta_s$, $\vartheta_s = $ const is the surface temperature, $\vartheta_0(t)$ is the temperature at the ball center, and d_0 is the ball diameter.

We assume that the diameter of a conventional cell of the pebble bed (involving the ball itself and the surrounding gas) is equal to the doubled diameter of the ball, $d_0 =$ 8.8 mm. We also assume that the relaxation time is the time interval during which the ball center is heated by 95%: $\vartheta_0 = 0.95\vartheta_s$, $\Delta = 0.05$. Replacing the thermometric conductivity of the ball α by its effective value for the cell of the inhomogeneous medium, we get $t_r \approx 0.107 d_0^2/\alpha_e$. For the above experiment conditions [60], we get $t_r \approx 10.4$ s, which is close to the measured value. The estimates give evidence that the values of the relaxation time obtained in the experiments in [60] are correct.

A transient thermal process in a polymethylmethacrylate plate at its sudden contact with liquid volume was experimentally studied in [62]. It was shown that at the sudden immersion of the plate in hot distilled water, the temperature in its center does not change during some time. After this, the temperature starts to increase, and the measured rate of its change was found to be much lower than predicted by the solution of the HHCE with third kind BCs. In [62], the experimental data were processed using the heat conduction equation derived by Luikov [63] from the system of differential Onsager equations.

In comparison with the standard HHCE, this equation involves an additional term with a mixed third derivative of the temperature (in time and in the transverse coordinate) and the new physical constant (the temperature damping time t_T). For a detailed account of the Luikov equation and its modifications, see, in particular, [64]. With the use of the Luikov equation for the processing of experimental data, Kirsanov et al. [62] found that $t_r = 3.5$ s and $t_T = 2.45$ s. For the current analysis, it is important that the solution t_r obtained in [62] is close to the values from [60] for the cases of sand and biological tissue.

3.15 General Analytical Solution of the Conjugate Problem

3.15.1 Integral Laplace Transform

The method of integral transform (Laplace transform, Fourier transform, Hankel transform, and Meijer transform) is an efficient method for solving many problems of mathematical physics. This method, which is also known the method of operational calculus, is capable in a number of cases of solving fairly involved mathematical problems via simple rules. In this method, the function under consideration (the original function) is replaced by a different function (the transform image) obtained by certain rules. Under this approach, operations on original functions are replaced by more simple operations on the transform image.

The application of the integral transform (in the frames of any method of operational calculus) to each term of the equation, and also to boundary-value conditions, results in the corresponding equations and boundary-value conditions with respect to the transform image. Under this approach, the original partial differential equation is transformed into an ordinary differential equation. Another application of the integral transform transforms this equation into the algebraic equation. Note that the operational method applies only to linear heat conduction equations with constant coefficients and linear BCs.

The most widely used of these is the Laplace transform, which is transformed for excluding the time coordinate from the differential equation [65]. If the original equation contains two independent variables, then we get an ordinary differential equation with respect to the image. Having found the solution of the first equation (i.e., the expression for the image) and applying to it the inverse Laplace transform, we get the solution of the original problem.

The inverse Laplace transform is applied to find the original of the function from its image. To this aim, the image should be written in the Heaviside form by using the necessary form of expansion of a linear fractional function. The resulting sum of simple partial fractions is subject to the Laplace transform. To this end, one can use tables of Laplace transforms, which contain the images of many time functions. The Laplace transform is a basic tool for finding analytical solutions to boundary value problems of transient heat conduction in canonical domains (plate, cylinder, ball, etc.). Below, the integral Laplace transform will be used for solving HHCEs.

For boundary value problems of the hyperbolic heat conduction model, the following dilemma is typical: notwithstanding its mathematical simplicity, they are incapable of delivering a simple analytical solution in most cases. Hence for HHCEs, it is much more difficult to find exact solutions than in the case of parabolic equations. The use of operation methods in this case results as a rule in involved functional constructions in the (Fourier or Laplace) image spaces. At present, only a relatively small part of such constructions was investigated (the available originals can be found in handbooks on operational calculus). In the majority of cases, such originals cannot be found analytically, and hence the solution is written down as involved quadratures. In this regard, of great value are rare cases in which an analytical solution can be found.

Below, we give analytical solutions of HHCEs obtained via the integral Laplace transform. They are used for the construction of the general solution of the conjugate problem. The following well-known analytical solution to the HHCE for a semi-infinite body with stationary first kind BC $x = 0 : \vartheta(0, t) = \vartheta_0 = \text{const}$ was obtained in [66]. Below, we extend this solution to the case when the surface temperature of the body is an arbitrary function $\vartheta_0(t)$ of time. Consider the homogeneous ICs: $\vartheta(x, 0) = 0$, $\frac{\partial\vartheta(x,0)}{\partial t} = 0$, $q(x, 0) = 0$. The decay condition of temperature perturbations should be satisfied at infinity: $x \to \infty : \vartheta(\infty, t) = 0$. The integration of Eq. (3.101) with due account of Cattaneo–Vernotte's hypothesis (3.100) gives the following expression

$$q(x,t) = -\frac{k}{t_r}\int\limits_0^t \frac{\partial \vartheta(x,\xi)}{\partial x}\exp\left(-\frac{t-\xi}{t_r}\right)d\xi.$$

From this quadrature, using the known temperature $\vartheta(x,t)$, one can evaluate the heat flux at any point of the body at an arbitrary time.

According to [65], the Laplace transform maps the original $f(x,t)$ into the image $F(p,t)$ via integration with the weight function

$$F(s,t) = \int_0^\infty f(x,t)\mathrm{e}^{-st}dt.$$

This will be symbolically denoted as $f(x,t) \Rightarrow F(p,t)$. So, for the Laplace transforms of the temperature and the BCs, we shall write

$$\vartheta(x,t) \Rightarrow T(x,s),\ \vartheta_0(t) \Rightarrow T_0(s).$$

The solution in the image domain reads as

$$T(x,s) = T_0(s)\exp\left(-\varepsilon x\sqrt{s^2 + s/t_r}\right).$$

Using the tables of inverse Laplace transform, we find the original and write the solution to the problem in the form

$$\vartheta(x,t) = H(\gamma)\left\{\vartheta_0(\gamma)\exp\left(-\frac{x}{2\sqrt{\alpha t_r}}\right) + \frac{x}{2\sqrt{\alpha t_r}}J_1\right\}. \qquad (3.102)$$

Here, we used the notation

$$H(\gamma) = \begin{cases} 0, x < 0 \\ \frac{1}{2}, x = 0 \\ 1, x > 0 \end{cases}$$

is the Heaviside unit function

$$J_1 = \int\limits_0^\gamma T_0(\tau)\frac{I_1(\chi/2)}{\chi}\exp\left(-\frac{t-\xi}{2t_r}\right)d\xi,\ \gamma = t - \varepsilon x,\ \varepsilon = \sqrt{\frac{t_r}{a}},\ \chi = \sqrt{(t-\xi)^2 - \varepsilon^2 x^2}.$$

Relation (3.102) is a generalization of the original solution of [66] to the case of a nonstationary BC of the first kind. The author has not succeeded in finding in the literature the solution of the problem for BC of the second kind (even for its stationary variant). Below, we give such a solution for the general case of a nonstationary BC $x = 0 : q(0, t) = q_0(t)$. At infinity of the body, the temperature gradient decay condition should be satisfied $x \to \infty : \frac{\partial \vartheta(\infty,t)}{\partial t} = 0$. For the convenience of further presentation, we replace this condition with the equivalent condition

$$-\lambda \frac{\partial \vartheta(0, t)}{\partial x} = q_0^*(t), q_0^*(t) = q_0(t) + t_r \frac{\partial q_0(t)}{\partial t}.$$

In this case, we use the Laplace transforms

$$q(x, t) \Rightarrow Q(x, s), q_0(t) \Rightarrow Q_0(s), q_0^*(t) \Rightarrow Q_0^*(s).$$

The solution in the image domain reads as

$$T(x, s) = Q_0(s)\exp\left(-\varepsilon x\sqrt{s^2 + s/t_r}\right)/\left(\lambda\varepsilon\sqrt{s^2 + s/t_r}\right).$$

Returning to the originals via the inverse Laplace transform, we get the following expression for the temperature field in the body

$$\vartheta(x, t) = H(\gamma)\left\{\frac{t_r}{\varepsilon k} q_0(\gamma)\exp\left(-\frac{x}{2\sqrt{\alpha t_r}}\right) + \frac{1}{2\varepsilon k} J_2\right\}. \tag{3.103}$$

The second term in curly brackets contains the following integral

$$J_2 = \int_0^\gamma q_0(\tau)\left[I_1(\chi/2) + (t - \tau)\frac{I_2(\chi/2)}{\chi}\right]\exp\left(-\frac{t - \xi}{2t_r}\right)d\xi.$$

3.15.2 Relative Heat Transfer Coefficient

Thus, we obtained general analytical solutions for two classical BCs written as quadratures (3.102), (3.103). Let us now transform the corresponding stationary BCs and introduce the HTC, which relates the heat flux and the temperature on the surface of the body

$$h(t) = \frac{q(0, t)}{\vartheta(0, t)}. \tag{3.104}$$

From (3.102)–(3.104), we now get

$$\left.\begin{aligned} \vartheta_0 = const : h_\vartheta &= \sqrt{\tfrac{\rho k c_p}{t_r}} \exp\left(-\tilde{t}\right) I_0\left(\tilde{t}\right), \\ q_0 = \text{const} : h_q &= \sqrt{\tfrac{\rho k c_p}{t_r}} \exp\left(\tilde{t}\right) \left\{ I_0\left(\tilde{t}\right) + 2\,\tilde{t}\left[I_0\left(\tilde{t}\right) + I_1\left(\tilde{t}\right)\right]\right\}^{-1}. \end{aligned}\right\} \quad (3.105)$$

Here, $I_0\left(\tilde{t}\right)$, $I_1\left(\tilde{t}\right)$ are the modified Bessel functions of the first kind, respectively, of zero and first order [67], and $\tilde{t} = t/2t_r$ is the dimensionless time.[2]

If the relaxation time is negligibly small ($t_r \to 0$), we get $\tilde{t} \to \infty$, and formulas (3.105) describe the HTC $h_{\vartheta *}$, h_{q*} for the parabolic case

$$\left.\begin{aligned} h_{\vartheta *} &= \sqrt{\tfrac{\rho k c_p}{\pi t}}, \\ h_{q*} &= \tfrac{\sqrt{\pi}}{2}\sqrt{\tfrac{\rho k c_p}{t}}. \end{aligned}\right\} \quad (3.106)$$

Based on the reference values (3.106), we introduce the relative HTC, using which one can express the effect of finite heat transfer rate

$$\left.\begin{aligned} \Psi_\vartheta &= \tfrac{h_\vartheta}{h_{\vartheta *}}, \\ \Psi_q &= \tfrac{h_q}{h_{q*}}. \end{aligned}\right\} \quad (3.107)$$

For the classical BCs, from (3.105), (3.106) we get

$$\left.\begin{aligned} \Psi_\vartheta &= \sqrt{2\pi\,\tilde{t}} \exp\left(-\tilde{t}\right) I_0\left(\tilde{t}\right), \\ \Psi_q &= \sqrt{\tfrac{8}{\pi}\,\tilde{t}} \exp\left(\tilde{t}\right) \left\{ I_0\left(\tilde{t}\right) + 2\,\tilde{t}\left[I_0\left(\tilde{t}\right) + I_1\left(\tilde{t}\right)\right]\right\}^{-1}. \end{aligned}\right\} \quad (3.108)$$

For small values of the relaxation time, both relative HTCs coincide and reduce to the reference parabolic case

$$\tilde{t} \to 0 : \Psi_\vartheta = \Psi_q = 1$$

In the other limit case, (3.108) gives

$$\tilde{t} \to \infty : \Psi_\vartheta \to \sqrt{2\pi\,\tilde{t}},\ \Psi_q = \sqrt{\frac{8}{\pi}\,\tilde{t}}.$$

These hyperbolic asymptotic formulas describe the maximal decrease of intensity of thermal conductivity due to the inertial delay of the heat flux. The results of calculations via (3.108) are shown in Fig. 3.18.

[2] For convenience, as a scale we use the doubled relaxation time $2t_r$.

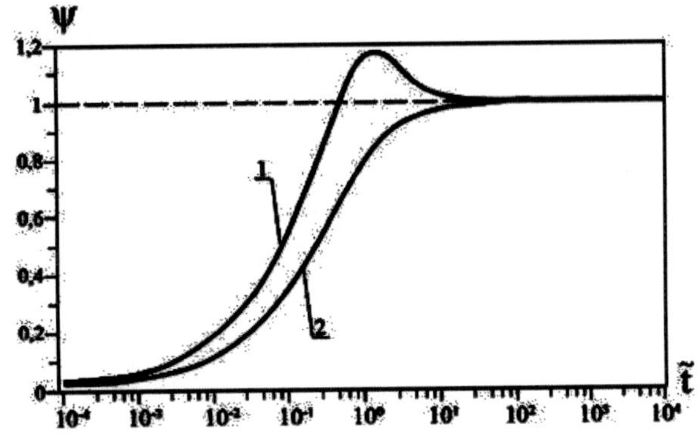

Fig. 3.18 The relative HTC. 1 Boundary condition of the first kind and 2 boundary condition of the second kind

The figure shows that with increasing $\tilde{t}$, the function Ψ_q increases monotonically from 0 to 1. At the same time, the dependence Ψ_ϑ has a local maximum with coordinates $\tilde{t} = 1.58$, $\Psi_\vartheta = 1.175$. From (3.105), it follows that initially the HTC for both BCs is described by the relation

$$t = 0 : h_\vartheta = h_q = \sqrt{\frac{\rho k c_p}{t_r}}. \tag{3.109}$$

Comparison of the limit cases (3.106), (3.109) shows the substantial difference between the two models of thermal conductivity:

- $\vartheta = \text{const}$. In the parabolic case, the initial heat flux is infinitely large, while in the hyperbolic case it remains finite.
- $q = \text{const}$. At the beginning of heating, for the PHCE the temperature differential at the surface is zero, while for the HHCE it increases in steps to some fixed value.

3.15.3 Dimensionless Heat Transfer Coefficients

We introduce the dimensionless relaxation time: $\tau_r = t_r / t_0$ and change in (3.105) to the dimensional HTC defined by (3.95)

$$\left.\begin{aligned} \tilde{h}_\vartheta &= \tau_r^{-1/2} \exp\left(-\tilde{t}\right) I_0\left(\tilde{t}\right), \\ \tilde{h}_q &= \tau_r^{-1/2} \exp\left(\tilde{t}\right) \left\{ I_0\left(\tilde{t}\right) + 2\,\tilde{t}\left[I_0\left(\tilde{t}\right) + I_1\left(\tilde{t}\right) \right] \right\}^{-1}. \end{aligned}\right\} \tag{3.110}$$

Relations (3.110), which involve two dimensionless times, are inconvenient for further analysis. With an error up to 1%, they can be approximated as follows

$$\left.\begin{aligned}\tilde{h}_\vartheta &= \left(\frac{3\tau_r+\tau}{3\tau_r^2+4\tau_r\tau+3\tau^2}\right)^{1/2},\\ \tilde{h}_q &= \left(\frac{3(3\tau_r+\tau)}{9\tau_r^2+12\tau_r\tau+4\tau^2}\right)^{1/2}.\end{aligned}\right\} \tag{3.111}$$

Here, we introduced the unified dimensionless values: $\tau = t/t_0$, $\tau_r = t_r/t_0$. In the limit as $t \to \infty$, the relative HTCs have the parabolic form, which depends only on the current time

$$\tau_r = 0 : \tilde{h}_\vartheta = \frac{1}{\sqrt{3\tau}}, \tilde{h}_q = \sqrt{\frac{3}{4}\tau}.$$

In turn, in the hyperbolic limit as $t \to 0$, these quantities are functions only of the relaxation time

$$\tau_r \to \infty : \tilde{h}_\vartheta = \tilde{h}_q = \frac{1}{\sqrt{\tau_r}}.$$

We can now give a construction of the analytical solution of the conjugate problem, which generalizes the parabolic variant (3.96)

$$\psi_* = \left\{1 + \left[\sqrt{3}\tilde{h}_\vartheta + \left(\sqrt{\frac{8}{3}}\tilde{h}_q - \sqrt{3}\tilde{h}_\vartheta\right)\varphi_*\right]\right\}^{-1}. \tag{3.112}$$

Here,

$$\varphi_* = \left(1 + \tau_*^{1/2}\right)^{-1/2} \tag{3.113}$$

is the transient function containing the a priori unknown dimensionless time τ_*. We shall search this quantity as a generalization of the current time to the case of HHCE. Analysis of (3.111) gives two possible variants: either $\tau_* \approx 1/3\tilde{h}_\vartheta^2$, or $\tau_* \approx 3/4\tilde{h}_q^2$. Calculations performed in the range $0 < \tau_r < 1$ for the above two quantities show that they differ by a small fraction of a percent. Hence for definiteness, we shall use below the transient function for the BC $\vartheta = \text{const}$

$$\varphi_* = 3^{1/4}\left(\frac{\tilde{h}_\vartheta}{1 + \sqrt{3}\tilde{h}_\vartheta}\right)^{1/2}. \tag{3.114}$$

Relations (3.110)–(3.113) constitute an analytical solution of the conjugate "heat conductivity–evaporation" problem for the general case of finite heat propagation rate. The generalized dependences $\psi_*(\tau)$ for various values of τ_r are given in Fig. 3.19, which shows that these nonlinear generalized dependences are extended flatter parts near $\tau = 0$ that gradually transform into the reference dependence (3.88) for $t_r = 0$. An important feature of the general solution for the HHCE is the fact that

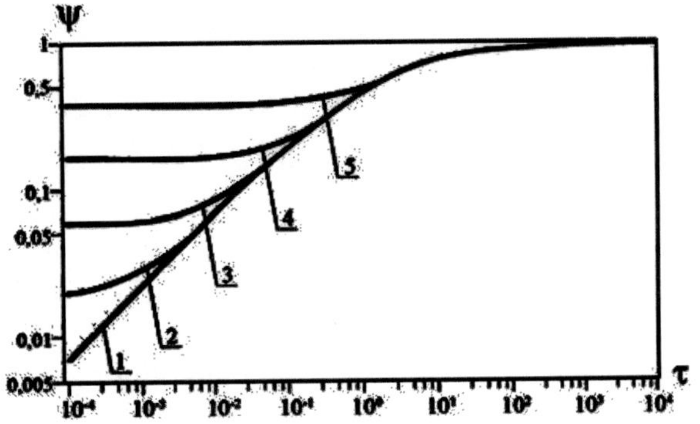

Fig. 3.19 The evaporation parameter versus time for various values of the relaxation time. 1 $\tau_r = 0$, 2 $\tau_r = 10^{-3}$, 3 $\tau_r = 10^{-2}$, 4 $\tau_r = 10^{-1}$, and 5 $\tau_r = 10^0$

for $\tau = 0$, the quantity ψ_* does not vanish, as in the particular case of the PHCE, but rather approaches a fixed finite value (a hyperbolic shelf)

$$\psi_{*\min} = \left(1 + \sqrt{\frac{3}{\tau_r}}\right)^{-1}. \tag{3.115}$$

Figure 3.20 shows the dependence $\psi_{*\min}(\tau_r)$. For $\tau_r = 0$, we have the classical case described by the PHCE: $\psi_{*min} = 0$. With increasing relaxation time, the quantity $\psi_{*\min}$ increases monotonically and as $\tau_r \to \infty$ it reaches the limit value: $\psi_{*\min} = 1$. This asymptotics corresponds to the complete switch-off of the heat conduction mechanism: $\psi_* = \psi_{*\min} = 1$. Physically, this means that over the entire interval $0 < \tau < \infty$, the whole heat radiation flux is spent completely for evaporation: $q_u \to q = \text{const}$.

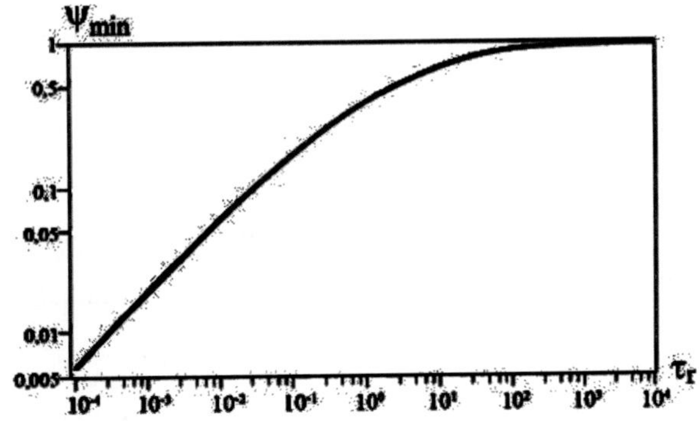

Fig. 3.20 The maximum value of the evaporation parameter versus the relaxation parameter

3.15.4 Two-Zone Approximation of the Solution

For approximate physical estimates, it is expedient to approximate the solution (3.112) by two-zone curves. To this end, we replace the gradual parts in Fig. 3.19 with horizontal shelves of width

$$\tau_0 = \frac{2}{15}\frac{\tau_r(5+3\sqrt{\tau_r})}{(1+\sqrt{\tau_r})} \tag{3.116}$$

in the range $0 \le \tau \le \tau_0$. The quantity τ_0 is the delay time during which a hyperbolic shelf of the dependence $\psi_*(\tau)$ is realized. As follows from (3.115), τ_0 is proportional to the relaxation time: $\tau_0 = 2/3\tau_r$ for $\tau_r = 0$, $\tau_0 = 2/5\ \tau_r$ as $\tau_r \to \infty$. According to (3.115), the height of the hyperbolic shelf of the dependence $\psi_*(\tau)$ is uniquely determined by the relaxation parameter

$$\tau_r \equiv \frac{t_r}{t_0} = \frac{3}{2\pi}\frac{\beta}{1-0.4\beta}\frac{r^4 t_r}{\alpha R_g c_p^2 T_s^3}.$$

As a result, we can write down the two-zone approximation of the analytical solution in the form (Fig. 3.21)

$$\psi = \begin{cases} \psi_{*\min},\ 0 < \tau < \tau_0, \\ \psi(\tau),\ \tau_0 \le \tau < \infty. \end{cases} \tag{3.117}$$

Figure 3.21 gives a physically transparent picture of how the relaxation parameter effects the evaporation intensity. Consideration of the finite heat propagation velocity cuts off a segment of length τ_0 from the limit curve described by formula (3.88). So, in the thermal balance with $\tau = 0$, there is a surge in the evaporation flow from zero to $q_u = \psi_{*min} q$. Throughout the hyperbolic zone $0 < \tau < \tau_0$, the thermal balance remains unchanged, the height of the hyperbolic shelf increasing with increased relaxation time. Next, starting from the moment $\tau = \tau_0$, a classical parabolic zone

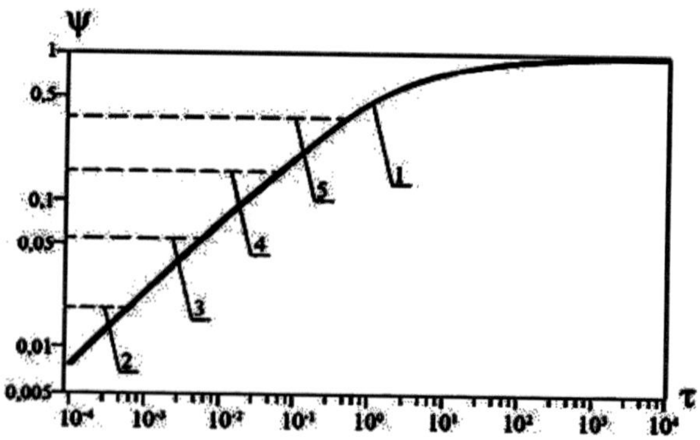

Fig. 3.21 Two-zone approximation of the solution (3.112). 1 $\tau_r = 0$, 2 $\tau_r = 10^{-3}$, 3 $\tau_r = 10^{-2}$, 4 $\tau_r = 10^{-1}$, and 5 $\tau_r = 10^0$

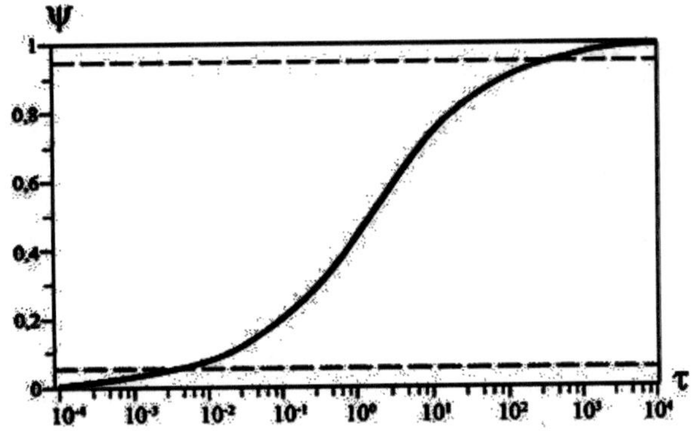

Fig. 3.22 Two-zone approximation. 1 $\tau_r = 0$, 2 $\tau_r = 10^{-3}$, 3 $\tau_r = 10^{-2}$, 4 $\tau_r = 10^{-1}$, and 5 $\tau_r = 10^0$

complying with (3.88) is implemented. Such an approach reflects the trend in the degeneration of the effect of the relaxation time on the heat transfer.

Figure 3.22 shows the limit cases of two-zone approximation. So, to the value $\psi_{*min} = 0.05$, there corresponds the delay time $\tau_0 \approx 5.5 * 10^{-3}$. Here, we have an extended parabolic zone during which the evaporation flux gets larger. To the value $\psi_{*min} = 0.95$, there corresponds $\tau_0 \approx 400$. Here, the thermal conductivity process becomes to have practically no effect on the heat transfer, so that throughout the entire second zone the evaporation flux increases by 5%.

The above universal solution (3.110)–(3.114) is compact and convenient for calculations. However, for the transparent detection of the effect of the condensation coefficient on the evaporation parameter, it is expedient to write the solution in the two-parameter form. To this end, in place of (3.87) we introduce the new time scale

$$t_1 = \frac{2\pi}{3} \frac{\alpha R_g c_p^2 T_s^3}{r^4}. \tag{3.118}$$

In view of (3.118), we redefine the relaxation parameter $\tau_{r1} = t_r / t_1$ and the dimensionless time $\tau_1 = t_1 / t_0$, and the problem receives in addition the second parameter β. Figure 3.23 shows that with the same value of the relaxation parameter and if β is reduced, the curves $\psi_*(\tau)$ move down and to the right. Physically, this means two interrelated trends: (a) transition of the conjugate problem from the hyperbolic to the parabolic form, (b) delay of the evaporation process.

Note that the experimental and theoretical results of [56–60] are debatable and are required special studies beyond the scope of the present work. As a promising direction in this way, we can mention the book [68], which analyzes the construction of the solution to the kinetic Boltzmann equation in the case of a multiscale DF. Using the well-known Chapman–Enskog method, one can obtain modified equations for the transfer of the molecular flows of heat and energy involving additional relaxation terms. Consideration of such additional terms gives the corresponding relaxation terms in expressions for the viscous stress tensor and the heat flux vector. The above results show the importance of the consideration of the heat flux relaxation effect, which is an integral part of the mechanism of nonstationary heat conduction.

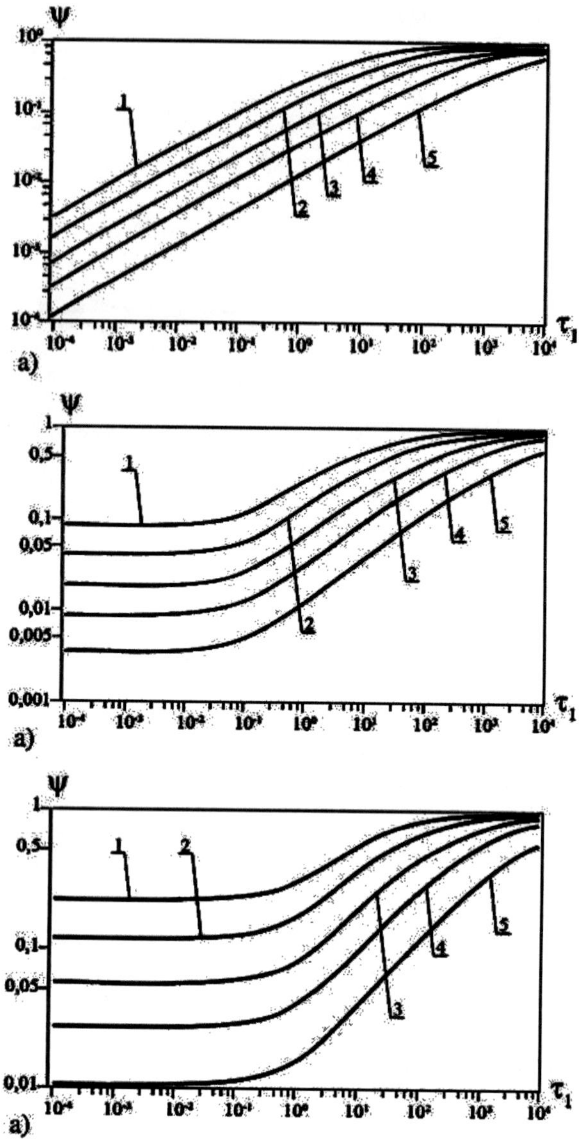

Fig. 3.23 The influence of the condensation coefficient on the evaporation parameter: 1 $\beta = 1$, 2 $\beta = 0.6$, 3 $\beta = 0.3$, 4 $\beta = 0.15$, *and* 5 $\beta = 0.06$. **a** $\tau_{r1} = 0$, **b** $\tau_{r1} = 0.1$, and **c** $\tau_{r1} = 1$

3.15.5 *Algorithm of Solution of the Conjugate Problem*

For the systematization of the above transformations and calculations, we give a short algorithm for the construction of an analytical solution of the conjugate problem.

(1) The "thermal conductivity–evaporation" conjugate problem is considered. A time-constant heat flow q = const is specified on the boundary of a semi-infinite body with a homogeneous initial distribution of temperatures (Fig. 3.8). The classical PHCE (3.49) is solved together with the relations of the linear kinetic theory of evaporation.

(2) The process is split into two parts. Throughout the thermal conductivity phase, the irradiated liquid body is heated to the saturated temperature with given pressure in the system. Next, the evaporation process from the surface is triggered and the conjugation stage is onset. The heat flux impinging on the surface involves two components: the heat flux extracted inward the body by the heat conduction mechanism and the component spent for evaporation. As a result, we get the solution (3.81) describing the dependence of the evaporation parameter on the dimensionless time and the kinetic parameter (3.79) (Figs. 3.9–3.13). The time scale is given by formula (3.67).

(3) We consider the particular case when the heat flux falls on the surface of a semi-infinite body uniformly heated up to the saturated temperature (Fig. 3.8). At the initial moment of irradiation, the conjugation stage is realized bypassing the thermal conductivity stage. As a result, we get the universal solution (3.88) describing the dependence of the evaporation parameter only on the dimensionless time (Fig. 3.16). The time scale is given by formula (3.87). We introduce the notion of the HTC that relates the heat flux and temperature on the body surface. Using this, we represent the solution (3.88) in the form (3.96) convenient for further calculations.

(4) Once the heat flux relaxation time is introduced, we transform it to the hyperbolic heat conduction Eq. (3.101). With the help of the integral Laplace transform, we construct its solutions for time-dependent boundary conditions: of BC of the first kind (formula (3.102)) and BC of the second kind (formula (3.103)). Next, we transform to stationary BC and evaluate the relative HTC for the classical BC (relations (3.108), Fig. 3.18).

(5) The analytical solution of the hyperbolic conjugate problem (3.112) is written down (Fig. 3.19). This relation, which extends the parabolic variant (3.96), involves the transient function (3.114), the dimensionless relative HTC (3.111), and the dimensionless time. Note that initially the evaporation parameter does not vanish (as in the parabolic case), but it describes a fixed finite value (3.115) (a hyperbolic shelf), Fig. 3.20.

6) We write down a two-zone approximation of solution (3.116) convenient for physical estimates of the conjugate problem. The approximation involves the initial hyperbolic shelf and the asymptotic parabolic branch (Fig. 3.21). The width of the hyperbolic branch (the delay time) is proportional to the relaxation time and is described by formula (3.115).

7) To find the effect of the condensation coefficient on the evaporation parameter, the two-zone approximation is written in the parametric form. To this end, the new time scale (3.117) is used (Fig. 3.23).

Thus, we have a simple physically transparent solution of the conjugate problem for the hyperbolic heat conduction equation.

3.16 Conclusions

An approximate kinetic analysis of the problem of intense evaporation was carried out. The approach developed in [4, 10] was supplemented with the condition of equality of mass velocities on the mixing surface and on the boundary between the Knudsen layer and the Navier–Stokes region. The obtained analytical solutions for temperatures, pressures, and mass velocities of vapor agree well with the available numerical and analytical solutions. The limiting mass flux of vapor flow in evaporation was calculated. The mechanism of reflection of molecules from the condensed-phase surface was analyzed. The effect of the condensation coefficient on the conservation equations of molecular flows of mass, momentum, and energy, and also on the thermodynamic state of the resulting vapor was studied. Analytical approximations of kinetic surges of temperature and pressure on the condensed-phase surface were obtained. The time-dependent "thermal conductivity in target–intensive evaporation" conjugate problem at atmospheric pressure was calculated. The solution to the chain of fundamental equations was put forward in the linear and nonlinear approximations. The asymptotic behavior of the solutions in terms of the key parameters of the systems was obtained and analyzed from the physical viewpoint. The results obtained can be used in the calculation of the combustion problem of liquid fuel in the context of the conjugate problem.

The conjugate problem for the hyperbolic heat conduction equation was considered. The integral Laplace transform was applied to find an analytical solution of the hyperbolic heat conduction equation in the general case when the temperature and the heat flux on the body surface are arbitrary time functions. A general solution to the problem is constructed using the concept of the relative HTC.

The relaxation time of the heat flow is estimated. This estimate compares favorably with the available experimental data for bodies of spatially inhomogeneous structure. A two-zone approximation of the solution was given, using which the following characteristic parameters of the conjugate problem were identified: the delay time, the height of the hyperbolic shelf, and the hyperbolic and parabolic zones of the evaporation process. It was shown that starting from a certain value of time delay, the heat conductivity of the irradiated body ceases to have any significant effect on the process of evaporation from its surface. With the decaying of the condensation coefficient, the dependence of the evaporation parameter on time is shown to involve two interrelated trends: transition of the conjugate problem from the hyperbolic to the parabolic form and delay of the evaporation process. Using the general analytical solution obtained above, one can study the trends of the effect of physical parameters on the parameters of the process of evaporation from the surface of the irradiated body. We also take into account the response delay effect of a heat flow owing to the finite heat propagation rate.

References

1. Kogan MN (1969) Rarefied gas dynamics. Plenum, New York
2. Muratova TM, Labuntsov DA (1969) Kinetic analysis of evaporation and condensation processes. High Temp 7(5):959–967
3. Labuntsov DA (2000) Physical foundations of power engineering. Selected works, Moscow Power Energetic Univ. (Publ.). Moscow (In Russian)
4. Labuntsov DA, Kryukov AP (1977) Processes of intense evaporation. Therm Eng 4:8–11
5. Bobylev AV (1987) Exact and approximate methods in the theory of nonlinear and kinetic boltzmann and landau equations. Keldysh Instutute Preprints, Moscow (In Russian)
6. Aristov VV, Zabelok SA (2010) Application of direct methods of solving boltzmann equations for modeling non-equilibrium phenomena in gases. Computing Center of the Russian Academy of Sciences, Moscow (In Russian)
7. Pao YP (1971) Temperature and density jumps in the kinetic theory of gases and vapors. Phys. Fluids 14:1340–1346
8. Pao YP (1973) Erratum: Temperature and density jumps in the kinetic theory of gases and vapors. Phys Fluids 16:1650
9. Anisimov SI (1968) Vaporization of metal absorbing laser radiation. Sov. Phys. JETP 27(1):182–183
10. Labuntsov DA, Kryukov AP (1979) Analysis of intensive evaporation and condensation. Int J Heat Mass Transf 2(7):989–1002
11. Zhakhovskii VV, Anisimov SI (1997) Molecular-dynamics simulation of evaporation of a liquid. J Exp Theor Phys 84(4):734–745
12. Frezzotti A (2007) A numerical investigation of the steady evaporation of a polyatomic gas. Eur J Mech B: Fluids 26:93–104
13. Gusarov AV, Smurov I (2002) Gas-dynamic boundary conditions of evaporation and condensation: Numerical analysis of the Knudsen layer. Phys Fluids 14:4242–4255
14. Rose JW (2000) Accurate approximate equations for intensive sub-sonic evaporation. Int J Heat Mass Transf 43:3869–3875
15. Hertz H (1882) Über die Verdünstung der Flüssigkeiten, inbesondere des Quecksilbers, im luftleeren Räume. Ann Phys Chem 17:177–200
16. Knudsen M (1934) The kinetic theory of gases. Methuen, London
17. Cercignani C (1981) Strong evaporation of a polyatomic gas. In: Fisher SS (ed) Rarefied gas dynamics, Part 1, vol 1. AIAA, pp 305–310
18. Skovorodko PA (2001) Semi-empirical boundary conditions for strong evaporation of a polyatomic gas. In: Proceedings of the AIP conference. American Institute of Physics, New York, pp 585, 588–590
19. Kogan MN, Makashev NK (1971) On the role of Knudsen layer in the theory of heterogeneous reactions and in flows with surface reactions. Izv. Akad. Nauk SSSR. Mekh Zhidk Gaza 6:3–11 (In Russian)
20. Anisimov SI, Rakhmatullina A (1973) Dynamics of vapor expansion on evaporation into vacuum Zh. Eksp Teor Fiz 64(3):869–876 (in Russian)
21. Landau LD, Lifshits EM (1987) Fluid mechanics. Butterworth-Heinemann
22. Zudin YB (2015) Approximate kinetic analysis of intense evaporation. J Eng Phys Thermophys 88(4):1015–1022
23. Zudin YB (2015) The approximate kinetic analysis of strong condensation. Thermophys Aeromech 22(1):73–84
24. Zudin YB (2016) Linear kinetic analysis of evaporation and condensations. Thermophys Aeromech 23(3):437–449
25. Maxwell JC (1860) Illustrations of the dynamical theory of gases: Part I. On the motions and collisions of perfectly elastic spheres. Philos Mag 19:19–32
26. Maxwell JC (1860) Illustrations of the dynamical theory of gases: Part II. On the process of diffusion of two or more kinds of moving particles among one another. Philos Mag 20:21–37

27. Langmuir I (1913) Chemical reactions at very low pressures. II. The chemical cleanup of nitrogen in a tungsten lamp. J Am Chem Soc 35:931–945
28. Gerasimov DN, Yurin EI (2017) Potential energy distribution function and its application to the problem of evaporation. J Phys Conf Ser 891:012005
29. Marek R, Straub J (2001) Analysis of the evaporation coefficient and the condensation coefficient of water. Int J Heat Mass Transf 44:39–53
30. Kryukov AP, Levashov VY (2011) About evaporation-condensation coefficients on the vapor-liquid interface of high thermal conductivity matters. Int J Heat Mass Transf 54(13–14):3042–3048
31. Bardos C, Golse F, Sone Y (2006) Half-space problems for the boltzmann equation. Surv J Stat Phys 124(2–4):275–300
32. Bond M, Struchtrup H (2004) Mean evaporation and condensation coefficient based on energy dependent condensation probability. Phys Rev E 70:061605
33. Hołyst R, Litniewski M, Jakubczyk D (2015) A molecular dynamics test of the Hertz-Knudsen equation for evaporating liquids. Soft Matter 11(36):7201–7206. https://doi.org/10.1039/c5sm01508a
34. Zudin YB (2018) Non-equilibrium evaporation and condensation processes: analytical solutions. Springer, Heidelberg, Dordrecht, London, New York
35. Zeldovich YB, Raizer YP (2002) Physics of shock waves and high-temperature hydrodynamic phenomena. Courier Corporation, North Chelmsford
36. Ytrehus T (1977) Theory and experiments on gas kinetics in evaporation. In: Potter JL (ed) Rarefied gas dynamics, vol 51(2). N.Y., pp 1197—1212
37. Khight CJ (1979) Theoretical modeling of rapid surface vaporization with dack pressure. AIAA J 17(5):519–523
38. Khight CJ (1982) Transient vaporization from a surface into vacuum. AIAA J 20(7):950–955
39. Kelly R, Miotello A (1994) Laser-pulse sputtering of atoms and molecules. Part II. Recondensation Effects Nucl Instr Meth B 91:682–691
40. Jeong H, Greif R, Russo RE (1999) Shock wave and material vapour plume propagation during excimer laser ablation of aluminium samples. J Phys D 32:2578–2585
41. Chen KR, Leboeuf JN, Wood RF, Geohegan DB, Donato JM, Liu CL, Puretzky AA (1996) Mechanisms affecting kinetic energies of laser-ablated materials. J Vacuum Sci Technol Vacuum, Surf Films 14(3):1111–1114
42. Bulgakov AV, Bulgakova NM (1998) Gas-dynamic effects of the interaction between a pulsed laser-ablation plume and the ambient gas: analogy with an underexpanded jet. J Phys D 31:693–703
43. Ho JR, Grigoropoulos CP, Humphrey JAC (1995) Computational study of heat transfer and gas dynamics in the pulsed laser evaporation of metals. J Appl Phys 78(6):4696–4709
44. Gusarov AV, Gnedovets AG, Smurov I (2000) Gas dynamics of laser ablation: influence of ambient atmosphere. J Appl Phys 88:4352–4364
45. Goodman TR (1964) Application of integral methods in transient non-linear heat transfer. In: Irvine TF, Hartnett JP (eds) Advances in heat transfer, vol. 1. Academic Press, New York, pp 51–122
46. Öziúik NM (1989) Boundary value problems of heat conduction. Dover Publications Inc., New York
47. Langford D (1973) The heat balance integral method. Int J Heat Mass Transf 16:2324–2328
48. Tsoi PV (2001) Problem of heat transfer with allowance for axial heat conduction for the flow of a liquid in tubes and channels. J Eng Phys Thermophys 74(1):183–191
49. Incropera FP, Dewitt DP (1996) Fundamentals of heat and mass transfer. J Wiley
50. Kreith FB, Mark S (2001) Principles of heat transfer. Brooks/Cole pub
51. Schlichting H (1974) Boundary layer theory. McGraw-Hill, New York
52. Wang L, Zhou X, We X (2007) Heat conduction: mathematical models and analytical solutions. Springer, Berlin, Heidelberg
53. Cattaneo C (1958) A form of heat conduction equation which eliminates the paradox of instantaneous propagation. Compt Rend 247:431–433

54. Vernotte P (1958) Les paradoxes de la théorie continue de i'equation de la chaleur. Compt Rend 246:3154–3155
55. Zudin YB (2017) Theory of periodic conjugate heat transfer, 3rd edn. Springer, Berlin, Heidelberg
56. Kaminski W (1990) Hyperbolic heat conduction equation for materials with a non-homogeneous inner structure. ASME J Heat Transfer 112:555–560
57. Mitra K, Kumar S, Vedavarz A, Moallemi MK (1995) Experimental evidence of hyperbolic heat conduction in processed meat J Heat Transfer. Trans ASME 117:568–573
58. Graßmann A, Peters F (1999) Experimental investigation of heat conduction in wet sand. Heat Mass Transf 35:289–294
59. Herwig H, Beckert K (2000) Experimental evidence about controversy concerning Fourier or non-Fourier heat conduction in Materials with nonhomogeneus inner structure. Heat Mass Transf 36:387–392
60. Roetzel W, Putra N, Das SK (2005) Experiment and analysis for non-Fourier conduction in materials with non-homogeneous inner structure. Int J Thermal Sci 42:541–552
61. Antimonov MS, Kudinov VA, Stefanyuk EV (2008) Analytical solutions to the heat conduction problems for a cylinder and a ball based on determining the temperature perturbation front. Comput Math Math Phys 48:648–658
62. Kirsanov YA, Kirsanov AY, Yudakhin AE (2017) Measurement of thermal relaxation and temperature damping time in a solid. High Temp 55:114–119
63. Luikov AV (1969) Analytical heat diffusion theory. Academic Press, New York
64. Kudinov VA, Kudinov IV (2013) Studying heat conduction taking into account the finite rate of heat propagation. High Temp 51:268–276
65. Schiff JL (1999) The Laplace transform theory and applications. Springer
66. Baumeister KJ, Hamill TD (1969) Hyperbolic heat-conduction equation: a solution for the semi-infinite body problem. ASME J Heat Transfer 91:543–548
67. Watson GN (1995) A treatise on the theory of Bessel functions. Cambridge University Press, Cambridge
68. Struchtrup H (2005) Macroscopic transport equations for rarefied gas flows approximation methods in kinetic theory. Springer, Berlin, Heidelberg

Chapter 4
Semi-Empirical Model of Strong Evaporation

Abbreviations

BC	Boundary condition
CPS	Condensed-phase surface
DF	Distribution function
KL	Knudsen layer
KSW	Kinetic shock wave

Symbols

c_p	Isobaric heat capacity
c_{v}	Isochoric heat capacity
h	Specific enthalpy
f	Distribution function
J_i	Molecular flux
j	Mass flux
I_i	Dimensionless molecular flux
k	Wave number
m	Molecular mass
M	Mach number
p	Pressure
$\tilde{p}$	Pressure ratio
S	Entropy
s	Speed ratio
T	Temperature
$\tilde{T}$	Temperature ratio
u_∞	Hydrodynamic velocity
v	Thermal velocity

Greek Letter Symbols

β	Condensation coefficient

Y. B. Zudin, *Non-equilibrium Evaporation and Condensation Processes*,
Mathematical Engineering, https://doi.org/10.1007/978-3-030-67553-0_4

ω	Frequency
ρ	Density

Subscripts

m	State at mixing surface
w	State at condensed-phase surface
∞	State at infinity

4.1 Strong Evaporation

Knowledge of laws of strong evaporation is instrumental for the solution of a number of applied problems: the effect of laser radiation on materials [1], calculation of the parameters of discharge into vacuum of a flashing coolant [2], etc. Strong evaporation also plays an important role in the fundamental problem of simulation of the inner cometary atmosphere. According to the modern view [3], the intensity of icy cometary nucleus varies, as a function of the distance to Sun, in a very substantial range and may reach very large values.

Mathematical modeling of strong evaporation requires setting Boundary Conditions (BCs) on the Condensed-Phase Surface (CPS) for the gas-dynamic equations in the exterior flow region (which is also called the Navier–Stokes region). The gas-dynamic laws become inapplicable in the layer adjacent to the CPS Knudsen layer, whose thickness is of the order of the mean free path of molecules. The standard concepts of the continuous medium (the density, temperature, and pressure) lose their phenomenological sense in the non-equilibrium Knudsen Layer (KL). In this setting, rigorous calculation of gas parameters can be carried out only by solving the kinetic Boltzmann equation [4], which describes the variation of the Distribution Function (DF) of molecules in terms of velocities.

Exact solutions of the very involved integro-differential Boltzmann equation are known only in certain special cases with spatial-homogeneous distributions of parameters [5]. Various approximate methods are employed for solving problems even with simple geometry (for example, the problem of evaporation of gas in the half-space): for example, reduction of the Boltzmann equation to a system of moment equations [6, 7], changing the Boltzmann equation by simplified equations (the relaxation Krook equation [1, 6, 7], the model Case equation [8]), etc. At present, strong evaporation is modeled, as a rule, using various numerical methods [1, 9].

In the case, when the gas velocity u_∞ is much smaller than the sonic velocity, the kinetic analysis is capable of giving the solution in the form of a system of nonlinear algebraic equations [6, 7] or quadratures [8]. Analytical representation of a solution is obtained by some or other approximation. Kinetic molecular structure of phase transitions is characterized by the following main peculiarities.

(1) Solution of the problem for the Boltzmann equation in the rigorous (microscopic) statement, which determines the DF in the KL, involves huge mathematical difficulties. For its part, the approximate (macroscopic) problem for the equation of the system in the Navier–Stokes region is considered traditional. However, even for this problem one has to specify the BC, which is determined from the solution of the microscopic problem. The relation between these different-scale problems is a subtle and intriguing point in the analysis. This is the "boundary region" in which widely differing ideas were raised for over the century. Some of them did not stand the test of time, but some formed a basis for impressive "breakthroughs", which are highly important in applications.

(2) Assume that we have a hypothetic exact solution of the Boltzmann equation for the DF. Hence, as a corollary, we have exact formulas for the gas-dynamic parameters in the Navier–Stokes region. As a result, extrapolation of these dependences to the CPS will result in the appearance on it fictitious values of temperature, density, and pressure of the gas. These values are not equal to the corresponding true values and form macroscopic jumps of temperature and pressure on the CPS.

(3) The DF of molecules emitted from the CPS is completely determined by its temperature, and hence, it has an isotropic equilibrium character (the classical Maxwell distribution function). For its part, the flow of the molecules incident to the CPS is formed as a consequence of their collisions between each other away from the CPS along the entire length of the KL. Its spectrum reflects some averaged state of vapor in the surface region. As a corollary, the DF has a discontinuity on the CPS, which is monotonically smoothed within the KL and disappears when reaching the boundary of the Navier–Stokes region. In a sense, the above microscopic jump is the ultimate cause of macroscopic jumps of the parameters obtained in the extrapolation of the dependences of temperature and pressure on the CPS.

(4) With some simplification, one may assert that there is a fundamental idea running like a golden thread through all available approximate models: find the DF without solving the Boltzmann equation, thereby solving the problem of setting the BC for the system of gas-dynamic equations in the Navier–Stokes region.

The author allows himself to compare the situation described above and the "breakthrough" in the description of real gases made in 1873 by Van der Waals. At his time, the theory of perfect gas considered molecules as noninteracting material points. Besides, the existence of molecules was not generally accepted at this time. In his doctoral thesis, Van der Waals put forward two bold assumptions: he assumed that each molecule occupies a finite volume and introduced the force of attraction between molecules (not clarifying its nature). This resulted in the equation of state for a real gas—the classical Van der Waals equation. Since then a swarm of modifications of the Van der Waals equation had appeared. These new equations contained some or other refinements, however, they did not substantially change their pre-image. The above demonstrative historical example is aimed at pointing out the value of innovative ideas in theoretical domains with a purpose to "circumvent" the master equations (the Boltzmann equation in this setting).

The intensity of evaporation is known to be characterized by the speed ratio $s_\infty = u_\infty / \mathrm{v}_\infty$, which is proportional to the Mach number

$$\mathrm{M} = \frac{u_\infty}{\sqrt{(c_p/c_\mathrm{v}) R_g T_\infty}}.$$

Here, c_p, c_v are, respectively, the isobaric and isochoric specific heat capacities of gas, $\mathrm{v}_\infty = \sqrt{2R_g T_\infty}$ is the thermal velocity (the mean square molecular velocity) of molecules, R_g is the individual gas constant, $u_\infty > 0$ is the velocity of vapor flow (the "hydrodynamic velocity"), T_∞ is the gas temperature, the index "∞" means the conditions in the Navier–Stokes region.

The theoretical basis for the study of non-equilibrium evaporation process is the linear kinetic analysis, which describes small departures of gas-dynamic parameters from the equilibrium:$s_\infty \ll 1$. The linear kinetic theory, residing on the solution of the linearized Boltzmann equation, may be found in [6, 7] in its complete form. If the velocity of gas egression is comparable with the sonic velocity (M $\approx$ 1) and if the kinetic jumps of parameters are comparable with its absolute values in the Navier–Stokes region, then one speaks about strong evaporation. There is a great number of studies on the numerical investigations of strong evaporation. We mention, in particular, [1] (a monatomic gas, the relaxation Krook equation) and [9] (a polyatomic gas, the Boltzmann equation, the Monte Carlo method).

The Fig. 4.1 shows that the true values of the parameters on both sides from the CPS are not equal to each other. The solid lines in the figure show the dependences of the true (statistically averaged) parameters: the density ρ, the temperature T, and the hydrodynamic velocity u, as calculated for the ratios of pressure $p_\infty / p_w = 0.3$. The dotted lines are the results of extrapolation of the dependences from the gas-dynamic inside the KL, they separate the macroscopic jumps parameters on the CPS. In the figure, the abscissa is the transverse coordinate z, as normalized by the length of free path of molecules l.

Figure 4.1 clearly shows two different levels of kinetic molecular description of the strong evaporation. In the rigorous (microscopic) approach [1, 6–9], the DF is determined from the solution of the Boltzmann equation, and then the DF is used as a weight function to calculate the moment of temperature, density, and pressure of the

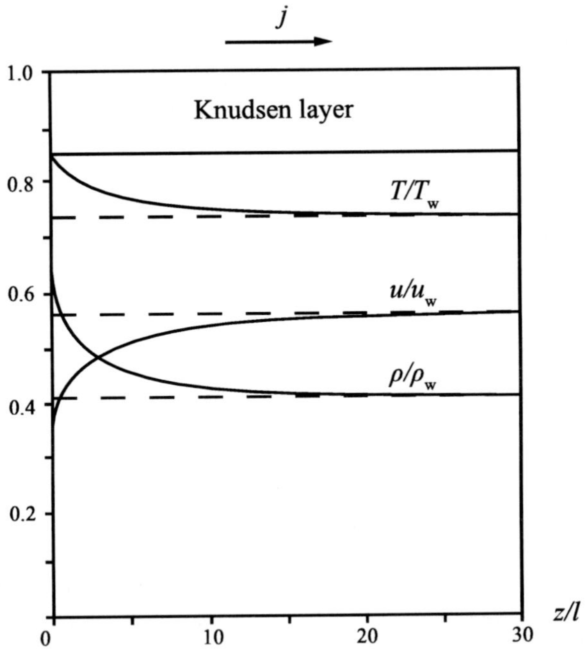

Fig. 4.1 Dependences of the true parameters on the transverse coordinate in the Knudsen layer

egressing gas. The microscopic approach is capable of providing full information about the KL and hence to ascertain both the true and extrapolated values of the parameters on the CPS.

The purpose of the approximate (macroscopic) analysis [10–16] is to specify the BC for the gas-dynamic equations in the Navier–Stokes region. For this purpose, the DF is approximated with free parameters, which are defined from the solution of the system of moment equations. The macroscopic approach is related with a substantial simplification of the mathematical description. However, the solutions obtained under this approach are very bulky, and in turn, calls for a numerical approximation. This therefore suggests a further simplification of the macroscopic description, which would enable one to express analytically the sought-for extrapolated jumps. Such a simplified approach will be presented below in the form of a semi-empirical model of strong evaporation.

4.2 Approximate Analytical Models

Let us now go back to the above problem of the relation between the macroscopic and microscopic descriptions. We formulate it as a concrete problem: to what extent one may simplify the mathematical description in order not to "spoil" too much the BC for the gas-dynamic equations in the Navier–Stokes region? We shall be based on the generally accepted point of view from applications to the effect that the details of the behavior of the DF (and hence, of the true distributions of the parameters in the KL) are of purely theoretical interest [4]. For applications, it suffices in most cases to know only the values of the extrapolated jumps of parameters on the CPS.

One should also not forget that the Boltzmann integro-differential equation is considered up to now as a "tough row to hue" for numerical studies. A usual approach here is to replace the collision integral with the simplified relaxation relation. This being so, a numerical realization of the microscopic approach underlying the Boltzmann equation contains in essence the macroscopic component. Finally, even considering the rapid development of computers in our computer age, it is still impossible to believe that the direct numerical modeling is capable to "cover" the entire range of practical applications.

The first approximate analytical model of strong evaporation was proposed in 1936 by Crout [17]. As distinct from the case of isotropic equilibrium distribution, Crout [17] considered anisotropic molecular spectrum of gas in the KL. Use was made of the ellipsoidal approximation of the DF, which differs from the Maxwell DF by the presence of different measures of thermal velocity in the longitudinal and transverse directions.

The anisotropic DF contained four free parameters: the longitudinal and transverse temperature, the density, and the velocity. Three such parameters were defined from the requirement that the molecular flows of mass, momentum, and energy, as calculated from the given DF, would be equal to the flows that are transferred by the molecules emitted from the CPS. The fourth parameter and the required characteristics of the evaporation processes were determined to form the laws of conservation of the molecular flows, as written for the molecular CPS and the Navier–Stokes region.

Thus, Crout had obtained a complete and qualitatively correct solution to the problem of evaporation of arbitrary intensity. The quantitative results of [17] match well the numerical results of [1, 9]. However, [17] has the following drawback: the adopted approximation of the distribution function on the surface is adapter to the BC on the surface only in the mean (in the terminology of the book [4]). Besides, this solution proves to be inaccurate in the region of low process intensity, quantitatively, it poorly matches the results of the linear theory. Very unfortunately, Crout's pioneer work, which was apparently far ahead of its time, is still left aside even now.

The next breakthrough in the macroscopic approach toward the problem of strong evaporation was made in 1968 by Anisimov [10]. His method was based on the approximation of the DF with one free parameter (the density of the molecular flow incident to the CPS). Next, he solved the system of equations of conservation of the molecular flows of mass and the normal component of the momentum and energy, which were the first three equations of the moment chain of equations [4]. In [10], he put forward a solution in the case of sonic evaporation ($M = 1$). The small two-page Anisimov's note [10] opened a line of research on the strong evaporation on the basis of the mass, momentum, and energy conservation laws. In [11, 12], the original one-parameter model [10] was extended to the general case of gas flow with arbitrary subsonic velocity ($0 \leq M \leq 1$).

The author of the present book proposed a two-parameter approximation of the DF [13–15], where the velocity of molecular flow flying toward the CPS was considered as an additional free parameter. In order to close the mathematical description, the system of three conservation equations was augmented with the "mixing condition" in some sections inside the KL. This two-parameter model was used to obtain approximate analytical solutions to problems of intense phase transition–evaporation [13] and condensation [14]. A linear kinetic analysis of evaporation and condensation, which is an asymptotic variant of the calculation method of [13, 14], was performed in [15].

The results of [13–15] were found to be in a good accord both with theoretical analytical results [6, 7] and with the numerical results of [1, 9], which were obtained for intensive phase changes. It is worth pointing out that the introduction of the intermediate conditional surface in the KL has certain common points with Crout's model. However, here there are principle differences: in Crout's model, one approximates the DF of the emitted flow, whereas in the mixing model one approximates the DF of the molecular flow that flies toward the CPS.

Rose [16] proposed a one-parameter approximation of the DF, where the displacement of the DF over the molecular velocity in the direction of the evaporation flow was considered as free parameter. It seems that such an approximation (of which no justification was given in [16]) is empirical, as distinct from the physically legitimate macroscopic models (the one-parametric one in [10–12] and the two-parametric one in [13–15]).

Nevertheless, the calculated results of [16] were found to match well with those of [11–13]. Comparison of the approximate results of [11–13, 16, 17] with each other and also with the numerical results of [1, 9] provides a satisfactory fit. The maximal deviation of gas parameters, as calculated by various methods, is as follows $\approx 1\%$ for the pressure $p_\infty \approx 2\%$ for the mass flux $J \approx 5\%$ for the temperature T_∞. It is remarkable, that as distinct from all other studies, the analytical curve $T_\infty(\mathrm{M})$ of the old paper [17] matches practically perfectly the results of numerical studies of [1, 9].

It is worth pointing out that approximate models used various (sometimes very different) approximations of the DF. The aforementioned agreement of the results suggests that the macroscopic description of strong evaporation is conservative with respect to the method of introduction of the free parameter into the DF. In this connection, we quote Gusarov and Smurov [1]: "…even rough approximation to the distribution function in terms of velocities in the KL may give satisfactory description of the gas-dynamic evaporation conditions…".

4.3 Analysis of the Available Approaches

This approach depends on the numerical simulation by the Monte Carlo method [9], numerical solution of the relaxation Krook equation [1], etc. Solving numerically the Boltzmann equation, one ascertains the DF, which is later used as a weight function in the corresponding integrals ("summational invariants" [4]). As a result, one determines the moments of the DF in the Navier–Stokes region: the temperature T_∞, pressure p_∞, density ρ_∞, and the gas mass flux $\rho_\infty u_\infty$.

Numerical methods are known to be a continuously improving powerful machinery for calculation of parameters of strong evaporation. However, their efficiency may be hindered by the calculation time, and the accuracy may decrease due to the presence of statistical noise. Numerical difficulties also arise near the regime of sonic evaporation ($\mathrm{M} = 1$). For example, in [1] the last calculation point was obtained with $\mathrm{M} \approx 0.86$, and in [9], with $\mathrm{M} \approx 0.96$. Finally, in the framework of numerical methods, it is impossible to secure the limiting process as $\mathrm{M} \to 0$. In particular, in [1, 9] the first calculation points were obtained with $\mathrm{M} \approx 0.1$.

The DF is determined from the solution of the linearized Boltzmann equation or its approximate analogues [6–8]. In short, the linearization procedure is as follows.

- the absolute values of gas parameters on the CPS (the subscript "w") and in the Navier–Stokes region (the subscript "∞") are assumed to be equal.
- the purpose of calculation is to obtain small (linear) differences of the temperature jump ($T_w - T_\infty \ll T_\infty$) and the pressure jump ($p_w - p_\infty \ll p_\infty$) c) linear analysis gives an asymptotic behavior as $\mathrm{M} \to 0$, hence it is in principal impossible to assess the precise.

The DF is given as an equilibrium Maxwell distribution with one free parameter [10–12, 16]. Solving the system of conservation equations of molecular flows of mass, the normal component of the momentum and energy, we find the temperature and pressure (or the temperature and density) of egressing gas, as well as the free parameter. If one defines a DF with two free parameters [13], then the method of solution remains the same, but the system of equations is augmented with the additional equation. Unlike numerical methods, the approximate approach is capable of obtaining the solution in the entire range of variation of Mach numbers, $0 \le \mathrm{M} \le 1$.

Approximations of the DF, which are used in models [11–13], lead to a system of nonlinear transcendent algebraic equations, which are poorly fit for numerical calculations. This suggests a further simplification of the mathematical description of the problem of strong evaporation. The semi-empirical approach proposed in the sequel may be looked upon as a means for constructing analytical approximations to solutions obtained in the framework of the approximate analytical approach.

4.4 The Semi-Empirical Model

Semi-empirical model of strong evaporation is based on the following two assumptions.

- the linear kinetic theory [6, 7] adequately describes the physical mechanism of kinetic molecular phenomena for evaporation.

- for a transition to the problem of strong evaporation, the linear kinetic jumps must be augmented with the quadratic terms, which describe the discontinuity surface[1] [18].

Let us write down the solutions of [6, 7] to the linear (with the superscript "I") jumps parameters: the linear pressure difference

$$\Delta p^I = F p_w s_w \tag{4.1}$$

and the linear differential of temperature

$$\Delta T^I = \frac{\sqrt{\pi}}{4} T_w s_w. \tag{4.2}$$

Here, $s_w = u_w / \mathrm{v}_w$ is the speed ratio, u_w is the gas velocity, p_w, T_w, ρ_w are, respectively, the pressure, temperature, and gas density, $\mathrm{v}_w = \sqrt{2R_g T_w}$ is the thermal velocity, the index "w" denotes the conditions on the CPS. The right-hand side of (4.1) involves the function of the condensation coefficient β

$$F(\beta) = 2\sqrt{\pi}\frac{1 - 0.4\beta}{\beta}, \tag{4.3}$$

which was introduced in [7].

According to [4], the total gas flow in the KL is formed as a result of interaction of two molecular flows: the one emitted by the surface and the discontinuity surface flowing toward from the Navier–Stokes region. The condensation coefficient β is defined as the ratio of the mass flow of the molecules adsorbed by the interphase boundary and the total mass flow of molecules incident to the CPS. The quantity β depends on the physical nature of the interphase surface and may vary in the range $0 \leq \beta \leq 1$. A survey of various approaches to the calculation and measurement of the condensation coefficient may be found in [19].

We shall be searching quadratic supplements (with the superscript "II") to the linear jumps of parameters describing nonlinear laws of evaporation. To this aim, we shall assume that on the CPS there exists a discontinuity of gas-dynamic parameters, which is described by the Rankine–Hugoniot equations [18]. In the actual fact, this hypothesis means that the linear kinetic jumps are superimposed on the rarefaction shock wave. In this case, we have the wave pressure difference

$$\Delta p_\infty^{\mathrm{II}} = \rho_\infty u_\infty^2 - \rho_w u_w^2 \tag{4.4}$$

and the wave differential of temperature

[1] For a gas discharge to the region of reduced pressure, the discontinuity surface is front of a rarefaction shock.

$$\Delta T_{\infty}^{\mathrm{II}} = \frac{1}{2c_p}\left(u_{\infty}^2 - u_w^2\right). \tag{4.5}$$

According to the initial concepts of the kinetic analysis [7], on the liquid side of a CPS molecules are in the state of chaotic thermal motion with zero mean velocity. When transferring to the gas side of the CPS, molecules accelerate discontinuously to form the flow of gas egressing from the phase interface. Based on this, the wave differential of parameters (4.4), (4.5) should be augmented with the terms that take into account the acceleration of gas flow escaping from the CPS

$$\Delta p_0^{\mathrm{II}} = \frac{1}{2}\rho_w u_w^2, \tag{4.6}$$

$$\Delta T_0^{\mathrm{II}} = \frac{1}{2}\frac{u_w^2}{c_p}. \tag{4.7}$$

Formula (4.6) is a consequence of the Bernoulli equation, formula (4.7) follows from the definition of the enthalpy of stagnation of a perfect gas. From (4.4)–(4.7), we get the total nonlinear differentials of the parameters (transition through a shock rarefaction wave plus acceleration of the flow) the nonlinear pressure jump

$$\Delta p^{\mathrm{II}} \equiv \Delta p_{\infty}^{\mathrm{II}} + \Delta p_0^{\mathrm{II}} = \rho_{\infty} u_{\infty}^2 - \frac{1}{2}\rho_w u_w^2 \tag{4.8}$$

and the nonlinear temperature jump

$$\Delta T^{\mathrm{II}} \equiv \Delta T_{\infty}^{\mathrm{II}} + \Delta T_0^{\mathrm{II}} = \frac{1}{2}\frac{u_{\infty}^2}{c_p}. \tag{4.9}$$

Summing the linear and nonlinear jumps parameters, we get equations for the resulting differences of pressure

$$\Delta p \equiv p_w - p_{\infty} = \Delta p^I + \Delta p^{\mathrm{II}}$$

and the temperature

$$\Delta T \equiv T_w - T_{\infty} = \Delta T^I + \Delta T^{\mathrm{II}}.$$

In view of relations (4.1), (4.2), (4.8), (4.9), we finally obtain

$$\Delta p = F p_w s_w + \rho_{\infty} u_{\infty}^2 - \frac{1}{2}\rho_w u_w^2, \tag{4.10}$$

$$\Delta T = \frac{\sqrt{\pi}}{4} T_w s_w + \frac{1}{2}\frac{u_{\infty}^2}{c_p}. \tag{4.11}$$

Formulas (4.10), (4.11) are the resulting relations for the above semi-empirical model. The remaining calculations are technical in nature. Let us introduce the dimensionless values of the gas parameters in the Navier–Stokes region: the pressure $\tilde{p} = p_\infty / p_w$, the temperature $\tilde{T} = T_\infty / T_w$, and the density $\tilde{\rho} = \rho_\infty / \rho_w$. These quantities are related by the state equation for a perfect gas

$$\tilde{p} = \tilde{\rho} \tilde{T} \, . \tag{4.12}$$

Let us write down expressions for the isochoric c_v and isobaric c_p specific heat capacities of a perfect polyatomic gas

$$c_v = \frac{i}{2} R, \, c_p = \frac{i+2}{2} R. \tag{4.13}$$

Here, i is the number of degrees of freedom of gas molecules: $i = 3$ for a monatomic gas, $i = 5$ for a diatomic gas, $i = 6$ for a polyatomic gas. Expressing from (4.10), (4.11) the temperature T_∞ and the pressure p_∞ of gas in the Navier–Stokes region and changing to the dimensionless form, it follows by (4.12), (4.13) that

$$\tilde{p} + F s_w + \left(\frac{2}{\tilde{\rho}} - 1 \right) s_w^2 - 1 = 0, \tag{4.14}$$

$$\tilde{T} + \frac{\sqrt{\pi}}{4} s_w + \frac{2}{i+2} \frac{s_w^2}{\tilde{\rho}^2} - 1 = 0. \tag{4.15}$$

We assume that the mass flow of gas discharging from the CP is defined as[2] $j_w \equiv \rho_w u_w$. We introduce the physically plausible assumption on the equality of mass flows on the CPS and in the Navier–Stokes region: $j_w = j_\infty$. It implies the relation of the speed ratios on the CPS (s_w) and in the gas1dynamic region (s_∞)

$$s_w = s_\infty \, \tilde{\rho} \, \sqrt{\tilde{T}}. \tag{4.16}$$

As an independent variable, we shall use the Mach number in the Navier–Stokes region

$$\mathrm{M} = \sqrt{\frac{2i}{i+2}} s_\infty. \tag{4.17}$$

In view of (4.3), the system of Eqs. (4.12)–(4.17) provides the closed description of the semi-empirical model of strong evaporation. From this equation, one may also

[2]Strictly speaking, physically sensible on the CPS are the temperature T_w and density ρ_w. The pressure p_w and velocity u_w are reference values.

find the dimensionless mass flow of gas discharging from the KL

$$\tilde{J} \equiv 2\sqrt{\pi} s_w = \frac{j_\infty}{j_{\mathrm{M}}}. \tag{4.18}$$

Here,

$$j_{\mathrm{M}} = \frac{1}{2\sqrt{\pi}} \rho_w v_w \tag{4.19}$$

is the molecular flow of mass emitted by the surface (which is also called the "one-way Maxwell flow"). According to [4], the classics of the early kinetic theory posed the problem of evaporation into vacuum: $j_\infty = j_{\mathrm{M}}$, $\tilde{J} = 1$. Later, the same authors refined the statement of the problem, taking into account the stagnation of the Maxwell flow by the presence of a "shielding vapor cloud" near the phase interface: $j_\infty < j_{\mathrm{M}}$, $\tilde{J} < 1$.

4.5 Validation of the Semi-Empirical Model

4.5.1 *Monatomic Gas*

As a criterion of efficiency of the semi-empirical model, one may consider the degree of agreement of the results obtained under this approach and the available solutions. Below, we shall be concerned with the results of such comparison in three parameters: the Mach number M, the condensation coefficient β, and the number of degrees of freedom of gas molecules i.

In Fig. 4.2, we match the results of calculation by (4.12)–(4.18) with those obtained in [12] for the case of evaporation of a monatomic gas with $\beta = 1$. The curves of the dependences of the dimensionless mass flux $\tilde{J}$ on the Mach number (Fig. 4.2a) are seen to be practically identical. A small difference (less than 2%) near the point $\mathrm{M} = 1$ may appear from the fact that by [12] the maximum of the dependence $\tilde{J}$ (M) is attained with $\mathrm{M} \approx 0.879$, whereas in our calculations it is attained at the point of sonic evaporation $\mathrm{M} = 1$. The departure of the curves for the dimensionless vapor temperature in the Navier–Stokes region (Fig. 4.2b) is at most 3% (2% for the corresponding curves of the dimensionless pressure, see Fig. 4.2c. Note that calculation by the semi-empirical model in fact replicates that by the mixing model, which was made by the author of the present book in [13–15]. Table 4.1 matches these results with sonic evaporation ($\mathrm{M} = 1$).

Increasing Mach number results in the following qualitative trends

Fig. 4.2 Parameters in the Navier–Stokes region vs. the Mach number. 1 Calculation by Eqs. (4.12)–(4.18), 2 results of [12]. **a** The dimensionless mass flux, **b** the dimensionless temperature, **c** the dimensionless pressure

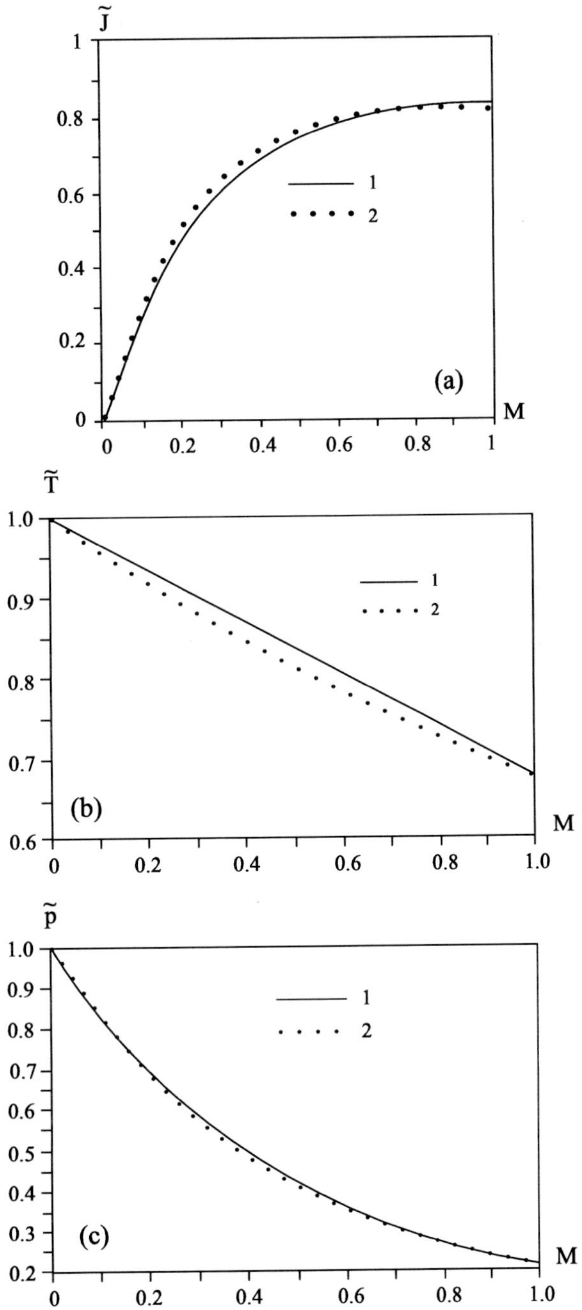

Table 4.1 Sonic evaporation (M = 1). Calculations by the semi-empirical model versus those of [13]

	Semi-empirical model	Paper [13]
$\tilde{T}_\infty$	0.672	0.657
$\tilde{p}_\infty$	0.209	0.208
$\tilde{j}$	0.826	0.829

- mass flux emitted by the CPS grows more slowly, reaching its maximum with M = 1,
- the temperature of the egressing vapor decreases nearly linearly (by one third in the limit),
- the vapor pressure markedly decreases with some delayed (approximately by five times in the limit).

It is of interest to estimate the relative contributions of the two components in the differences of pressure and temperature (Fig. 4.3): the linear component

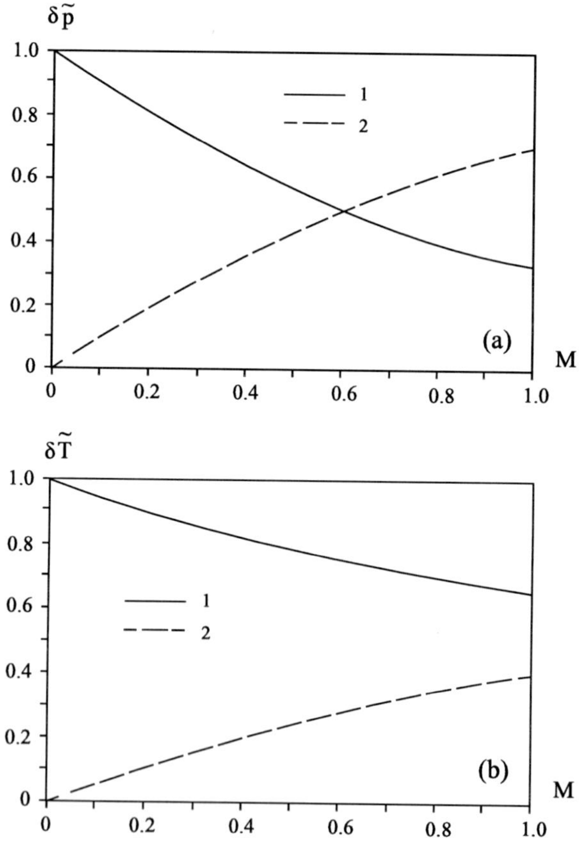

Fig. 4.3 Relative contributions of the linear (1) and the nonlinear (2) kinetic jumps. **a** Pressure jump, **b** temperature jump

$$\delta p^I = \frac{\Delta p^I}{\Delta p}, \delta T^I = \frac{\Delta T^I}{\Delta T} \tag{4.20}$$

and the nonlinear components

$$\delta p^{\mathrm{II}} = \frac{\Delta p^{\mathrm{II}}}{\Delta p}, \delta T^{\mathrm{II}} = \frac{\Delta T^{\mathrm{II}}}{\Delta T}. \tag{4.21}$$

Figure 4.3a shows that the pressure differences δp and δp^{II} are commensurable even in the regime of sonic evaporation $\mathrm{M} = 1$. The linear component of the temperature difference δT^I is seen to substantially increase the nonlinear term δT^{II} in the entire range of variation of the Mach number (Fig. 4.3b). Figures 4.4 depicts the graphs of dependences of $\tilde{J}, \tilde{T}, \tilde{p}$ on the Mach number for three values of β.

The effect of the condensation coefficient is taken into account by the parameter F from the linear pressure jump (4.1). A decrease in β is seen to result in a decrease of the mass flux (Fig. 4.4a) and the pressure (Fig. 4.4c) and in an increase of temperature in the gas-dynamic region (Fig. 4.4b). A decrease in the condensation coefficient has the following qualitative trends.

- the scale of the dependence $\tilde{J}$ (M) markedly decreases,
- the dependence $\tilde{T}$ (M) becomes more and more convex,
- the dependence $\tilde{p}$ (M) fails down catastrophically.

4.5.2 Sonic Evaporation

Figure 4.5 shows the dependences of $\tilde{J}, \tilde{T}, \tilde{p}$ (respectively, a, b, c) on the condensation coefficient, as calculated in the case of sonic evaporation ($\mathrm{M} = 1$). Solid lines correspond to calculation by (4.12)–(4.18), the dotted lines show calculations by the method od [12]. The dependences for $\tilde{J}_1 \equiv \tilde{J}\Big|_{\mathrm{M}=1}$, $\tilde{p}_1 \equiv \tilde{p}\Big|_{\mathrm{M}=1}$ are seen to be practically identical. However, the dependences of $\tilde{T}_1 \equiv \tilde{T}\Big|_{\mathrm{M}=1}$ on β differ even qualitatively: according to our calculations, a decrease in the condensation coefficient results in a linear growth of the gas temperature in the Navier–Stokes region, whereas $\tilde{T}_1 = $ idem by [12]. A decrease of the condensation coefficient has the following qualitative effects: (a) the mass flux emitted from the CPS decreases nearly linearly to zero, (b) the "sonic temperature" remains practically the same, (c) the "sonic pressure" decreases to zero.

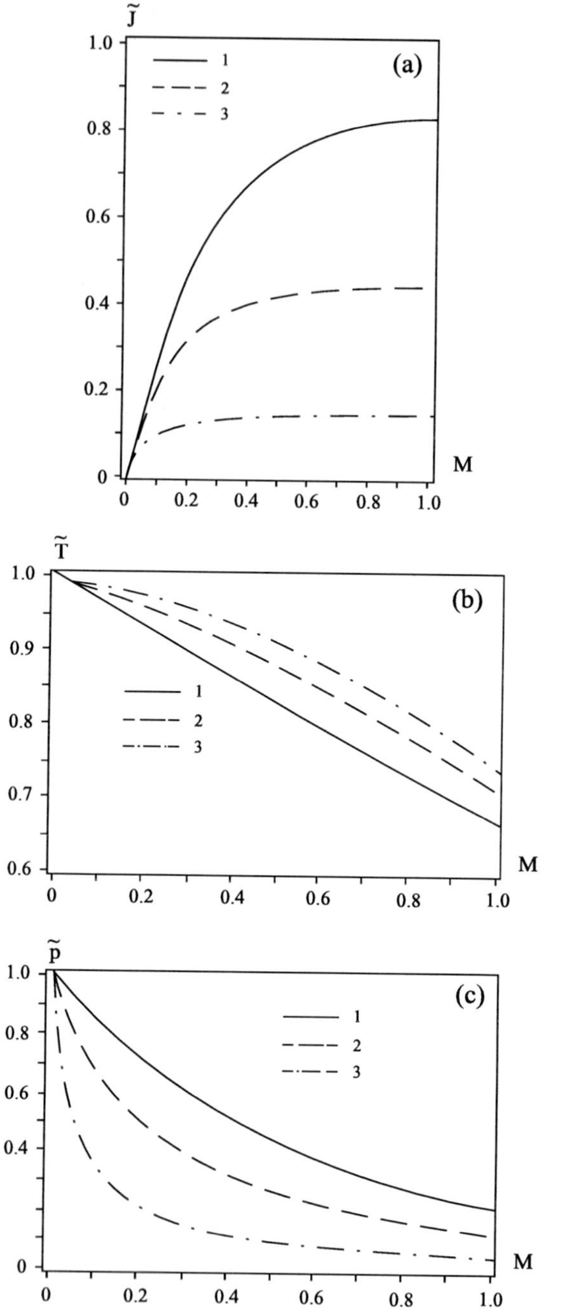

Fig. 4.4 Parameters in the Navier–Stokes region vs. the Mach number for various values of the condensation coefficient. 1 $\beta = 1$, 2, $\beta = 0.5$, and 3 $\beta = 0.15$. **a** The dimensionless mass flux, **b** the dimensionless temperature, **c** the dimensionless pressure

Fig. 4.5 Sonic evaporation. Parameters in the Navier–Stokes region vs. condensation coefficient. 1 Calculation by Eqs. (4.12)–(4.18), 2 results of [12]. **a** Dimensionless mass flux, **b** dimensionless temperature, **c** dimensionless pressure

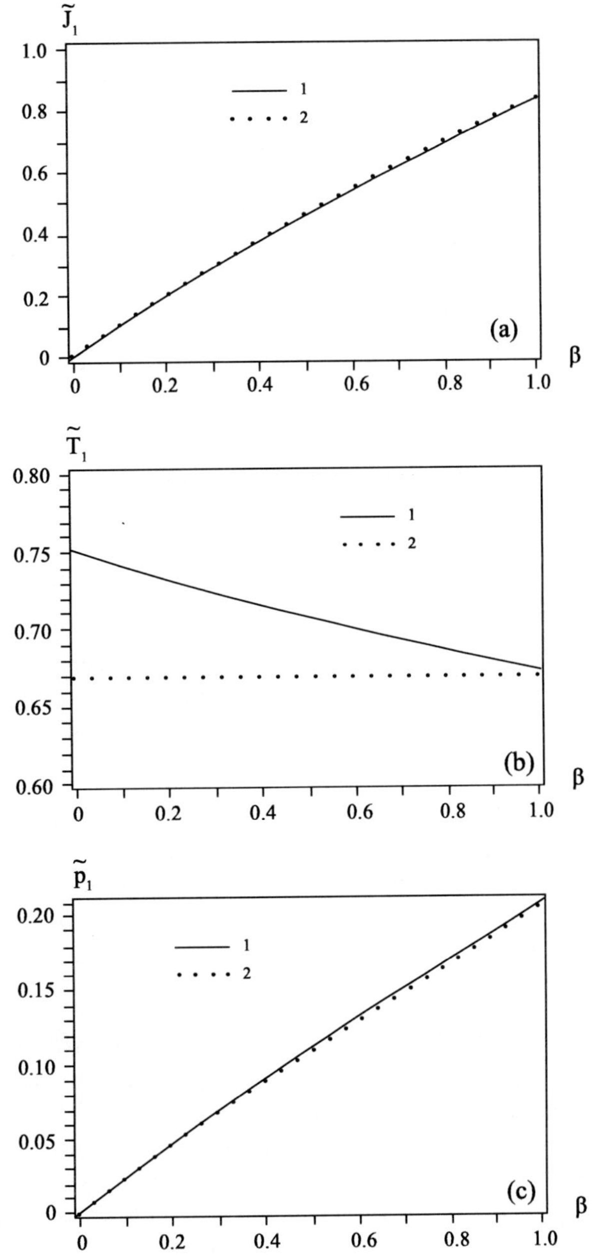

4.5.3 Polyatomic Gas

The problem of strong evaporation for a polyatomic gas was first solved by Cercignani [20] using the moment method. Frezzotti [9] studied numerically the above by the Monte Carlo method in the interval $0.1 \leq \mathrm{M} \leq 0.96$. In the framework of the semi-empirical model, the effect of the number of degrees of freedom of gas molecules is governed by the specific heat capacity of a perfect gas (formulas (4.13)). Figures 4.6 shows the dependences of $\tilde{J}$, $\tilde{T}$, $\tilde{p}$ on the Mach number in the case of a polyatomic gas.

It seems that the mass flux of gas decreases as the degree of freedom of gas molecules increases (Fig. 4.6a), while the temperature and pressure in the Navier–Stokes region both increase (Fig. 4.6b, c). Table 4.2 compares the results of calculation by the semi-empirical model with the results of [9] for cases $i = 3, 5, 6$ with $\mathrm{M} = 0.96$.

4.5.4 Maximum Mass Flow

The author of [21] was the first to realize that the theoretical dependences for $\tilde{J}$ (M) of the paper [20] with $i = 3, 5, 6$ feature a nonphysical maximum in the interval $0.8 \leq \mathrm{M} \leq 0.9$. Sone and Sugimoto [21] proposed a semi-empirical for correction of the near sonic parameters of gas, their methods resided on the conservation laws for the flows of mass, momentum, and energy in the KL. The correction was based on the numerical results of [22] for the Boltzmann equation in a monatomic gas. As a result, Sone and Sugimoto [21] obtained physically based values of the mass flux, temperature, and pressure of gas with sonic evaporation.

Table 4.3 compares the results of calculation by the semi-empirical model with $\mathrm{M} = 1$ with the results of [21] for cases $i = 5, 6$. The above translation of the coordinate of the maximum mass flux into the region $\mathrm{M}_{\max}$ is also characteristic of the approximate models of [12, 13] (Table 4.4).

By contrast, in the first model of [23], the use of the composite DF resulted in the translation of the coordinate of maximum into the region: $\mathrm{M}_{\max}$. Mazhukin et al. [23] reached the required alignment of maxima for the dependences $\tilde{J}$ (M) with the point $\mathrm{M}_{\max}$ in their second model "…by using additional correction parameters…", acknowledging in the meantime that "…this choice of the correction coefficients is by no means unique…" (Table 4.4). Here, it is worth pointing out that the semi-empirical model is capable of uniquely forecasting the maxima of the dependences $\tilde{J}$ (M) with $\mathrm{M}_{\max}$ for an arbitrary number of degrees of freedom of gas molecules i.

Figure 4.7 illustrates the sonic maximum of the dependences of the mass flux on the Mach number. It seems that this important property of the model follows from the conclusions of the theory of rarefaction shock [18]: in a supersonic flow,

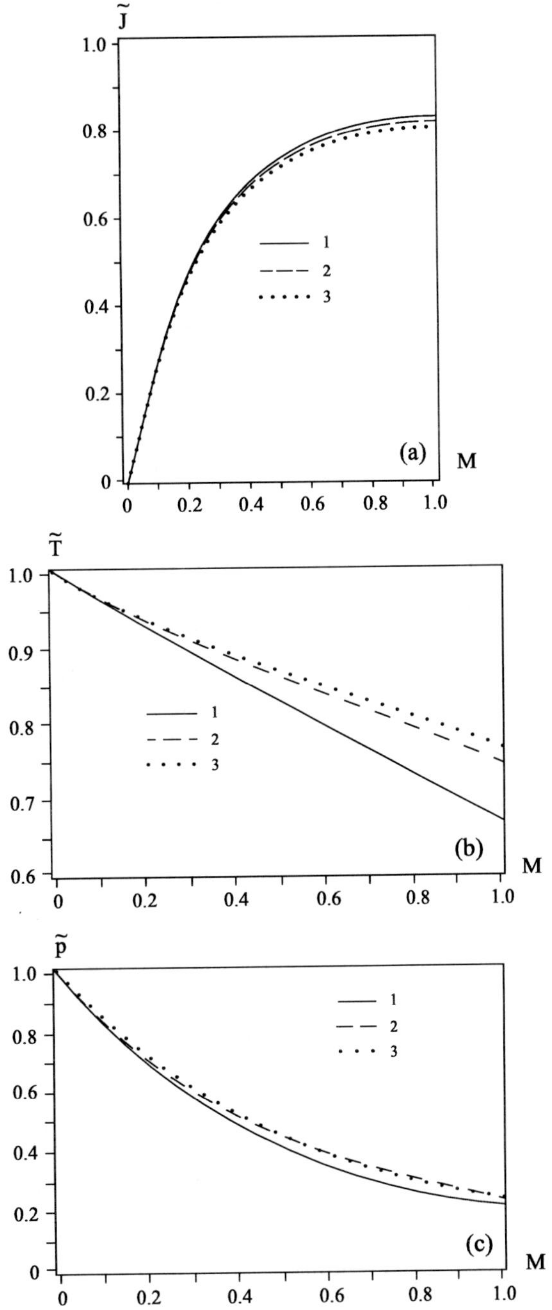

Fig. 4.6 Polyatomic gas. Parameters in the Navier–Stokes region vs. the Mach number for various values of the degree of freedom of gas molecules. 1 $i = 3$, 2 $i = 5$, and 3 $i = 6$. **a** The dimensionless mass flux, **b** the dimensionless temperature, **c** the dimensionless pressure

Table 4.2 M = 0.96. Calculations by the semi-empirical model versus those of [9]

	Semi-empirical model	Paper [9]	Semi-empirical model	Paper [9]	semi-empirical model	Paper [9]
	$i = 3$		$i = 5$		$i = 6$	
$\tilde{T}_\infty$	0.685	0.667	0.757	0.763	0.779	0.793
$\tilde{J}$	0.826	0.836	0.809	0.807	0.804	0.798

Table 4.3 Sonic evaporation (M = 1). Calculations by the semi-empirical model versus those of [21]

	Semi-empirical model	Paper [21]	Semi-empirical model	Paper [21]
	$i = 5$		$i = 6$	
$\tilde{T}_\infty$	0.749	0.758	0.771	0.791
$\tilde{p}_\infty$	0.236	0.2366	0.244	0.245
$\tilde{J}$	0.809	0.805	0.804	0.796

Table 4.4 Calculated values of the Mach number in various models in which the maximum of vapor mass flux is attained

Paper	[17]	[12]	[13]	[23] (first model)	[23] (second model)	Semi-empirical model
M_{max}	0.954	0.879	0.928	1.11	1.0	1.0

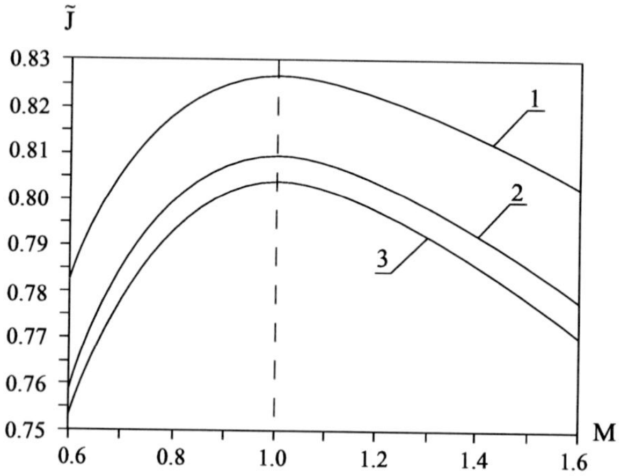

Fig. 4.7 Polyatomic gas. Demonstration of the sonic maximum of the dependences of the mass flux on the Mach number. 1 $i = 3$, 2 $i = 5$, and 3 $i = 6$

any perturbation is referenced to the surface. If the region of supersonic flow exists ab initio, then it is unstable and should separate from the surface.

Correspondingly, the region M > 1 in Fig. 4.7 is physically unrealizable, its role is to illustrate the existence of maxima of the dependences $\tilde{J}$ (M) with M = M_{max}. It is worth pointing out that the conclusion on the nonexistence of supersonic evaporation was first made in the early paper [17], and was later justified by Crout [10]. This being so, the semi-empirical model had enabled one to describe with good accuracy the effect of the following parameters on the parameters of strong evaporation: the Mach number, the condensation coefficient, and the number of degrees of freedom of gas molecules. This suggests that the physical assumptions of the model are capable of adequately represent (in the large) the kinetic laws of strong evaporation.

4.6 Hydrodynamic Instability

4.6.1 Small Disturbance Method

Theoretically any boundary-value problem for the Navier–Stokes equations under known stationary BC should have an exact stationary solution. This condition is necessary, but not sufficient. Each solution of the hydrodynamics equations should in addition satisfy the stability condition with respect to small perturbations. Physically, this is the requirement that the perturbations should decay in time (otherwise the motion will be unstable and cannot be effected in practice.

The mathematical study of the stability with respect to small perturbations follows the traditional approach of [24]. Assume there is a stationary flow in which the distributions of the pressure p_0 and the velocity vector $\mathbf{v}_0$ satisfy the stationary Navier–Stokes equations

$$(\mathbf{v}_0\nabla)\mathbf{v}_0 = -\frac{\nabla p_0}{\rho} + \nu\Delta\mathbf{v}_0,\ \mathrm{div}\mathbf{v}_0 = 0. \tag{4.22}$$

In accordance with the small disturbance method, the stationary parameters are subject to nonstationary small perturbations $\mathbf{v}'$, p'. The resulting summary flow

$$\mathbf{v} = \mathbf{v}_0 + \mathbf{v}',\ p = p_0 + p' \tag{4.23}$$

should satisfy the nonstationary Navier–Stokes equations

$$\frac{\partial \mathbf{v}}{\partial t} + (\mathbf{v}\nabla)\mathbf{v} = -\frac{\nabla p_0}{\rho} + \nu\Delta\mathbf{v},\ \mathrm{div}\mathbf{v} = 0. \tag{4.24}$$

Equations (4.24) are linearized: only the terms linear in the perturbation are retained and the terms of higher orders of smallness are dropped out

$$\frac{\partial \mathbf{v}'}{\partial t} + (\mathbf{v}_0\nabla)\mathbf{v}' = -\frac{\nabla p'}{\rho} + \nu\Delta\mathbf{v}',\ \mathrm{div}\mathbf{v}' = 0. \tag{4.25}$$

So, $\mathbf{v}'$ satisfies the system of homogeneous differential equations in which the coefficients depend only on the coordinates. By linearity, the general solution of (4.25) is a superposition of particular solutions, in which $\mathbf{v}'$ depends on time via the factor $\exp(-i\omega t)$. In general, the frequencies ω of the perturbations are complex quantities and can be determined by solving Eqs. (4.25) with the corresponding external conditions. If, in addition, there appear frequencies with positive imaginary parts, the factor $\exp(-i\omega t)$ will exponentially increase in time and the motion will be unstable with respect to small perturbations. It follows that a necessary condition that the motion be stable is that the imaginary part of ω be negative. In this case, the appearing perturbations will exponentially decrease in time.

4.6.2 *Kelvin–Helmholtz Instability*

The mass, energy, and momentum conservation laws underlying the equations of dynamics of an inviscid non-heat-conductive gas can be applied not only to continuous gas-dynamic quantities, but also to the domains in which they have a discontinuity. Since the length scale is absent in the dynamics of an ideal gas, it may contain arbitrary thin transition layers. In the limit case, we have a discontinuity: finite jumps of gas-dynamic quantities through the infinitely thin layer. On the discontinuity surface, the compatibility conditions should be satisfied, which secure the continuity of the gas parameters upstream the discontinuity (the index "1") and downstream the discontinuity (the index "2"). Let the square brackets denote the difference of values of some quantity from both sides of the discontinuity surface.

Let us write down the continuity conditions for the mass flow

$$[\rho u] = 0, \tag{4.26}$$

the momentum flow

$$\left[p + \rho u^2\right] = 0, \tag{4.27}$$

and the energy flow

$$\left[u^2/2 + h\right] = 0. \tag{4.28}$$

Here, u is the component of the velocity vector $\mathbf{v}$ orthogonal to the surface. Equations (4.26)–(4.28) form a closed system of compatibility conditions. From these equations, a conclusion can be made of the possibility of existence of two types of discontinuity surfaces. Let us first consider the most simple case when the mass flux through the discontinuity is absent: $\rho u = 0$. Since the densities from both sides of the discontinuity are nonzero, the velocities should vanish $u = 0$. Here, conditions (4.26), (4.28) are satisfied identically, and condition (4.27) guarantees that the pressures from both sides of the discontinuity are equal. So, in this case, the velocities and the pressure of the gas are continuous on the discontinuity surface

$$[u] = 0, [p] = 0. \quad (4.29)$$

If there exist tangents to the discontinuity surface, these tangents may also have a jump, which is called the "tangential discontinuity". The tangential discontinuity may lead to the development of the instability, which is known as the Kelvin–Helmholtz instability after Helmholtz Thomson (Lord Kelvin), who pioneered the study of this instability in 1871 (see [26]) and in 1868 (see [26). It is a hydrodynamic instability in which incompressible and inviscid fluids are in relative and irrotational motion. In the Kelvin–Helmholtz instability, the velocity and density profiles are uniform in each fluid layer, but they are discontinuous at the plane interface between the two incompressible fluids. This discontinuity in the tangential velocity induces the vorticity at the interface.

Let us give an analysis of the stability of tangential discontinuity for the case of an incompressible ideal liquid [27]. Assume that two layers of liquid slide over one another, forming a jump of the tangent with respect to the velocity of the discontinuity surface. Without loss of generality we can assume that an arbitrary small part of the discontinuity surface is flat and that the liquid from one side of the discontinuity is at rest.[3] Let $\mathbf{v}$ be the velocity vector of the liquid from the moving side of the discontinuity. Let the x-axis be directed along the vector $\mathbf{v}$, and let the z-axis be directed along the normal vector to the surface.

Assume that the discontinuity surface is subject to a perturbation of small amplitude, under which all hydrodynamic parameters (the coordinates of the surface, the velocity, and the pressure of the liquid) are proportional to $\exp(i(kx - \omega t))$. Putting $\nu = 0$ in the Navier–Stokes Eq. (4.25), we arrive at the Euler equations, which describe the flow of an ideal liquid with $\mathbf{v}_0 = \mathbf{v} = \text{const}$. So, given a constant shear velocity, we have the following system of equations for the perturbation $\mathbf{v}'$

$$\text{div}\mathbf{v}' = 0, \frac{\partial \mathbf{v}'}{\partial t} + (\mathbf{v}\nabla)\mathbf{v}' = -\frac{\nabla p'}{\rho}. \quad (4.30)$$

Here, $\mathbf{v}'$ is the pulsation of the tangent velocity. Since $\mathbf{v}$ is directed along the x-axis, the second equation in (4.30) can be rewritten as

$$\frac{\partial \mathbf{v}'}{\partial t} + \mathrm{v}\frac{\partial \mathbf{v}'}{\partial x} = -\frac{\nabla p'}{\rho}, \quad (4.31)$$

where $\mathrm{v} \equiv |\mathbf{v}|$.

Taking the divergence of both sides of (4.31) and using Eq. (4.30), we get zero at the left, so that p' should satisfy the Laplace equation

[3] This can always be achieved by a corresponding choice of the system of coordinates.

$$\Delta p' = 0. \tag{4.32}$$

Let $\zeta = \zeta(x, t)$ denote the deviation along the z-axis of the discontinuity surface under the perturbation. Then, $\partial\zeta/\partial t$ the drift velocity of the surface coordinate with a given coordinate x. The component w_z' of the liquid velocity which is normal to the discontinuity surface is equal to the motion velocity of the surface itself, and so in the effective linear approximation we have

$$\frac{\partial\zeta}{\partial t} = w_z' - \mathrm{v}\frac{\partial\zeta}{\partial x}. \tag{4.33}$$

Let us search for the expression for the pressure pulsations p' in the form

$$p' = f(z)\exp(i(kx - \omega t)),$$

where k is the wavenumber of the perturbation. Substituting into (4.32), we get the following equation for $f(z)$

$$\frac{d^2 f}{dx^2} - k^2 f = 0,$$

which has the solution $f \sim \exp(\mp kz)$. Assume that the space from the side of the moving discontinuity surface corresponds to the positive values of z. Then, we should write $f \sim \exp(-kz)$. Hence, we get the expression for the pressure pulsations

$$p' = \text{const}\exp(i(kx - \omega t))\exp(-kz). \tag{4.34}$$

Substituting this expression into the z-component of Eq. (4.31), we get

$$w_z' = \frac{kp'}{i\rho_1(k\mathrm{v} - \omega)}. \tag{4.35}$$

Now from (4.33), we obtain the pulsation of the normal velocity on the surface

$$w_z' = i\zeta(k\mathrm{v} - \omega).$$

This together with (4.35) shows that

$$p' = -\zeta\frac{\rho_1(k\mathrm{v} - \omega)^2}{k}. \tag{4.36}$$

The pressure from the side at rest is expressed by the same formula, but with $\mathrm{v} = 0$ and with the opposite sign (since in this region $z < 0$ and since all the quantities should be proportional to $\exp(kz)$. So, we have

$$p' = \zeta \frac{\rho_2 \omega^2}{k}. \tag{4.37}$$

Here, the different densities ρ_1, ρ_2 are introduced with the aim at describing the case of the interface region between two different immiscible liquids. From the condition $p_1' = p_2'$ on the discontinuity surface, we get

$$\rho_1 (k\mathrm{v} - \omega)^2 = -\rho_2 \omega^2,$$

from which follows the required dependence between ω and k

$$\omega = k\mathrm{v} \frac{\rho_1 \mp i\sqrt{\rho_1 \rho_2}}{\rho_1 + \rho_2}. \tag{4.38}$$

From (4.38), it follows that the frequency of pulsations is a complex quantity and there always exist values of ω with positive imaginary part. This means that the tangential discontinuity is unstable with respect to small perturbations for any values of the frequency and the wave number. Such form of instability is called "absolute".

With increasing discontinuity of the velocity, the increment of development of the Kelvin–Helmholtz instability will be also increasing. If the discontinuity velocities are comparable with the sound velocity in liquid, the above classical theory becomes inapplicable, the compressibility of the medium plays an essential role. Landau and Lifshitz [27] were the first to draw attention to the necessity of consideration of compressibility of liquid for large discontinuity velocities. In particular, they showed that, in this case, the dependence of the perturbation frequency on the wave number is more involved than in (4.38). However, in this setting, the main theoretical conclusions on absolute instability of tangential discontinuity remain valid.

4.6.3 *Kinetic Shock Wave*

Consider now the case when the discontinuity surface crosses a finite mass flux. In this case, the density and pressure (and hence, other thermodynamic quantities), as well as the normal velocity component, have discontinuities described by Eqs. (4.26)–(4.28), which in this setting are written in the form

$$[\rho u] = 0, \left[p + \rho u^2\right] = 0, \left[\rho u^2/2 + h\right] = 0. \tag{4.39}$$

The discontinuity type described by conditions (4.39) is known as the "shock wave" [18]. If now we change back to the fixed coordinate system, then u should be always replaced by the difference between the velocity component u_n normal to the discontinuity surface and the velocity u_s of the surface itself directed along the normal vector to it

$$u = u_n - u_s. \tag{4.40}$$

Here, the velocity u is the velocity of gas motion with respect to the discontinuity surface. Condition (4.40) can be reformulated as follows: the velocity of propagation of the discontinuity surface itself with respect to the gas at rest is $-u = u_s - u_n$. Here, it should be recalled that in the general case of density discontinuity, this velocity of gas is different on both sides of the surface.

As in the above case of tangential discontinuity, the shock wave will be stable if an arbitrary initial perturbation on the discontinuity surface will decay with time. If the perturbations increase unboundedly, the developing instability will generate perturbations of the gas-dynamic quantities in the region downstream of the shock wave. Let us investigate in this relation the above semi-empirical model of strong evaporation.

We recall that this model is based on the principle of superposition of a gas-dynamic rarefaction shock wave [18] and linear kinetic discontinuities [6, 7]. From physical considerations, it is natural to assume that the stability of the shock wave depends on the boundary-value conditions on the discontinuity surface. Based on this formula, we put forward the hypothesis: the system of Eqs. (4.12)–(4.17) describes the standing rarefaction wave. This specific shock wave, whose front coincides with the Condensed-Phase Surface (CPS), will be called the "Kinetic Shock Wave" (KSW). To justify this hypothesis and investigate the KSW stability, we shall consider the thermodynamic aspect of the problem of strong evaporation.

4.6.4 *Thermodynamic System*

Einstein's high appreciation of thermodynamics is given in the following quotation [28]: "A law is more impressive the greater the simplicity of its premises, the more different are the kinds of things it relates, and the more extended its range of applicability. Thermodynamics is the only physical theory of universal content, which I am convinced, that within the framework of applicability of its basic concepts will never be overthrown". In the narrow sense, this thermodynamic problem calls for the prediction and description of that finite state which should be attained by a given system under the superimposed external conditions. A thermodynamic system evolves in the course of sequential spontaneous transformations taking place inside it. One of the principal provisions of the classical thermodynamics is the proof of a stable thermodynamic equilibrium for a finite state of a closed system. The complete equilibrium state is characterized by the minimum of the corresponding thermodynamic potential Π

$$\mathrm{d}\Pi = 0, \mathrm{d}^2\Pi > 0.$$

As a thermodynamic potential, one may consider, in particular, the free Helmholtz or Gibbs energy [29]. The possibility of evolution of a closed system to a state with one minimum of the thermodynamic potential follows from the second law of thermodynamics: the entropy S of an isolated system involving irreversible processes is always increasing. It is worth pointing out that the process of evolution of the system itself cannot in principle be described in the framework of the classical thermodynamics, because it does not involve the time factor.

In open systems such evolution is possible, but not always. Depending on the BC, a spontaneous variation of properties of the system may result in some stable state, which is general is not an equilibrium. It is believed that a system is at a dynamical equilibrium if the BC imposed on this system are "compatible with equilibrium". A characteristic example of this situation is the stationary distribution of the temperatures in the system. The natural limit case of the dynamical equilibrium is a stable thermodynamic equilibrium. In the absence of force fields, it is characterized by the isotropy in temperature, pressure, and the chemical potential of all the components existing in the system. This means that inside the system the gradients of the quantities are zero and that mass and energy fluxes are absent. In turn, a system which did not attain the complete thermodynamic equilibrium state (a non-equilibrium system) is characterized by the gradients of the corresponding parameters, and as a rule, by the presence of mass and energy fluxes.

A typical example is given by a two-stage evolution, when a non-equilibrium system tends to a dynamical equilibrium and (once it is reached) to the complete equilibrium. The study of such a process and a finite state to which the system not capable of reaching a complete equilibrium is tending forms a part of the subject of the thermodynamics of irreversible processes (or the non-equilibrium thermodynamics) [30]. The non-equilibrium thermodynamics is based on the concept of a thermodynamic force X_i, which is equivalent to that of an irreversible process. Thermodynamic forces appear under the spatial inhomogeneity of a system with respect to the temperature, concentration, and pressure. The principal problems under consideration here is the identification of local states of the system and the analysis of their stability. In special cases, the ultimate result of the development of instability of stationary states of an open system is the formation of organized dissipative structures.

A variation of the entropy of an open system may proceed in one of the two ways.

- In the course of an inadvertent progress of thermodynamically irreversible processes, the entropy inside the system is always increasing ($d_i S > 0$).
- Due to the energy exchange of the system with the external environment, the entropy may either increase or decrease ($d_e S > 0, d_e S < 0$).

In the non-equilibrium thermodynamics, it is postulated that the components $d_i S$ and $d_e S$ are independent, and the total variation of the entropy of an open system dS is equal to their sum

$$dS = d_i S + d_e S.$$

This relation also holds for an arbitrary time, and so we have

$$dS/dt = dS/dt = d_i S/dt + d_e S/dt.$$

It follows that the rate of change of the entropy of the system dS/dt is equal to the sum of the entropy production rate inside the system $d_i S/dt$ and the rate of entropy exchange between the system and the environment $d_e S/dt$. For thermodynamically reversible processes, the entropy remains constant: $d_i S = 0$. For any irreversible variations, we have $d_i S > 0$. For an isolated system, the second law of thermodynamics implies that $d_e S \equiv 0$, $dS = d_i S \geq 0$.

Let us introduce the energy dissipation rate in the form of heat for a temperature-isotropic system

$$\boldsymbol{P} = T d_i S/dt = T\sigma,$$

where $\sigma \equiv d_i S/dt$ is the entropy production rate in the system. Here, the following cases are possible.

- $\sigma > 0$, $\boldsymbol{P} > 0$. Internal irreversible processes take place in the system.
- $d_e S/dt > 0$. The entropy of the system is increasing due to the exchange of matter and energy with the external environment.
- $d_e S/dt < 0$. The outflow of positive entropy from the system to the external environment (for example, due to the outflow of heat or a part of matter from the system) exceeds the external heat gain.

Here, the derivative $d_e S/dt$ can be positive or negative. Hence, irrespective of the necessary observance of the inequality $d_i S/dt > 0$, the total entropy of the open system can be increasing or decreasing. So, the two components of the entropy variation $d_i S$ and $d_e S$ demonstrate the principal differences in the thermodynamic properties of open and isolated systems.

4.6.5 *Onsager Reciprocal Relation*

A system near a thermodynamic equilibrium is characterized by a weak effect of external forces[4] and small velocities of processes. In this case, there is linear relation between the flow J and the drive power X with the coefficient L

$$J = LX.$$

In general, several interrelated processes take place in the system. As a result, the flow of each thermodynamic parameter will depend not only on the "proper" thermodynamic force, but also on the drive forces of the remaining processes. A qualitative and quantitative study of this mutual influence is an important property in non-equilibrium thermodynamics.

[4]By thermodynamic forces we mean, as a rule, the gradient of intensive thermodynamic parameters.

In 1931, Onsager [31] put forward the principle of interaction of thermodynamic processes. We denote by a_j the intensive thermodynamic parameters (T, p, μ). Then, the thermodynamic forces will be determined by the relation of the form $X_j \sim \nabla a_j$. According to Onsagers's theory, flows of interacting irreversible processes are described by linear differential equations with constant coefficients

$$J_i = \sum_j L_{ij} \nabla a_j.$$

In the case when the coefficients L_{ij} are independent of X_j, this implies the fundamental Onsager reciprocal relation using which one can describe the entropy production rate in involved systems

$$J_i = \sum_j L_{ij} X_j.$$

In this setting, interacting thermodynamic processes are called "conjugate processes". The Onsager reciprocal relation also holds for a wide class of simultaneously occurring processes: diffusion and heat transfer, electron current and ion diffusion, the cases when many chemical reactions take place simultaneously, etc. The simplest case of interaction of two processes 1 and 2 can be described using the cross-effect coefficients L_{11}, L_{12}

$$\left.\begin{aligned} J_1 &= L_{11} X_1 + L_{12} X_2, \\ J_2 &= L_{21} X_1 + L_{22} X_2. \end{aligned}\right\}$$

Using the invariance principle of the motion laws of particles under time reversal, Onsager formulated the principle of commutation of the coefficients
$L_{ij} = L_{ji}$.

An important postulate of the non-equilibrium thermodynamics is the Curie symmetry principle [32], which was formulated in 1894 (Curie, P. Sur la symétrie des phénomènes physiques: symétrie d'un champ électrique et d'un champ magnétique, Journal de Physique, 1894, 3e série, 393–415). Later "the Curie symmetry principle" was further developed and supplemented and can be formulated in modern terms as follows (see [33]).

- If the causes of phenomena have equal symmetry elements, then they are also preserved in the symmetry of consequences.
- The number of symmetry elements possessed by the causes is always smaller than or equal to the number of elements of the action they cause.
- The action of several causes of which each has a proper symmetry gives a result, which preserves only the coincident symmetry elements of their causes. The consequences may have a higher symmetry than their causes.
- If the result has certain degradation in symmetry (a dissymmetry), then the same dissymmetry should also be manifested in its causes.

In particular, in accordance with the Curie symmetry principle, the cross-effect coefficients L_{ij} can be nonzero only in the cases when the interacting thermodynamic forces X_j have the same spatial dimension (for example, they are either scalar or vector quantities, and so on).

4.6.6 Prigogine Theorem

Consider the entropy production rate in involved thermodynamic systems. In the case when several processes take place simultaneous, we have, for the energy dissipation rate

$$\boldsymbol{P} = T\sum_k J_k X_k > 0.$$

From the Onsager reciprocal relation, it follows that

$$\boldsymbol{P} = \sum_k \sum_j L_{kj} X_k X_j > 0.$$

This expression is known as the Rayleigh–Onsager dissipation function. This function is an analogue of the Rayleigh dissipation function, which was introduced in mechanics in 1878 by Lord Rayleigh [34]. If a system involves irreversible processes, then $\boldsymbol{P} > 0$. Hence in addition to the Onsager reciprocal relation for the thermodynamic system, the following relations should also hold

$$L_{ii} \geq 0,\ L_{ij} > -\left(L_{ii} X_i^2 + L_{jj} X_j^2\right)/2X_i X_j.$$

Here, the sign of the off-diagonal reciprocity coefficients can be positive. In the linear approximation, the entropy production rate (or the dissipation of energy) can be also expressed as the quadratic flow function J_i

$$\boldsymbol{P} = \sum_k \sum_j \widetilde{L}_{kj} J_k J_j > 0.$$

Here, the coefficients $\widetilde{L}_{kj}$ are also called the Onsager coefficients and have the same properties as the coefficients L_{kj}.

So, the representation of the function $\boldsymbol{P}$ (or its derivative $d_i S/dt$) in the form of flows is equivalent to its representation in terms of forces. Under stationary external conditions, a finite stationary state may settle in a non-equilibrium system. Its specific feature is the stationarity of the internal parameters of the systems, and as a rule, the absence of flows of internal thermodynamic parameters. The latter depend on the course of the irreversible processes exited by internal thermodynamic forces. The total entropy of an open system in a stationary state is also constant, i.e.

$$dS/dt = dS/dt = d_i S/dt + d_e S/dt.$$

However, the terms $d_i S/dt$ and $d_e S/dt$ can be nonzero. The non-equilibrium thermodynamics is also capable of answering the question about the stability of non-equilibrium stationary states with respect to external perturbations and inadvertent internal fluctuations. This stability can be investigated by analyzing the variation of the quantity of the entropy production rate as the system leaves the stationary state.

In 1940s, Prigogine [35], when considering the variation of $d_i S/dt$ in time, described the evolution of the system to a stationary state. This problem is close in spirit to the problem of classical thermodynamics about the direction of an inadvertent irreversible process in an isolated system. According to the second law of thermodynamics, these variations occur only in the direction of the entropy increase. Prigogine proved that in a finite equilibrium state the entropy attains its maximum value.

Consider an open system involving m simultaneously occurring irreversible processes. The energy dissipation rate and the corresponding entropy production rate in this system

$$\boldsymbol{P} = \sum_{i=1}^{m} J_i X_i > 0.$$

According to Prigogine's hypothesis, an establishment of a stationary state with $J_i = 0$ for a flow of internal parameters a_i is consequent of variations of the internal thermodynamic forces X_i. Let us investigate the character of the dependence of the quantity $\boldsymbol{P}$ on the force X_i. To this end, we take the partial derivative $\partial \boldsymbol{P}/\partial X_i$ with constant $X_i (j \neq i)$. The Rayleigh–Onsager dissipation function can be written as

$$\partial \boldsymbol{P}/\partial X_i = 2 \sum_k L_{ik} X_k = 2 J_i.$$

It follows that the condition of attainment of a stationary state in some internal variable ($J_i = 0$) and the condition $\partial \boldsymbol{P}/\partial X_i = 0$ are equivalent. Since the quantity $\boldsymbol{P}$ should be a quadratic form in the entire range of variation of the independent variable X_i, the condition $\partial \boldsymbol{P}/\partial X_i = 0$ serves simultaneously as the condition for minimality of $(X_1, X_2, \ldots, X_m)$ in the variable X_i. In a similar way, one may obtain stationarity conditions also for flows of different internal variables. So, the conditions $J_k = 0$ and $\partial \boldsymbol{P}/\partial X_k = 0$ are also equivalent. The same equivalence conclusion also holds for an arbitrary system involving any number of processes related with each other by the Onsager reciprocal relation. This implies the following important property of irreversible processes: "Under fixed external conditions in a stationary state of an open system close to a thermodynamic equilibrium, the entropy production rate is constant, positive, and minimal".

This conclusion is the main result of Prigogine's theorem of the minimum entropy production [36]—this is a theorem of the thermodynamics of non-equilibrium processes according to which the minimum level of production of the entropy in a system, under conditions preventing the attainment of the equilibrium state, corresponds to the steady state of this system. If there are no such hindrances, then the production of entropy reaches its absolute minimum (which is zero). The theorem was derived in 1947 by I. Prigogine from the Onsager reciprocal relations. It is valid only if the kinetic coefficients of the Onsager relations are constant, for real systems, the principle is merely an approximation. Thus, the minimization of the production of entropy for the steady state is not as general a principle as the maximization of entropy for the equilibrium state.

Prigogine theorem also holds for any open systems near a thermodynamically stable state. It is also essential that only a monotone decrease in the quantity $\partial \boldsymbol{P}/\partial X_k = 0$ is possible as the system inadvertently evolves to a stationary state, i.e., the inequality $\partial \boldsymbol{P}/\partial t < 0$ always holds. Hence, the minimum principle of the entropy production rate is a quantitative criterion for determining the direction of inadvertent changes in an open system (i.e., a criterion of its evolution). In an isothermic system, this principle is equivalent to the principle of the minimum of the energy dissipation rate, which was formulated by Onsager in 1930s when he considered electrodynamic problems.

In the course of the evolution of a system from some initial state, the quantity $\boldsymbol{P} = T d_i S/dt$ decreases monotonically. However, the value $\boldsymbol{P}$ remains positive and tends monotonically to the minimal constant positive value, which corresponds to the attainment of a finite stationary state. So, Prigogine's theorem provides a criterion of the evolution of a system to a stationary state near a thermodynamic equilibrium.

4.6.7 Change of the Entropy

As was already pointed out, the entropy of the system characterizes the direction of the development of inadvertent thermodynamic processes in this system. The existence and properties of the entropy as a function of the state of a thermodynamic system follow from the second law of thermodynamics. In essence, its various formulations available in the literature are particular cases of the general principle that the entropy is nondecreasing [37]: "In an adiabatically isolated thermodynamic equilibrium the entropy cannot decreases—it is either conserved (if in the system only reversible processes take place) or is increasing (if the system involves at least one irreversible process)".

It follows that in an equilibrium state the entropy of an isolated thermodynamic system should always have the greatest possible value. Based on the general thermodynamic laws studied above, we consider the aforementioned hypothesis on the KSW. Let us find the difference of the entropies of an ideal gas when transiting through the KSW front

$$\Delta S = c_p \ln\frac{T_\infty}{T_w} - R\ln\frac{p_\infty}{p_w}.$$

Changing to the dimensionless form, we get

$$\Delta\, \tilde{S} = \frac{k+2}{2}\ln \tilde{T} - \ln \tilde{p}, \tag{4.41}$$

where $\Delta\, \tilde{S} = \Delta S/R$.

Figure 4.8 shows the dependencies of the entropy increment on the Mach number for various values of the condensation coefficient β, as calculated from (4.41). According to (4.41), a transition across the KSW front is characterized by the two important properties.

(1) A positive increment of the entropy. The inequality $\Delta\, \tilde{S} > 0$ holds in the interval $0 < \mathrm{M} < 1$. Moreover, in a sonic point, we have $\mathrm{M} = 1$, $\Delta\, \tilde{S} = 0$. This means that, for the entire range

$$0 < \mathrm{M} \leq 1,\ \Delta\, \tilde{S} \geq 0. \tag{4.42}$$

Therefore, by the second law of thermodynamics, a transition across the KSW front is an irreversible process. According to [18], a gas-dynamic rarefaction shock wave is unstable, because for it the inequality $\Delta\, \tilde{S} \leq 0$ always holds [18]. This result means that the superimposition of kinetic jumps to a rarefaction shock wave transforms it into the class of stable discontinuous flows. This stability property of the KSW, which we have just established, provides an evidence of the physical justification of the semi-empirical model of strong evaporation.

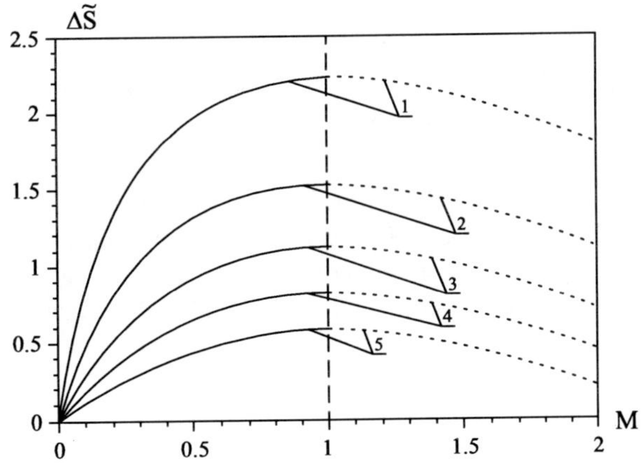

Fig. 4.8 Dependence of the entropy gain on the Mach number via (4.41). 1 $\beta = 1$, 2 $\beta = 0.8$, 3 $\beta = 0.6$, 4 $\beta = 0.4$, and 5 $\beta = 0.2$. Unfeasible supersonic regimes are shown by dotted lines

(2) The entropy production rate. The inequality $d\tilde{S}/d\mathrm{M} > 0$ holds in the interval $0 < \mathrm{M} < 1$. Moreover, at a sonic point

$$\mathrm{M} = 1, \frac{d\tilde{S}}{d\mathrm{M}} = 0, \frac{d^2\tilde{S}}{d\mathrm{M}^2} > 0. \tag{4.43}$$

This means that if the evaporation proceeds at the velocity of sound, the entropy increment is minimal. By Prigogine' theorem, this is indicative of the attainment of an equilibrium state. It is interesting to note that the dependence of the molecular mass flow on the Mach number of the molecular mass flow has properties identical to (4.43) (see Fig. 4.7). So, here we can speak about the similarity of the dependencies $\tilde{J}$ (M) and $\Delta\tilde{S}$ (M).

4.7 Inverse and Ill-Posed Problems

4.7.1 Conditions for Well-Posedness

The study of the direct problem of strong evaporation showed that the KSW is stable and satisfies the conclusion of Prigogine's theorem. Consider now the inverse problem of strong evaporation.

In mathematical physics, by "direct problems" one usually means problems in which physical processes or phenomena are modeled. In direct problems, it is required to find a function describing a physical process at each point of the domain under investigation at each time. For the solution of a direct problem, the following information is required.

(1) the domain in which the process is studied,
(2) the equation describing the given process,
(3) the initial conditions (if the process is nonstationary),
(4) the conditions on the boundary of the domain under consideration (if the domain is bounded).

In addition, the properties of the medium (i.e., the coefficients of the equations), the initial state of the process, and its properties on the boundary are assumed to be known a priori. However, in practical applications, the medium properties are generally unknown. This leads to inverse problems, in which it is required to find the unknown coefficients of the equations[5] and the initial and BC.

[5] As a rule, the sought-for coefficients are thermophysical properties of the medium under study.

In the majority of cases, the inverse problems are "ill-posed problems"—the well- and ill-posed problems differ only in the degree of certainty of their solutions. Hadamard [38] pioneered the study of such problems—in 1902, he formulated his famous three conditions which should be satisfied by a well-posed problem.

(1) Existence of the solution. The problem should have a solution for any admissible initial data. This means that among the initial data there are no contradictory conditions—the appearance of such conditions would exclude the possibility of the solution of the problem.
(2) Unique solvability of the problem. To each set of initial data, there corresponds a unique solution. In other words, there are sufficiently many initial data to guarantee the existence of a unique solution of the problem.
(3) Stability of the problem. Small perturbations in initial data give small deviations in the solution.

The first two conditions are called the conditions of mathematical definiteness of the problem. To clarify the third condition, consider the operator equation

$$Aq = f \tag{4.44}$$

with linear continuous operator A.

Assume it is required to find a solution q that corresponds to the given right-hand side f of the equation. Here, the sought-for physical characteristics q cannot be directly specified. On the other hand, one has only the data f related to q via the operator A (as a rule such data are obtained from a physical experiment). The operator equation of the form (4.44) is a typical mathematical model for many inverse problems. Moreover, condition (3) means that the inverse operator A^{-1} is continuous, i.e., to small variations of the right-hand side u there corresponds small variations of the solution z.

It is interesting to note that Hadamard himself proposed to deal only with well-posed problems. However, it was found later that the majority of the applied problems are ill-posed problems. So, any system of differential equations describes a real physical process only to some degree of accuracy. Moreover, any instrument for measuring some or other physical phenomenon is also imperfect. As a result, small perturbation in measurement of the initial data may cause considerable variations in the solutions of the equations. In this situation, one cannot guarantee that a selected mathematical model is capable of adequately describing the corresponding physical phenomenon. If, the other way around, small perturbations of the initial conditions result in small variations of the solution in the entire domain of their existence, then the corresponding mathematical model should be regarded as adequate.

This leads to the following question of great value in applications: under which conditions, the mathematical model described by the system of differential equations will be stable. So, the third Hadamard condition presents the greatest challenge [39]. Examples of some well- and ill-posed problems are given in Table 4.5, which shows that each ill-posed problem (see the right-hand column of the table) can be formulated as an inverse problem to some direct well-posed problem (the left-hand columns).

Table 4.5 Examples of well-posed and ill-posed problems

Well-posed problems		Ill-posed problems
Arithmetic		
Multiplication by a small number A $Aq = f$		Division by a small number A $q = A^{-1} f(A \ll 1)$
Algebra		
Multiplication by a matrix $Aq = f$		Solve the system $Aq = f$ with an ill-conditioned or degenerate matrix A
Calculus		
Integration $f(x) = f(0) + \int_0^z q(\xi)d\xi$	Differentiation $q(x)=f'(x)$ $q(x)$	
Differential equations		
The Sturm–Liouville problem $u''(x) - q(x)u(x) = \lambda u(x),$ $u(0) - hu'(0),$ $u(1) - Hu'(1)=0$		The inverse Sturm–Liouville problem $\{\lambda_n, \|u\|^2\} \to q(x)$ Find $q(x)$ From the spectral data $\{\lambda_n, \|u\|\}$
Integral equations		
Volterra and Fredholm equation of the second kind $q(x) + \int_0^x K(x, \xi)q(\xi)d\xi=f(x)$ $q(x) + \int_a^b K(x, \xi)q(\xi)d\xi=f(x)$		Volterra and Fredholm equation of the first kind $\int_0^x K(x, \xi)q(\xi)d\xi=f(x)$ $\int_a^b K(x, \xi)q(\xi)d\xi=f(x)$

The inverse and ill-posed problems share in common the important property of instability of the solution with respect to small errors in the specification of the initial data. It is important to note that in the majority of cases any inverse problem is also an ill-posed problem. In turn, as a rule, an ill-posed problem can be formulated as an inverse problem with respect to some direct (well-posed) problems.

Historically, the inverse and ill-posed problems were formulated and studied independently and in parallel, and so now both terms are used in scientific literature. In the formal language, one can say that the solutions of inverse and ill-posed problems are aimed at the study of properties and methods of regularization of unstable problems [40]. More specifically, this means the investigation of stable methods of approximation of unstable maps. In the language of linear algebra, this is the search of approximate methods of finding normal pseudosolutions of systems linear algebraic equations with rectangular, degenerate, or ill-conditioned matrices. From the point of view of the information theory, the inverse and ill-posed problems deal with properties of maps of compact sets with large epsilon-entropy to tables with small epsilon-entropy [41].

4.7.2 Numerical Methods

Numerical study of inverse problems is traditionally based on gradient methods, which are briefly listed below.

- The gradient descent method. The principal issue of the method is the determination of the step size involved in the descent.
- The steepest descent method. In contrast to the previous method (in which the step size of the descent should be manually specified before proceeding with the iterations), this algorithm involves an automatic step selection mechanism at each iteration.
- The coordinate descent method. The crux of the method is that each time the descent is effected only in one variable (the internal iteration). Once all the variables have been dealt with (the external iteration), the algorithm returns to the first one.

Another important issue is the convergence of gradient methods. It is in principle possible that the level sets of a function under study are very elongated in one direction and compressed in a different direction. The traditional gradient methods demonstrate poor convergence on this class of functions. In particular, if the matrix A in (4.44) is ill-conditioned, then the level surfaces of the minimized function may be highly nonspherical—they are elongated along some surface called the "ravine bottom". Functions with such property are called "ravine functions".

Low efficiency of gradient methods in minimization of ravine functions stimulated the development of heuristic procedures based on the separation of motion with respect to "fast" and "slow" variable. Such procedures, which do not provide high precision of the solution, are aimed at improving an approximation point to the solution of the problem in a relatively small number of steps (for a successive use of this problem by a different method). As a typical example of such an approach, we can mention the classical Gel'fand–Tseitlin ravine function method [42]. The principal difficulty of this method lies in the heuristic selection of its parameters. If one is clumsy in the choice of such parameters, the ravine function method may degenerate into one of ineffective variants of the gradient method.

4.8 Direct and Inverse Problems of Strong Evaporation

4.8.1 Strong Evaporation

In kinetic analysis, the spectrum of emitted molecules is described by the equilibrium Maxwell distribution

$$f_w^+ = \frac{n_w}{\pi^{3/2} \mathrm{v}_w^3} \exp\left(-\frac{c_x^2 + c_y^2 + c_z^2}{\mathrm{v}_w^2} \right). \tag{4.45}$$

Here, $n = \rho/m$ is the numerical density of the molecules, m is the mass of a molecule. The function f_w^+ is described in the intervals of molecular velocities: $-\infty < c_x < \infty$, $-\infty < c_y < \infty$, $0 < c_z < \infty$.

The DF of the molecules flying from the gas-dynamic region to the surface is described by the Maxwell distribution displaced along the velocity axis c_z by the quantity of the evaporation rate u_∞

$$f_\infty^- = \frac{n_\infty}{\pi^{3/2} \mathrm{v}_\infty^3} \exp\left(-\frac{c_x^2 + c_y^2 + (c_z - u_\infty)^2}{\mathrm{v}_\infty^2} \right). \tag{4.46}$$

In the process of their motion, the molecules interact between each other and with the counter-current flow, hence the surface is reached by the molecular flow with deformed "negative half" of the DF: $f_w^- \neq f_\infty^-$. As a result, the total DF on the surface will consist of the positive and negative halves formed under different conditions

$$\left. \begin{aligned} 0 \le c_z < \infty : f_w = f_w^+, \\ -\infty < c_z \le 0 : f_w = f_w^-. \end{aligned} \right\}$$

It follows that the spectrum of the molecules on the CPS is discontinuous (Fig. 4.9). Away from the surface, this discontinuity should become more smooth due to intermolecular collisions.

As a result, on the interphase surface, there coexist two oppositely directed (the emitted J_i^+ and the incident J_i^-) molecular flows. In case of a disbalance $J_i^+ > J_i^-$, evaporation flows appear on the external region to the surface: the mass flow ($i = 1$)

$$J_1^+ - J_1^- = \rho_\infty u_\infty, \tag{4.47}$$

the momentum flow ($i = 2$)

$$J_2^+ - J_2^- = \rho_\infty u_\infty^2 + p_\infty \tag{4.48}$$

and the energy flow ($i = 3$)

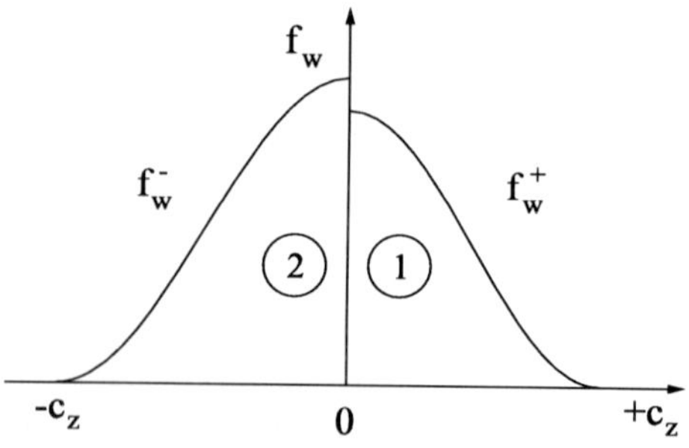

Fig. 4.9 Form of the velocity distribution function of molecules on the condensed-phase surface in evaporation. 1 The positive half of the distribution function, 2 the negative half of the distribution function

$$J_3^+ - J_3^- = \frac{\rho_\infty u_\infty^3}{2} + \frac{5}{2} p_\infty u_\infty. \tag{4.49}$$

Here, ρ_∞ is the density, p_∞ is the pressure.

Below we shall consider the case of an absolutely permeable CPS: it adsorbs all the molecules arriving from the gas region and emits all the molecules arriving from the depth of the condensed phase. Concrete expressions for the emitted and incident mass fluxes are determined via integration of the DF over the three-dimensional field of molecular velocities. The integration intervals over the velocities parallel to the surface are equal for the positive and negative halves of the DF: $-\infty < c_x < \infty$, $-\infty < c_y < \infty$. The difference is manifested for integration along the velocity orthogonal to the surface, where the domains of the functions should be taken into account: $0 < c_z < \infty$ for f_w^+, $-\infty < c_z < 0$ for f_w^- (Fig. 4.9). As a result, we get the following expressions for the emitted flows

$$\left.\begin{aligned} J_1^+ &= m\int_{-\infty}^{\infty}\left\{\int_{-\infty}^{\infty}\left(\int_0^{\infty} f_w^+ c_z dc_z\right)dc_y\right\}dc_x, \\ J_2^+ &= m\int_{-\infty}^{\infty}\left\{\int_{-\infty}^{\infty}\left(\int_0^{\infty} f_w^+ c_z^2 dc_z\right)dc_y\right\}dc_x, \\ J_3^+ &= \frac{1}{62} m\int_{-\infty}^{\infty}\left\{\int_{-\infty}^{\infty}\left(\int_0^{\infty} f_w^+ c_z c^2 dc_z\right)dc_y\right\}dc_x, \end{aligned}\right\} \tag{4.50}$$

and for the incident flows

$$\left.\begin{aligned} J_1^- &= m\int_{-\infty}^{\infty}\left\{\int_{-\infty}^{\infty}\left(\int_{-\infty}^{0} f_w^+ c_z dc_z\right)dc_y\right\}dc_x, \\ J_2^- &= m\int_{-\infty}^{\infty}\left\{\int_{-\infty}^{\infty}\left(\int_{-\infty}^{0} f_w^+ c_z^2 dc_z\right)dc_y\right\}dc_x, \\ J_3^- &= \frac{1}{2} m\int_{-\infty}^{\infty}\left\{\int_{-\infty}^{\infty}\left(\int_{-\infty}^{0} f_w^+ c_z c^2 dc_z\right)dc_y\right\}dc_x. \end{aligned}\right\} \tag{4.51}$$

Substituting f_w^+ from (4.45) into the integrals (4.50), we have, for the emitted flows J_i^+

$$J_1^+ = \frac{1}{2\sqrt{\pi}} \rho_w \mathrm{v}_w,$$

$$J_2^+ = \frac{1}{4} \rho_w \mathrm{v}_w^2,$$

$$J_3^+ = \frac{1}{2\sqrt{\pi}} \rho_w \mathrm{v}_w^3.$$

To solve this problem, it suffices to find the negative half of the DF. Below, we will describe the approach in which f_w^- is specified in some class of functions with unknown free parameters. In order to find these parameters, one should solve the conservation Eqs. (4.47)–(4.49). Such an approach can be called "macroscopic".

4.8.2 Macroscopic Models of Strong Evaporation

Let us briefly discuss the main provisions of the existing macroscopic models of intense evaporation [12]. In the entire region filled by gas, the equation of state for an ideal gas should be satisfied

$$p = \frac{1}{2}\rho \mathrm{v}^2, \tag{4.52}$$

where $\mathrm{v} = \sqrt{2R_g T}$ is the thermal velocity of the molecules.

It is assumed that during the motion to the CPS the negative half of the DF is deformed due to intermolecular collisions, hence, on some intermediate "mixing surface" in the KL it assumes the form

$$f_m^- = \frac{n_m}{\pi^{3/2}\mathrm{v}_m^3}\exp\left(-\frac{c_x^2 + c_y^2 + (c_z - u_m)^2}{\mathrm{v}_m^2}\right).$$

On the mixing surface, the following conservation laws should be satisfied: the mass flux conservation law

$$\rho_w \mathrm{v}_w - \rho_m \mathrm{v}_m I_1 = 2\sqrt{\pi}\rho_\infty u_\infty, \tag{4.53}$$

the momentum flux conservation law

$$\rho_w \mathrm{v}_w^2 - \rho_m \mathrm{v}_m^2 I_2 = 4\rho_\infty u_\infty^2 + 2\rho_\infty \mathrm{v}_\infty^2, \tag{4.54}$$

and the energy flux conservation law

$$\rho_w \mathrm{v}_w^3 - \rho_m \mathrm{v}_m^3 I_3 = \sqrt{\pi}\rho_\infty u_\infty^3 + \frac{5}{2}\sqrt{\pi}\rho_\infty \mathrm{v}_\infty^2 u_\infty. \tag{4.55}$$

Equations (4.53)–(4.55) involve the dimensionless incident fluxes

$$I_1 = \exp(-s_m^2) - \sqrt{\pi}s_m \mathrm{erfc}(s_m),$$

$$I_2 = \frac{2}{\sqrt{\pi}}s_m \exp(-s_m^2) - (1 + 2s_m^2)\mathrm{erfc}(s_m),$$

$$I_3 = \left(1 + \frac{s_m^2}{2}\right)\exp(-s_m^2) - \left(\frac{5\sqrt{\pi}}{4}s_m + \frac{\sqrt{\pi}}{2}s_m^3\right)\mathrm{erfc}(s_m).$$

Here, $\mathrm{erfc}(s_m)$ is the complementary error function, the subscript "w" denotes the condensed-phase surface, "m" corresponds to the mixing surface, and "∞" denotes the external gas-dynamic region.

Assume that the parameters on the mixing surface and in the gas-dynamic are related by the following coefficients

$$\rho_m = \alpha_\rho \rho_\infty, \; \mathrm{v}_m = \alpha_\mathrm{v} \mathrm{v}_\infty, \; s_m = \alpha_s s_\infty.$$

Here, $s_\infty = u_\infty/\mathrm{v}_\infty$ is the velocity ratio related to the Mach number as $\mathrm{M} = \sqrt{6/5} s_\infty$.

In the standard setting of the problem, the density ρ_w and the temperature T_w are given. This means that on the CPS the thermal velocity v_w and the velocity ratio s_∞ (or the related Mach number) are specified. In this case, the system of Eqs. (4.53)–(4.55) will contain the following five unknowns: the density ρ_m, the thermal velocity v_m, the evaporation rate u_m (on the mixing surface), the density ρ_∞, and the thermal velocity v_∞ (in the external gas-dynamic region). In the one-parameter model [43], we adopt the assumption: $\mathrm{v}_m = \mathrm{v}_\infty$, $s_m = s_\infty$, which means that $\alpha_\mathrm{v} = \alpha_s = 1$. Hence, the DF will contain the unique free parameter α_ρ, and the number of unknowns is reduced to three. The model of [43] is a four-dimensional approximation to the solution of the Boltzmann equation [4]. In a more general case, only one assumption is used: $\mathrm{v}_m = \mathrm{v}_\infty$, or $\alpha_\mathrm{v} = 1$. Such approximation of the negative half of the DF is more flexible and involves two free parameters: α_ρ, α_s. In this case, system (4.53)–(4.55) will contain four unknowns. For its closure, one adopts the physical assumption about the continuity of the evaporation flux as it moves from the mixing surface to the external gas-dynamic region

$$\rho_m u_m = \rho_\infty u_\infty.$$

This equality is called the "mixing condition", and the two-parameter model of [12], the "mixing model". The latter model is a five-momentum approximation to the solution of the Boltzmann equation [4].

In macroscopic models [12, 43], in accordance with the kinetic theory, the incident fluxes J_i^- are evaluated via the integrals (4.51). One can also mention some less rigorous approaches. So, for example, in [17] in place of the Maxwell spectrum of emitted molecules, one uses an anisotropic DF involving two different temperatures. In the integrands in (4.51), Rose [44] used a correction function, which distorts the classical Maxwell distribution (4.46).

So, the direct problem of strong evaporation involves the following stages.

- Define the DF as a displaced Maxwell distribution with free parameters.
- Use the kinetic integrals to determine the incident molecular fluxes J_i^-.
- The quantities J_i^- are substituted into the system of conservation Eqs. (4.53)–(4.55), whose solution is used to evaluate the free parameters (one in the one-parameter model and two in the mixing model), as well as the extrapolated BC of interest.

It is worth pointing out that in models of [12, 17, 43, 44] the authors used widely differing approximations of the DF and a different number of equations in the mathematical description. Nevertheless, the results obtained in these papers are consistent with each other and in good agreement with numerical studies [1, 9]. This remarkable fact suggests the following hypothesis: the macroscopic description of strong evaporation is conservative with respect to the method of approximation of the DF.

4.8.3 Solution of Inverse Problem

Consider now the inverse problem of recovery of the free parameters $\alpha_\rho, \alpha_v, \alpha_s$ from the given ρ_∞, v_∞ (or p_∞, T_∞). The system of transcendent Eqs. (4.53)–(4.55) involving complementary error functions was solved using the fsolve function in Maple. The following basic efficient principles of numerical methods are well known [45]: the universality, the simplicity of computation management, the accuracy provision, and the convergence rate. In our setting, the difficulty is that (in contrast to the case of nonlinear algebraic equations), there does not exist a universal algorithm for the analysis of the convergence and verification of the correctness of the solution of transcendent equations.

Figure 4.10 shows that the results of the solution for the mixing model and for the KSW model are quite close to each other. Calculations show that the coefficients α_ρ, α_v vary in a relatively small range. However, the coefficient α_s manifests the abnormal behavior: with increasing Mach number it passes through zero and changes its sign. This means that the DF displacement has the form typical of the inverse direction of the vapor flow (i.e., for the condensation problem).

Consider now the inverse problem with the initial data obtained from numerical solutions [1, 9].

Figure 4.11 shows that the coefficients of thermal velocity α_v and displacement of the distribution function α_s vary relatively slow. However, the density coefficient α_ρ in the range $0.6 \leq M \leq 1$ increases by three orders. Since there are no universal recommendations for the solution of transcendent equations, we shall start with the following assumptions in our analysis of inverse problems.

(1) Assumption 1. A negative displacement of the DF (Fig. 4.10a) stems from the ravine character of the level surface of the sought-for functions. As was already pointed out, the ravine character (i.e., elongation of the level lines along one direction) induces an abrupt aggravation of the convergence of gradient methods for searching extrema of functions [45]. Ravine properties of a system of equations reduce considerably the efficiency of traditional iterative algorithms of numerical solution. In this regard, refinement calculations using the modified Newton–Raphson method [46] were performed, this method is a root-finding algorithm, which produces successively better approximations to the roots (or zeroes) of a real-valued function. The results of the refinement calculations were found to be practically equal to those obtained via the *fsolve* function in Maple. The conclusion is made that the anticipated ravine features of the equations have no effect on the solution of the inverse problem.

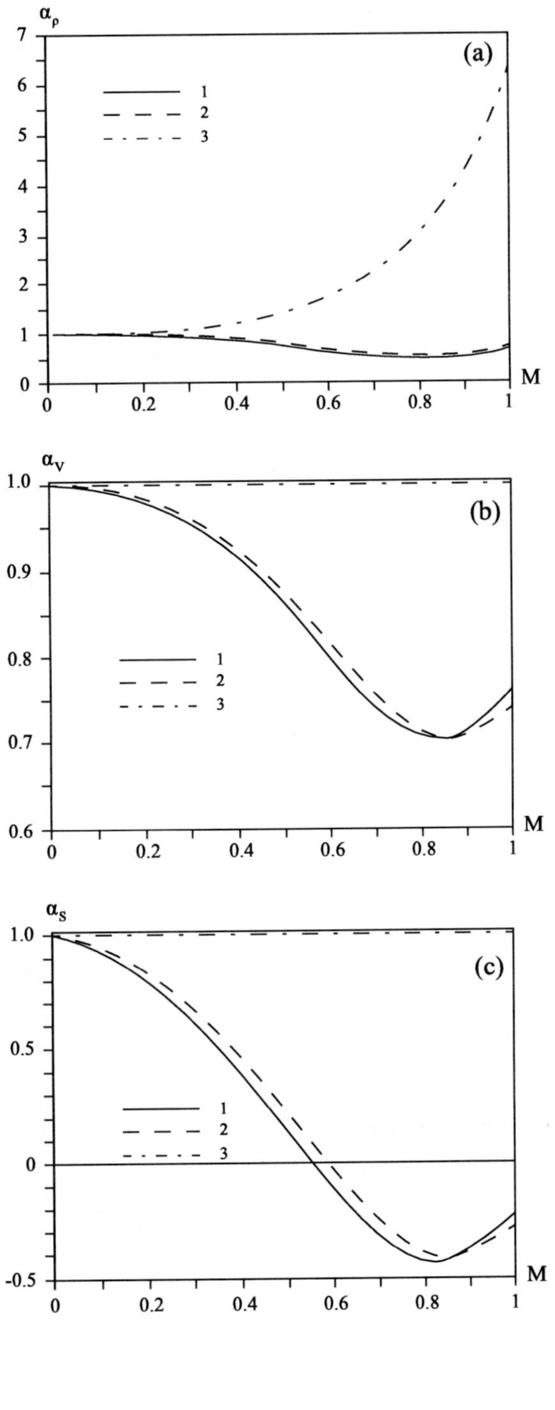

Fig. 4.10 Dependence of the coefficients of inverse problem on the Mach number for various macroscopic models. ———— The mixing model, — — — the model shock evaporation wave, —·—·—· the one-parameter model. **a** Density coefficient, **b** thermal velocity coefficient, **c** displacement factor of the distribution function

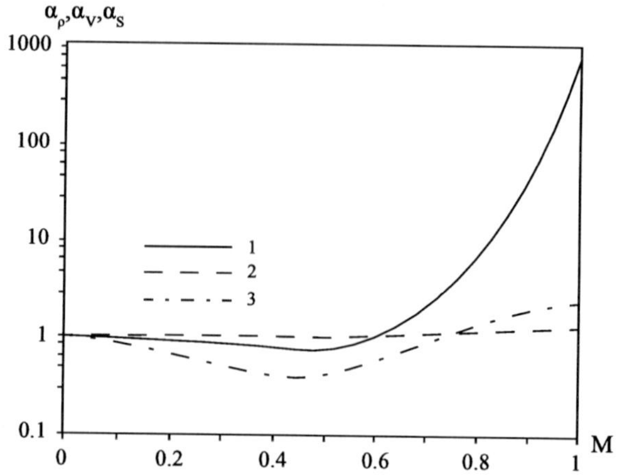

Fig. 4.11 Dependence of the coefficients of inverse problem on the Mach for numerical results. ———— The density coefficient, — — — the thermal velocity coefficient, — · — · — · the displacement factor of the distribution function

(2) Assumption 2. The reason for the abnormal behavior of α_s (Fig. 4.11) is the instability of the solution of the inverse problem. We recall that a problem is unstable if small errors in the original quantity result in considerable errors in calculations [40, 41]. A test of the numerical method for evaluation of the density coefficient α_ρ involved the following stages.

- Evaporation in the transonic regime (M = 1) was considered.
- A maximal range for the available solutions of the direct problem for both analytical methods was specified: one-parameter and two-parameter methods.
- Some specific parameter of the function (the pressure or the temperature) was varied with the fixed value of the second parameter.
- The value of α_ρ was determined from the solution of the inverse problem.

It was shown that, with fixed temperature and variable pressure, this quantity changes by almost five orders. Conclusion: the solution of the inverse problem is unstable. Above it was shown that in the direct problems the use of considerably different methods gives pretty close results. In contrast, the recovery of free parameters in inverse problems from the given BC is strongly unstable. This means that the direct and inverse problems of strong evaporation have opposite properties.

In this relation, it is worth mentioning a conceptually close result from high mathematics [47]. It is known that strong variation of the parameters of an integrand results in (with the exception of special cases) a weak variation of the primitive function. The other way round, even for a small variation of a differentiable function its derivative may change considerably. The above estimates support the above hypothesis on the conservatism of the direct problem of strong evaporation. The principal result of the present chapter is the development of a model of a kinetic shock evaporation wave. This result greatly simplifies the kinetic analysis of the problem. The role and place of this model among the available ones is illustrated in Table. 4.6.

Table 4.6 Comparison of characteristics of various approaches to the solution of the problem of strong evaporation

	The Boltzmann equation and its analogues	Distribution function of the incident flux	Conservation equation of molecular fluxes	Results of solution
Rigorous approach [1, 9]	Numerical solution	Found from the solution of the Boltzmann equation	Become identities	Detailed description of the distribution function
Macroscopic approach [12, 17, 43, 44]	Not used	Given as a Maxwell distribution with free parameters	Solution is searched	Extrapolated boundary conditions
Shock evaporation wave (the present chapter)	Not used	Not used	Replaced by a superposition of known solutions	Extrapolated boundary conditions

4.9 Conclusions

The results of calculation of the parameters of strong evaporation in the framework of macroscopic models are shown to be conservative with regard to the means of approximation of the distribution function. We proposed a semi-empirical model of strong evaporation based on the linear kinetic theory. Extrapolated jumps of density and temperature on the condensed-phase surface are obtained by summing the linear and quadratic components. The expressions for the linear jumps are taken from the linear kinetic theory of evaporation. The nonlinear terms are calculated from the relations for a rarefaction shock wave with due account of the corrections for the acceleration of the egressing flow of gas. Analytical dependences of the vapor parameters in the gas-dynamic region on the Mach number, the condensation coefficient, and the number of degrees of freedom of gas molecules are put forward. The results of calculation by the semi-empirical model match well the results of the available analytical and numerical studies. The semi-empirical model is shown as being precisely satisfying the condition for the maximum of the dependence of the evaporation mass flux on the Mach number as the evaporating gas reaches the sonic velocity. The model proposed can be used for calculations of the extrapolated jumps of pressure and temperature on the condensed-phase surface with strong evaporation.

We give an analysis of macroscopic models based on various approximations of the distribution function of the molecular fluxes incident on the surface. For a kinetic shock evaporation wave, its model is proposed based on the composition of classical results of the linear kinetic evaporation theory and the theory of gas-dynamic shock rarefaction wave. A kinetic shock wave is shown to be stable based on the second law of thermodynamics. It is proved that the maximum entropy principle, which follows from Prigogine's theorem, is satisfied during transonic evaporation. A highly unstable solution of the inverse problem of intense evaporation was obtained. It is shown that the solution of the intense evaporation problem is conservative with respect to the method of approximation of the distribution function. The principal result of this section is the development of the model of a kinetic shock wave, which substantially simplifies the kinetic analysis of the intense evaporation problem.

References

1. Gusarov AV, Smurov I (2002) Gas-dynamic boundary conditions of evaporation and condensation: numerical analysis of the Knudsen layer. Phys Fluids 14(12):4242–4255
2. Larina IN, Rykov VA, Shakhov EM (1996) Evaporation from a surface and vapor flow through a plane channel into a vacuum. Fluid Dyn 31(1):127–133
3. Crifo JF (1994) Elements of cometary aeronomy. Curr Sci 66(7–8):583–602
4. Kogan MN (1995) Rarefied gas dynamics. Springer
5. Bobylev AV (1984) Exact solutions of the nonlinear Boltzmann equation and of its models. Fluid Mech Soviet Res 13(4):105–110
6. Labuntsov DA (1967) An analysis of the processes of evaporation and condensation. High Temp 5(4):579–647
7. Muratova TM, Labuntsov DA (1969) Kinetic analysis of the processes of evaporation and condensation. High Temp 7(5):959–967
8. Latyshev AV, Yushkanov AA (2008) Analytical methods in kinetic theory methods in kinetic theory. MGOU, Moscow (in Russian)
9. Frezzotti AA (2007) A numerical investigation of the steady evaporation of a polyatomic gas. Europ J Mech B: Fluids 26:93–104
10. Anisimov SI (1968) Vaporization of metal absorbing laser radiation. Sov Phys JETP 27(1):182–183
11. Labuntsov DA, Kryukov AP (1977) Intense evaporation processes. Therm Eng 4:8–11
12. Labuntsov DA, Kryukov AP (1979) Analysis of intensive evaporation and condensation. Int J Heat Mass Transf 2:989–1002
13. Zudin YB (2015) Approximate kinetic analysis of intense evaporation. J Engng Phys Thermophys 88(4):1015–1022
14. Zudin YB (2015) The approximate kinetic analysis of strong condensation. Thermophys Aeromechanics 22(1):73–84
15. Zudin YB (2016) Linear kinetic analysis of evaporation and condensation. Thermophys Aeromechanics 23(3):437–449
16. Rose JW (2000) Accurate approximate equations for intensive sub-sonic evaporation. Int J Heat Mass Transf 43:3869–3875
17. Crout PD (1936) An application of kinetic theory to the problems of evaporation and sublimation of monatomic gases. J Math Phys 15:1–54
18. Zeldovich YB, Raizer YP (2002) Physics of shock waves and high-temperature hydrodynamic phenomena. Courier Corporation
19. Kryukov AP, Levashov VY, Pavlyukevich NV (2014) Condensation coefficient: definitions, estimations, modern experimental and calculation data. J Engng Phys Thermophys 87(1):237–245
20. Cercignani C (1981) Strong evaporation of a polyatomic gas. In: Rarefied gas dynamics, International symposium, 12th, Charlottesville, VA, 7–11 July 1980, Technical papers. Part 1, American Institute of Aeronautics and Astronautics, New York, 1981, pp 305–320
21. Skovorodko PA (2000) Semi-empirical boundary conditions for strong evaporation of a polyatomic gas. In: Bartel T, Gallis M (eds) Rarefied gas dynamics, 22th international symposium. Sydney, Australia, 9–14 July 2000. In: AIP conference proceedings 585. American Institute of Physics, Melville, NY, 2001, pp 588–590

22. Sone Y, Sugimoto H (1993) Kinetic theory analysis of steady evaporating flows from a spherical condensed phase into a vacuum. Phys Fluids A 5:1491–1511
23. Mazhukin VI, Prudkovskii PA, Samokhin AA (1993) About gas-dynamical boundary conditions on evaporation front. Matematicheskoe Modelirovanie 5(6):3–10 (in Russian)
24. Drazin PG, Reid WH (2010) Hydrodynamic stability (Cambridge texts in applied mathematics). 2nd edn. Cambridge University Press
25. Helmholtz H (1868) Über discontinuierliche Flüssigkeits-Bewegungen. Monatsberichte Der Königlichen Preussische Akademie Der Wissenschaften Zu Berlin 23:215–228
26. Kelvin L, (Thomson W), (1871) Hydrokinetic solutions and observations. Phil Mag 42:362–377
27. Landau LD, Lifshits EM (1987) Fluid mechanics. Butterworth-Heinemann
28. Einstein A (1949) Autobiographical notes. In: Schilpp PA (ed) Albert Einstein, Philosopher-Scientist. The Library of Living Philosophers, Open Court, La Salle IL
29. Sonntag RE, Borgnakke C, Van Wylen GJ (2002) Fundamentals of thermodynamics. 6th ed. Wiley, New Jersey
30. De Groot SR, Mazur P (2011) Non-equilibrium thermodynamics. Dover Publications Inc., New York
31. Onsager L (1931) Reciprocal Relation in Irreversible Processes. I Phys Rev Am Phys Soc (APS) 37(4):405–426
32. Curie P (1894) Sur la symétrie des phénomènes physiques: symétrie d'un champ électrique et d'un champ magnétique. J de Phys 3e série 393–415
33. Castellani E, Jenann I (2016) Which curie's principle? Philos Sci 83(5):1002–1013
34. Goldstein H (1980) Classical mechanics, 2nd edn. Reading, MA, Addison-Wesley
35. Prigogine I (1945) Modération et transformations irréversibles des systèmes ouverts: bulletin de la classe des sciences. Académie R de Belg 31:600–606
36. Glansdorff P, Prigogine I (1971) Thermodynamic theory of structure, stability, and fluctuations. Wiley-Interscience, New York
37. Bazarov IP (1964) Thermodynamics Pergamon Press, New York
38. Hadamard J (1902) Sur les problèmes aux dérivées partielles et leur signification physique. Princeton University Bulletin, pp 49–52
39. Ramm A (2005) Inverse problems mathematical and analytical techniques with applications to engineering New York, Springer Science
40. Heinz WE, Hanke M, Neubauer A (1996) Regularization of inverse problems. Kluwer, Dordrecht
41. Hanke M (2017) A taste of inverse problems: basic theory and examples. SIAM, Philadelphia
42. Gelfand M (1966) The ravine method in the problems of the X-Ray structural analysis. Nauka, Moscow [in Russian]
43. Zudin YB (2019) Non-equilibrium evaporation and condensation processes: analytical solutions, 2nd edn. Springer, Heidelberg
44. Rose JW (2000) Accurate approximate equations for intensive subsonic evaporation. Int J Heat Mass Transf 43:3869–3875
45. Gill PE, Murray W, Wright MH (1981) Practical optimization. Academic Press, London
46. Süli E, Mayers D (2003) An introduction to numerical analysis. Cambridge University Press
47. Smirnov VI, Lohwater AJ (2014) A course of higher mathematics: integration and functional analysis, vol 5, Pergamon

Chapter 5
Approximate Kinetic Analysis of Strong Condensation

Abbreviations

BC	Boundary condition
CPS	Condensed-phase surface
KL	Knudsen layer

Symbols

$\mathbf{c}$	Molecular velocity vector
f	Distribution function
J	Molecular flux
j	Mass flux
m	Molecular mass
M	Mach number
n	Molecular gas density
p	Pressure
$\tilde{p}$	Pressure ratio
T	Temperature
$\widetilde{T}$	Temperature ratio
$\mathbf{u}_\infty$	Hydrodynamic velocity vector
u_∞	Hydrodynamic velocity
$\widetilde{u}_\infty$	Speed ratio
v	Thermal velocity

Greek Letter Symbols

ρ	Density
β	Condensation coefficient
η	Pressure factor

Subscripts

δ	State at mixing surface
w	State at condensed-phase surface

Y. B. Zudin, *Non-equilibrium Evaporation and Condensation Processes*, Mathematical Engineering, https://doi.org/10.1007/978-3-030-67553-0_5

∞ State at infinity

5.1 Introduction

In recent years, there has been a growing interest in new fundamental and application problems focused on the study of strong phase transitions like evaporation and condensation. Problems of this kind arise in the study of many processes. In applying laser methods for material treatment, it is crucial to know the laws of evaporation process (thermal laser ablation from the target surface) and condensation process (for expanding vapor cloud interacting with the target) [1]. The power industry faces the risks of accidental situations caused by rapid contacts between volumes of cold liquid and hot vapor. The shock impact of two phases produces a pulse of rarefaction wave in vapor, this wave is followed by drastic changes in pressure and strong condensation [2]. When solar emission reaches the comet surface, strong evaporation of ice core produces atmosphere. The intensity of the evaporation mass flux varies, with the distance from the comet to the sun, in a wide range and can reach high levels. The time-variable nature of evaporation has a strong impact on comet atmosphere density and atmosphere flow [3].

In mathematical simulation for strong phase transitions, the Boundary Conditions (BC) at the condensed-phase boundary are determined by the solution to Boltzmann kinetic equation [4]. The Boltzmann equation describes the flow inside the Knudsen Layer (KL), attached to the surface from the gas side, which has the layer thickness about several molecular free paths. The linear kinetic theory for describing evaporation (condensation) at low velocities and which is based on linearized version of Boltzmann equation was presented in final outlay in [5, 6]. The phase transitions when the vapor velocity crossing the surface is comparable to the sonic velocity are graded as strong evaporation (or strong condensation). For strong phase transitions, the influence of viscosity and thermal conductivity on the heat and fluid flows deteriorates, so the flow in the external "Navier–Stokes region" behind the KL is described by a system of gas equations [4].

Strong evaporation (strong condensation) is described by macro-scaled jumps in parameters: temperature, density, and pressure of gas. These parameters are used as BC for continuum medium (different from the values at the surface). In general case, the molecules emitted by the Condensed-Phase Surface (CPS) have a spectrum different from the incoming molecules of gas phase. As a result, the molecular velocity distribution exhibits a discontinuity (microscopic jump) which dies down gradually within the KL and disappears when reaching the boundary of the Navier–Stokes region.

The complexity of kinetic analysis is explained by the need for considering the interlinked problems of different scales—for Boltzmann equation within the KL, and a macroscopic boundary-value problem for solving the equations in the Navier–Stokes region. Meanwhile, the extrapolated BC for the second problem are taken from the solution of the first problem. It has been proved [4–7] that accurate kinetic analysis is possible only on the basis of the Boltzmann equations. A lesson can be taught from [8], where extrapolated BC for the Euler equations were derived without using any Boltzmann equations. The author of [8] used a sophisticated mathematical procedure involving transformation of linearized Boltzmann equation into integro-differential Wiener–Hopf equation, then transformed it into a matrix form, factorized the matrix equation, and, finally, studied this equation using the Gohberg–Krein theorem about self-conjugated matrix. However, in the next paper [9], the author of [8] admitted that the previous results were erroneous.

The infeasibility of microscopic analysis without solving the Boltzmann equation does not deny the trails in macro-scale methods of conjugation of the Navier–Stokes region with the surface of condensed phase. The objective of applied kinetic analysis is the extrapolation toward the surface of Euler equation solutions. The simplified problem statement opens ways for integral form of distribution function instead of detailed study [4]. On the other hand, generation of high-accuracy numerical solutions of Boltzmann equation [10, 11] makes possible to find the required parameters of the distribution function. This creates the opportunity for applying the approximate solutions for efficient validation of approximate solutions. The goal of this research is approximate analytical solution for strong condensation problem.

5.2 Macroscopic Models

The subject for kinetic analysis is a three-dimensional molecular velocity distribution $f = f(\mathbf{c})$, which varies from the equilibrium Maxwellian distribution within the Navier–Stokes region

$$f_\infty = \frac{n_\infty}{\pi^{3/2}\mathrm{v}_\infty^3}\exp\left(-\left(\frac{\mathbf{c}-\mathbf{u}_\infty}{\mathrm{v}_\infty}\right)^2\right) \tag{5.1}$$

to the discontinuity distribution function at the surface of condensed phase

$$c_z > 0 : f_w = f_w^+, \tag{5.2}$$

$$c_z < 0 : f_w = f_w^-. \tag{5.3}$$

The distribution function of surface-emitted molecules f_w^+ is assigned as equilibrium half-Maxwellian distribution at surface temperature T_w and vapor saturation pressure at this temperature $p_w(T_w)$

$$f_w^+ = \frac{n_w}{\pi^{3/2}\mathrm{v}_w^3}\exp\left(-\left(\frac{\mathbf{c}}{\mathrm{v}_w}\right)^2\right). \tag{5.4}$$

Here, $n_\infty = p_\infty/k_B T_\infty$, $n_w = p_w/k_B T_w$ are the molecular densities of gas at infinity and at the CPS, correspondingly, $\mathbf{c}$, $\mathbf{u}_\infty$ are the vectors of molecular and hydrodynamic velocities, c_z is the normal vector to surface component of molecular velocity, $\mathrm{v}_\infty = \sqrt{2k_B T_\infty/m}$, $\mathrm{v}_w = \sqrt{2k_B T_w/m}$ are the thermal velocities of molecules at infinity and CPS.

Note that physically plausible relation (5.4) lacks any theoretical basis. For example, the authors of [12] wrote "Any kind of fundamental derivation procedure for this kind of BC is unknown for us". In work [12], the spectrum of surface-emitted molecules to vacuum was simulated with molecular dynamic methods. This research allowed to draw a conclusion: "Thus, for the case of low density vapor, the use of half-Maxwellian distribution as a BC for solving gas dynamic problems looks as a reasonable approximation".

Let us consider a problem of evaporation (condensation) for half-space of motionless vapor (ideal monatomic gas). For the one-dimensional variant, the vector of hydrodynamic velocity u_∞ degenerates into scalar velocity of evaporation (condensation) u_∞. Under steady conditions, the molecular flows of mass, momentum, and energy via any plane parallel to the surface are equal. If we use the BCs (5.1), this helps in expressing the fluxes in terms of the flow parameters at infinity and in writing the molecular fluxes of mass

$$\int_c mc_z f\,\mathbf{dc} = \rho_\infty u_\infty, \tag{5.5}$$

of the momentum

$$\int_c mc_z^2 f\,\mathbf{dc} = \rho_\infty u_\infty^2 + p_\infty, \tag{5.6}$$

and of the energy

$$\frac{1}{2}\int_c mc^2 c_z f\,\mathbf{dc} = \frac{1}{2} u_\infty \left(\rho_\infty u_\infty^2 + 5p_\infty\right). \tag{5.7}$$

Here, ρ_∞ is the density, p_∞ is the pressure, u_∞ is the hydrodynamic velocity, the index "∞" denotes the conditions at infinity, $c^2 = c_x^2 + c_y^2 + c_z^2$ is the squared modulus of molecular velocity, c_x and c_y are projections of molecular velocity vector on axes x and y lying in a plane parallel to the surface, c_z is the normal component of the molecular velocity.

Integration of left-hand sides in Eqs. (5.5)–(5.7) is carried out over the entire three-dimensional molecular velocities: $-\infty < c_x < \infty$, $-\infty < c_y < \infty$, $-\infty < c_z < \infty$. When the relations between flow parameters (at infinity) included into the right-hand sides of Eqs. (5.5)–(5.7) are questioned, it is enough to know the velocity distribution function at the surface. Since the distribution function of surface-reflected molecules f_w^+ is known already from the BC (5.4), then finding the macroscopic BC requires the distribution function of molecules to fall on the surface f_w^-. Let us rewrite the set of Eqs. (5.5)–(5.7) in a more convenient form

$$J_1^+ - J_1^- = \rho_\infty u_\infty, \tag{5.8}$$

$$J_2^+ - J_2^- = \rho_\infty u_\infty^2 + p_\infty, \tag{5.9}$$

$$J_3^+ - J_3^- = \frac{1}{2}\rho_\infty u_\infty^3 + \frac{5}{2}p_\infty u_\infty. \tag{5.10}$$

Here, J_i^+ and J_i^- are the outcoming and incoming molecular fluxes to the surface, $i = 1, 2, 3$.

One can see from Eqs. (5.8)–(5.10) that the disbalance of the molecular mass fluxes ($i = 1$), the momentum fluxes ($i = 2$), and the energy fluxes ($i = 3$) at the surface (left-hand sides of equations) creates in the Navier–Stokes region (right-hand sides of equations) the macroscopic fluxes of evaporation ($u_\infty > 0$ with $J_i^+ > J_i^-$) or condensation ($u_\infty < 0$ with $J_i^+ < J_i^-$). The values J_i^+ are calculated in a standard manner [4] while substitution of functions $f = f_w^+$ from the BC (5.4) in the integrands in left-hand sides of Eqs. (5.5)–(5.7)

$$\left.\begin{aligned} J_1^+ &= \tfrac{1}{2\sqrt{\pi}}\rho_w \mathrm{v}_w, \\ J_2^+ &= \tfrac{1}{4}\rho_w \mathrm{v}_w^2, \\ J_3^+ &= \tfrac{1}{2\sqrt{\pi}}\rho_w \mathrm{v}_w^3, \end{aligned}\right\} \tag{5.11}$$

where ρ_w is the density.

We should emphasize a general difference of microscopic and macroscopic approaches. For the first approach, the solution of Boltzmann equation with BC (5.1)–(5.4) defines the accurate distribution function that converts the conservation equations into identities. For the second approach, the set of Eqs. (5.5)–(5.7) has unknown Distribution function. This means that value f_w (actually, the negative half f_w^-) is found from model ideas.

5.3 Strong Evaporation

Let us assume that the distribution function for input molecules is described by a half-Maxwellian distribution (5.1) within the Navier–Stokes region

$$f_w^- = f_\infty^- \equiv f_\infty|_{c_z<0}. \tag{5.12}$$

The assumption (5.12) has the physical meaning that the spectrum of incoming molecules does not change throughout the KL. The set of Eqs. (5.8)–(5.10) with account for (5.11), (5.12) after some transformation can be represented in the form

$$\frac{\sqrt{\tilde{T}}}{\tilde{p}} - I_1^- = 2\sqrt{\pi}\tilde{u}_\infty, \tag{5.13}$$

$$\frac{1}{\tilde{p}} - I_2^- = 2 + 4\tilde{u}_\infty^2, \tag{5.14}$$

$$\frac{1}{\sqrt{\tilde{T}\tilde{p}}} - I_3^- = \sqrt{\pi}\tilde{u}_\infty^3 + \frac{5\sqrt{\pi}}{2}\tilde{u}_\infty. \tag{5.15}$$

Here, $\tilde{u}_\infty \equiv u_\infty/\mathrm{v}_\infty$ is the speed ratio related to the Mach number in the Navier–Stokes region $\mathrm{M}_\infty \equiv u_\infty(5k_BT_\infty/3m)^{-1/2}$ through the relation: $\tilde{u}_\infty = \sqrt{5/6}\mathrm{M}_\infty$. The dimensionless incident molecular fluxes at the surface of condensed phase I_i^- are written in the form

$$\left.\begin{aligned} I_1^- &= \exp\left(-\tilde{u}_\infty^2\right) - \sqrt{\pi}\tilde{u}_\infty \mathrm{erfc}\left(\tilde{u}_\infty\right), \\ I_2^- &= \tfrac{2}{\sqrt{\pi}}\tilde{u}_\infty \exp\left(-\tilde{u}_\infty^2\right) - \left(1 + 2\tilde{u}_\infty^2\right)\mathrm{erfc}\left(\tilde{u}_\infty\right), \\ I_3^- &= \left(1 + \tfrac{1}{2}\tilde{u}_\infty^2\right)\exp\left(-\tilde{u}_\infty^2\right) - \tfrac{\sqrt{\pi}}{2}\tilde{u}_\infty\left(\tfrac{5}{2} + \tilde{u}_\infty^2\right)\mathrm{erfc}\left(\tilde{u}_\infty\right). \end{aligned}\right\} \tag{5.16}$$

Here, $\mathrm{erfc}\left(\tilde{u}_\infty\right)$ is an additional integral of possibilities.

The evaporation problem gives us the dependencies of the two factors on $\tilde{u}_\infty$: for the temperature ratio $\tilde{T} = T_\infty/T_w$, and the pressure ratio $\tilde{p} = p_\infty/p_w$ This means that the set of Eqs. (5.13)–(5.15) is an overdetermined problem. Thus, assumption (5.12) rendering the idea of zero change of molecular spectrum within the KL is too strict, and therefore, is incorrect. The macroscopic theory of strong evaporation was formulated for the first time in the paper [13]. The author of [13] had analyzed the limiting case of evaporation when the hydrodynamic velocity of flow in the Navier–Stokes region is equal to sound velocity and assumed that the value f_w^- is proportional to the negative half-function of distribution in the Navier–Stokes region (5.1)

$$f_w^- = \alpha f_\infty^- \equiv \alpha f_\infty|_{c_z<0}. \tag{5.17}$$

The physical meaning of hypothesis (5.17) is that the molecular spectrum in the KL gains changes due to molecular collisions. Hence, the paper [13] presented for the first time the theoretical calculation for parameters of sonic evaporation for the case: $\mathrm{M}_\infty = 1$, $\tilde{T} = 0.669$, $\tilde{p} \approx 0.206$, and $\alpha \approx 6.29$. The advanced concept of [13] discovered a new class of kinetic problems and initiated a follow-up bunch of publications. A similar approach was applied in [14] and (independently) in [15] for study of the entire range of evaporation mass flux $0 < \mathrm{M}_\infty \le 1$. This research was continued in a series of publications where the simulation methods differed only in the tools for the approximation of the function f_w^-.

5.4 Strong Condensation

Unlike the case of evaporation, the temperature ratio $\widetilde{T} = T_\infty / T_w$ for condensation problem is not unknown (it is only a parameter). The goal in the condensation problem is in finding the relation $\widetilde{p}\left(\widetilde{u}_\infty\right)$ (or the similar function $\widetilde{p}$ (M_∞)) at $\widetilde{T}=$ idem. In this case, the over-specification of the set of Eqs. (5.13)–(5.15) needs assignment of two free parameters. In [16], the authors used the simulation method for the study of strong condensation: the method depends on a simplified integration in the space of molecular velocities. This approach is helpful in finding the connections between parameters T, p, u_∞, but it does not reproduce the distribution function (only few momentums). In the papers [17, 18], their authors used the simulation method for the study of the Boltzmann equation with a detailed analysis of the distribution function, the results gave the profiles of density, temperature, and pressure in the KL.

The assignment of the temperature ratio $\widetilde{T}$ as a parameter ensures splitting of strong condensation problem into two independent physical problems. In the first case, the "normal" condensation of hot gas upon a cold surface $\left(\widetilde{T} > 1\right)$ meets the standard macroscopic ideas about physical process. As for the second problem, the "anomalous" condensation of cold gas at the hot surface $\left(\widetilde{T} < 1\right)$ is allowed via mathematical description of the problem. The anomalous condensation occurs, in particular, while evaporation of substance into vacuum during pulse laser ablation [19]. After the end of laser pulse, the surface is overheated, and the vapor undergoes adiabatic expansion into vacuum and cools down rapidly. Meanwhile, the condensed phase is cooled rather slowly, so the low-temperature vapor starts condensing on the high-temperature surface.

5.5 The Mixing Model

The idea of the above model of strong condensation is based upon elementary kinetic concepts. It is known from [4] that conservation equations for molecular fluxes are the result of integrating the distribution function over the entire vector space of molecular velocities with the weights $1, \mathbf{c}, \mathbf{c}^2$, respectively. At the outcome, the left-hand sides of Eqs. (5.5)–(5.7) contain the values with dimensionality linear in density: they are as follows: $\rho \mathrm{v}$, $\rho \mathrm{v}^2$, $\rho \mathrm{v}^3$. Thus, the original hypothesis (5.17) can be interpreted as introduction of conditional mixing surface "$\delta - \delta$" inside the KL. The density ρ_δ at this surface is related to the density ρ_∞ at the outer boundary of the KL through a proportionality coefficient $\rho_\delta = \alpha_\rho \rho_\infty$. Herein we develop the integration of approaches formulated in [13, 14] for the case of strong condensation. We assume that all three macroscopic parameters at the mixing surface are related to their values at infinity through appropriate proportionality coefficients

$$\rho_\delta = \alpha_\rho \rho_\infty,\ \mathrm{v}_\delta = \alpha_\mathrm{v} \mathrm{v}_\infty,\ \tilde{u}_\delta = \alpha_u \tilde{u}_\infty. \tag{5.18}$$

Here, $\tilde{u}_\delta = u_\delta/\mathrm{v}_\delta$, $\mathrm{v}_\delta = \sqrt{2k_B T_\delta/m}$ is the speed ratio and thermal velocity at the surface "$\delta - \delta$", $\alpha_\rho, \alpha_\mathrm{v}, \alpha_u$ are the coefficients defined below. When the density and the temperature are known, the pressures at those surfaces are found from the ideal gas equation

$$p_\infty = \frac{\rho_\infty k_B T_\infty}{m}, \tag{5.19}$$

$$p_\delta = \frac{\rho_\delta k_B T_\delta}{m}. \tag{5.20}$$

Thus, we obtain the four unknowns: $\tilde{p}, \alpha_\rho, \alpha_\mathrm{v}, \alpha_u$ for three Eqs. (5.13)–(5.15). The enclosing mathematical description of this problem is found through the additional relation between parameters at surface "$\delta - \delta$". Let us fix the three-dimensional distribution function (5.1) at the KL boundary "$\infty - \infty$" (at infinity) for the planes $c_x = c_y = 0$ and normalize it to a maximal value. This normalized 1D distribution function takes the form

$$\tilde{f}_\infty = \exp\left[-\left(\tilde{c}_z - \tilde{u}_\infty\right)^2\right]. \tag{5.21}$$

Here, $\tilde{c}_z = c_z/\mathrm{v}_\infty$ is a dimensionless normal velocity of molecules. We assume also that inside the KL, the negative half-function $\tilde{f}$ is deformed into ellipsoid-type distribution

$$\tilde{\tilde{f}}^{-} \equiv \tilde{f}\Big|_{c_z<0} = \exp\left(\left[-\left(\varepsilon\tilde{c}_z - \tilde{u}\right)^2\right]\right). \tag{5.22}$$

Here, we introduced the following local values: ε is the ellipse ratio, $\tilde{c}_z = c_z/\mathrm{v}$, $\tilde{u} = \mathrm{v}$, where c_z and u are the molecular and hydrodynamic velocities, $\mathrm{v} = \sqrt{2kT/m}$ is the molecular thermal velocity corresponding to local temperature T.

Figure 5.1 presents the relations $\tilde{f}^{-}\left(\tilde{c}_z\right)$ on the CPS taken from [17].

The same paper presents the calculated values for ellipsoidal parameter: for the anomalous zone $\left(\tilde{T} = 0.2, \tilde{p} = 2\right)$: $\varepsilon_w = 2.11$, $\tilde{u}_w = -0.801$, for the standard zone $\left(\tilde{T} = 4, \tilde{p} = 2\right)$: $\varepsilon_w = 0.5$, $\tilde{u}_w = 0.45$. According to Eq. (5.21), the boundary of

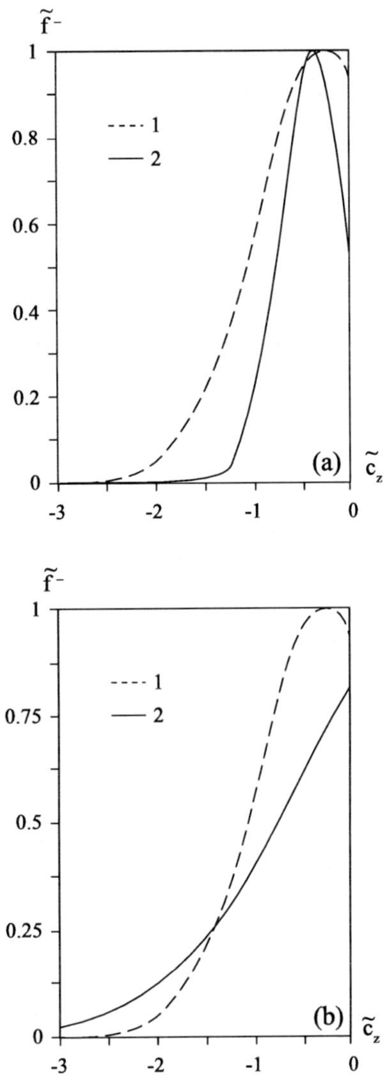

Fig. 5.1 Distribution function from [17] versus the dimensionless molecular velocity. 1 Mixing surface, 2 condensed-phase surface. **a** Anomalous zone of condensation, **b** standard zone of condensation

KL demonstrates $\varepsilon_\infty = 1$ for the entire temperature range $\left(0 < \widetilde{T} < \infty\right)$. The value $\widetilde{u}_\infty \equiv \sqrt{5/6}\mathrm{M}_\infty$ in Eq. (5.21) was found graphically from [17] for given values $\widetilde{p}$ and $\widetilde{T}$. The velocity obtained for the anomalous zone was $\widetilde{u}_\infty \approx -0.26$ and for the standard zone, $\widetilde{u}_\infty \approx -0.274$. The distribution function (5.21) is also plotted in Fig. 5.1 (curve 1).

Figure 5.1 demonstrates the growth in non-equilibrium of the spectrum of the incident molecular flux as it approaches through the KL toward the surface. The deformation of distribution function occurs differently for different branches of condensation. For $\tilde{T} < 1$ the function $\tilde{f}^-$ becomes more steep while retaining the initial hydrodynamic shift $\tilde{u} < 0$ (Fig. 5.1a), at $\tilde{T} > 1$ we observe the opposite pattern: function $\tilde{f}^-$ becomes more "smeared", and the shift changes the sign $\tilde{u} > 0$ (Fig. 5.1b), which is typical of the evaporation process [13–15].

The additional "mixing condition" is found through formulating the gas mass flux at infinity j_∞, which is available in the right-hand side of Eq. (5.5)

$$j_\infty \equiv \rho_\infty u_\infty = u_\infty \int_{\substack{-\infty < c_x < \infty \\ -\infty < c_y < \infty \\ -\infty < c_z < \infty}} m f_\infty d\mathbf{c}. \tag{5.23}$$

Integration in formula (5.23) is carried out in the limits from "$-\infty$" to "$+\infty$" for all three components of molecular velocity $\boldsymbol{c}$: components parallel to the surface c_x, c_y, and normal to the surface c_z It is enough to integrate for the negative half-space of the normal velocity component c_z ($-\infty < c_z < 0$) and integration for components c_x and c_y is left as formulated. This gives us the condensation flux—a component of molecular flux directed to the condensed phase

$$j_\infty^- = u_\infty \int_{\substack{-\infty < c_x < \infty \\ -\infty < c_y < \infty \\ -\infty < c_z \le 0}} m f_\infty^- d\mathbf{c}. \tag{5.24}$$

The integrand in (5.24) is a negative half distribution function that takes at infinity the value f_∞^- defined by ratio (5.1) written for area $c_z < 0$ We have a similar definition procedure for condensation flux at the surface "$\delta - \delta$"

$$j_\delta^- = u_\delta \int_{\substack{-\infty < c_x < \infty \\ -\infty < c_y < \infty \\ -\infty < c_z \le 0}} m f_\delta^- d\mathbf{c}. \tag{5.25}$$

Here, the negative half-function f_δ^- at the mixing surface is found from the formula

$$f_\delta^- \equiv f_\delta|_{c_z<0} = \frac{n_\delta}{\pi^{3/2} \mathrm{v}_\delta^3} \exp\left(-\left(\frac{\mathbf{c} - \boldsymbol{u}_\delta}{\mathrm{v}_\delta}\right)^2\right). \tag{5.26}$$

The closing equation for complete mathematical description of problem is obtained from a simple physical assumption : the process of molecular spectrum redistribution does not change the condensation flux, it remains steady. This produces a required mixing condition that links the molecular fluxes at the surfaces "$\infty - \infty$" and "$\delta - \delta$"

$$j_{\infty}^{-} = j_{\delta}^{-}. \tag{5.27}$$

Let us summarize the key moments of mixing model.

- strict assignment in the Navier–Stokes region for distribution function (5.12) overdetermination for the system of equations describing molecular flux conservation, therefore, the assumption (5.12) is incorrect,
- hypothesis (5.17) introduces a free parameter into mathematical description of the problem: this enables solving the evaporation problem [13, 14],
- the mixing model based on hypothesis (5.18) is a generalization for approach [13] for the condensation case,
- the problem closing is achieved with the conservation condition of condensation flux (5.27).

5.6 Solution Results

The system of Eqs. (5.8)–(5.10), (5.18), (5.27) are a closed description of strong condensation problem. After several transformations, this system is transformed into the following form

$$\frac{\sqrt{\tilde{T}}}{\tilde{p}} - \alpha_\rho \alpha_v K_1^- = 2\sqrt{\pi}\,\tilde{u}_\infty, \tag{5.28}$$

$$\frac{1}{\tilde{p}} - \alpha_\rho \alpha_v^2 K_2^- = 2 + 4\tilde{u}_\infty^2, \tag{5.29}$$

$$\frac{1}{\sqrt{\tilde{T}\,\tilde{p}}} - \alpha_\rho \alpha_v^3 K_3^- = \sqrt{\pi}\,\tilde{u}_\infty^3 + \frac{5\sqrt{\pi}}{2}\tilde{u}_\infty, \tag{5.30}$$

$$\operatorname{erfc}\left(\tilde{u}_\infty\right) = \alpha_\rho \alpha_v \alpha_u \operatorname{erfc}\left(\alpha_u \tilde{u}_\infty\right), \tag{5.31}$$

where we can find coefficients α_ρ, α_v, and α_u in relation (5.18). The dimensionless molecular fluxes K_i^- $(i = 1, 2, 3)$ incident to the mixing surface are written in the form

$$\left.\begin{aligned} K_1^- &= exp\left(-\tilde{u}_\delta^2\right) - \sqrt{\pi}\tilde{u}_\delta erfc\left(\tilde{u}_\delta\right), \\ K_2^- &= \frac{2}{\sqrt{\pi}}\tilde{u}_\delta exp\left(-\tilde{u}_\delta^2\right) - \left(1 + 2\tilde{u}_\delta^2\right) erfc\left(\tilde{u}_\delta\right), \\ K_3^- &= \left(1 + \frac{1}{2}\tilde{u}_\delta^2\right) exp\left(-\tilde{u}_\delta^2\right) - \frac{\sqrt{\pi}}{2}\tilde{u}_\delta\left(\frac{5}{2} + \tilde{u}_\delta^2\right) erfc\left(\tilde{u}_\delta\right). \end{aligned}\right\} \quad (5.32)$$

The function K_i^- is derived from the functions I_i^- defined through Eqs. (5.16), after the replacement $\tilde{u}_\infty \Rightarrow \tilde{u}_\delta \equiv \alpha_u \tilde{u}_\infty$. Equations (5.28)–(5.30) are obtained after calculations of the molecular fluxes in the left-hand sides of Eqs. (5.8)–(5.10) using appropriate distribution functions. The surface-emitted fluxes J_i^+ are calculated with the equilibrium half-Maxwellian distribution f_w^+ written for the CPS (formula (5.4)). The incident flows to the surface J_i^- are calculated with a shifted variant of half-Maxwellian distribution f_δ^- at the mixing surface (formula (5.26)).

The system of Eqs. (5.28)–(5.32) was solved using Maple computer algebra package. It should be taken into account during simulation that for the case of condensation problem, the Mach number in the Navier–Stokes region varies in the range $-1 \leq \text{M}_\infty < 0$ This corresponds to variation of speed ratio in the range $-\sqrt{5/6} \leq \tilde{u}_\infty < 0$. Figure 5.2 compares this simulation with results of simulation study from [17]. It is worth noting that simulated curves $\tilde{p}^{-1}\left(\tilde{T}\right)$ pass through a maximum in the vicinity of conjugation point $\tilde{T} = 1$ for two condensation branches—standard and anomalous ones (with a slight deviation to the left from this point).

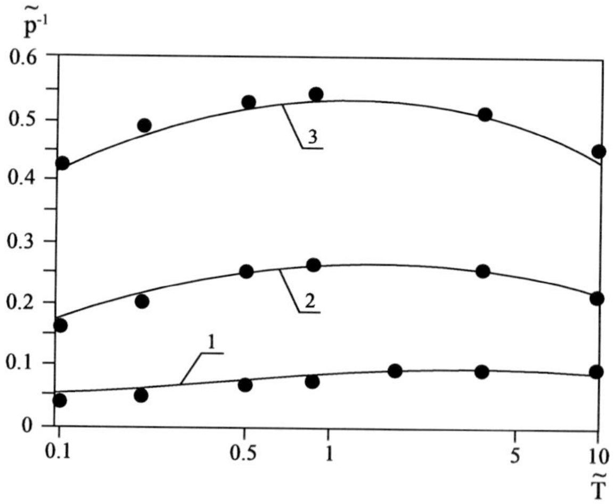

Fig. 5.2 Dependence for reverse pressure ratio on temperature ratio. Symbols correspond to the simulation data [17] calculations from the set of Eqs. (5.28)–(5.31). 1 $\text{M}_\infty = -1$, 2 $\text{M}_\infty = -0.6$, and 3 $\text{M}_\infty = -0.3$

At the microscopic level, the mixing model (not based on Boltzmann equation) does not reproduce the numerical evolution of the distribution function even at qualitative level (Fig. 5.1). However, at the macroscopic level, we observe a good correspondence between the approximate analytical formula $\tilde{p}\left(\tilde{T}, \mathrm{M}_\infty\right)$ with its accurate numerical version. Smoothing of microscopic errors in the distribution function while transition to extrapolated BC confirms the efficiency of theoretical method formulated in [13]. The same conclusion is claimed in [17]: "even a rough approximation for velocity distribution function in the KL might ensure satisfactory analytical description of gas dynamic conditions".

5.7 Sonic Condensation

Let us consider in detail the regime of sonic condensation (curve 1 in Fig. 5.2). As one can see from Fig. 5.3, the calculation is close to the dependency $\tilde{p}^{-1}\left(\tilde{T}\right)$ for the standard branch $\tilde{T} > 1$ but remarkably higher than the value $\tilde{p}^{-1}$ for the anomalous branch $\tilde{T} < 1$. The possible reason for this discrepancy might be the undercount of the choking effect for vapor flow as it reaches the sonic speed at the surface "$\delta - \delta$".

After calculating the Mach number $\mathrm{M}_\delta \equiv u_\delta (5k_B T_\delta / 3m)^{-1/2}$ at the mixing surface for the case of sonic condensation, the following results have been obtained (Fig. 5.4).

- for standard condensation the flow is subsonic $\left(\tilde{T} > 1 : |\mathrm{M}_\delta| < 1\right)$,
- for anomalous condensation the flow is supersonic $\left(\tilde{T} < 1 : |\mathrm{M}_\delta| > 1\right)$,

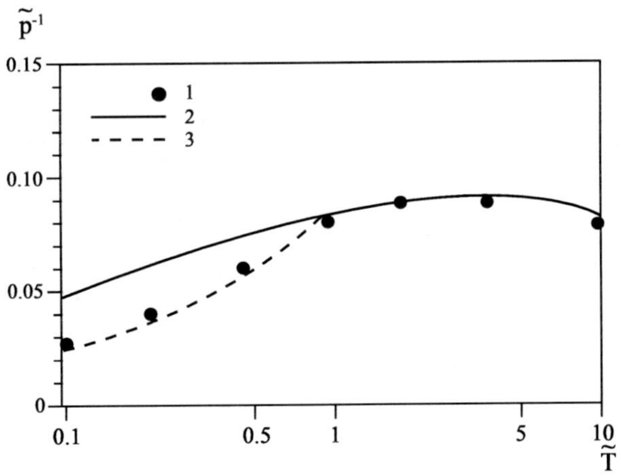

Fig. 5.3 Dependence for the reverse pressure ratio for the case of sonic flow condensation ($\mathrm{M}_\infty = -1$) on temperature ratio. 1 Simulation data [17], 2 calculations from the set of Eqs. (5.28)–(5.31), 3 calculation by (5.35)

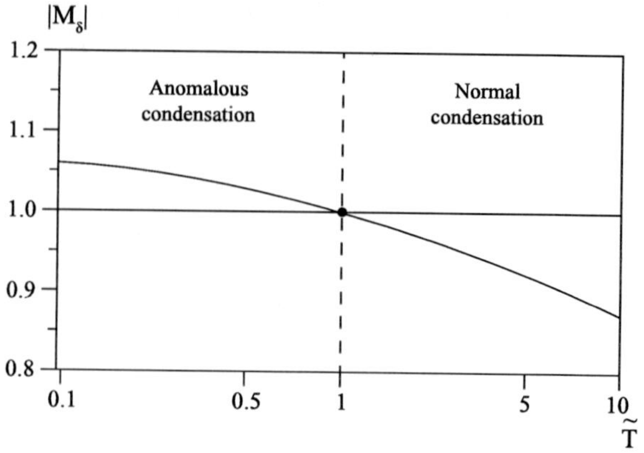

Fig. 5.4 Dependence of the Mach number on the temperature ratio on the mixing surface for the case of sonic condensation ($M_\infty = -1$)

- at the point of connection of the two branches of condensation the flow is sonic $\left(\tilde{T}=1 : |M_\delta| = 1\right)$.

This means that the vapor flow while moving from the KL boundary to the mixing surface: lows down and becomes subsonic at $\tilde{T}>1$, accelerates and becomes supersonic at $\tilde{T}<1$, remains sonic flow at $\tilde{T}=1$. The problem of stability for stationary supersonic condensation remains controversial issue. Thus, the authors of papers [16, 17] admit the possibility of supersonic condensation for certain parameter range, and in the papers [20, 21], it was stated that supersonic condensation ahead of the surface produces a shock wave, which returns the flow from supersonic mode to the sonic mode. We accept that the hydrodynamic velocity at the mixing surface cannot exceed the sonic speed: $|M_\delta| \leq 1$. Physically, this assumption is reduced to the condition of flow choking for the anomalous condensation branch

$$\tilde{T} \leq 1 : M_\infty = M_\delta = -1. \tag{5.33}$$

It follows from Eq. (5.33) that for anomalous zone $\tilde{T} \leq 1$, all parameters of incident molecular flux between the KL boundary and the mixing surface remain steady

$$\alpha_\rho = \alpha_v = \alpha_u = 1. \tag{5.34}$$

The use of conditions (5.33), (5.34) brings the following results for anomalous condensation: conservation equation for the condensation flux (5.31) becomes an equality. Conservation equations for the molecular flux of mass (5.28), momentum (5.29), and energy (5.30) give three different dependencies $\tilde{p}\left(\tilde{T}\right)$ The degenerate equation of energy (5.30) gives the following relation

$$\tilde{p} \approx \frac{11.7}{\sqrt{\tilde{T}}}. \quad (5.35)$$

As obvious from Fig. 5.3, the dependency $\tilde{p}^{-1}\left(\tilde{T}\right)$ calculated from formula (5.35) is perfect for describing the results of sonic condensation at $\tilde{T} < 1$. It is important that in the simulation study [17], the actual sonic regime was not achieved. The upmost subsonic simulated mode corresponded to $M_\infty \approx -0.95$. The points for $M_\infty = -1$ for the interval $0.1 < \tilde{T} < 10$ in Fig. 5.3 were obtained in [17] through extrapolation of smoothed simulated curves (this can be the origin of errors). It is interesting to note that in [18] (mostly reproducing the original analysis in [17]) for the case of polyatomic gas, the calculations were performed for subsonic mode with $M_\infty \approx -0.97$. Thus, the mixing model represented by a system of Eqs. (5.28)–(5.35) gives a good qualitative description of the pressure ratio $\tilde{p}$ on temperature ratio $\tilde{T}$ for the case of strong condensation.

The further development of model is based on construction of incident molecular flows at fixed points within the KL. More specifically, this procedure is reduced to a step-by-step build-up of mixing surfaces with a certain step until the process reaches the CPS. For the first stage, it is reasonable to transform the Maxwellian distribution function (5.26) into ellipsoid-type distribution (5.21), this will bring out the ellipsoidal parameter $\varepsilon \neq 1$. Obviously, this idea of process arrangement is close to the simulation techniques for the discretization of Boltzmann equation for the transverse coordinate performed in [22]. It is known that using the standard difference schemes with the second order of approximation leads to problems related to reproduction of discontinuity of the distribution function at the surface. On the opposite, the mixing model, although approximate one, is free of this shortcoming. Fulfillment of this program requires focused and time-consuming research with nonobvious outcome.

5.8 Approximate Solution

The analytical solution (5.28)–(5.31) is very cumbersome and can hardly be represented by formulas. This suggests the problem of finding an approximation to this solution in a form convenient for applied calculations. The study of the asymptotic behavior of the general solution as a function of the temperature factor suggests the following form of approximation

$$\tilde{T} \to 0:\ \tilde{p} \equiv \tilde{p}_0 = 1 + f_0 \tilde{T}^{-1/2}, \quad (5.36)$$

$$\tilde{T} \to \infty:\ \tilde{p} \equiv \tilde{p}_\infty = 1 + f_\infty \tilde{T}^{1/2}. \quad (5.37)$$

Here, $\tilde{p} \equiv \eta^{-1} = p_\infty/p_w$ is the ratio of pressures, $\eta \equiv p_w/p_\infty$—"pressure factor", $f_0(|\mathrm{M}|), f_\infty(|\mathrm{M}|)$ are some functions of the Mach number.

In constructing approximations, we pose the condition that in the limit as $|\mathrm{M}| \to 0$ the approximate solution should converge to the analytical solution in the linear approximation, which was obtained in [9]

$$\tilde{p} = 1 + \left(1.01\tilde{T}^{1/2} + 0.929\tilde{T}^{-1/2}\right)|\mathrm{M}|. \tag{5.38}$$

Omitting the intermediate operations, we write the approximate solution as

$$\left.\begin{aligned} \tilde{p} = 1 + \left(\left(\tilde{p}_0 - 1\right)^{1/n} + \left(\tilde{p}_\infty - 1\right)^{1/n}\right)^n, \\ n = 1 + \chi_n. \end{aligned}\right\} \tag{5.39}$$

Here, χ_0, χ_∞, χ_n are polynomials

$$\chi = k_1|\mathrm{M}| + k_2|\mathrm{M}|^2 + k_3|\mathrm{M}|^3 + k_4|\mathrm{M}|^4 + k_5|\mathrm{M}|^5 + k_6|\mathrm{M}|^6, \tag{5.40}$$

whose coefficients are given in Table 5.1.

Figure 5.5 shows that the dependences of the pressure factor on the temperature factor, as calculated from (5.39), accord well with the analytical solution (5.28)–(5.31) (the maximal relative departure is at most 4%). It is well known that for a kinetic analysis of condensation one usually assumes that the pressure in the Navier–Stokes region always exceeds the saturation pressure at CPS temperature: $\tilde{p} > 1, p_\infty > p_w$ [5, 6, 8–13]. At the same time, the ratio of densities ρ_∞/ρ_w can be smaller or greater than 1. We now fix the temperature factor $\tilde{T} = \tilde{T}_*$ at the point of maximum of the dependence $\eta\left(\tilde{T}\right)$ (i.e., at the point of minimum of the dependence $\tilde{p}\left(\tilde{T}\right)$). Then, for each Mach number, we get the minimal possible value of the ratio of densities: ρ_∞/ρ_w increases irrespective of whether $\tilde{T}$ becomes greater or smaller than $\tilde{T}_*$. For the isothermal case $\tilde{T}_* = 1$, $T_* = T_w$, which separates the abnormal and normal and branches of condensation, we have $|\mathrm{M}|_{min} = 0.2437$.

Table 5.1 Coefficients of the polynomials in (5.40)

	χ_n	χ_0	χ_∞
k_1	1.3	0.9293	1.011
k_2	2.8	1.028	−0.002559
k_3	−10.1	−5.18	1.222
k_4	11.1	19.19	−3.612
k_5	−4.1	−25.055	8.042
k_6	0	13.05	−4.967

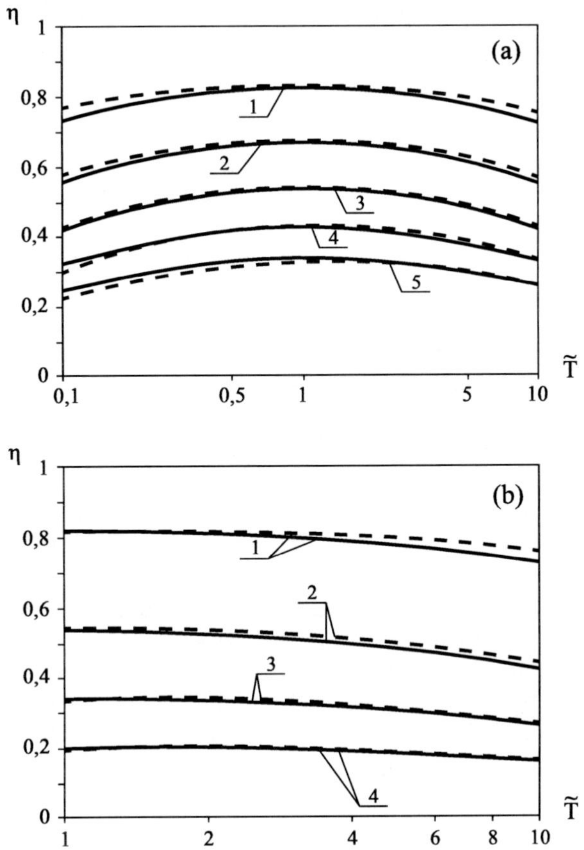

Fig. 5.5 Pressure factor versus the temperature factor. *Solid line* calculated from system of Eqs. (5.28)–(5.32), *dashed line* calculated from (5.39). **a** Anomalous and standard zones of condensation, 1 $|M| = 0.1$, 2 $|M| = 0.2$, 3 $|M| = 0.3$, 4 $|M| = 0.4$, and 5 $|M| = 0.5$. **b** standard zone of condensation, 1 $|M| = 0.1$, 2 $|M| = 0.3$, 3 $|M| = 0.5$, and 4 $|M| = 0.7$

Hence, in the framework of the kinetic analysis, in order to get a stable isothermal condensation, it is required to exactly specify the above Mach number and produce the pressure $p_\infty \approx 1.647 p_w$ in the condensing vapor. In this case, the number of molecules incident on the surface is $\approx$ 1.647 times smaller than that emitted from the CPS at the same temperature $T_\infty = T_w$. At the same time, under the phenomenological approach, the equality of the CPS and vapor temperatures should result in a complete extinction of condensation.[1] The above example clearly illustrates the nontrivial kinetics of strong condensation.

[1]As regards the abnormal condensation branch, it should not take place at all according to the standard approaches.

5.9 Supersonic Condensation Regimes

5.9.1 Introduction

In terms of applications, the kinetic analysis of the strong condensation problem is required for specification of boundary conditions on the CPS for flow equations of a compressible gas in the Navier–Stokes region. The principal difference between the subsonic and supersonic condensation regimes is in the possibility or impossibility of upstream propagation of the flow of perturbations of density (the acoustic wave). For subsonic condensation ($|\mathrm{M}| < 1$), the acoustic wave propagates toward the condensing gas. Hence, one parameter (for example, p_∞) should be the sought-for one, and the two remaining ones (for example, $|\mathrm{M}|$, T_∞) should be specified. For supersonic condensation, density perturbations do not take place upstream the flow. In this case, CPS ceases to have an effect on the flow, and so the three parameters in the Navier–Stokes region ($|\mathrm{M}|$, T_∞, p_∞) can be specified arbitrarily.

However, such a "self-similarity" of the outer flow by no means indicates that the setting of the supersonic condensation problem for the Boltzmann equation is senseless per se. In contrast, the solution to this problem is necessary for the answer to the following question: for which parameters of the incident flow and of the CPS the supersonic condensation is possible and for which it is impossible. In the first case, one should assume that the CPS receives all the incident molecular flow and has no effect on it. In the second case, a subsonic condensation BC should be specified on the surface. Such a statement of the problem results in a radical change of the flow: near the CPS the gas will be decelerated to a subsonic velocity. A shock wave will propagate upstream the supersonic incident flow. As a result, it can be assumed that under stable supersonic condensation the KL merges with a standing shock wave. So, the kinetic analysis of supersonic condensation is required not for the solution of "self-similar" gas dynamic equations, but rather for qualitative "diagnostic of the flow".

The impossibility of preliminary prediction of the domain of existence of the solution for supersonic condensation necessitates the use of the "cut-and-try method": one constructs a concrete solution, studies its stability, and only eventually becomes able to conclude about the existence of this flow. Conceptually, this peculiarity of condensation corresponds to the following well-known theorem from calculus: termwise differentiation of an infinite function series is only possible once the uniform convergence of the series composed of the derivatives of the functions is established [23].

By now, only a limited number of theoretical investigations of the supersonic condensation is available. Numerical analysis [16] shows that there exist domains of parameters in which stationary supersonic condensation is impossible. In [20], Mott-Smith's method was used to evaluate supersonic condensation regimes in the form of a shock wave in front of CPS, which transforms a supersonic motion into a subsonic one. In [24, 25], it was numerically obtained that stable stationary solutions exist in this setting only for some definite combinations of the parameters. In [26], the direct Monte Carlo statistical modeling method was used to show that in the subsonic regime, for each value of |M|, there exits some limit bottom curve $\tilde{p}\left(\tilde{T}\right)_{\lim}$ (in our notation, the limit upper curve $\eta\left(\tilde{T}\right)_{\lim}$). This curve determines the boundary between the domain in which p_∞, T_∞ can be arbitrary and the domain of parameters for which supersonic condensation is impossible. In [26], the theoretical dependences for $\tilde{p}\left(\tilde{T}\right)_{\lim}$ with $|M| = \text{idem}$ have a minimum (like the dependences $\tilde{p}\left(\tilde{T}\right)$ in the case of subsonic condensation).

5.9.2 *Calculation Results*

We now employ the "mixing model" of [5] to analyze the supersonic condensation. The following results were obtained with the help of Eqs. (5.28)–(5.31).

(1) In the domain $1 \leq |M| \leq 1.16$, the dependences $\eta\left(\tilde{T}\right)$ have the same form as in the subsonic condensation setting: with increasing Mach number these curves move down and the points of their maxima move to the right along the $\tilde{T}$-axis (Fig. 5.6a).

(2) For $|M| \approx 1.16$, the dependence $\eta\left(\tilde{T}\right)$ exhibits an opposite behavior: in the domain $1.16 < |M| \leq 1.375$ the abnormal branch of $\eta\left(\tilde{T}\right)$ continues to move down, whereas the normal branch of the solution starts to move up. With increasing Mach number, the maximum value of the dependence $\eta\left(\tilde{T}\right)$ is increasing (Fig. 5.6b). Correspondingly, the minimum value of the dependence $\tilde{p}\left(\tilde{T}\right)$ is decreasing. A qualitative indication of the existence of such an inversion of $\eta\left(\tilde{T}_*\right)_{\max}$ is also contained in the paper [26].[2]

(3) At the point $|M| \approx 1.375$ the solution bifurcates: the last single-valued solution is replaced by the first two-valued solution (Fig. 5.7). We note that in [20, 21] in the domain $|M| \geq 1.3$ the following peculiarity was pointed out: disappearance of a shock wave-type solution.

(4) In the domain $1.375 < |M| \leq 2.24$ only two-valued solutions are realized: to each value of the temperature factor there correspond two values of pressure in the Navier–Stokes region. Here, the branch of abnormal condensation $\tilde{T} < 1$ first disappears $|(M| = 1.375)$, and later, as the Mach number increases, it is "filled" by the two-valued domain.

(5) In turn, the domain $1.375 < |M| \leq 2.24$ can be subdivided into two subdomains. In the range $1.375 \leq |M| \leq 1.5$, the curves exhibit self-intersections

[2] In [26], the inversion was predicted when the flow exactly reaches the sound speed condensation $|M| = 1$.

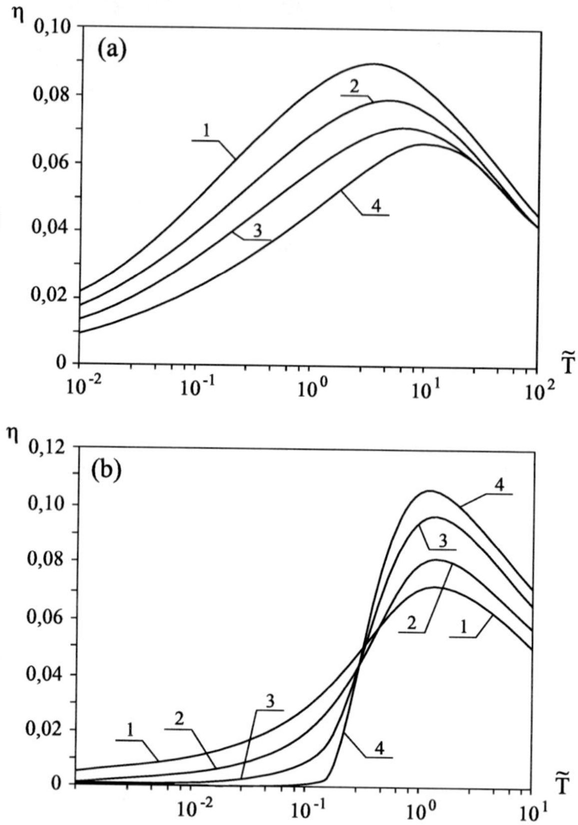

Fig. 5.6 The pressure factor versus the temperature factor for the supersonic condensation domain. **a** Domain of Mach number $1 \leq |M| \leq 1.16$, 1 $|M| = 1$, 2 $|M| = 1.05$, 3 $|M| = 1.1$, 4 $|M| = 1.16$, **b** domain of Mach number $1.25 \leq |M| \leq 1.375$, 1 $|M| = 1.25$, 2 $|M| = 1.3$, 3 $|M| = 1.35$, 4 $|M| = 1.375$

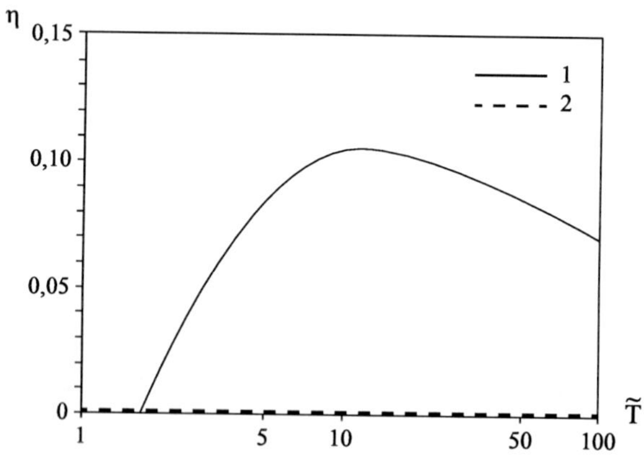

Fig. 5.7 Bifurcation domain for the solution. *Solid line* upper curve, *dashed line* lower curve

and hence are difficult to interpret. In the range $1.5 \leq |\mathrm{M}| \leq 2.24$, the two-valued curves $\eta\left(\tilde{T}\right)$ become ordered. Each of these pairs of curves can be separated into the "lower curve" (smaller than η, greater than $\tilde{p}$) and the "upper curve" (greater than η, smaller than $\tilde{p}$). As the boundary of the lower and upper curves, we take the point at which the condition $\frac{d\eta}{d\tilde{T}} = \infty$ is satisfied. Calculations show that for the entire range $1.5 \leq |\mathrm{M}| \leq 2.24$ this condition is satisfied approximately for the same pressure factor $\eta \approx 0.6$ (see Fig. 5.8). This figure shows that with increasing Mach number both branches move to the left along the $\tilde{T}$-axis and expand more and more.

The following principal question arises when considering the two-valued domain of solutions: which of the branches of the dependences $\eta\left(\tilde{T}\right)$ is actually realized? In this connection, we recall that in the frameworks of the model under consideration, the "mixing surface" is in fact the "outer surface" of the CPS. We shall assume that with fixed parameters $\tilde{T}$, $|\mathrm{M}|$ the regime with smaller value of the Mach number is

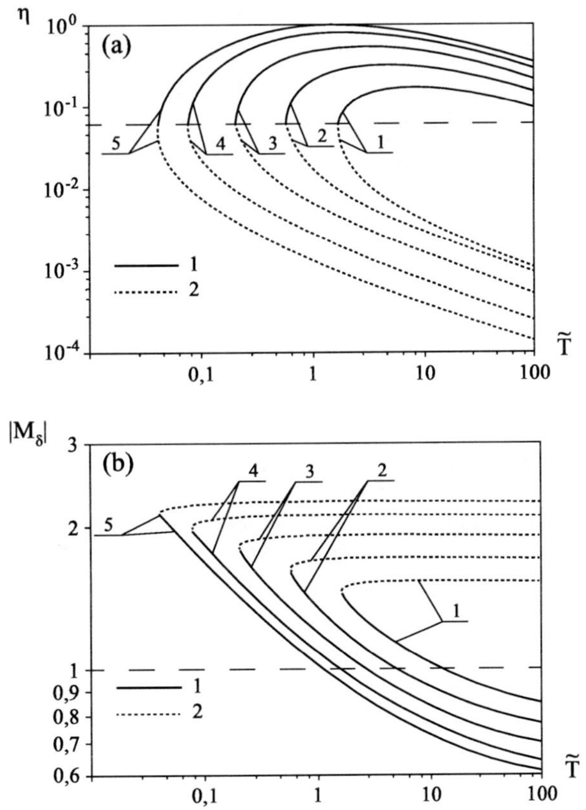

Fig. 5.8 The domain of two-valued solutions. *Solid line* upper curve, *dashed line* lower curve. **a** Pressure factor versus the temperature factor. 1 $|\mathrm{M}| = 1.5$, 2 $|\mathrm{M}| = 1.7$, 3 $|\mathrm{M}| = 1.9$, 4 $|\mathrm{M}| = 2.1$, and 5 $|\mathrm{M}| = 2.24$, **b** $|\mathrm{M}_\delta|$ versus the temperature factor, 1 $|\mathrm{M}| = 1.5$, 2 $|\mathrm{M}| = 1.7$, 3 $|\mathrm{M}| = 1.9$, 4 $|\mathrm{M}| = 2.1$, and 5 $|\mathrm{M}| = 2.24$

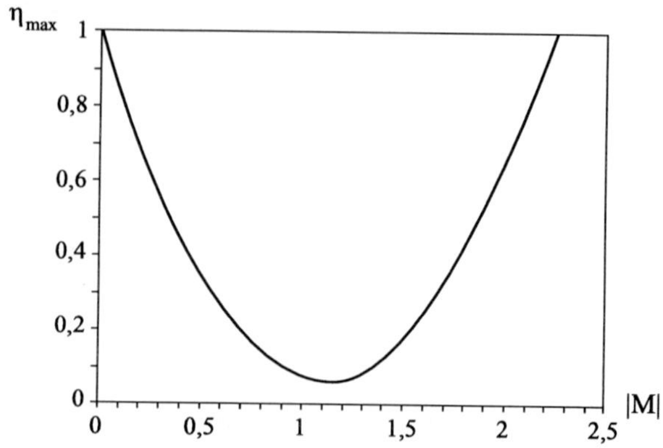

Fig. 5.9 Maxima of the pressure factor versus the Mach number

realized on the mixing surface. Figures 5.8b shows that for the upper branch $\eta\left(\widetilde{T}\right)$ in Fig. 5.8a, the quantity $|M_\delta|$ will always be smaller than for the lower branch.

Physically this means that, in the two-valued domain of supersonic condensation, regimes with smaller gas pressure are "chosen". On this basis, in what follows the upper curves $\eta\left(\widetilde{T}\right)$ in Fig. 5.8a will be called "stable". With increasing temperature factor, the quantity $|M_\delta|$ for each "stable" curve decreases and for certain values of $\widetilde{T}$ it becomes smaller than 1. Hence, these regimes can be interpreted as a shock wave transforming a supersonic motion into a subsonic one. For $|M| \approx 2.24$, the isobaric condensation was attained at the maximum point of the dependence $\eta\left(\widetilde{T}\right)$. In our analysis, this case was considered as a limit one, because otherwise one has to assume the existence of nonphysical regimes with inversion of pressures on the CPS and in the condensing gas.

Calculations show that the envelop of the family of maxima of functions of the pressure factor of the Mach number for the entire domain of existence of condensation $0 < |M| \le 2.24$ (Fig. 5.9) has a minimum at the point of inversion $|M| \approx 1.16$. An isobaric equilibrium case is attained as $|M| \to 0$: with zero condensation flow the pressure in gas is equal to the saturation pressure at the CPS temperature. The problem is posed with a fixed temperature factor $\widetilde{T} \neq 1$, and hence the isothermality condition does not hold in a general case hold. Above it was shown that the isobaric condensation is also attained on the upper boundary of existence of the domain of supersonic condensation $|M| \approx 2.24$, i.e., in the limit non-equilibrium case. Let us now consider the coordinates of the maxima of the pressure factor. To this aim, we calculated the dependence on the Mach number of the inverse temperature factor $\tau_* \equiv \widetilde{T}_*^{-1}$, which corresponds to the envelop of the family of maxima $\eta\left(\widetilde{T}\right)$.

Figure 5.10 shows that the dependence $\tau_*(|M|)$ has a minimum for $|M| \approx 1.29$. It is worth pointing out that as $|M| \to 0$ the quantity τ_* tends not to 1, but rather to the value ≈ 1.0886. This result was obtained in [9] in the study of the linear condensation problem. In turn, on the upper boundary of regimes of supersonic condensation, we

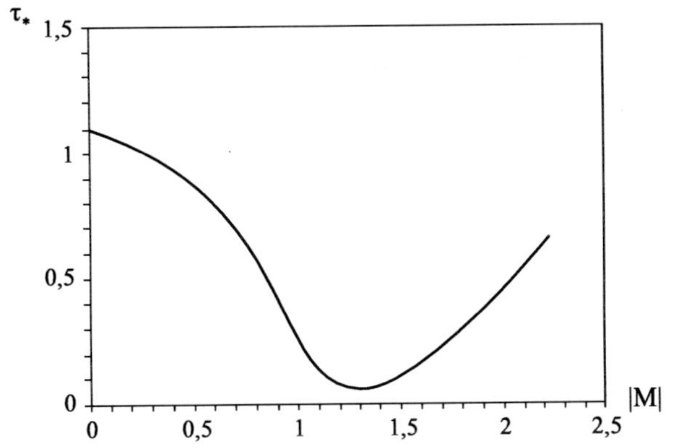

Fig. 5.10 Coordinates of maxima of the pressure factor versus the Mach number

have $|\mathrm{M}| \approx 2.24$, $\tau_* \approx 0.64$. So, as distinct from the dependence $\eta(|\mathrm{M}|)_{\max}$, the limit value τ_* is distinct from 1 on both boundaries of the domain of existence of condensation $0 < |\mathrm{M}| \leq 2.24$.

5.10 Diffusion Scheme of Reflection of Molecules

The previous arguments apply to the case when a surface absorbs all the molecules incident on it from the Navier–Stokes region ("an absolutely permeable CPS"). In a general case, the CPS has a limited permeability and transmits only a part of the molecular flow passing through it. This part is controlled by the condensation coefficient β. The quantity β, which reflects the state of the surface and the physical nature of the condensed phase, varies in the range $0 < \beta \leq 1$.

Let us consider the effect of the condensation coefficient on the equations of conservation of molecular flows. We denote by J_i^+ the flows of mass ($i = 1$), momentum ($i = 2$), and energy ($i = 3$) emitted by an absolutely permeable CPS. Assume now that the surface transmits not the entire flow incident from the liquid phase to the gaseous phase, but only its part $J_{i\beta}^+ = \beta J_i^+$. In its turn, we assume that only the part β of the molecular flow J_i^- incident on the CPS from Navier–Stokes region is captured by it, while the remaining part $(1 - \beta)J_i^-$ is reflected from it. Then, the total molecular flow outgoing from the surface is as follows

$$J_{i\beta}^+ = \beta J_i^+ + (1 - \beta)J_i^-, \tag{5.41}$$

$i = 1, 2, 3$. From (5.41) we can find the macroscopic flows $J_{i\infty}$ in the Navier–Stokes region, which are defined as the difference between the emitted and the incident molecular flows

$$J_{i\infty} = J_{i\beta}^+ - J_i^- = \beta\left(J_i^+ - J_i^-\right). \tag{5.42}$$

Let us introduce the "permeability coefficients" defined as the ratio of the flow emanating from the CPS (the case $\beta < 1$) to the flow emitted by an absolutely permeable CPS (the case $\beta = 1$)

$$\psi_i = \frac{J_{i\beta}^+}{J_i^+}. \tag{5.43}$$

Using (5.41), (5.42) in (5.43), we find that

$$\psi_i = 1 - 2\sqrt{\pi}\frac{1-\beta}{\beta}\frac{J_{i\infty}}{J_i^+}. \tag{5.44}$$

The quantities J_i^+ are determined by (5.11). The macroscopic flows in the Navier–Stokes region can be found from Eqs. (5.8)–(5.10)

$$J_{1\infty} = \rho_\infty u_\infty, \tag{5.45}$$

$$J_{2\infty} = \rho_\infty u_\infty^2 + p_\infty, \tag{5.46}$$

$$J_{3\infty} = \frac{1}{2}\rho_\infty u_\infty^3 + \frac{5}{2}p_\infty u_\infty. \tag{5.47}$$

From Eqs. (5.43)–(5.47), we can find expressions for the permeability coefficients for each of the three conservation equations

$$\psi_1 = 1 - 2\sqrt{\pi}\frac{1-\beta}{\beta}\frac{\tilde{u}_\infty}{\eta\sqrt{\tilde{T}}}, \tag{5.48}$$

$$\psi_2 = 1 - 2\frac{1-\beta}{\beta}\frac{1+2\tilde{u}_\infty^2}{\eta}, \tag{5.49}$$

$$\psi_3 = 1 - \frac{5\sqrt{\pi}}{2}\frac{1-\beta}{\beta}\frac{\sqrt{\tilde{T}}\tilde{u}_\infty\left(1+2/5\tilde{u}_\infty^2\right)}{\eta}. \tag{5.50}$$

So, in the general case $\beta \leq 1$, the equations of conservation of molecular flows (5.8)–(5.10) assume the form

$$\psi_1 J_1^+ - J_1^- = J_{1\infty}, \tag{5.51}$$

$$\psi_2 J_2^+ - J_2^- = J_{2\infty}, \tag{5.52}$$

$$\psi_3 J_3^+ - J_3^- = J_{3\infty}. \tag{5.53}$$

In the kinetic analysis, it is assumed that all three permeability coefficients are equal, $\psi_1 = \psi_2 = \psi_3$. Such a simplified scheme of reflection of molecules is known as the "diffusion scheme". This scheme was first proposed in [5] in the linear approximation frameworks. In [27], this scheme was modified for the case of intensive evaporation condensation. Later, more involved schemes of interaction of molecules of the incident flow with the CPS were used [28].

It is worth mentioning that the diffusion scheme is self-inconsistent.

- In the frameworks of the diffusion scheme only the permeability coefficient of the flow of mass ψ_1 is determined. Physically this means that only the balance of the mass flow ($i = 1$) is achieved, whereas the flows of the momentum normal component ($i = 2$) and the energy ($i = 3$) through the CPS remain unbalanced. In other words, the diffusion scheme involves only formula (5.48) and ignores formulas (5.49), (5.50).
- The kinetic analysis deals only with the gaseous phase with specified BC on the surface. However, in the frameworks of the diffusion scheme, it is assumed that the CPS also "filters out" the molecules incoming from the liquid phase. This implicitly includes the liquid phase in the motion mechanism of molecular flows, which contradicts the original process picture.

However, the studies based on modeling the evaporation/condensation processes by the molecular dynamics method are free of such self-inconsistencies [29]. This method, which appeared in the mid-1950s, deals with relatively small systems. The principal idea of the molecular dynamics method is the study of various properties of substances by modeling the motion and interaction of separate particles (atoms or molecules). The interaction of particles is described with the help of various potentials: the hard-sphere potential, the Lennard–Jones potential, continuous potentials, etc. Studies of intensive evaporation by the molecular dynamics method showed the important role of fluctuations of the bonding energy in the surface layer of liquid. It was shown that a considerable contribution to the molecular flow comes from the molecules whose kinetic energy has the same order with the energy of heat motion.

However, it is worth noting that by now in the framework of the molecular dynamics method no correlation dependences for parameters were provided which are suitable for applied calculations. So, below we shall use the diffusion scheme of reflection of molecules tested in a large number of studies. Let us proceed with the calculation of the permeability coefficient ψ_1. According to (10), we have

$$J_1^+ = \rho_w \sqrt{\frac{kT_w}{2\pi m}}. \tag{5.54}$$

Taking into account that when changing from the general case $\beta \leq 1$ to the limit case $\beta = 1$ the CPS temperature remains the same ($T_w = \text{idem}$), we have from (5.43), (5.45)

$$\psi_1 = \frac{\rho_{w\beta}}{\rho_w}. \tag{5.55}$$

So, is the diffusion scheme of reflection the actual density of vapor saturation at the surface temperature ρ_w is replaced by its modified value $\rho_{w\beta}$

$$\rho_w \to \rho_{w\beta} = \rho_w\left(1 - 2\sqrt{\pi}\frac{1-\beta}{\beta}\frac{\tilde{u}_\infty}{\eta\sqrt{\tilde{T}}}\right). \tag{5.56}$$

As was already pointed out, the molecular flow emitted from the surface has an equilibrium Maxwell spectrum defined by relation (5.4). In the frameworks of the diffusion scheme, it is assumed that after the interaction with the CPS the reflected flow "forgets" its original spectrum and also acquires a Maxwell distribution.

5.11 The General Case of Boundary Conditions

Using (5.56), we write the system of equations of conservation of molecular flows (5.8)–(5.10) in the following form

$$\frac{1}{2\sqrt{\pi}}\left(\rho_{w\beta}\mathrm{v}_w - \rho_\delta \mathrm{v}_\delta I_1^-\right) = 2\sqrt{\pi}\,\rho_\infty u_\infty, \tag{5.57}$$

$$\frac{1}{4}\left(\rho_{w\beta}\mathrm{v}_w^2 - \rho_\delta \mathrm{v}_\delta^2 I_2^-\right) = \rho_\infty u_\infty^2 + p_\infty, \tag{5.58}$$

$$\frac{1}{2\sqrt{\pi}}\left(\rho_{w\beta}\mathrm{v}_w^3 - \rho_\delta \mathrm{v}_\delta^3 I_2^-\right) = \frac{1}{2}\rho_\infty u_\infty^3 + \frac{5}{2}p_\infty u_\infty. \tag{5.59}$$

We see that the form of equations remains the same if ρ_w is replaced by $\rho_{w\beta}$. Now, the original system of Eqs. (5.8)–(5.10) can be written as

$$\sqrt{\tilde{T}}\eta_\beta - \alpha_\rho\alpha_\mathrm{v} I_1^- = 2\sqrt{\pi}\tilde{u}_\infty, \tag{5.60}$$

$$\eta_\beta - \alpha_\rho\alpha_\mathrm{v}^2 I_2^- = 2 + 4\tilde{u}_\infty^2, \tag{5.61}$$

$$\frac{1}{\sqrt{\tilde{T}}}\eta_\beta - \alpha_\rho\alpha_\mathrm{v}^3 I_3^- = \sqrt{\pi}\tilde{u}_\infty^3 + \frac{5}{2}\sqrt{\pi}\tilde{u}_\infty, \tag{5.62}$$

$$\mathrm{erfc}\left(\tilde{u}_\infty\right) = \alpha_\rho\alpha_\mathrm{v}\alpha_u \mathrm{erfc}\left(\alpha_u\tilde{u}_\infty\right). \tag{5.63}$$

Here, η_β is the modified pressure factor related to the true value η by the relation

$$\eta = \eta_\beta + 2\sqrt{\pi}\frac{1-\beta}{\beta}\frac{\tilde{u}_\infty}{\sqrt{\tilde{T}}}. \tag{5.64}$$

Let us express the speed ratio in terms of the Mach number for a monoatomic gas, $\tilde{u}_\infty = \sqrt{5/6}\mathrm{M} = -\sqrt{5/6}|\mathrm{M}|$.[3] Now from Eq. (5.64), we find the required pressure factor

$$\eta = \eta_\beta - A. \tag{5.65}$$

Here,

$$A = \sqrt{\frac{10\pi}{3}}\frac{1-\beta}{\beta}\frac{|\mathrm{M}|}{\sqrt{\tilde{T}}}.$$

is the generalized parameter describing the mechanism of diffuse reflection, which involves the condensation coefficient, the temperature factor, and the Mach number. Relation (5.65) is of key importance for calculation of the effect of β on the process of condensation.

Let us consider various cases of variation of the parameter A.

- The modified pressure factor is calculated from the system of Eqs. (5.60)–(5.63) as a function of the temperature factor $\tilde{T}$ and the Mach number $|\mathrm{M}|$, $\eta_\beta = \eta_\beta\left(\tilde{T}, |\mathrm{M}|\right)$. The limit case $\beta = 1$ corresponds to an absolutely permeable surface, $A = 0$, $\eta = \eta_\beta$.
- For $\beta < 1$, from (5.65) we have $A > 0$, $\eta < \eta_\beta$. This means that the surface pressure factor is decreasing with decreasing permeability, i.e., the pressure in the Navier–Stokes region is increasing. With a fixed β this trend is expressed the more the smaller is the temperature factor and the larger is the Mach number.
- To each combination of the parameters $\tilde{T}$, $|\mathrm{M}|$, there corresponds some minimal value of the condensation coefficient, which can be found from the equality $A = 0$

$$\beta_{\min} = \sqrt{\frac{3}{10\pi}}\frac{\eta_\beta\sqrt{\tilde{T}}}{|\mathrm{M}|}. \tag{5.66}$$

It follows that condensation is possible only in the range $\beta_{\min} < \beta \leq 1$, whereas the domain

[3] Recall that in the above case of consideration the speed ratio and the Mach number are negative.

$0 < \beta \leq \beta_{\min}$ is "locked" for condensation. Figure 5.11 shows that the pressure factor versus the $\beta_{\min}$.

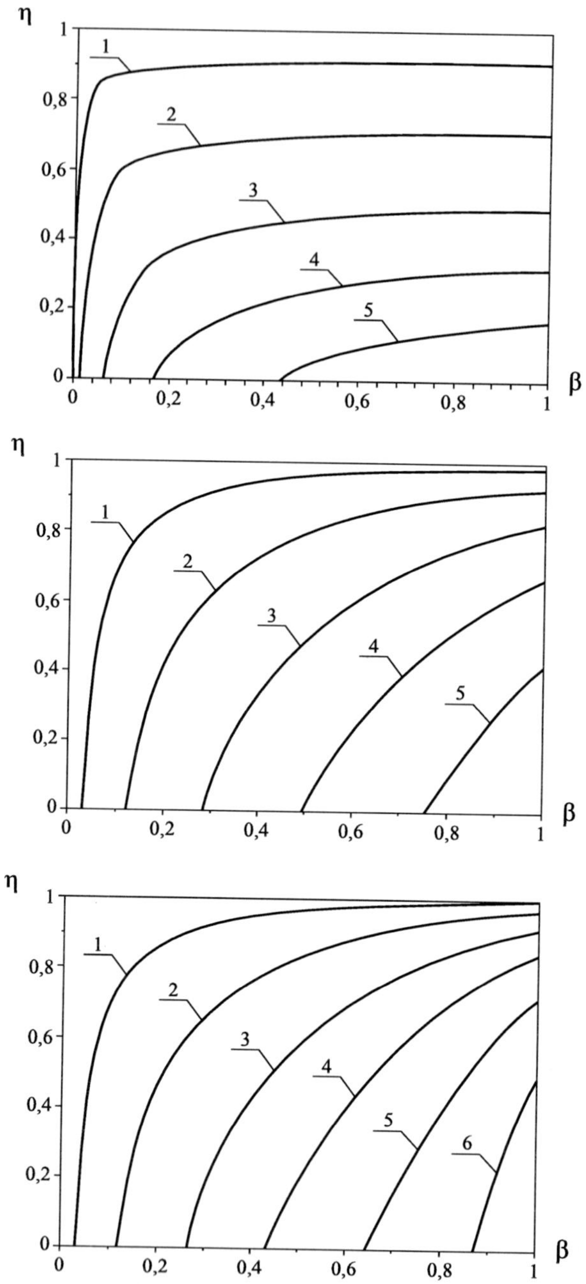

Fig. 5.11 The pressure factor versus the β. **a** $\widetilde{T} = 10$, $1\ |\mathrm{M}| = 1 * 10^{-2}$, $2\ |\mathrm{M}| = 4 * 10^{-2}$, $3\ |\mathrm{M}| = 1 * 10^{-1}$, $4\ |\mathrm{M}| = 2 * 10^{-1}$, $5\ |\mathrm{M}| = 4 * 10^{-1}$, **b** $\widetilde{T} = 1$, $1\ |\mathrm{M}| = 1 * 10^{-2}$, $2\ |\mathrm{M}| = 4 * 10^{-2}$, $3\ |\mathrm{M}| = 1 * 10^{-1}$, $4\ |\mathrm{M}| = 2 * 10^{-1}$, $5\ |\mathrm{M}| = 4 * 10^{-1}$, **c** $\widetilde{T} = 0.1$, $1\ |\mathrm{M}| = 1 * 10^{-3}$, $2\ |\mathrm{M}| = 4 * 10^{-3}$, $3\ |\mathrm{M}| = 1 * 10^{-2}$, $4\ |\mathrm{M}| = 2 * 10^{-2}$, $5\ |\mathrm{M}| = 4 * 10^{-2}$, $6\ |\mathrm{M}| = 1 * 10^{-1}$

5.12 The Effect of β on the Condensation Process

Figures 5.12, 5.13 and 5.14 given the locked value of the condensation coefficient versus the temperature factor for subsonic (Fig. 5.12) and supersonic (Figs. 5.13 and 5.14) condensation with the Mach number as a parameter.

The figures show that the pressure factor decreases from some maximal value to zero as β decreases from $\beta_{\min}$ to 1. This means that with decreasing β and a fixed value of p_w, the pressure in the Navier–Stokes region increases, so that as $\beta \to \beta_{\min}$ the "condensation is locked", $p_\infty \to \infty$. So, each curve M = idem intersects the abscissa axis at the point $\beta = \beta_{\min}$. The region to the right of this point is "allowed" for condensation, the domain to the left is "locked".

Figures 5.12, 5.13 and 5.14 shows that with decreasing $\tilde{T}$ the condensation "lock threshold" increases monotonically, assuming the value 1 as $\tilde{T} \to 0$, which makes condensation impossible. With increasing Mach number the curve $\beta_{\min}\left(\tilde{T}\right)$ moves

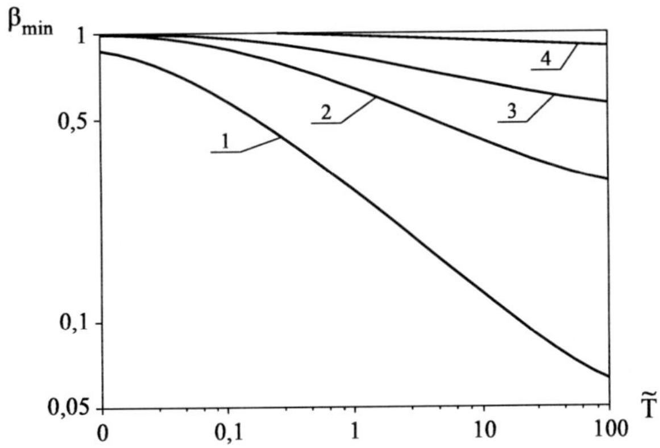

Fig. 5.12 Locked value of the condensation coefficient versus the temperature factor for $|M| \leq 1$. 1 $|M| = 0.1$, 2 $|M| = 0.3$, 3 $|M| = 0.5$, and 4 $|M| = 1$

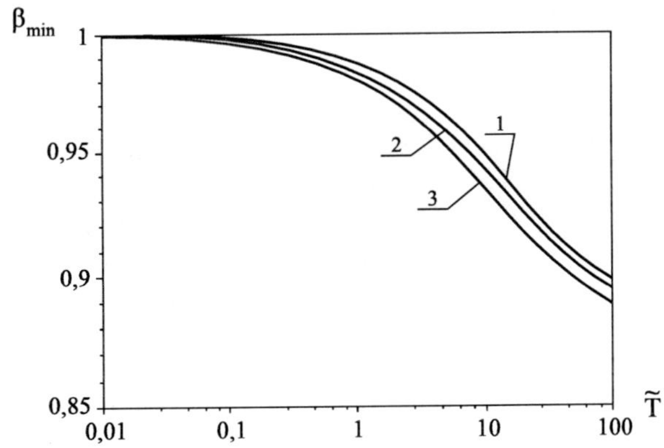

Fig. 5.13 Locked value of the condensation coefficient versus the temperature factor for $|M| \geq 1$.
1 $|M| = 1.05$, 2 $|M| = 1.1$, and 3 $|M| = 1.16$

Fig. 5.14 Locked value of the condensation coefficient versus the temperature factor for $|M| \geq 1$.
1 $|M| = 1.2$, 2 $|M| = 1.3$, and 3 $|M| = 1.375$

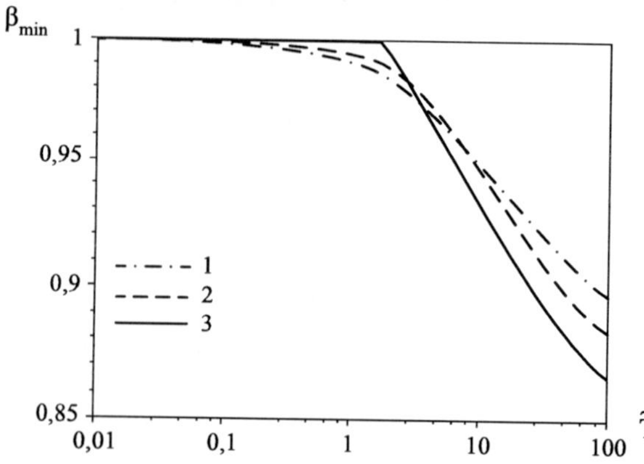

more and more toward 1, thereby increasing the "lock threshold" and decreasing the "allowed domain" (Fig. 5.12). In the single-valued domain of supersonic condensation $1 < |M| \leq 1.16$, the subsonic trends of the solution are preserved. Moreover, with increasing Mach number the domain of existence of condensation becomes more and more narrow (Fig. 5.13).

The inversion of $\beta_{\min}\left(\tilde{T}\right)$ occurs in the range $1.16 < |M| \leq 1.375$ (Fig. 5.14). In the abnormal domain $\left(\tilde{T} < 1\right)$, this curve continues to increase with increasing $|M|$, and in the normal domain $\left(\tilde{T} > 1\right)$ this curve starts to move down. In the two-valued domain of supersonic condensation, the unstable branches become practically equal to 1. As a result, such branches are hardly probable already for two reasons: supersonic Mach numbers on the mixing surface and condensation lock owing to decreasing CPS permeability. At the same time, stable branches feature much broader "allowable" domain.

5.13 Conclusions

A model of strong condensation (mixing model) is developed: it is based on the conservation equations for molecular fluxes of mass, momentum, and energy within the Knudsen layer. The closing relationship for this model is the condition of condensation flux conservation between the Knudsen layer and the mixing surface. The approximate analytical solution for strong condensation problem was obtained in the form of pressure ratio versus temperature ratio (with Mach number as a parameter). Analysis of condensation with the sonic gas flow uses the gas flow choking condition at the mixing surface. The analytical solution thus obtained was compared with available simulation data. The direction for further elaboration of the analytical model is outlined. Based on the previously obtained analytical solution of the problem of strong condensation its approximation is derived suitable for calculations of the entire range of subsonic condensation. The dependence of the pressure factor on the temperature factor with the Mach number as a parameter was an unknown magnitude. An equilibrium isothermal case was considered. This case separates the abnormal and the normal condensation branches.

The previously developed "mixing model" was used to calculate regimes of supersonic condensation. Peculiarities of supersonic condensation with increased Mach number are studied: the inversion of the solution, bifurcation of the solution, transition to two-valued solutions, the limit Mach number, for which a solution exists. A classification of two-valued solutions was obtained capable of singling out stable branches. For the entire range of variation of the condensation intensity, the envelope of the family of maxima of pressures was constructed. The effect of the condensation coefficient on the conservation equations for mass, the normal component of the momentum, and the energy of molecular flows was studied. The "condensation lock" phenomenon due to reduced permeability of the condensed-phase surface was examined. Calculations of the "allowed" condensation regimes were carried out.

References

1. Mazhukin VI, Mazhukin AV, Demin MM, Shapranov AV (2013) The dynamics of the surface treatment of metals by ultra-short high-power laser pulses. In: Sudarshan TS, Jeandin M, Firdirici V (eds) Surface modification technologies XXVI (SMT 26), vol 26, pp 557–566
2. Lezhnin SI, Kachulin DI (2013) The various factors influence on the shape of the pressure pulse at the liquid-vapor contact. J Eng Thermophys 22(1):69–76
3. Zakharov VV, Crifo JF, Lukyanov GA, Rodionov AV (2002) On modeling of complex non-equilibrium gas flows in broad range of Knudsen numbers on example of inner cometary atmosphere. Math Models Comput Simul 14(8):91–95
4. Kogan MN (1995) Rarefied gas dynamics. Springer
5. Labuntsov DA (1967) An analysis of the processes of evaporation and condensation. High Temp 5(4):579–647
6. Muratova TM, Labuntsov DA (1969) Kinetic analysis of the processes of evaporation and condensation. High Temp 7(5):959–967
7. Cercignani C (1990) Mathematical methods in kinetic theory. Springer, US
8. Pao YP (1971) Temperature and density jumps in the kinetic theory of gases and vapors. Phys Fluids 14:1340–1346
9. Pao YP (1971) Erratum: temperature and density jumps in the kinetic theory of gases and vapors. Phys Fluids 16:1650
10. Aristov VV, Panyashkin MV (2011) Study of spatial relaxation by means of solving a kinetic equation. Comput Math Math Phys 51(1):122–132
11. Tcheremissine FG (2012) Method for solving the Boltzmann kinetic equation for polyatomic gases. Comput Math Math Phys 52(2):252–268
12. Zhakhovskii VV, Anisimov SI (1997) Molecular-dynamics simulation of evaporation of a liquid. J Exp Theor Phys 84(4):734–745
13. Anisimov SI (1968) Vaporization of metal absorbing laser radiation. Sov Phys JETP 27(1):182–183
14. Labuntsov DA, Kryukov AP (1979) Analysis of intensive evaporation and condensation. Int J Heat Mass Transf 2(7):989–1002

15. Ytrehus T (1977) Theory and experiments on gas kinetics in evaporation. In: Potter JL (ed) Rarefied gas dynamics: technical papers selected from the 10th international symposium on rarefied gas dynamics. Snowmass-at-Aspen, CO, July 1976. In: Progress in astronautics and aeronautics, vol 51. American Institute of Aeronautics and Astronautics, pp 1197–1212
16. Aoki K, Sone Y, Yamada T (1990) Numerical analysis of gas flows condensing on its plane condensed phase on the basis of kinetic theory. Phys Fluids 2:1867–1878
17. Gusarov AV, Smurov I (2002) Gas-dynamic boundary conditions of evaporation and condensation: numerical analysis of the Knudsen layer. Phys Fluids 14(12):4242–4255
18. Frezzotti A, Ytrehus T (2006) Kinetic theory study of steady condensation of a polyatomic gas. Phys Fluids 18(2):027101–027101-12.
19. Gusarov AV, Smurov I (2001) Target-vapour interaction and atomic collisions in pulsed laser ablation. J Phys D Appl Phys 34(8):1147–1156
20. Kuznetsova IA, Yushkanov AA, Yalamov YI (1997) Supersonic condensation of monoatomic gas. High Temp 35(2):342–346
21. Kuznetsova IA, Yushkanov AA, Yalamov YI (1997) Intense condensation of molecular gas. Fluid Dyn 6:168–174
22. Vinerean MC, Windfäll A, Bobylev AV (2010) Construction of normal discrete velocity models of the Boltzmann equation. Nuovo Cimento C 33(1):257–264
23. Smirnov VI (1964) A course in higher mathematics. Pergamon Press, Oxford, Addison-Wesley, Reading, Mass.
24. Abramov AA, Kogan MN (1984) The supersonic regime of gas condensation. Akademiia Nauk SSSR, Doklady 278(5):1078–1081 (In Russian)
25. Abramov AA, Butkovskii AV (2008) The effect of the flow-to-wall temperature ratio on strong condensation of gas. High Temp 46(2):229–233
26. Bardos C, Golse F, Sone Y (2006) Half-space problems for the Boltzmann equation. Surv J Stat Phys 124(2–4):275–300
27. Kogan MN, Makashev NK (1971) On the role of Knudsen layer in the theory of heterogeneous reactions and in flows with surface reactions. Izv Akad Nauk SSSR Mekh Zhidk Gaza 6:3–11 (In Russian)
28. Bond M, Struchtrup H (2004) Mean evaporation and condensation coefficient based on energy dependent condensation probability. Phys Rev E 70:061605
29. Zhakhovsky VV, Kryukov AP, Levashov VY, Shishkova IN, Anisimov SI (2018) Mass and heat transfer between evaporation and condensation surfaces: atomistic simulation and solution of Boltzmann kinetic equation. In: Proceedings of the national academy of sciences, Apr 2018. 201714503. https://doi.org/10.1073/pnas.1714503115

Chapter 6
Linear Kinetic Analysis of Evaporation and Condensation

Abbreviations

CPS Condensed-phase surface
DF Distribution function
KL Knudsen layer

Symbols

c Molecular velocity vector
c Molecular velocity
j Mass flux
J_i Molecular flux
f Distribution function
F Temperature factor
M Mach number
m Molecular mass
p Pressure
s Speed ratio
T Temperature
v Thermal velocity
$\mathbf{u}_\infty$ Hydrodynamic velocity vector
u_∞ Hydrodynamic velocity

Greek Letter Symbols

β Condensation coefficient
ρ Density

Subscripts

δ State at mixing surface
w State at condensed-phase surface
∞ State at infinity

Y. B. Zudin, *Non-equilibrium Evaporation and Condensation Processes*,
Mathematical Engineering, https://doi.org/10.1007/978-3-030-67553-0_6

6.1 Introduction

Non-equilibrium evaporation and condensation are important aspects of numerous fundamental and applied problems. Designing heat screens for space vehicles includes simulating the events of the depressurization of the protection shell of nuclear power units. This problem requires calculation of parameters for strong evaporation of coolant during ejection into vacuum [1]. The film boiling heat transfer of superfluid helium is very intensive due to very low thermal resistance, therefore the non-equilibrium effects at the interface become of utmost importance [2]. The contact of hot vapor with cold liquid in the steam volume creates a pulsed wave of rarefaction pressure followed by a pressure jump (and strong condensation) [3].

Computation of non-equilibrium processes during evaporation/condensation requires solving a system of conservation laws for gas in a remote (away from the interface) Navier–Stokes region. The flow in this zone is governed by the thermal velocity $\mathrm{v}_\infty = \sqrt{2k_B T_\infty/m}$ and by the hydrodynamic velocity u_∞ ($u_\infty > 0$ for evaporation, $u_\infty < 0$ for condensation). The intensity of phase transitions is described by the speed ratio

$$s \equiv \frac{u_\infty}{\mathrm{v}_\infty} = \frac{u_\infty}{\sqrt{2k_B T_\infty/m}},$$

which is related to the Mach number[1]

$$\mathrm{M} \equiv u_\infty (5k_B T_\infty/3m)^{-1/2}$$

via the ratio

$$s = \sqrt{\frac{5}{6}}\mathrm{M}.$$

Here, m is the molecular mass, k_B is the Boltzmann constant, and T_∞ is the gas temperature in the Navier-Stokes region.

The equations of the continuous medium become inapplicable for the Knudsen layer (KL), which attaches the interface, this layer has thickness about the molecular free path. Since the KL is out of equilibrium, the concepts of density, temperature, and pressure lack their original phenomenological meaning. The state of gas within the KL is defined by the interaction of opposite molecular flows: the difference between the flow emitted by the condensed-phase surface (CPS) and the flow entering the layer from the gas zone.

The molecular emission from the CPS depends on the surface temperature and independent of the Navier–Stokes region, the spectrum of molecules incident to the interface is formed due to molecular collisions in the remote layers of gas. The

[1] We shall be concerned with the case of single-atom gas.

overlapping of quite different molecular flows creates the discontinuity in the Distribution Function (DF) of molecular velocities within the KL. The discontinuity in the DF declines and becomes monotonically more smooth away from the CPS, it disappears at the outer side of the KL (where the molecular velocity spectrum takes the Maxwellian form). The conjunction conditions for the condensed and gaseous phases can be found by solving the Boltzmann equation that describes the DF in the KL [4].

The accurate solution of this extremely complicated integrodifferential Boltzmann equation was known for special cases with uniform distribution of parameters [5]. For boundary-value problem (the gas fills a half-space and is limited with two surfaces), it seems impossible to exactly solve the Boltzmann equation. Therefore, researchers use the kinetic analysis with approximate analogs of the Boltzmann equation: the relaxation Krook equation [4], the Case model equation [6], the chain of moment equations [7, 8], etc. The specifics of the kinetic analysis resides in the necessity of solving a complex conjugate problem—a macroscopic boundary-value problem for the gas dynamics equations in the region of continuous medium flow (also called the Navier–Stokes region) and a microscopic problem for the Boltzmann equation in the KL.

The theoretical foundation of the study of non-equilibrium processes of evaporation/condensation is the linear kinetic analysis, which describes small deviations of gas-dynamic parameters from their equilibrium level. The linear kinetic theory was based on solving the linearized Boltzmann equation and was developed in the papers [7, 8]. The systematic outline of the results of [7, 8] is available in [9]. The analogues of the Boltzmann equation: the chain of moment equations and the relaxation Krook equation. Recently, the evaporation/condensation problem in the form of linear kinetic tasks was accomplished using the theory of distributions [6] and the methods developed in the theory of functions of complex variable [10]. For the later purposes, it is worth noting that within the linear analysis [7–9], the processes of evaporation and condensation are assumed symmetric, they differ only in the opposite directions of vapor flow.

The mathematical description of non-equilibrium evaporation/condensation is simplified if we skip the task of DF simulation in the KL. In this situation, there is no need to assign the true Boundary Conditions (BC) on the CPS. Instead, we have to define the extrapolated BC for gas-dynamic equations in the Navier–Stokes region. The extrapolation of gas-dynamic parameters to the interface produces kinetic jumps at the interface: the extrapolated values of temperature, density, and pressure of gas are not equal to true values.

The phase transition with the almost sonic velocity of gas flow, depending on the vapor flow direction, is called the strong evaporation (for $u_\infty > 0$) or strong condensation (for $u_\infty < 0$). The analytical study of strong evaporation was initiated in [11]. The author of the present book had approximated the spectrum of molecules approaching the interface using reasonable physical assumptions, the limiting case of strong evaporation was studied when the hydrodynamic velocity equals the sonic velocity (the Mach number equals one). In the following papers [12, 13], the original method of [11] was extended to the entire range of the Mach number. Labuntsov and Kryukov [12, 13] obtained an analytical solution for strong evaporation, which enables a correct limiting transition to the classical linear theory [8, 9]. It is critical to emphasize that Labuntsov and Kryukov [13], and Yano [14] demonstrated the asymmetry of strong evaporation and strong condensation.

At a fixed temperature of CPS, the BC in the Navier–Stokes region calls for only one parameter (the pressure, for example), and for the case of condensation, we need to assign two parameters (pressure and temperature, for example). In this version of analysis [13, 14], there was no restriction on the Mach number while it tends to zero. Therefore, we assume that asymmetry of strong evaporation/condensation should be kept even for the linear approximation of the problem. Thus, we have two quite different approaches for the description of non-equilibrium phase transitions: the symmetric linear [6–9] and the asymmetric nonlinear [12–14] ones. The purpose of this chapter is to analyze the linear problem for evaporation/condensation with account for asymmetry. We apply the analytical "mixing model" proposed earlier in [15].

6.2 Conservation Equations

The subject for the kinetic analysis is the three-dimensional molecular velocity distribution $f = f(\mathbf{c})$, which varies from equilibrium Maxwellian distribution within the Navier–Stokes region

$$f_\infty = \frac{n_\infty}{\pi^{3/2}\mathrm{v}_\infty^3}\exp\left(-\left(\frac{\mathbf{c}-\mathbf{u}_\infty}{\mathrm{v}}\right)^2\right) \tag{6.1}$$

to the discontinuous DF at the CPS

$$c_z > 0 : f_w = f_w^+,\, c_z < 0 : f_w = f_w^-,$$

herein $\mathbf{c}$ and $\mathbf{u}_\infty$ are the vectors of molecular and hydrodynamic velocity.

Let us consider the standard case: the CPS captures completely the input molecular flux, there is no secondary emission in the form of reflected molecules. Thus, the emitted molecular spectrum takes the form of equilibrium semi-Maxwellian distribution for the surface temperature of T_w and known saturated vapor pressure at the same temperature $p_w(T_w)$

$$f_w^+ = \frac{n_w}{\pi^{3/2}\mathrm{v}_w^3}\exp\left(-\left(\frac{\mathbf{c}}{\mathrm{v}_w}\right)^2\right). \tag{6.2}$$

Here, $n_\infty = p_\infty/k_B T_\infty$, $n_w = p_w/k_B T_w$ is the molecular gas density at infinity and the CPS, respectively, c_z is the velocity component normal to the CPS, $\mathrm{v}_w = \sqrt{2k_B T_w/m}$ is the thermal velocity at the CPS.

The particularly remarkable commonly used ratio (6.2) looks like a reasonable physical hypothesis. The paper [16] deals with the spectrum of molecules emitted from the CPS: evaporation to vacuum was simulated using the method of molecular dynamics. It was demonstrated that for the case of low vapor density, the employment of the semi-Maxwellian distribution (6.2) as a BC is a correct assumption.

We consider the problem of evaporation/condensation for a half-space for the vapor steady at infinity (situation of single-atomic gas). For the one-dimensional case, the vector of hydrodynamic velocity $\mathbf{u}_\infty$ degenerates into a scalar velocity for evaporation/condensation u_∞. Under steady conditions, molecular fluxes of mass, momentum, and energy through any surface parallel to the CPS are the same. If we use the BC (6.1), this formulates these fluxes through the flow parameters at infinity, so we obtain the conservation laws for the mass flux

$$\int_{c} mc_z f d\mathbf{c} = \rho_\infty u_\infty, \tag{6.3}$$

the momentum flux

$$\int_{c} mc_z^2 f d\mathbf{c} = \rho_\infty u_\infty^2 + p_\infty, \tag{6.4}$$

and the energy flux

$$\int_{c} mc^2 c_z f d\mathbf{c} = u_\infty\left(\frac{1}{2}\rho_\infty u_\infty^2 + \frac{5}{2}p_\infty\right). \tag{6.5}$$

Here, ρ_∞ is the density, p_∞ is the pressure, u_∞ is the hydrodynamic velocity, the index "∞" denotes the conditions at infinity, $c^2 = c_x^2 + c_y^2 + c_z^2$ is the modulus of molecular velocity squared, c_x and c_y are the projections of the molecular velocity to x- and y-axes (they are oriented in a plane parallel to the CPS).

Integrating the left-hand sides of Eqs. (6.3)–(6.5) is performed for the entire three-dimensional space of molecular velocities: $-\infty < c_x < \infty, -\infty < c_y < \infty, -\infty < c_z < \infty$. To find relations between the flow parameters at infinity (included into right-hand sides of Eqs. (6.3)–(6.5)), it is sufficient to know the distribution function at the CPS. Since we know the positive half of this function f_w^+ from the BC (6.2), the definition of BC requires finding the negative component f_w^-. Let us rewrite the set of Eqs. (6.3)–(6.5) in a more attractive form

$$J_1^+ - J_1^- = \rho_\infty u_\infty, \tag{6.6}$$

$$J_2^+ - J_2^- = \rho_\infty u_\infty^2 + p_\infty, \tag{6.7}$$

$$J_3^+ - J_3^- = \frac{1}{2}\rho_\infty u_\infty^3 + \frac{5}{2}p_\infty u_\infty, \tag{6.8}$$

where J_i^+ and J_i^- are the outcoming and incoming molecular fluxes from to the CPS, $i = 1, 2, 3$.

One can see from Eqs. (6.6)–(6.8) that the disbalance of the molecular mass fluxes ($i = 1$), the momentum fluxes ($i = 2$), and the energy fluxes ($i = 3$) at the CPS (see the left-hand sides of the equations) produces macroscopic flows of evaporation ($u_\infty > 0$ for $J_i^+ > J_i^-$) or condensation ($u_\infty < 0$ for $J_i^+ < J_i^-$). In the Navier–Stokes region (described by the right-hand parts). Disregarding the nonlinear terms in the equations right parts (6.6)–(6.8), we obtain

$$J_1^+ - J_1^- = \rho_\infty u_\infty, \tag{6.9}$$

$$J_2^+ - J_2^- = p_\infty, \tag{6.10}$$

$$J_3^+ - J_3^- = \frac{5}{2} p_\infty u_\infty. \tag{6.11}$$

Here, the symbol "$\infty - \infty$" stands for the outer boundary of the KL, with the Navier–Stokes region behind this boundary.

Here, we have the continuous medium equations. According to the physical model formulated in [15], we introduce an intermediate surface denoted as "$\delta - \delta$" (the mixing surface) with parameters p_δ, T_δ, and u_δ placed between the surfaces "$w - w$" and "$\infty - \infty$". Thus, the KL is split into two subzones—the internal and external ones (Fig. 6.1). Then we write the condition of mass flux conservation between the surfaces "$\delta - \delta$" and "$\infty - \infty$"—"the mixing condition" [15]

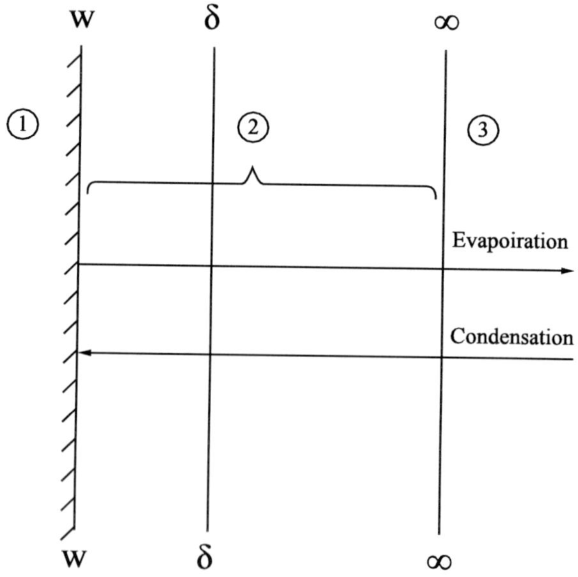

Fig. 6.1 Diagram of mixing model. 1 The condensed-phase, 2 the Knudsen layer, and 3 the Navier–Stokes region

$$\rho_\delta u_\delta = \rho_\infty u_\infty = \text{const.} \tag{6.12}$$

On the surface "$\delta - \delta$", the spectrum of molecules moving toward the interface is described by the DF shifted relative the zero by the magnitude of hydrodynamic velocity u_δ

$$f_\delta^- = \frac{p_\delta}{k_B T_\delta}\left(\frac{m}{2\pi k_B T_\delta}\right)^{3/2} \exp\left(-\frac{m(\boldsymbol{c}_\delta - u_\delta)^2}{2k_B T_\delta}\right). \tag{6.13}$$

Here, $\boldsymbol{c}_\delta$ is the vector of molecular velocity at the mixing surface.

Using the ideal gas equations of state for the surfaces "$w - w$", "$\delta - \delta$", "$\infty - \infty$", we obtain the relation between the thermodynamic parameters

$$\frac{p_w}{p_\infty} = \frac{\rho_w}{\rho_\infty}\frac{T_w}{T_\infty}, \quad \frac{p_\delta}{p_\infty} = \frac{\rho_\delta}{\rho_\infty}\frac{T_\delta}{T_\infty}. \tag{6.14}$$

The expressions for emitting molecular fluxes J_i^+ in the system of Eqs. (6.9)–(6.11) are calculated by the known method through substituting the function $f = f_w^+$ from the BC (6.2) into the sub-integral expressions in the left parts of Eqs. (6.3)–(6.5) [4]

$$\left.\begin{aligned} J_1^+ &= \tfrac{1}{2\sqrt{\pi}}\rho_w \mathrm{v}_w, \\ J_2^+ &= \tfrac{1}{4}\rho_w \mathrm{v}_w^2, \\ J_3^+ &= \tfrac{1}{2\sqrt{\pi}}\rho_w \mathrm{v}_w^3. \end{aligned}\right\} \tag{6.15}$$

The equations for molecular fluxes approaching from the Navier-Stokes region J_i^+ are formulated in the following view

$$\left.\begin{aligned} J_1^- &= \tfrac{1}{2\sqrt{\pi}}\rho_\delta \mathrm{v}_\delta I_1, \\ J_2^- &= \tfrac{1}{4}\rho_\delta \mathrm{v}_\delta^2 I_2, \\ J_3^- &= \tfrac{1}{2\sqrt{\pi}}\rho_\delta \mathrm{v}_\delta^3 I_3. \end{aligned}\right\} \tag{6.16}$$

Here, I_i is the corresponding dimensionless fluxes determined from the integration of the negative semi-Maxwellian distribution (6.13) over the three-dimensional field of molecular velocities, $i = 1, 2, 3$.

The functions $I_i(s_\delta)$ in the general form were presented in [15]. Here, $s_\delta \equiv u_\delta/\mathrm{v}_\delta$ is the speed ratio, $\mathrm{v}_\delta = \sqrt{2k_B T_\delta/m}$ is the thermal velocity, all the values are related to the mixing surface. The specific expressions for I_i in the linearized form are shown below. If there is no phase transition ($s_\delta = s = 0$), this normalization is valid

$$I_1 = I_2 = I_3 = 1. \tag{6.17}$$

In view of (6.14)–(6.16), the system of Eqs. (6.9)–(6.11) is represented in the form

$$\frac{p_w}{p_\infty}\sqrt{\frac{T_\infty}{T_w}} - \frac{p_\delta}{p_\infty}\sqrt{\frac{T_\infty}{T_\delta}} I_1 = 2\sqrt{\pi} s, \tag{6.18}$$

$$\frac{p_w}{p_\infty} + \frac{p_\delta}{p_\infty} I_2 = 2, \tag{6.19}$$

$$\frac{p_w}{p_\infty}\sqrt{\frac{T_w}{T_\infty}} - \frac{p_\delta}{p_\infty}\sqrt{\frac{T_\delta}{T_\infty}} I_3 = \frac{5\sqrt{\pi}}{2} s. \tag{6.20}$$

6.3 Equilibrium Coupling Conditions

Here we consider the case of phase equilibrium with no phase transition ($s = 0$). Accordingly, the set of Eqs. (6.18)–(6.20) with account for conditions (6.17) gives the expressions

$$\frac{p_w^0}{p_\delta^0} = \sqrt{\frac{T_w^0}{T_\delta^0}}, \ \frac{p_w^0}{p_\infty^0} + \frac{p_\delta^0}{p_\infty^0} = 2, \ \frac{p_w^0}{p_\delta^0} = \sqrt{\frac{T_\delta^0}{T_w^0}}.$$

This produces the equilibrium conditions for matching of the condensed and gaseous phases: the isobaric condition for pressure

$$p_w^0 = p_\delta^0 = p_\infty^0, \tag{6.21}$$

and the isothermal condition for temperature

$$T_w^0 = T_\delta^0. \tag{6.22}$$

Here, the superscript "0" stands for the equilibrium.

The isobaric condition for the KL (6.21) is physically obvious: the thermodynamic equilibrium inside the gaseous zone denies any steady jump in pressure. Meanwhile, the isothermal condition (6.22) is valid not for the entire KL, but only for the inner area which is limited by surfaces denoted as "$w - w$" and "$\delta - \delta$". This means that for the equilibrium situation, the temperature field (in a general situation) may have a discontinuity. This discontinuity can be characterized by (non-unit) "temperature factor" (Fig. 6.2)

$$F \equiv T_\infty^0 / T_w^0 \neq 1.$$

We emphasize that the discontinuity in the temperature field occurs deeply in the KL: macroscopic laws do not work here. Indeed, for the solid medium, condition

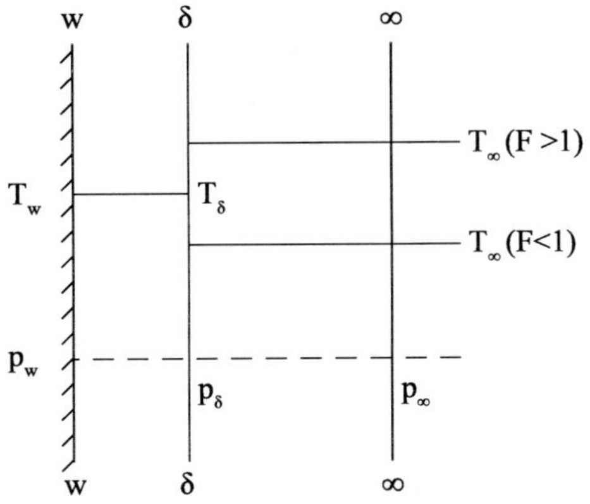

Fig. 6.2 Distributions for temperature and pressure inside the Knudsen layer

(6.22) cannot be satisfied: the Fourier law states that the discontinuity in the one-dimensional temperature distribution is smoothed due to unsteady thermal diffusion [17]. Thus, the mixing model tells us that the gas state at the surface "$\delta - \delta$" is not uniform. On one side, here we can apply the macroscopic concepts—the mass flux (6.12) and the equation of ideal gas (6.14). On the other hand, the gradient law of Fourier is not valid at the mixing surface (this means that the temperature field discontinuity is possible). This combination of macroscopic and microscopic properties allows one to use the temperature factor F as a BC both for the situation of phase equilibrium ($s = 0$), and for the situation of phase transition ($s \neq 0$).

6.4 Linear Kinetic Analysis

6.4.1 Linearized System of Equations

We linearize the set of Eqs. (6.18)–(6.20) over a small parameter $|s| << 1$ According to [7, 8], the analysis of linear processes of evaporation/condensation allows one to arbitrarily choose basic values for counting the parameter jumps. For simplicity, we take the fixed values of pressure and temperature at the outer margin of the KL: $p_\infty \equiv p^0_\infty = \text{const}$, $T_\infty \equiv T^0_\infty = \text{const}$. From these values, the expression of the temperature factor is written in the form

$$F \equiv \frac{T_\infty}{T^0_w}. \tag{6.23}$$

The functions I_i in the right-hand sides of (6.16) are the dimensionless molecular fluxes directed to the CPS [12, 13]. The linearization of dimensionless fluxes over the parameter $|s_\delta|$ gives the following set of parameters

$$\left.\begin{aligned} I_1 &= 1 - \sqrt{\pi} s_\delta, \\ I_2 &= 1 - \tfrac{4}{\sqrt{\pi}} s_\delta, \\ I_3 &= 1 - \tfrac{5\sqrt{\pi}}{4} s_\delta. \end{aligned}\right\}$$

With the mixing condition (6.12), the equation of state (6.14) and the equilibrium condition for matching (6.21) and (6.22) gives the ratio

$$\frac{|s_\delta|}{|s|} = \frac{s_\delta}{s} \equiv \frac{u_\delta / \mathrm{v}_\delta}{u_\infty / \mathrm{v}_\infty} = \frac{\rho_\infty}{\rho_\delta}\sqrt{\frac{T_\infty}{T_\delta}} = \frac{p_\infty}{p_\delta}\sqrt{\frac{T_\delta}{T_\infty}}.$$

Since the values $|s_\delta| \ll 1$ and $|s| \ll 1$ are proportional to the linear deviations of these parameters from equilibrium, we can use $p_\delta \cong p_\delta^0$ and $T_\delta \cong T_\delta^0$ in the last chain of ratios. In turn, using (6.21) we have $p_\delta^0 = p_\infty^0 \equiv p_\infty$. This produces a relation between the speed ratio at the two surfaces "$\delta - \delta$" and "$\infty - \infty$"

$$\frac{s_\delta}{s} = \sqrt{\frac{T_w^0}{T_\infty}} \equiv G^{-1},$$

where $G \equiv \sqrt{F} = \sqrt{T_\infty / T_w^0}$. Now we present the values of the pressures p_δ and p_w and the temperatures T_δ and T_w in the new linearized form

$$p_\delta = p_\infty(1 + \eta_\delta s),\ p_w = p_\infty(1 + \eta_w s),\ T_\delta = T_\infty F^{-1}(1 + \tau_\delta s),\ T_w = T_\infty F^{-1}(1 + \tau_w s).$$

The set of Eqs. (6.18)–(6.20) after linearization transforms into

$$\eta_w - \eta_\delta + \frac{1}{2}(\tau_\delta - \tau_w) = \sqrt{\pi} G^{-1}, \tag{6.24}$$

$$\eta_w + \eta_\delta = \frac{4}{\sqrt{\pi}} G^{-1}, \tag{6.25}$$

$$\eta_w - \eta_\delta + \frac{1}{2}(\tau_w - \tau_\delta) = \frac{5\sqrt{\pi}}{2} G - \frac{5\sqrt{\pi}}{4} G^{-1}. \tag{6.26}$$

From (6.24)–(6.26), we obtain the ratios for coefficients of kinetic jumps at surfaces "$w - w$" and "$\delta - \delta$"

$$\eta_w = \frac{5\sqrt{\pi}}{8} G + \left(\frac{2}{\sqrt{\pi}} - \frac{\sqrt{\pi}}{16}\right) G^{-1}, \tag{6.27}$$

$$\eta_\delta = -\frac{5\sqrt{\pi}}{8} G + \left(\frac{2}{\sqrt{\pi}} + \frac{\sqrt{\pi}}{16}\right) G^{-1}, \tag{6.28}$$

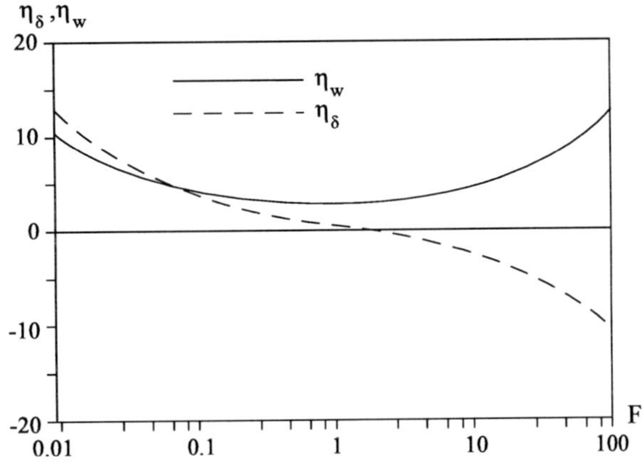

Fig. 6.3 Pressure jump coefficient on the condensed-phase surface and on the mixing surface as a function of the temperature factor

$$\tau_w - \tau_\delta = \frac{5\sqrt{\pi}}{2}G - \frac{9\sqrt{\pi}}{4}G^{-1}. \tag{6.29}$$

From the point of view of applications, the most interesting are the jump coefficients for the pressure and the temperature at the CPS η_w, τ_w and the corresponding values at the mixing surface η_δ, τ_δ, which are internal model parameters.[2] The system of three Eqs. (6.27)–(6.29) comprises the four unknowns η_w, η_δ, τ_w, τ_δ. The first two unknowns are obtained from relations (6.27), (6.28). Figure 6.3 demonstrates the coefficient of pressure jumps versus the temperature factor. The function $\eta_w(F)$ has a minimum with coordinates $F_{\min}$. This dependence on the mixing surface $\eta_\delta(F)$ is the sign-alternating one (Fig. 6.3): this function is positive at small values of temperature factor and negative at high values of temperature factor: it passes the zero at $F = 1.119$.

6.4.2 *Symmetric and Asymmetric Cases*

It is worth noting that formulas (6.27) and (6.28) for pressure jumps are symmetric: they both apply to process of evaporation ($s > 0$) and to condensation ($s < 0$). For the next stage of analysis—finding the coefficient of the temperature jump—we will observe the asymmetry for evaporation/condensation. For the problem of condensation, the values of temperature in cross sections "$w - w$" and "$\infty - \infty$" are strictly related via the temperature factor (6.23). Thus, at $T_\infty =$ const, we have $T_w = T_w^0 =$ const and equality (6.23) is rewritten in the form $F = T_\infty / T_w =$ const. This means that the CPS has no linear temperature jump: $\tau_w = 0$ In turn, the coefficient of temperature jump at the mixing surface is described by the relation

[2]Herein the concept of "jump" is used only referring to the desired jumps on the condensed-phase surface.

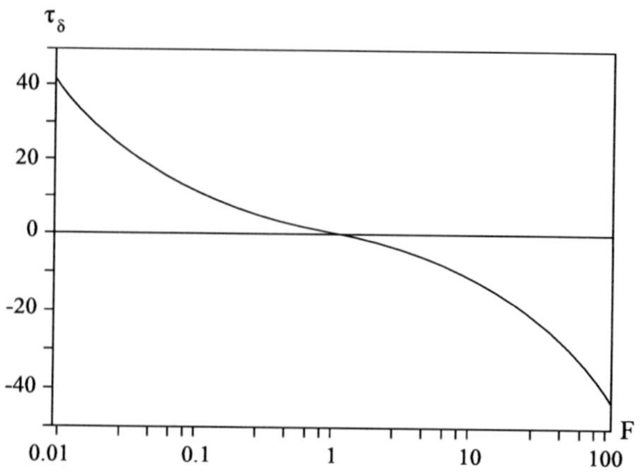

Fig. 6.4 Dependence of temperature jump coefficient on the mixing surface on the temperature factor

$$\tau_\delta = \frac{5\sqrt{\pi}}{2}G + \frac{9\sqrt{\pi}}{4}G^{-1}. \tag{6.30}$$

As one can see in Fig. 6.4, the function $\tau_\delta(F))$ is similar to the function $\eta_\delta(F)$ (the dashed line in Fig. 6.3): this function is positive at low values and negative at high values of temperature factor and crosses the zero level at $F = 0.9$. As known from [15, 18], the condensation problem is divided into two separate tasks: (a) the "anomalous" condensation of cold vapor on a hot surface (the area $F < 1$), (b) the "normal" condensation of hot vapor on a cold surface (area $F > 1$). It is interesting to note that the margins of the anomalous and normal areas ($F = 1$) almost coincide with the coordinate of minimum of dependence $\eta_w(F)$, as calculated from relation (6.27) (solid line in Fig. 6.3).

Thus, the linear problem of condensation is *asymmetric*.

- The temperatures in the cross sections "$w - w$" and "$\infty - \infty$" are related through the temperature factor (6.23), which takes any values in the range $0 < F < \infty$.
- The condition $F = \text{const}$ gives the condition $\tau_w \equiv 0$.
- The desired parameter is the pressure jump that depends on F and becomes zero for the condition of equilibrium: $\eta_w \to 0$ as $s \to 0$.
- For the case of phase equilibrium inside the KL, the isobaric condition is fulfilled, but this is not true for isothermicity (asymmetry property).

In contrast to the above situation, the problem of evaporation is *symmetric*.

- The desired parameters are jumps of pressure and temperature, meanwhile $\eta_w \to 0$, $\tau_w \to 0$ as $s \to 0$.
- The assigning of the temperature factor F as the incoming parameter is out of the problem statement.
- The values η_w and τ_w are independent of the temperature factor, and therefore, we take this factor as equivalent to unit: $F = 1$.

- For the situation of phase equilibrium inside the KL, the conditions of isobaric and isothermal behavior take place (property of symmetry).

Unlike the condensation problem, evaporation is not a closed problem. Thus, we need to find the fourth variable (the temperature jump), and this requires an additional condition to the set of Eqs. (6.27)–(6.29). Within this model, this closing relation is difficult to find, so we are satisfied with semi-empirical reasoning derived from the structure of Eq. (6.29). The definition of the problem of condensation gives us the condition $\tau_w = 0$. Thus, the plausible hypothesis for the evaporation problem is an alternative condition: $\tau_\delta = 0$. Thus, we assume in (6.27)–(6.29), $G = F = 1$, and this gives us the relation for jumps in pressure and temperature

$$\eta_w = 2.125,\ \tau_w = 0.4431. \tag{6.31}$$

The coefficient for the pressure jump at the mixing surface is the following: $\eta_\delta = 0.1314$.

6.4.3 Kinetic Jumps

The fundamental study [8] presented calculations for kinetic-type jumps for a linear problem of evaporation. There were considered 12 possible variants of a solution: 5 of them were based on a chain of moment equations and 7 variants were based on the relaxation Krook equation. The author of [9] had offered the "best" variant of solutions from [8]

$$\eta_w = 2.13,\ \tau_w = 0.454. \tag{6.32}$$

Relations (6.32) are taken from the later monograph [9]. As one can see from the comparison of (6.31) and (6.32), their maximal relative difference is only 2.4%. Now we present our solutions for kinetic jumps (in a more detailed form). The problem of evaporation has the following relations

$$u_\infty > 0,\ \frac{p_w - p_\infty}{p_\infty} = 2.125\frac{u_\infty}{\mathrm{v}},\ \frac{T_w - T_\infty}{T_\infty} = 0.4431\frac{u_\infty}{\mathrm{v}_\infty}. \tag{6.33}$$

Using (6.33) one finds that: $p_w > p_\infty$, $T_w > T_\infty$, i.e., the pressure and temperature for the vapor emitted from the CPS is higher than those parameters in the Navier–Stokes region. The problem of condensation has the following relations

$$u_\infty < 0,\ \frac{p_w - p_\infty}{p_\infty} = \left(1.108\sqrt{\frac{T_\infty}{T_w}} + 1.018\sqrt{\frac{T_w}{T_\infty}}\right)\frac{u_\infty}{\mathrm{v}_\infty}. \tag{6.34}$$

According to (6.34), $p_w < p_\infty$, and this means that vapor absorbed by the interface has the pressure lower than the incoming vapor from the Navier–Stokes region. And the vapor temperature at infinity distance may be either higher than the temperature of CPS ($T_\infty > T_w$, normal condensation) or lower ($T_\infty < T_w$, anomalous condensation). If the temperatures are equal $T_w \approx T_\infty$, the pressure difference $p_w - p_\infty$ reaches a minimum. The essential result of our chapter is relation (6.34) describing the dependence of the pressure jump on the temperature factor for the condensation problem. Up to our best knowledge, the literature for today has no accurate solution to the linear problem of condensation that would be similar to the problem of evaporation [8, 9]. Therefore, the validation of solution (6.34) was carried out via comparison with the simulation results for strong condensation [18]. Gusarov and Smurov [18] calculated the pressure ratio $\tilde{p}_w \equiv p_w/p_\infty$ for three values of the temperature factor: $F = 0.1, 0.2, 0.5, 1, 4$, and 10. Each of the calculated regimes comprised from 7 to 9 calculation points in the interval $0 < |\mathrm{M}| < 1$.

Figure 6.5 presents the dependence $\tilde{p}_w(|\mathrm{M}|)$, which was obtained for the case $F = 0.2$ [18]. As for determining the pressure jumps from numerical results of [18], the procedure was as follows. As the base point, we take the initial point with coordinates $\tilde{p}_w|_{\mathrm{M}=0} = 1$, which correspond to the equilibrium state $s = 0$. Then we add two nearest points with the lowest value of $|\mathrm{M}|$. These three points give the straight line $\tilde{p}_w = 1 - \alpha|\mathrm{M}|$. By expressing the Mach number $\mathrm{M} \equiv u_\infty(5k_B T_\infty/3m)^{-1/2}$ in terms of the speed ratio $s \equiv u_\infty(2k_B T_\infty/m)^{-1/2}$, we obtain $|\mathrm{M}| = \sqrt{6/5}|s|$. The linearized relation for the dimensionless pressure at the CPS gives the ratio $\tilde{p}_w = 1 + \eta_w s = 1 - \eta_w|s|$. Comparing the two last expressions for $\tilde{p}_w$ produces the coefficient of linear pressure jump, which is extrapolated to $|\mathrm{M}| = 0$: $\eta_w = \sqrt{6/5}\alpha$. For the chosen regime with $F = 0.2$ (Fig. 6.5), we obtain $\eta_w = 2.77$. Note here that for the regimes with $F = 1$ and 10 considered in [18], the second non-zero point was calculated for $|\mathrm{M}| > 0.1$. Herein, the error for extrapolation from three points becomes too high. Therefore, we took a linear extrapolation for two points—for the basic initial point and the first calculated point.

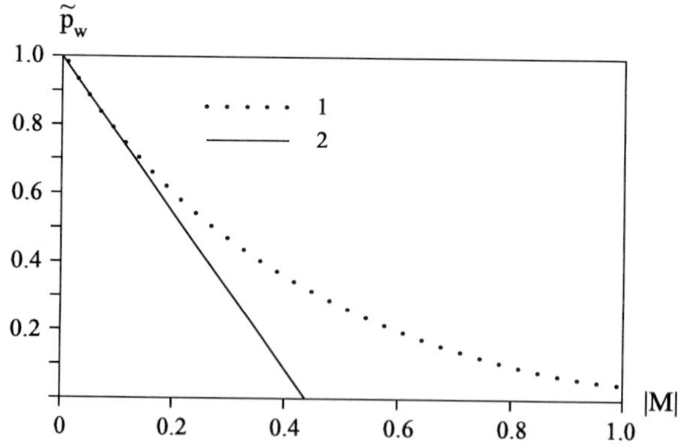

Fig. 6.5 Linear jumps of pressure for the condensation problem. 1 Simulation data from [18], 2 linear approximation

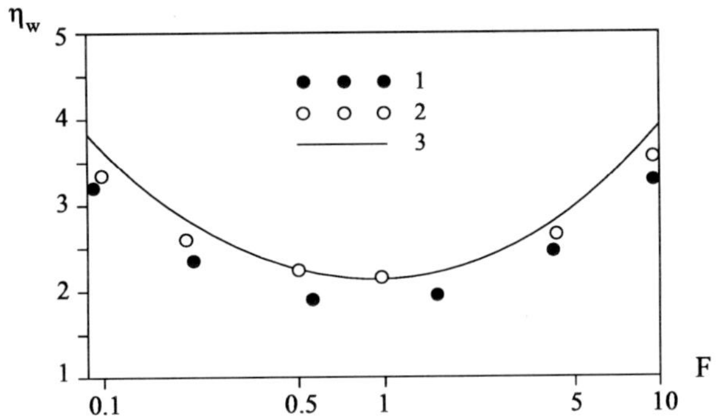

Fig. 6.6 Simulation results from [18]. 1 Simple extrapolation, 2 refined extrapolation, 3 calculation by formula (6.27)

The so obtained analytical formula $\eta_w(F)$ is in qualitative compliance with the numerical results (total 6 points), and this reproduces the minimum at $F \approx 1$ (Fig. 6.6). However, the quantitative agreement between the two approaches is much worse: the maximum for the relative deviation is $\approx$16%. As one can see from Fig. 6.6, this type of "simple" extrapolation produces a systematic deviation between the theory and simulation. We can use the "improved" extrapolation by using a specific feature of the mixing model: for $F = 1$, the coefficients of pressure jump η_w for the problems of evaporation and condensation are the same and equal 2.125 (see the first formula (6.31)). Note also that this value almost coincides with the results of accurate formulation of linear theory of evaporation [8, 9]: η_w (see the first formula (6.32)). Thus, we take the value $\eta_w|_{F=1} = 2.125$ as an exact solution of the linear problem of condensation. Next, we compare this with the extrapolated value $\eta_w = 1.95$ [18] and estimate the systematic error for the situation of "simple" approximation as $\approx$9%. Multiplying of all extrapolated values η_w by the increasing coefficient 1.09, we obtain the refined extrapolation. This procedure reduces the deviation between the analytical and simulation curves to a reasonable level of $\approx$8% (Fig. 6.6), and the difference still looks like a systematic error.

6.4.4 *Effect of Condensation Coefficient*

The previous analysis was concerned with the limit case when all the molecules incident on the surface are completely absorbed by it. Assume now that not all the flow incoming from the liquid to the gaseous phase is passed through the surface, but only a part of it. In turn, assume that only the part β of the flow of the molecules J^- incident from the Navier–Stokes region on the CPS is captured by it (the remaining part $(1-\beta)J^-$ is reflected from it.[3] This part is determined by the "condensation coefficient". The quantity β, which reflects the state of the surface and the physical nature of the condensed-phase, may, in general, vary in the range $0 < \beta \leq 1$. Then

[3]Below for convenience we shall drop the subscript and write: $J_1^+ \equiv J^+$, $J_1^- \equiv J^-$.

the total molecular flow outgoing from the surface is

$$J_\beta^+ = \beta J^+ + (1-\beta)J^-, \tag{6.35}$$

$i = 1, 2, 3$. From (6.35), it is possible to find the macroscopic flows J_∞ in the Navier–Stokes region, which are defined as the difference of the emitted and incident molecular flows

$$J_\infty = J_\beta^+ - J^- = \beta\left(J^+ - J^-\right). \tag{6.36}$$

Let us introduce the "permeability coefficients" as the ratio of the flow outgoing from the CPS (the case $\beta < 1$) and the flow emitted by an absolutely permeable CPS (the case $\beta = 1$)

$$\psi = \frac{J_\beta^+}{J^+}. \tag{6.37}$$

Hence, using (6.35), (6.36) in (6.37) gives

$$\psi = 1 - \frac{1-\beta}{\beta}\frac{J_\infty}{J^+}. \tag{6.38}$$

In the linear approximation, we have $\rho_w = \rho_\infty$, $\mathrm{v}_w = \mathrm{v}_\infty$. Hence, in view of (6.38)

$$\psi = 1 - 2\sqrt{\pi}\frac{1-\beta}{\beta}\frac{s}{\sqrt{F}}. \tag{6.39}$$

So, for the diffuse reflection the density ρ_w of the emitted molecular flow should be replaced by $\rho_w^* = \rho_w \psi$, where the factor $\psi < 1$ is given in (6.39). Imitating the earlier discussion of the "reference" case $\beta = 1$, we get the generalized relations for the kinetic pressure jumps of pressure and temperature on the CPS.

Evaporation problem. The temperature jump is independent of β and is determined by the formula

$$\tilde{\tau}_w = 0.454. \tag{6.40}$$

The dependence of the pressure jump on the condensation coefficient is described by the relation

$$\tilde{\eta}_w = 2\sqrt{\pi}\frac{1-0.4\beta}{\beta}. \tag{6.41}$$

The resulting expressions for the kinetic jumps in the evaporation problem are practically equal to those obtained in the framework of the classical linear theory [8, 9]

$$\tilde{\eta}_w = 2\sqrt{\pi}\frac{1 - 0.399\beta}{\beta}, \ \tilde{\tau}_w = 0.443. \tag{6.42}$$

Condensation problem: $\tilde{\tau}_w = 0$. The pressure jump is determined by the relation

$$\tilde{\eta}_w = \left(\frac{2}{\sqrt{\pi}} - \frac{\sqrt{\pi}}{16}\right)F^{-1/2} + \frac{5\sqrt{\pi}}{8}F^{1/2} + B, \tag{6.43}$$

where $B = 2\sqrt{\pi}\frac{1-\beta}{\beta}$.

Figure 6.7 shows the dependences $\tilde{\eta}_w(F)$ for various values of the condensation coefficient. The characteristic of these dependencies is the presence of a minimum, whose coordinates depend on β. The figure shows that $\tilde{\eta}_w$ increases monotonically with decreasing β. Besides, the values of the temperature factor $F_{\min}$, for which the curves $\tilde{\eta}_w(F)$ pass through the minimum, move more and more inside the normal condensation domain ($F > 1$). The dependencies $F_{\min}(\beta)$, $\tilde{\eta}_{w\min}(\beta)$ are shown in Figs. 6.8 and 6.9. It is worth pointing out that according to (6.43) the minimum of the dependence $\tilde{\eta}_w(\beta)$ in the "equilibrium case" ($T_w = T_\infty$, $F = 1$) is moved from the point $\beta = 1$

$$\beta_{\min}|_{F=1} \approx 0.975. \tag{6.44}$$

Relation (6.44) is another manifestation of the asymmetry of processes of evaporation and condensation. For the reference case ($\beta = 1$, $B = 0$), relation (6.43) specializes in formula (6.27). It is interesting to note that the jumps of pressure and temperature on the mixing surface are independent of the condensation coefficient and are described by the previous relations (6.28), (6.29).

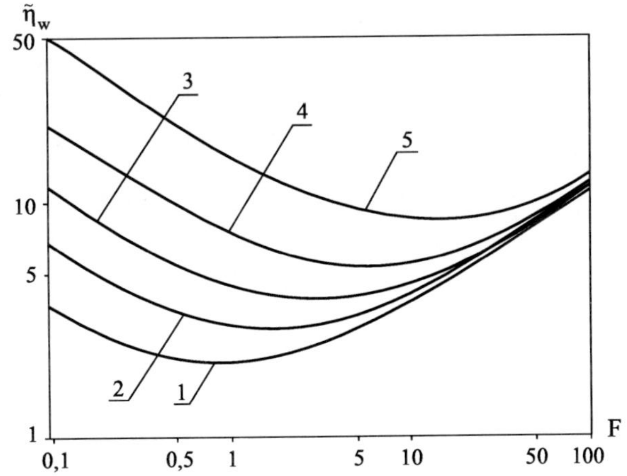

Fig. 6.7 Pressure jump versus the temperature factor. 1 $\beta = 1$, 2 $\beta = 0.8$, 3 $\beta = 0.6$, 4 $\beta = 0.4$, 5 $\beta = 0.2$

Fig. 6.8 Dependence $F_{\min}(\beta)$. **a** $10^{-3} \leq \beta \leq 1$, **b** $\beta \sim 1$

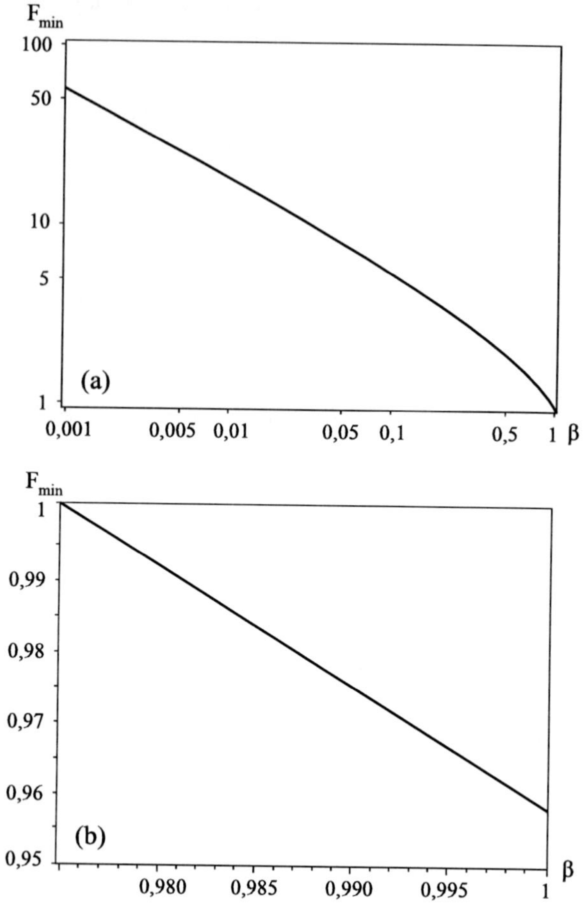

Fig. 6.9 Dependence $\tilde{\eta}_{w\min}(\beta)$

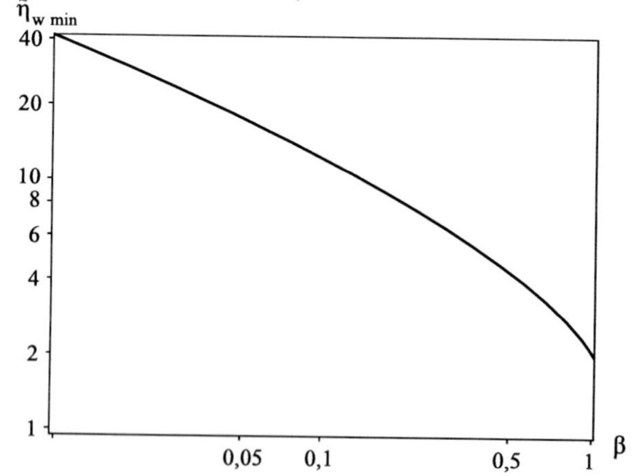

6.4.5 *Short Description*

Now we give a short description of the used algorithm within the mixing model framework [15].

- For the mixing surface (within the KL), we define the DF for the molecule flux moving to the interface, this distribution is semi-Maxwellian shifted due to the flow of evaporation/condensation.
- The system of equations describing the conservation laws for the mass flux (6.3), the momentum flux (6.4), the energy flux (6.5), augmented with the mixing condition (6.12), and this set after linearization transforms into the set of Eqs. (6.27)–(6.29).
- The pressure jump coefficients are given by Eqs. (6.27) and (6.28). The defining of temperature jump coefficient requires the additional condition for Eq. (6.29).
- By equating to zero the temperature jump coefficient at the CPS (this is merely a definition), we formulate the condensation problem. For this situation, the temperature factor is taken as a BC.
- By equating to zero the temperature jump coefficient at the mixing surface (a reasonable hypothesis), we formulated the evaporation problem. This makes the temperature factor equal to one.

In general, the use of the mixing model [15] gives a compliant description of linear processes of evaporation/condensation with an account of asymmetry of those processes. The model is modified by building of DF for molecular flux moving toward the CPS using the certain cross sections inside the KL. This enables a constriction of several mixing surfaces with a certain step until we reach the interface. As a result, one can calculate the integral characteristics (for example, kinetic jumps of parameters) and ascertain a certain type of information about the behavior of DF inside the KL. The mixing model was developed in the papers of the author of the present book [15, 19, 20].

6.5 Conclusions

The previously developed "mixing model" laid the foundation for the analysis of linear kinetic problems of evaporation and condensation. Using this model, we introduced the "mixing surface" within the Knudsen layer: here the state of gas is of a mixed nature. On the one hand, we used macroscopic concepts: the mass flux and the ideal gas equation. On the other hand, the mixing surface may have a discontinuity of the steady temperature field. This combination of microscopic and macroscopic properties of gas within the Knudsen layer gives the final formulas describing both the evaporation (one boundary condition is assigned) and the condensation (two boundary conditions). In doing so, the asymmetry of evaporation/condensation is proved for the linear asymptotic. The expressions for pressure and temperature jumps were obtained for the evaporation problem: these results almost coincide with those of the classical linear theory. The key result of this research is an analytical dependence of the pressure jump on the temperature factor (the condensation problem). It was demonstrated that this dependence has a minimum near the margin between the anomalous and normal regimes of condensation. The directions for further development of this analytical model were proposed. The previous results are extended to the case of diffusion reflection of molecules from the phase boundary.

References

1. Larina IN, Rykov VA, Shakhov EM (1996) Evaporation from a surface and vapor flow through a plane channel into a vacuum. Fluid Dyn 1:127–133
2. Kryukov AP, Yastrebov AK (2003) Analysis of the transfer processes in a vapor film during the injection of a highly heated body with a cold liquid. High Temp 41(5):680–687
3. Lezhnin SI, Kachulin DI (2013) The various factors influence on the shape of the pressure pulse at the liquid-vapor contact. J Eng Thermophys 22(1):69–76
4. Kogan MN (1969) Rarefied gas dynamics. Plenum, New York
5. Bobylev AV (1987) Accurrate and approximate methods in the theory of the Boltzmann and Landau nonlinear kinetic equations. Keldysh Institute Preprints, Moscow (In Russian)
6. Latyshev AV, Yushkanov AA (2008) Analytical methods in kinetic theory. Moscow State Regional University, Moscow (In Russian)
7. Labuntsov DA (1967) An analysis of the processes of evaporation and condensation. High Temp 5(4):579–647
8. Muratova TM, Labuntsov DA (1969) Kinetic analysis of the processes of evaporation and condensation. High Temp 7(5):959–967
9. Labuntsov DA (2000) Physical foundations of power engineering. Selected works, Moscow Power Energetic Univ (Publ), Moscow (In Russian)
10. Siewert CE (2003) Heat transfer and evaporation/condensation problems based on the linearized Boltzmann equation. Eur J Mech B Fluids 22:391–408
11. Anisimov SI (1968) Vaporization of metal absorbing laser radiation. Sov Phys JETP 27(1):182–183
12. Labuntsov DA, Kryukov AP (1977) Processes of intense evaporation. Therm Eng 4:8–11
13. Labuntsov DA, Kryukov AP (1979) Analysis of intensive evaporation and condensation. Int J Heat Mass Transf 2(7):989–1002
14. Yano T (2008) Half-space problem for gas flows with evaporation or condensation on a planar interface with a general boundary condition. Fluid Dyn Res 40(7–8):474–484
15. Zudin YB (2015) The approximate kinetic analysis of strong condensation. Thermophys Aeromech 22(1):73–84
16. Zhakhovskii VV, Anisimov SI (1997) Molecular-dynamics simulation of evaporation of a liquid. J Exp Theor Phys 84(4):734–745
17. Carslaw HS, Jaeger JC (1986) Conduction of heat in solids. Clarendon, London
18. Gusarov AV, Smurov I (2002) Gas-dynamic boundary conditions of evaporation and condensation: numerical analysis of the Knudsen layer. Phys Fluids 14:4242–4255
19. Zudin YB (2015) Approximate kinetic analysis of intense evaporation. J Eng Phys Thermophys 88(4):1015–1022
20. Zudin YB (2016) Linear kinetic analysis of evaporation and condensation. Thermophys Aeromech 23(3):437–449

Chapter 7
Binary Schemes of Vapor Bubble Growth

Symbols

α	Thermal diffusivity
c_p	Isobaric heat capacity
Ja	Jakob number
k	Thermal conductivity
m	Growth modulus
p	Pressure
q	Heat flux
R	Bubble radius
L	Heat of phase transition
S	Stefan number
T	Temperature
t	Time

Greek Letter Symbols

β	Evaporation–Condensation coefficient
μ	Dynamic viscosity
ν	Kinematic viscosity
ρ	Density
$\mathbb{R}$	Thermal resistance

Subscripts

b	Vapor bubble
e	State at energy spinodal
l	Liquid
max	Maximum
min	Minimum
v	Vapor
s	Saturation state
∞	State at infinity

Y. B. Zudin, *Non-equilibrium Evaporation and Condensation Processes*,
Mathematical Engineering, https://doi.org/10.1007/978-3-030-67553-0_7

$*$ State at blocking point

7.1 Introduction

In applications related to the physics of boiling, one has to know the dependence of the bubble growth rate at a heated surface on the thermophysical properties of a liquid and vapor, capillary, viscous, and inertial forces, as well as on the kinetic molecular laws operating at an interface [1]. The problem of bubble growth in the rigorous formulation is described by partial differential equations taken separately for liquid and vapor phases and supplemented with compatibility conditions at the interface. In the general case, the solution of such multiparametric problem can be only numerical. At the same time, to model the problems of the physics of boiling, approximate analytical solutions for the growth of a bubble are needed to find the general laws governing the influence of various parameters.

The basis of the physical modeling of the process of boiling is an idealized problem on a spherically symmetric growth of a vapor bubble in an infinite volume of a uniformly superheated liquid [2]. The first reference to the dynamics of a vapor cavity in a liquid was made in 1859 in [3]. The problem of bubble growth in a superheated liquid was extensively studied in experiments and theory; see the surveys in [4, 5]. Nevertheless, within the framework of this thoroughly researched problem, there are "blank pages" of great practical importance that were never considered before. So, we shall be concerned with the problem of the spherically symmetric growth of a vapor bubble in an infinite volume of a uniformly superheated liquid. A description of the limiting schemes of bubble growth according to the Labuntsov classification is given, supplemented with regard to the energetic thermal scheme. The attempts involving the "universal approximation" of the classical Scriven solution were shown to be futile.

7.2 Limiting Schemes of Growth

Labuntsov [6] was the first to suggest considering the influence of each effect within the framework of the following four limiting schemes (Fig. 7.1).

(a) dynamic viscous scheme,
(b) dynamic inertial scheme,
(c) energetic kinetic molecular (non-equilibrium) scheme,
(d) energetic thermal scheme.

The real picture of a bubble growth is determined in the general case by the total effect of the enumerated factors. With the effect of two (or several) factors, the bubble growth rate will always be lower than the least of the values calculated within

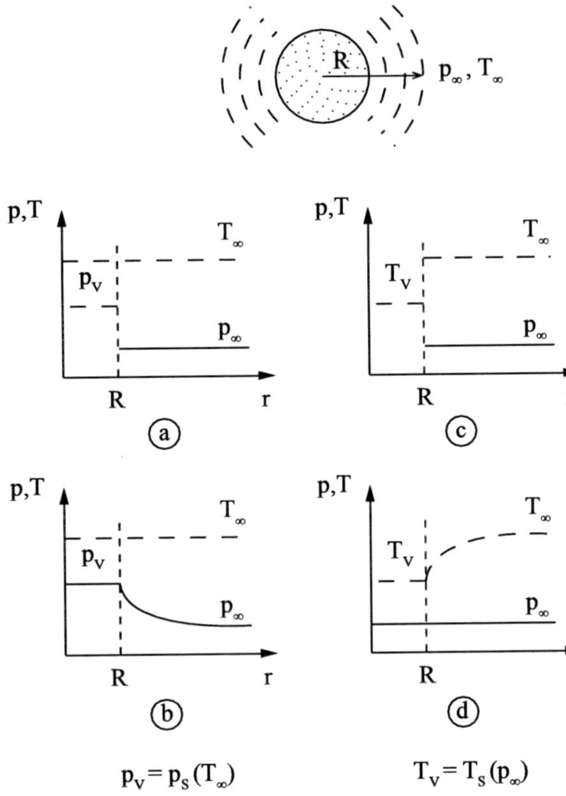

Fig. 7.1 Limiting schemes of vapor bubble growth. **a** Dynamic viscous scheme, **b** dynamic inertial scheme, **c** energetic molecular-kinetic scheme, and **d** energetic thermal scheme

the framework of limiting schemes. Analytical solutions for the bubble growth rate within the framework of each scheme are presented in [6]. The basic relation for the process of vapor bubble growth in a superheated liquid is the heat balance equation

$$q = \rho_V L \dot{R}. \tag{7.1}$$

In the overwhelming majority of cases, the bubble growth rate $\dot{R} = dR/dt$ is governed by the response of a liquid medium to the spherical expansion of a bubble (dynamic effects) and by the evaporation intensity on its boundary (energetic effects). The dynamic effects may be of a viscous and inertial nature. The energetic effects are determined by the non-equilibrium evaporation and by the mechanism of heat supply to the interface. The systematic approach suggested by Labuntsov allows one to find the relative contribution of each effect and also to single out the main factor that limits the bubble growth. Following [6], we will briefly consider the limiting schemes of bubble growth.

The dynamic viscous scheme [6] describes the case where the difference between the vapor pressure in a bubble and the liquid pressure away from the bubble are balanced by the normal component of the tensor of viscous stresses at the interface on the liquid phase side (Fig. 7.1a)

$$\Delta p = 4\frac{\mu \dot{R}}{R}. \tag{7.2}$$

Here, $\Delta p = p_v - p_\infty$, $p_v = p_s(T_\infty)$, is the saturation pressure at a temperature in the bulk of the liquid. Such scheme determines the growth rate of a bubble of a very small radius in a very viscous liquid. Here, the boundary of the Newtonian liquid region can be reached within the framework of which the linear coupling between the tensor of deformation rates and stress tensor is valid.

The dynamic inertial scheme [7] describes the growth of a bubble at a constant difference of pressures in the bubble and in the surrounding liquid ($p_v - p_\infty = \text{const}$) due to the inertial response of the liquid to its spherical repulsion (Fig. 7.1b). The temperature in both phases is considered constant and equal to T_∞. The velocity of the radial motion of liquid is defined by the Rayleigh equation [4, 5]

$$\frac{\Delta p}{\rho} = \frac{3}{2}\dot{R}^2 + R\ddot{R}, \tag{7.3}$$

where $\ddot{R} = d^2R/dt^2$. At $\Delta p = \text{const}$, Eq. (7.3) yields the classical Rayleigh formula

$$R = \sqrt{\frac{2}{3}\frac{\Delta p}{\rho}}t. \tag{7.4}$$

It was assumed in (7.4) that the vapor density is much lower than the liquid density ($\rho_v << \rho$). Such a scheme can occur in the case of an infinitely high thermal conductivity of liquid.

The energetic kinetic molecular (non-equilibrium) scheme is considered when non-equilibrium effects dominate at the interface on the side of the vapor phase. The pressure in both phases is assumed to be constant ($p_v = p_\infty = \text{const}$). A constant temperature difference is maintained at the interface: $\Delta T = T_\infty - T_v = \text{const}$ (Fig. 7.1c). Based on the kinetic theory of evaporation, the following expression for the bubble growth rate was obtained in [8]

$$\dot{R} = f\frac{p_\infty - p_s}{\rho_v\sqrt{2\pi R_g T_\infty}}. \tag{7.5}$$

Here, $p_s = p_s(T_\infty)$, and

$$f(\beta) = \frac{\beta}{1 - 0.4\beta}$$

is the function of the evaporation–condensation coefficient β. The value β is defined as the ratio of the mass flow of the molecules adsorbed by the interphase boundary and the total mass flow of molecules incident on the CPS. The quantity β depends on the physical nature of the interphase surface and may vary in the range $0 \leq \beta \leq 1$. Such a scheme can appear only with $\beta \ll 1$.

7.3 The Energetic Thermal Scheme

The energetic thermal scheme describes the growth of a bubble due to the heat supply from the superheated liquid to the interface by the mechanism of nonstationary heat conduction. The pressure in both phases is assumed to be constant ($p_v = p_\infty =$ const), and the temperature of the vapor in the bubble is equal to the saturation temperature at the pressure in the system $T_v = T_s(p_\infty)$, Fig. 7.1d. The bubble growth is determined by the self-similar thermal diffusion law

$$R = m\sqrt{\alpha t}, \tag{7.6}$$

where m is the "growth modulus" [1].

The studies of the energetic thermal scheme were initiated by German thermophysicists [9–11]. In 1930, for calculating the heat flux supplied to the interface, Bosnjakovic [9] used the solution of the problem of nonstationary heat conduction

$$q = \frac{k\Delta T}{\sqrt{\pi \alpha t}}. \tag{7.7}$$

From (7.7) using (7.1), (7.6), we find an expression for the growth modulus

$$m = \frac{2}{\sqrt{\pi}}\text{Ja}. \tag{7.8}$$

Here,

$$\text{Ja} = \frac{\rho c_p \Delta T}{L\rho_v} \tag{7.9}$$

is the Jakob number defined as the ratio of the enthalpy of liquid superheating to the phase transition enthalpy (both quantities are taken with respect to a unit volume), and $\Delta T = T_\infty - T_v$ is the temperature drop, $T_v = T_s(p_\infty)$.

As a result of later investigations of the thermal law of growth [10, 11], formulas close to (7.8) were derived. The earlier theoretical models in [9–11] had a quasi-stationary character as they ignored the influence of liquid medium motion on the bubble growth. Plesset and Zwick [12], with the aid of a very complex mathematical analysis and following the self-similarity law (7.6), have obtained, for the first time, an expression for the heat flux on the surface of a sphere during its expansion in a superheated liquid

$$q = \sqrt{\frac{3}{\pi}} \frac{k\Delta T}{\sqrt{\alpha t}}. \tag{7.10}$$

Relation (7.10) differs from its quasi-stationary analog (7.7) by the presence of the spherical factor $\sqrt{3}$. Based on the solution of the energy equation for a vapor bubble, the well-known Plesset–Zwick formula was obtained in [12] as

$$m = 2\sqrt{\frac{3}{\pi}}\mathrm{Ja}. \tag{7.11}$$

Plesset and Zwick [12] proceeded from the problem formulation in the most general form (with due account of the influence of the viscous and inertial resistance, capillary forces, and non-equilibrium condition of evaporation), and thereafter they successively invoked a series of simplifying assumptions. In contrast to this, Birkhoff et al. [13] initially assigned the diffusion law of growth (7.6) and showed that Plesset–Zwick's formula represented asymptotics of the general solution of the energetic thermal scheme with $\mathrm{Ja} \to \infty$. This limiting case, by analogy with the approach of [9], corresponds to the physical ideas about a thin thermal boundary layer in a liquid on the bubble surface, however, in contrast to the quasi-stationary approach of [9], it takes into account the spherical dynamics of the bubble.

An exact analytical solution of the problem within the framework of the energetic thermal scheme was obtained for the first time by Scriven in [14] in the form of a quadrature. Unfortunately, Scriven [14] presented the results of calculations in the form of a bulky table unsuitable for practical applications. It seems that because of this fact the pioneering work of Scriven in its original form was hardly used. The indisputable advantage of [14] is in the analytical asymptotics of the energetic thermal scheme that were obtained in that work and that are briefly considered in what follows. The asymptotics of slow bubble growth ($\mathrm{Ja} \to 0$)

$$m = \sqrt{2\mathrm{Ja}} \tag{7.12}$$

follows from the problem on stationary supply of heat to an isothermal spherical surface [15]

$$q = \frac{k\Delta T}{R}. \tag{7.13}$$

The asymptotics of rapid bubble growth ($\mathrm{Ja} \to \infty$) is written as

$$\mathrm{S} = \sqrt{\pi} m_* \exp\left(m_*^2\right) \mathrm{erfc}(m_*). \tag{7.14}$$

Here,

$$\mathrm{S} = \frac{c_p \Delta T}{L} \tag{7.15}$$

is the Stefan number defined as the ratio of the enthalpy of liquid superheating to the phase transition enthalpy (both quantities are referred to as unit mass), $m_* = \varepsilon m/\sqrt{12}$ is the modified growth modulus, erfc(m_*) is the additional probability integral, and $\varepsilon = \rho_v/\rho$ is the density phase ratio.

The solution of the form of Eq. (7.14) was obtained for the first time in 1860 by Neumann in describing the problem of solidification of liquid on a spherical surface [15]. This classical solution entered into the course of Neumann's lectures published in 1912 by Riemann and Weber. A detailed description of the problem of solidification (melting) for the equality of densities of both phases was given in 1890–1891 in Stefan's works. Since then, this classical problem of mathematical physics together with its multiple modern modifications [16] has been referred to in the literature under the name "Stefan problem".

In order to elucidate the connection between Eq. (7.14) and the Stefan problem, we will go over to the coordinate system fixed on the bubble boundary. Now the evaporation-induced decrease in the liquid layer adjacent to the interface ΔR can be described by the formula $\Delta R \equiv R\varepsilon = m\varepsilon\sqrt{at}$. If we multiply ΔR by the spherical factor $\sqrt{3}$, the value of m_* will be equal to the self-similar variable of the Stefan problem [15, 16].

In turn, the asymptotics of the rapid growth contains two asymptotics. With $m_* \to 0$, Plesset–Zwick's formula (7.11) follows from (7.14). Expanding the right-hand side of (7.14) in a series with $m_* \to \infty$, we obtain

$$m = \sqrt{\frac{6}{1-\mathrm{S}}}\frac{1}{\varepsilon}. \tag{7.16}$$

The growth modulus determined from formula (7.16) corresponds to the heat flux supplied to the bubble boundary

$$q = \sqrt{\frac{3}{2}}\frac{k\Delta T}{\sqrt{(1-\mathrm{S})\alpha t}}. \tag{7.17}$$

This asymptotic form describes the case of an infinitely high rate of growth: $\mathrm{S} \to 1$, $m \to \infty$. Its physical meaning is dictated by the specificity of the energetic thermal scheme [17]. When the excess enthalpy of liquid superheating $c_p\Delta T$ becomes equal to the value of the evaporation heat L, each elementary volume of liquid near the interface is easily converted into vapor, moreover, this does not require heat supply from outside, so that any limitations on the phase transformation rate disappear. In reality, the bubble growth rate at $\mathrm{S} = 1$ cannot become infinitely high; it is limited by other factors not accounted for in the analysis. At high growth rates, the inertial forces arising in the liquid and leading to the disturbance of the initial condition of constant pressure in the liquid and in the bubble become substantial. Moreover, the very process of evaporation cannot feature an infinitely high intensity because of the kinetic molecular trends inherent in it.

The asymptotics of a homogeneous medium ($\varepsilon = 1$) describes the problem of spherical solidification under the conditions of the exact equality between the densities of phases [15]

$$S = \frac{m^2}{2}\left[1 - \frac{\sqrt{\pi}}{2} m \exp\left(\frac{m^2}{4}\right) \operatorname{erfc}\left(\frac{m}{2}\right)\right]. \tag{7.18}$$

Relation (7.18) was obtained by Frank [18] in 1950 and up to now serves as a basis for the theoretical investigation of the problems of solidification/melting [19]. Within the framework of the energetic thermal scheme, this hypothetical case corresponds to the conditions of the thermodynamic critical point $\rho = \rho_V$. The Jakob number (7.9) becomes in this case equal to the Stefan number (7.15). With small and large m, Eq. (7.18) coincides with the corresponding asymptotic formulas (7.12) and (7.16) if we assume in them that $\varepsilon = 1$.

The case most important for applications is realized with the strong inequality between the densities of phases ($\varepsilon \ll 1$). To describe the Scriven solution under these conditions, Labuntsov et al. [20] approximated the tabulated data of [14] for the entire range of variation of the Jakob number by the empirical formula

$$m = 2\sqrt{\frac{3}{\pi}}\mathrm{Ja}\left[1 + \frac{1}{2}\left(\frac{\pi}{6\mathrm{Ja}}\right)^{2/3} + \frac{\pi}{6\mathrm{Ja}}\right]^{1/2}. \tag{7.19}$$

Let us consider the possible semi-empirical technique for approximating the Scriven solution using the analogy with the spherical problem of nonstationary heat conduction [15]. It follows from the solution of this problem that the heat flux on the surface of the sphere can be written as the superposition of two components

$$q = \frac{k\Delta T}{\sqrt{\pi \alpha t}} + \frac{k\Delta T}{R}. \tag{7.20}$$

With $t \to 0$, the first (nonstationary) term, coinciding with (7.7), dominates on the right-hand side of Eq. (7.20). With $t \to \infty$, the second (stationary) component, coinciding with (7.13), is predominant there. We replace now the nonstationary component in Eq. (7.20) by expression (7.10) obtained in [12] for a sphere expanding by the self-similar law (7.6). Now, using the heat balance Eq. (7.1)

$$\mathrm{Ja} = \frac{1}{2}\frac{m^2}{1 + \sqrt{3/\pi}m}. \tag{7.21}$$

Resolving the quadratic Eq. (7.21) for the growth modulus m, we obtain

$$m = \sqrt{\frac{3}{\pi}}\mathrm{Ja} + \sqrt{\frac{3}{\pi}\mathrm{Ja}^2 + 2\mathrm{Ja}}. \tag{7.22}$$

With Ja $\to$ 0, Eq. (7.22) yields the stationary asymptotics (7.12). With Ja $\to \infty$, relation (7.22) goes over into an asymptotic nonstationary form (7.11). Note that both approximations, i.e., the empirical (7.19) and semi-empirical (7.22) ones, agree up to 2% with the Scriven solution in the entire range of variation of the Jakob number ($0 < \mathrm{Ja} < \infty$). By its structure, formula (7.22) is identical to that proposed by Labuntsov and Yagov [1] for the case of bubble growth on a heating surface

$$m = 0.3\mathrm{Ja} + \sqrt{(0.3\mathrm{Ja})^2 + 12\mathrm{Ja}}. \tag{7.23}$$

It is important to note that relation (7.23) was obtained in [1] as a result of the summation of two heat fluxes supplied to the bubble surface: through the liquid microlayer from the wall surface (the lower part of the bubble) and from the superheated liquid layers (the upper part of the bubble). It seems that the indicated correspondence is an argument in favor of the above rule of the superposition of heat fluxes (7.20).

Straub [21] gives a description of unique experiments carried out with boiling in the region of microgravitation on a platform falling from a laboratory tower of height 110 m. A distinctive feature of the indicated experiment conducted within the framework of the ZARM program (Zentrum für Angewandte Raumfahrttechnick und Mikrogravitation) was modeling, in a "pure form", of spherical–symmetric growth of a bubble on a single nucleation site for a very long time (up to 5 s).

Figure 7.2 presents a comparison of the results of calculations by Eq. (7.22) with the experimental data of [22] obtained for the coolant R11 in the region $0.2 \leq \mathrm{Ja} \leq 4$. The results presented in Fig. 7.2 make it possible to reveal the noticeable contribution of the slow growth asymptotics (7.12) and also demonstrate the inapplicability of Plesset–Zwick's formula in the region of small Jakob numbers. This fact confirms

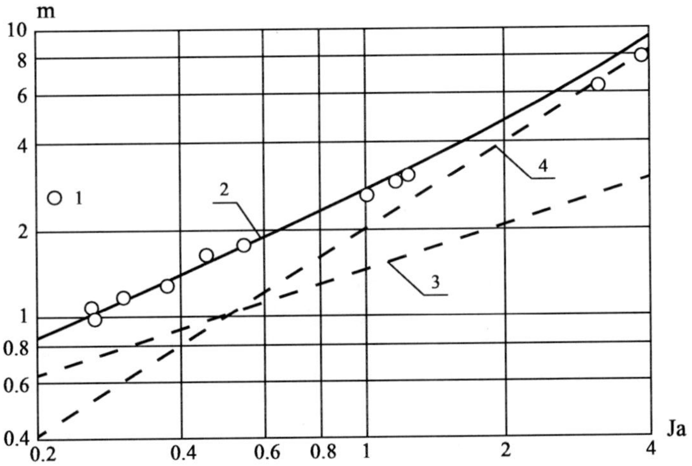

Fig. 7.2 Energetic thermal scheme of vapor bubble growth. 1 Experimental data of [22], 2 calculation by Eq. (7.22), 3 quasi-stationary asymptotics (7.12), and 4 calculation by the Plesset–Zwick formula (7.11)

the inference made in [13] about the asymptotic character of relation (7.11), which is strictly valid only for $\mathrm{Ja} \to \infty$.

In [23], an attempt was made to obtain an approximation of the Scriven solution that would be applicable for the entire resultant table of [14]. In [23], the following "universal approximate solution" was obtained (with the announced maximum relative error of 2.4%)

$$m = \sqrt{\frac{3}{\pi}}\mathrm{Ja}\psi + \sqrt{\frac{3}{\pi}(\mathrm{Ja}\psi)^2 + 2\mathrm{Ja}}\,, \tag{7.24}$$

where

$$\psi(\mathrm{S}) = 1 + \sqrt{\frac{\pi}{2}}\left(\frac{1}{\sqrt{1-\mathrm{S}}} - 1\right). \tag{7.25}$$

It is interesting to compare approximation (7.24) with the Frank exact solution (7.18), which is the sole analytical asymptotics of the Scriven quadrature in the entire range of variation of the growth modulus $0 < m < \infty$. For this limiting case, it is necessary to assume in (7.24) that $\varepsilon = 1 : \mathrm{Ja}\psi = \mathrm{S}$. The calculation shows that the maximum relative error of Eq. (7.24) exceeds 12% (with $\mathrm{S} \approx 0.75$ and $m \approx 3.9$). A selective comparison of Eq. (7.24) directly with Scriven's table shows that in the range of anomalous superheatings ($0.75 \le \mathrm{S} \le 1$), the error of Eq. (7.24) increases. Thus, the "universal approximation" suggested in [23] for the Scriven solution should be recognized as unsound. It allows one to take the next step and come to the binary schemes of growth that describe the simultaneous effect of two factors on the growth of a bubble.

7.4 Binary Schemes of Growth

7.4.1 The Viscous–Inertial Scheme

The case of the simultaneous effect of both dynamic factors—viscous and inertial—is described by the Rayleigh–Plesset equation [4, 5]

$$\Delta p = 4\frac{\mu\dot{R}}{R} + \frac{3}{2}\rho\dot{R}^2. \tag{7.26}$$

If the inertial forces are small $\left(\rho\dot{R}^2 << \mu\dot{R}/R\right)$, then the viscous scheme of growth (7.2) follows from (7.26). At $\Delta p = p_\mathrm{v} - p_\infty = \mathrm{const}$, Eq. (7.26) can be transformed into a quadratic equation whose solution is

$$\frac{dR^2}{dt} = \frac{8}{3}\left(\sqrt{\nu^2 + \frac{3}{8}\frac{\Delta p R^2}{\rho}} - \nu\right). \tag{7.27}$$

With $R \rightarrow 0$, Eq. (7.27) yields the viscous asymptotics (7.2) that determines the initial stage of the bubble growth. With $R \rightarrow \infty$, the bubble growth rate is determined by the inertial law (7.4). A comparative evaluation of the influence of viscosity and inertial forces can be made without resorting to the integration of differential Eq. (7.27). After some transformations, this equation can be presented in the form

$$\mathrm{Re}_0 = \frac{4}{3}\left(\sqrt{1 + \frac{9}{16}\mathrm{Re}_\infty^2} - 1\right). \tag{7.28}$$

It is seen from Eq. (7.28) that the binary viscous–inertial scheme of bubble growth is determined by two Reynolds numbers. The quantity $\mathrm{Re}_0 = R_0 \dot{R}_0/\nu$ is responsible for the initial viscous stage of growth: $\dot{R} = \dot{R}_0$. The quantity $\mathrm{Re}_\infty = R_0 \dot{R}_\infty/\nu$ determines the asymptotic inertial stage. Here, $\dot{R}_\infty = \sqrt{2/3(\Delta p/\rho)}$ is the Rayleigh rate of growth (7.4). With $\mathrm{Re}_\infty \rightarrow 0$, the viscous asymptotics $\mathrm{Re}_0 = 3/8\mathrm{Re}_\infty^2$ follows from Eq. (7.28). The bubble starts to grow following the viscous scheme (7.2) and with time passes into the inertial stage (7.4). With $\mathrm{Re}_\infty \rightarrow \infty$, the inertial asymptotics $\mathrm{Re}_0 = \mathrm{Re}_\infty$ is realized. There is no effect of viscous forces, and from the very beginning the bubble grows by the inertial scheme (7.4).

7.4.2 The Inertial–Thermal Scheme

The region of small superheatings $S \ll 1$. In the general case, the spherical expansion of a bubble in a liquid causes the inertial resistance of the liquid leading to a rise in pressure at the interface. The temperature of the saturated vapor in the bubble will increase in this case, whereas the temperature difference between the liquid and vapor will decrease. As a consequence, the intensity of the heat supply to the interface will decrease. As a result, the bubble growth rate will be lower than that predicted by the thermal energetic scheme.

Account for the variability of the bubble surface temperature in time leads to the binary inertial–thermal scheme of bubble growth. This scheme has been studied for the first time in [24] with the aid of very coarse assumptions (the linear approximation of the saturation curve, the vapor density in a bubble constant in time, etc.). In [25], an approximate analytical solution was obtained for the limiting case of the inertial–thermal scheme that corresponds to the region of high Jakob numbers

$$R = 1.2\left(\frac{kc_p}{\rho_v}\right)^{1/4} \frac{R_g^{3/4} T_s^{5/4} t^{3/4}}{L}. \tag{7.29}$$

The characteristic features of Eq. (7.29) are the absence in it of the liquid superheating ΔT and the non-self-similar "three quarters" growth law $R \sim t^{3/4}$ that differs from the diffusional relation $R \sim t^{1/2}$.

7.4.3 The Region of High Superheatings

As shown in the numerical investigation [26], when the vapor temperature is variable in time, it is possible to use, with acceptable accuracy, for the heat flux supplied to the bubble surface, relations obtained within the framework of a purely thermal scheme. Therefore, in analyzing the inertial–thermal scheme in the region of high superheatings, we will proceed from relation (7.17). It follows from this relation, in particular, that it is incorrect to use the Jakob number as a base parameter in the region of high superheatings. This may lead to the "impermissible intrusion" into the region $\mathrm{S} \equiv \varepsilon \mathrm{Ja} > 1$.

The author of [27] carried out a numerical investigation of the problem on the growth of a bubble in the region of high superheatings for the limiting case where the liquid is superheated to a temperature corresponding to the boundary of thermodynamic stability (of the spinodal [28]). It is shown in [27] that the experimentally observed increased rates of growth can be attributed to the long inertial stage in the course of which a substantial pressure difference is maintained between the liquid and the bubble.

We now consider the case where the initial Stefan number exceeds unity: $\mathrm{S} = \mathrm{S}_{\max} > 1$. Since relation (7.17) must hold during the entire period of bubble growth, it follows from this that the local Stefan number must always remain smaller than unity: $\mathrm{S} = \mathrm{S}(t) < 1$. Such matching of parameters in the physical sense is reduced to the mutual effect of the thermal and inertial mechanisms, which is called below the effect of "pressure blocking".

We consider the limiting case of a maximum possible Stefan number in the p, T diagram (Fig. 7.3). The state of saturated vapor in a bubble corresponds to the point $A\{p_{\min}\}$ on the binodal (saturation line), and the state of the superheated liquid, to the point $B\{p_{\min}\}$ on the spinodal (the line of the limiting superheating of liquid), AB is the isobar. The Stefan number is calculated by the formula

$$\mathrm{S} = \frac{1}{L} \int_{T_{\min}}^{T_{\max}} c_p dt. \tag{7.30}$$

Here, $T_{\max}$ is the spinodal temperature. Let us assume that the name "energy spinodal" is given in the p, T diagram the $T_e(p)$ curve for which the following condition is satisfied on each isobar

$$\mathrm{S}_e = \frac{1}{L} \int_{T_{\min}}^{T_e} c_p dt. \tag{7.31}$$

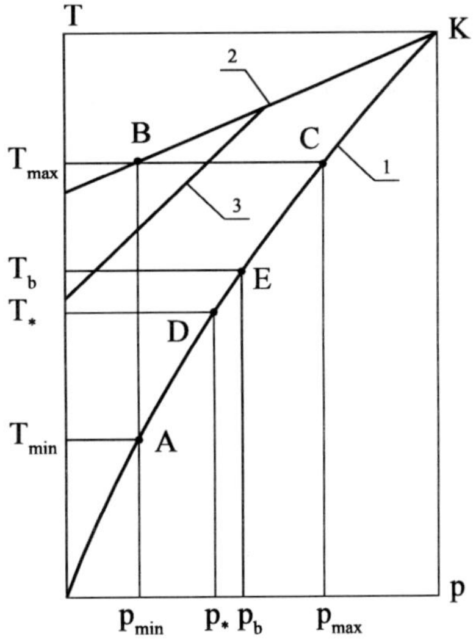

Fig. 7.3 Stages of vapor bubble growth in a superheated liquid with S_{max}. 1 Binodal, 2 spinodal, and 3 energy spinodal

The energy spinodal is described in Fig. 7.3 by line 3 lying between binodal 1 and spinodal 2 and intersecting with the latter. We will consider the stages of bubble growth in the p, T diagram.

(1) The initial state: $t = -0$. The vapor is at the point $A\{p_{min}, T_{min}\}$ on binodal 1, the liquid is at the point $B\{p_{min}, T_{max}\}$ on spinodal 2 and remains there at $t > 0$, and the Stefan number $S = S_{max}$.

(2) The initial state: $t = +0$. The inertial response of the liquid is engaged. The pressure in the bubble increases along the binodal from the point $A\{p_{min}, T_{min}\}$ to the point $C\{p_{max}, T_{max}\}$; the temperature drop and the Stefan number go to zero: $\Delta T = S = 0$. All these changes occur abruptly.

(3) The transition stage: $t > 0$. The bubble grows by the inertial–thermal scheme, the state of the vapor "drifts" along the binodal from the point $C\{p_{max}, T_{max}\}$ to the point $D\{p_*, T_*\}$, and the Stefan number increases $0 \le S(t) < 1$.

(4) The asymptotic stage: $t \to \infty$. The state of the vapor "hovers" in the vicinity of the point $D\{p_*, T_*\}$ that corresponds to the condition of the energy spinodal: $S(t) \to c_p(T_{max} - T_*)/L$. The temperatures T_{max} and T_* are on the "isobar of blocking": $p = p_* = \text{const}$. From this, it follows that with $t \to \infty$, the bubble will grow by the asymptotic inertial scheme (7.4), where $\Delta p = p_* - p_{min}$.

Thus, when a vapor bubble grows in the region S_{max}; the pressure blocking effect takes place in the vapor phase: $S(t) \to 1$, $p(t) \to p_* > p_{min}$, $T(t) \to T_* > T_{min}$. In reality, the vapor state in the bubble attains the pressure blocking point $D\{p_*, T_*\}$ only asymptotically with $t \to \infty$. Let us consider the law of bubble growth at some point $E\{T_b, p_b\}$ that "moves" to the point $D\{p_*, T_*\}$ and reaches it when $t \to \infty$. We introduce the notation for the temperature drop $\Delta T_b = T_{max} - T_b$ and pressure drop $\Delta p_b = p_b - p_{min}$ for the "vapor–liquid" system. The vapor pressure and temperature at the point $E\{T_b, p_b\}$ are higher than the vapor pressure and temperature at the point $D\{p_*, T_*\}$ by the values $\Delta p_s = p_b - p_*$ and $\Delta T_s = T_b - T_*$. These values are connected with the respective total drops by the relations $\Delta p_s + \Delta p_* = \Delta p_b$, $\Delta T_s = \Delta T_* - \Delta T_b$. Small (linear) pressure and temperature drops along the saturation curve are connected by means of the Clapeyron–Clausius equation

$$\frac{\Delta T_s}{T_*} = \frac{\Delta p_s}{L\rho_v}. \tag{7.32}$$

With regard to the adopted notation, from Eqs. (7.1), (7.3), (7.17), and (7.32) we obtain a biquadratic equation for the time dependence of the bubble growth rate in the region of pressure blocking. The asymptotics of its solution at $t \to \infty$ can be written in the form

$$\dot{R}^2 = \frac{2}{3}\frac{\Delta p_*}{\rho} + \sqrt{\frac{\rho}{\rho_v}\frac{\alpha L^2}{c_p T_*}\frac{1}{t}}. \tag{7.33}$$

According to (7.33), for $t \to \infty$ the bubble grows by the limiting Rayleigh law

$$\dot{R} = \sqrt{\frac{2}{3}\frac{\Delta p_*}{\rho}}. \tag{7.34}$$

On the decrease in the blocking pressure drop $\Delta p_* = p_* - p_{\min}$ (i.e., when the Stefan number tends to unity "from above"), the second (nonstationary) term prevails on the right-hand side of Eq. (7.3). The limiting case $\Delta p_* = 0$ describes the hypothetical equilibrium case with $\mathrm{S} = 1$

$$\dot{R} = \left(\frac{\alpha \rho L^2}{c_p \rho_v T_* t}\right)^{1/4}. \tag{7.35}$$

If we formally integrate (7.35) with the initial condition $R = 0$ a $t = 0$, we obtain the following "equilibrium" law of growth

$$R = \frac{4}{3}\left(\frac{\alpha \rho L^2}{c_p \rho_v T_*}\right)^{1/4} t^{3/4}. \tag{7.36}$$

It should be noted, however, that relations (7.35) and (7.36) correspond to an absolutely unstable equilibrium $\mathrm{S} = 1$. In reality, when $\mathrm{S} < 1$, the growth dynamics will "stall" into the inertial–thermal scheme [27] and at $\mathrm{S} > 1$, into the Rayleigh law (7.4) at $\Delta p = p_* - p_{\min}$. It is interesting to compare the "three quarters" growth laws obtained, respectively, for the condition of anomalous Jakob numbers (R_{Ja}, Eq. (7.29) and anomalous superheats ($\mathrm{R_S}$, Eq. (7.36))

$$\frac{R_S}{R_{Ja}} \approx \left(\frac{L^2}{R_g c_p T_s^2}\right)^{3/4}. \tag{7.37}$$

We take as an example the case of dodecane at $p_s = 0.01$ bar, when the combination of both limiting effects is possible theoretically: Ja $\gg 1$ and S $= 1$. Then, from Eq. (7.37) we obtain that $R_S/R_{Ja} \approx 5.6$.

7.4.4 The Non-equilibrium-Thermal Scheme

As is known [1], the evaporation–condensation coefficient designates the fraction of molecular flow directed from the vapor phase and adsorbed by the interface. The value of β depends on the state of the surface and on the physical nature of the condensed phase and in the general case can vary in the range $0 < \beta < 1$. A few experimental and theoretical data on the evaporation–condensation coefficient [29, 30] indicate that $\beta \approx 1$ under normal conditions.

Consider the case of a simultaneous effect of mechanisms of nonstationary thermal conductivity and non-equilibrium on the growth law of a vapor bubble. According to the limiting schemes of the growth concept, the heat flux on the bubble surface can be found by adding two thermal resistances

$$q_R = (\mathbb{R}_k + \mathbb{R}_t)\Delta T. \tag{7.38}$$

The expression for the kinetic molecular thermal resistance $\mathbb{R}_k$

$$\mathbb{R}_k = f\frac{T_\infty\sqrt{2\pi R_g T_\infty}}{\rho_v L^2} \tag{7.39}$$

can be found from (7.1) and (7.5).

In finding the energetic thermal resistance W_t, one should take into account that (7.38) holds if there is a thin thermal layer in the liquid. This means that in evaluating W_t, one should start from relation (7.10) for the heat flux. In this connection, it is worth noting that the approaches of [9–13], in spite of their differences, are based on physical interpretations about a thin thermal layer in liquid on a bubble surface through which a heat flux is supplied to the surface. Using Plesset–Zwick's formula (7.11), which holds for Ja $\gg 1$, we have

$$\mathbb{R}_t = \sqrt{\frac{\pi}{3}}\frac{\sqrt{\alpha t}}{k}. \tag{7.40}$$

From (7.1), (7.38)–(7.40), we get the binary bubble growth law

$$\tilde{R} = \sqrt{\tilde{t}} - \ln\left(1 + \sqrt{\tilde{t}}\right). \tag{7.41}$$

Here, $\tilde{R} = R/R_0, \tilde{t} = t/t_0$ are, correspondingly, the dimensionless radius of the bubble and time. The length and time scales read as

$$R_0 = \sqrt{\frac{18}{\pi} \frac{fkc_p\rho T_\infty \Delta T \sqrt{R_g T_\infty}}{\rho_v^2 L^3}}, \tag{7.42}$$

$$t_0 = 6 \frac{f^2 kc_p \rho R_g T_\infty^3}{\rho_v^2 L^4}. \tag{7.43}$$

For $\tilde{t} \ll 1$, the solution (7.41) becomes the kinetic molecular law $\tilde{R} = \tilde{t}$, and for $\tilde{t} \gg 1$, it becomes the asymptotic law of growth $\tilde{R} = \sqrt{\tilde{t}}$. In the dimensional form, the above asymptotic formulas are described by formulas (7.5) and (7.11), respectively.

In [21, 22], experiments on the influence of non-equilibrium effects on laws of bubble growth were carried out. These papers give results of unique experiments on nucleate boiling of freons (R11, R113) on a falling platform, which were conducted on a test column of height 110 m as a part of the ZARM program. The distinctive feature of these experiments is that in [21, 22], an actual modeling "in the pure form" of a spherically symmetric bubble growth on an isolated center of evaporation was achieved for a long time (up to 5 s).

A detailed numerical study of bubble growth on the basis of the system of conservation equations for both phases with due account of all possible factors (the surface tension, and the viscosity and compressibility of liquid and vapor) was also conducted in [22]. It should be noted, however, that such a detailed description of the problem seems excessive. So, in [1] it was shown that in the absence of shock gas-dynamic phenomena, there is no need to solve the system of equations for the vapor phase. The author of [22] gives a great number of the experimental data obtained by him for two liquids in the range $0.9 \leq \mathrm{Ja} \leq 32.6$ with the help of the regression analysis. As a result, the following approximation was obtained

$$R = 2.03\mathrm{Ja}(\alpha t)^{0.43}. \tag{7.44}$$

It is worth pointing out that formula (7.44) does not accord with the dimensional analysis and hence is purely empirical: the exponent which differs from "½" violates the self-similar character of the thermal diffusion law (7.6).

The processing of the experimental data from [22] revealed a considerable scatter in the values of the evaporation–condensation coefficient: $10^{-2} \leq \beta \leq 0.7$ for Freon 11 and $8.1 \times 10^{-3} \leq \beta \leq 1.0$ for Freon 113. The author of [22] explains this by a possible uncontrolled effect from high-boiling impurities (like Freons 13, 22, 23, 114) and mixtures of oils. Indeed, even a minute concentration of impurities (despite the fact that the cleanness guaranteed by the vendor of freons is 99.98%,) could in principle result in the shielding of the surface of a growing bubble. As a corollary, this could result in a significant decrease in its growth rate in comparison with the values predicted by the energetic thermal scheme.

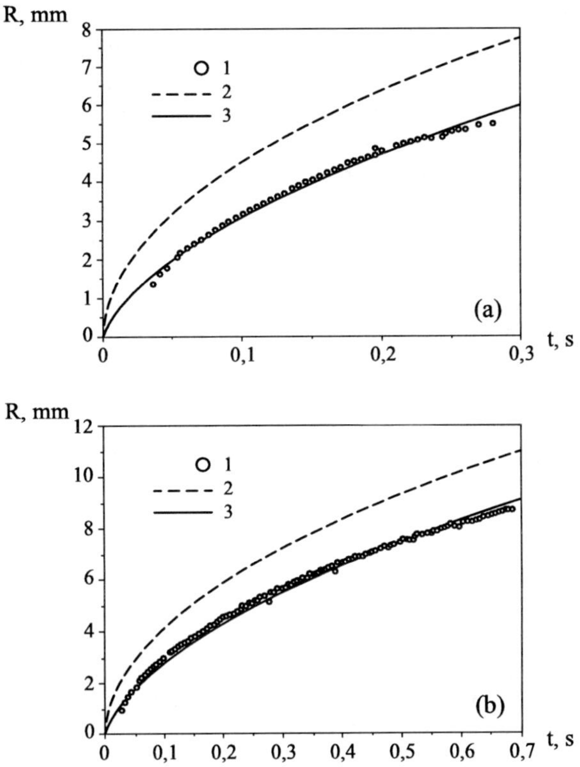

Fig. 7.4 Non-equilibrium effects versus the bubble growth rate. 1 Experimental data [22], 2 calculation by Plesset–Zwick's formula (7.11), and 3 calculations by the non-equilibrium-thermal scheme, formula (7.41). **a** Freon 113, Ja = 30.8, $\beta = 4.27 \times 10^{-3}$ and

In turn, according to (7.41) this corresponds to a strong decrease of β. Let us estimate the possible value of β from experiments of [22]. In these experiments, we obtained two experimental curves for the domain of applicability of the physical model under consideration: Ja $\gg$ 1. These arrays of a point were processed using formulas (7.41)–(7.43) and then compared with the theoretical dependence (7.41). Figure 7.4 shows that both regimes correspond to approximately the same value of the evaporation–condensation coefficient: $\beta \approx 4.3 \times 10^{-3}$.

7.5 Conclusions

Our main result here is the justification of the effect of the bubble pressure blocking in the region of anomalous superheats of liquid, as well as of the bubble growth law (7.33) following from this effect. It is shown that in the course of the growth of a vapor bubble in a liquid, the enthalpy of the superheating of which exceeds the phase transition heat, there appears a singularity in the mechanism of heat supply from a liquid to vapor. The elimination of this singularity within the framework of the binary inertial–thermal scheme must lead to the pressure blocking effect in the vapor phase. It is shown that due to this, the asymptotic bubble growth will occur by the inertial scheme. To qualitatively illustrate the pressure blocking effect, the notion of the "energy spinodal" was introduced, which corresponds to the condition that the liquid superheating enthalpy is equal to the phase transition heat. It is shown that in the region of high superheats of liquid, it is necessary to use the Stefan number as the determining parameter instead of the Jakob number. A qualitative analysis of the effect of the Stefan number on the bubble growth rate in the region of high superheats was carried out. The bubble growth law in the pressure blocking region was derived. A comparison was made of the limiting "three quarters" laws of bubble growth that describe the case of high Jakob numbers and high superheats, respectively. An analytical solution for the binary law of bubble growth taking into account the energetic and the non-equilibrium effects was obtained. The evaporation–condensation coefficient was estimated by comparing the theoretical solution with experimental data on the growth of a vapor bubble under reduced gravity conditions.

References

1. Labuntsov DA (2000) Physical foundations of power engineering. Selected works, Moscow Power Energetic Univ. (Publ.). Moscow (In Russian)
2. Labuntsov DA, Yagov VV (2007) Mechanics of two-phase systems. Moscow Power Energetic Univ. (Publ.), Moscow (In Russian)
3. Besant WH (2013) A treatise on hydrostatics and hydrodynamics. Forgotten Books, London
4. Prosperetti A, Plesset MS (1978) Vapor bubble growth in a superheated liquid. J Fluid Mech 85:349–368
5. Brennen CE (1995) Cavitation and bubble dynamics. Oxford University Press, Oxford
6. Labuntsov DA (1974) Current views on the bubble boiling mechanism. In: Heat transfer and physical hydrodynamics. Nauka, Moscow, pp 98–115 (In Russian)
7. Rayleigh L (1917) On the pressure developed in a liquid during the collapse of a spherical cavity. Philos Mag 34:94–98
8. Muratova TM, Labuntsov DA (1969) Kinetic analysis of the processes of evaporation and condensation. High Temp 7(5):959–967
9. Bosnjakovic F (1930) Verdampfung und Flüssigkeits Überhitzung. Technische Mechanik und Thermodynamik 1:358–362
10. Jakob M, Linke W (1935) Wärmeübergang beim Verdampfen von Flüssigkeiten an senkrechten und waagerechten Flächen. Phys Zeitschrift 36:267–280
11. Fritz W, Ende W (1936) Über den Verdampfungsvorgang nach kinematographischen Aufnahmen an Dampfblasen. Berechnung des Maximalvolumens von Dampfblase. Phys Zeitschrift 37:391–401
12. Plesset MS, Zwick SA (1954) The growth of vapor bubbles in superheated liquids. J Appl Phys 25:493–500
13. Birkhoff G, Margulis R, Horning W (1958) Spherical bubble growth. Phys Fluids 1:201–204
14. Scriven LE (1959) On the dynamics of phase growth. Chem Eng Sci 10(1/2):1–14
15. Carslaw HS, Jaeger JC (1986) Conduction of heat in solids. Clarendon, London
16. Mccue SW, Wu B, Hill JM (2008) Classical two-phase Stefan problem for spheres. Proc R Soc Lond, Ser A Math Phys Eng Sci 464(2096):2055–2076
17. Labuntsov DA, Yagov VV (1978) Mechanics of simple gas-liquid structures. Moscow Power Energetic Univ. (Publ.), Moscow (In Russian)

18. Frank FC (1950) Radially symmetric phase growth controlled by diffusion. Proc R Soc Lond, Ser A Math Phys Eng Sci 201(1067):586–599
19. Papac J, Helgadottir A, Ratsch C, Gibou FA (2013) Level set approach for diffusion and Stefan-type problems with Robin boundary conditions on quadtree/octree adaptive Cartesian grids. J Comput Phys 233:241–261
20. Labuntsov DA, Kol'chugin BA, Golovin VS, Zakharova EA, Vladimirova LN (1964) High-speed cine-photography investigation of the growth of bubbles in saturated water boiling in a wide range of pressures. High Temp 2(3):446–453
21. Straub J (2001) Boiling heat transfer and bubble dynamics in microgravity Adv. Heat Transf 35:157–172
22. Winter J (1997) Kinetik des Blasenwachstums. Dissertation. Technische Universität München, München
23. Avdeev AA (2014) Laws of vapor bubble growth in the superheated liquid volume (thermal growth scheme). High Temp 40(2):588–602
24. Mikic BB, Rosenow WM, Griffith P (1970) On bubble growth rates. Int J Heat Mass Transf 13:657–666
25. Yagov VV (1988) On the limiting law of growth of vapor bubbles in the region of very low pressures (high Jakob numbers). High Temp 26(2):251–257
26. Korabelnikov AV, Nakoryakov VE, Shraiber IR (1981) Taking account of non-equilibrium evaporation in the problems of the vapor bubble dynamics. High Temp 19(4):586–590
27. Aktershev SP (2004) Growth of a vapor bubble in an extremely superheated liquid. Thermophysics and Aeromechanics 12 (3): 445–457 Skripov VP (1974) Metastable Liquid. Wiley, New York
28. Debenedetti PG (1996) metastable liquids: concepts and principles. Princeton University Press, Princeton
29. Kryukov AP, Levashov VY (2011) About evaporation-condensation coefficients on the vapor-liquid interface of high thermal conductivity matters. Int J Heat Mass Transf 54(13–14):3042–3048
30. Kryukov AP, Levashov VY, Pavlyukevich NV (2014) Condensation coefficient: Definitions, estimations, modern experimental and calculation data. J Eng Phys Thermophys 87(1):237–245

Chapter 8
Pressure Blocking Effect in a Growing Vapor Bubble

Symbols

c_p	Isobaric heat capacity
Ja	Jakob number
k	Thermal conductivity
L	Heat of phase transition
m	Growth modulus
p	Pressure
q	Heat flux
R	Bubble radius
S	Stefan number
T	Temperature
t	Time

Greek Letter Symbols

α	Thermal diffusivity
ρ	Density

Subscripts

b	State in a vapor bubble
cr	State at critical point
e	State at energy spinodal
max	Maximum (on spinodal)
min	Minimum (on binodal)
s	Saturation state
v	Vapor
∞	State at infinity
$*$	State at pressure blocking point

Y. B. Zudin, *Non-equilibrium Evaporation and Condensation Processes*,
Mathematical Engineering, https://doi.org/10.1007/978-3-030-67553-0_8

8.1 Introduction

The phenomenon of gas (vapor) bubbles in a liquid, in spite of the fluctuation character of their nucleation and the short lifetime, has a wide spectrum of manifestations: underwater acoustics, sonoluminescence, ultrasonic diagnostics, decreasing friction by surface nanobubbles, nucleate boiling, etc. [1]. Such exotic manifestations of the bubble behavior as a micropiston injection of droplets in jet printing and the spiral rise path of bubbles in a liquid ("the Leonardo da Vinci paradox") permitted the authors of [2] to speak of "bubble puzzles".

The most important application of the bubble dynamics is the effervescence of a liquid superheated with respect to the saturation temperature. The liquid retains thereby the properties of the initial phase but becomes unstable (or metastable). The result of the demonstration of metastability of the liquid is the initiation and growth of nuclei of a new (vapor) phase in it. An ideal subject of investigation of this phenomenon is the spherically asymmetric growth of the vapor bubble in the volume of a uniformly superheated liquid. However, the experimental realization of such a process presents great challenges. One of the few exceptions is the experiments performed in Germany on the effervescence of a liquid in microgravity [3] (the parabolic flight path of an aircraft, a platform falling from a tower or into a shaft, a space flight).

The computation-theoretical part of this investigation is described in Picker's dissertation [4], where the vapor bubble growth was modeled numerically with account for practically all factors on which this process. Author of [4] solved the system of nonstationary differential mass, momentum, and energy equations for both the liquid and the vapor phase amplified by the compatibility conditions at the interphase boundaries. The results obtained in [4] reflect the achievements of computational mathematics that permit investigating the given problem for concrete ranges of parameters. However, the disadvantages of the utilitarian numerical approach show up thereby. On the one hand, in [4] a huge amount of experimental data was processed, and on the other hand, the calculated recommendations of the work are limited by the ranges of parameters of the numerical experiment.

Therefore, the analytical approach to the problem on the bubble growth based on the investigation of the influence of individual physical factors is still topical. In the present chapter, we propose an analytical solution to the problem on the bubble growth in a highly superheated liquid (the superheating enthalpy exceeds the phase transition heat). It has been shown that a peculiarity of the mechanism of heat transfer from the liquid to the interface leading to the effect of pressure blocking in the vapor phase can arise [5]. This fact was demonstrated by the results of the numerical calculation for the conditions of a concrete experiment.

8.2 The Inertial-Thermal Scheme

In his fundamental work [6], Labuntsov proposed a systematic approach to the problem of the vapor bubble growth in a superheated liquid. He showed that the growth rate in the general case is determined by the following four physical effects

(a) the viscous resistance of the medium displaced by the bubble,
(b) the inertial reaction of the liquid to the swelling of the bubble in it,
(c) non-equilibrium effects at the interface,
(d) the mechanism of heat transfer from the superheated liquid to the bubble boundary.

Taking into account the action of each of the factors under the assumption that the influence of the others is absent leads to limiting schemes of the bubble growth. Analysis of [6] led to an important conclusion: under the simultaneous action of two (or several) factors, the growth rate will always be lower than the least limiting value calculated from the point of view of the corresponding schemes. Each of the limiting schemes, which can differ widely in the physical content, is based on the thermal balance equation

$$q = \rho_v L\dot{R}. \tag{8.1}$$

With $\Delta p = \text{const}$, $\rho_v \ll \rho$ the dynamic inertial scheme described by the classical Rayleigh formula is realized

$$R = \sqrt{\frac{2}{3}\frac{\Delta p}{\rho}}t. \tag{8.2}$$

Here, $\Delta p = p_v - p_\infty$ is the "bubble–liquid" pressure drop, p_v is the saturation pressure at the temperature of the superheated liquid, and p_∞ is the pressure in the liquid at infinity.

The energy thermal scheme describes the bubble growth due to the heat transfer from the superheated liquid to the interface by the mechanism of nonstationary heat conduction. The pressure in both phases is assumed to be constant ($p_v = p_\infty = \text{const}$), and the vapor temperature in the bubble T_v is equal to the saturation temperature at the pressure in the system. The bubble growth rate is determined by the self-similar thermal diffusion law $R = m\sqrt{\alpha t}$, here m is the growth modulus. This implies the following expression for the thermal growth rate

$$\dot{R} = \frac{1}{2}m\sqrt{\frac{\alpha}{t}}. \tag{8.3}$$

The exact analytical solution for the energy thermal scheme was obtained by Scriven in [7] in the form of a quadrature. The results of the solution were systematized in the form of a table which can be given as a tabulated dependence of the form $m = f(\varepsilon, \mathrm{S})$, where ε is the density ratio between the vapor and the liquid phases, and S is the Stefan number defined as the ratio of the superheating enthalpy of the liquid to the phase transition enthalpy (both quantities are referred to as mass unit)

$$\varepsilon = \frac{\rho_v}{\rho}, \tag{8.4}$$

$$\mathrm{S} = \frac{c_p \Delta T}{L}, \tag{8.5}$$

where $\Delta T = T_\infty - T_v$ is the "liquid-vapor" temperature drop. The paper [7] also puts forward analytical asymptotic formulas for the solution by the diagnostic variables. Below we will use the asymptotics of [7] with $m_* \to \infty$

$$\mathrm{S} = \sqrt{\pi} m_* \exp(m_*^2) \mathrm{erfc}(m_*). \tag{8.6}$$

Here,

$$m_* = \frac{\varepsilon m}{\sqrt{12}} \tag{8.7}$$

defines the modified growth modulus, and $\mathrm{erfc}(m_*)$ is the additional integral of probabilities. In turn, (8.6) contains two asymptotics. Making $m_* \to 0$ we obtain $\mathrm{S} = \sqrt{\pi} m_*$ or

$$m = 2\sqrt{\frac{3}{\pi}} \mathrm{Ja}. \tag{8.8}$$

Here, Ja is the Jakob number, defined as the ratio of the liquid superheating enthalpy to the phase transition enthalpy (both quantities are referred to a volume unit)

$$\mathrm{Ja} = \frac{\rho c_p \Delta T}{L \rho_v}. \tag{8.9}$$

The Stefan number (8.5) and Jakob number (8.9) characterize, respectively, the mass and volume degrees of metastability of liquid. They are related as follows

$$\mathrm{S} = \varepsilon \mathrm{Ja}. \tag{8.10}$$

Relation (8.8), which was first obtained in [8] in the course of a very painstaking mathematical analysis, represents the well-known theoretical Plesset–Zwick formula. The implicit dependence $\mathrm{S}(m_*)$, as given by formula (8.6), can be conveniently written in the form of the approximate explicit dependence $m_*(\mathrm{S})$

$$m_* = \frac{1}{\sqrt{\pi}} \frac{\mathrm{S}\left(1 + \left(\sqrt{\pi/2} - 1\right)\mathrm{S}\right)}{\sqrt{1 + \sum_{i=1}^{n} \beta_i \mathrm{S}^i}} i = 1, 2, ...n. \tag{8.11}$$

Formula (8.11) enables one to generate approximations of the exact solution (8.6) with any required accuracy. For instance, as n increases from 1 to 7, the error in calculations by (8.11) decreases from 3 to 0.01% in the ranges of parameters: $0 < m_* \le 6, 0 < \mathrm{S} \le 0.98665$. The values of the polynomial coefficients in (8.11) at $n = 7$ accurate to four decimal places are as follows: $\beta_1 = -0.7604$, $\beta_2 = -0.4452$, $\beta_3 = 0.6153$, $\beta_4 = -1.5366$, $\beta_5 = 2.3369$, $\beta_6 = -1.7361$, $\beta_7 = 0.5261$. Using (8.7) from approximation (8.11), we obtain the generalized Plesset–Zwick formula

$$m = 2\sqrt{\frac{3}{\pi}}\mathrm{Ja}\,\psi, \tag{8.12}$$

where the factor ψ is determined by the relation

$$\psi(\mathrm{S}) = \frac{\left(1 + \left(\sqrt{\pi/2} - 1\right)\mathrm{S}\right)}{\sqrt{1 + \sum_{i=1}^{n} \beta_i \mathrm{S}^i}}. \tag{8.13}$$

Making $\mathrm{S} \to 0$ we have $\psi \to 1$, relation (8.12) becoming the classical Plesset–Zwick formula (8.8). From the series expansion on the right side of equality (8.6) with $m_* \to \infty$, we obtain an asymptotics, from which we have the expression for the growth modulus

$$m = \sqrt{\frac{6}{1 - \mathrm{S}} \frac{1}{\varepsilon}}. \tag{8.14}$$

Note that approximation (8.11) was initially constructed from the conditions of providing the passage to the limit in asymptotics (8.14)

$$\mathrm{S} \to 1, \psi \to \sqrt{\frac{\pi}{2}} \frac{1}{\mathrm{S}\sqrt{1 - \mathrm{S}}}, m \to \sqrt{\frac{6}{1 - \mathrm{S}} \frac{\mathrm{Ja}}{\mathrm{S}}}.$$

In view (8.10), this implies the asymptotic formula (8.14) describing the case of an infinitely high growth rate: $\mathrm{S} \to 1, m \to \infty$. The physical meaning of this asymptotics is dictated by the specificity of the energy thermal scheme [9]. When the liquid superheating enthalpy $c_p \Delta T$ becomes equal to the evaporation heat L, each elementary volume of the liquid near the interface is free to change into vapor, and no heat input from the outside is needed, so that all limitations for the phase transformation rate disappear. From relation (8.14), it follows that in the range of high superheating temperatures, one should use the Stefan number for the base parameter—in using the Jakob number, the inadmissible "getting into" the region of $\mathrm{S} \equiv \varepsilon \mathrm{Ja} > 1$ is possible.

In the overwhelming majority of cases, the bubble growth rate $\dot{R} = dR/dt$ will be determined by the simultaneous action of the inertial reaction of the liquid medium and the evaporation rate at the interface [9]. From the point of view of the "binary" growth inertial-thermal scheme of [5], the spherical expansion of a bubble in the liquid causes its dynamic reaction, leading to a pressure increase in the bubble. As a consequence, the saturated vapor temperature in the bubble increases and the "liquid-vapor" temperature drop decreases. The heat flow to the interface turns out to be smaller, the growth rate of the bubble is lower than that predicted by the energy thermal scheme.

Consequently, in the general case, it is necessary to take into account the time variation of the bubble surface temperature. The inertial-thermal scheme of the growth was first described theoretically in [10] in the case $S \ll 1$. A number of simplifying assumptions were made, the strongest of which are the following: the vapor density in the bubble is constant throughout the growth period and is equal to the saturated vapor density at pressure p_∞ in the system. The portion of the saturation curve relating to the pressure and the temperature in the vapor phase is approximated by the linear segment. In [11], the computing method of [10] (under the same strong assumptions) was used for the case $S \sim 1$. A detailed numerical investigation of the bubble growth for the range of high superheating temperatures of the liquid was carried out in [12].

8.3 The Pressure Blocking Effect

Since Gibb's time [13], it was known that the only physical restriction imposed on the liquid temperature in the metastable region follows from the condition of its thermodynamic stability: the upper limit of the existence of the liquid phase is the limiting superheating temperature (the spinodal temperature). Following [5], we consider the case of highly superheated liquid: $S = S_{max} > 1$. The starting point is the important practical conclusion drawn as a result of the numerical investigation [14]: at a time variable temperature of the vapor one can use simultaneously, with fair accuracy, both limiting laws of growth—the inertial and the thermal ones. This enables one to equate the right-hand sides of Eqs. (8.2) and (8.3)

$$\dot{R} = \sqrt{\frac{2}{3}\frac{p_v - p_\infty}{\rho}} = \frac{1}{2}m\sqrt{\frac{\alpha}{t}}. \tag{8.15}$$

From (8.15), we can obtain with the help of the generalized Plesset–Zwick formula (8.12) the equation for the growth rate in the implicit form

$$t = f(\dot{R}). \tag{8.16}$$

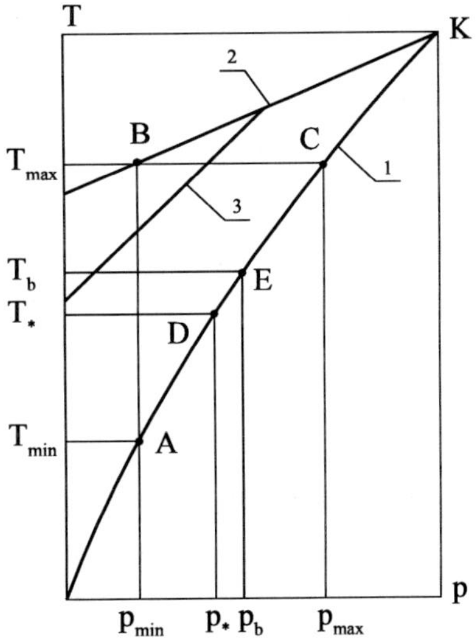

Fig. 8.1 Stages of vapor bubble growth in a superheated liquid with S_{max}. 1 Binodal, 2 spinodal, and 3 energy spinodal

The explicit representation of (8.16) must include the equation of the saturation curve and is very awkward. According to the asymptotics of the infinite growth rate (8.14), the local Stefan number should always be less than 1: $S = S(t) < 1$. The fulfillment of this physical condition leads to a specific inertial-thermal scheme of the vapor bubble growth at $S \to 1$.

Let us consider in the p, T diagram the process of bubble growth in the case of $S = S_{max}$ (Fig. 8.1). The state of the saturated vapor in the bubble corresponds to the point $A\{p_{min}, T_{min}\}$ on the binodal (saturation line), the state of the superheated liquid corresponds to the point $B\{p_{min}, T_{max}\}$ on the spinodal (limiting superheating line), and AB is an isobar. The Stefan number is calculated by the formula

$$S = \frac{1}{L}\int_{T_{min}}^{T_{max}} c_p dt, \tag{8.17}$$

where T_{max} is the spinodal temperature. According to [5], by the "energy spinodal" we shall mean the $T_e(p)$—curve in the p, T diagram for which at each isobar the condition

$$S_e = \frac{1}{L}\int_{T_{min}}^{T_e} c_p dt$$

is fulfilled. The energy spinodal is described by line 3 in Fig. 8.1 lying between binodal 1 and spinodal 2 and intersecting the latter. Thus, at deep penetration into the metastable region, the critical value of $S = 1$ can be attained theoretically before the spinodal is reached.

Let us consider in the p,T diagram the stages of bubble growth with $S_{max} > 1$ (Fig. 8.1).

(1) The initial state: $t = -0$. The vapor is at the point $A\{p_{min}, T_{min}\}$ on binodal 1, and the liquid is at the point $B\{p_{min}, T_{max}\}$ on spinodal 2 and remains there at $t > 0$, the Stefan number $S = S_{max}$.

(2) The initial state: $t = +0$. The inertial response of the liquid is engaged. The pressure in the bubble increases along the binodal from the point $A\{p_{min}, T_{min}\}$ to the point $C\{p_{max}, T_{max}\}$, the temperature drop and the Stefan number go to zero: $\Delta T = S = 0$. All these changes occur abruptly.

(3) The transition stage: $t > 0$. The bubble grows by the inertial-thermal scheme, the state of the vapor "drifts" along the binodal from the point $C\{p_{max}, T_{max}\}$ to the point $D\{p_*, T_*\}$, the Stefan number increases: $0 \le S(t) < 1$.

The asymptotic stage: $t \to \infty$. The state of the vapor "hovers" in the vicinity of the point $D\{p_*, T_*\}$ that corresponds to the condition of the energy spinodal: $S(t) \to c_p(T_{max} - T_*)/L$. The temperatures T_{max} and T_* are on the "isobar of blocking": $p = p_* = \text{const}$. From this, it follows that with $t \to \infty$ the bubble will grow by the asymptotic inertial scheme

$$\dot{R}_* = \sqrt{\frac{2}{3}\frac{(p_* - p_{min})}{\rho}}. \tag{8.18}$$

Thus, when the vapor bubble grows in the region of $S_{max} > 1$, the pressure blocking effect must take place in the vapor phase [5], $S(t) \to 1$, $p(t) \to p_* > p_{min}$, $T(t) \to T_* > T_{min}$

8.4 The Stefan Number in the Metastable Region

The rigorous calculation of the Stefan number (8.5) in the metastable region by formula (8.7) can only be performed on the basis of the equation of state of real gases [15]. Notably, by the time the well-known monograph [15] was written, more than 100 equations of state based on the classical Van der Waals equation had been published. Since that time their number continued to grow steadily, but the most suitable equations for engineering calculations were still the "old" equations, such as the Dieterici, Berthelot, Redlich–Kwong, and other equations [16]. For instance, from the Soave–Redlich–Kwong equation [16] the following approximation of the spinodal equation can be obtained

$$\frac{T_{max}}{T_{cr}} = 0.89 + 0.11\frac{p}{p_{cr}}. \tag{8.19}$$

Here, T_{cr} and p_{cr} are, respectively, the temperature and pressure at the critical point K (Fig. 8.1).

The traditional method of approximate calculation of the thermal properties in the metastable region is the method of temperature approximations [13]. To illustrate this method, let us consider the limiting case of spinodal superheatings (Fig. 8.1). At the instant of time $t = -0$ (effervescence), the temperature of the liquid at the point B at a given pressure in the system can be calculated by formula (8.19). At time $t = +0$ (initiation of the inertial reaction of the medium), the bubble pressure increases stepwise to the value of $p_{\max}$ (the point C). In this case, the isobar heat capacity on the binodal at its intersection with the isotherm $T = T_{\max}$ is equal to some value of $c_p(T_{\max})$. This value will be larger than that of the isobar heat capacity at the saturation temperature $c_p(T_{\min})$. According to the method of temperature approximations, the error in determining c_p in the metastable region cannot be verified in principle.

In his monograph [17], Novikov generalized the results of his investigations of many years in the area of the theory of phase transitions of the second kind. Developing the ideas of Gibbs and Landau, Novikov [17] proved the existence of an analogy between the singular behavior of the thermodynamic properties in the vicinity of the thermodynamic critical point and in the vicinity of the spinodal. In particular, according to [17], the function $c_p(T)$ in the metastable region with approach to the spinodal along the isobar obeys the universal scaling law

$$\frac{c_{p\min}}{c_p} = (1 - \theta)^{\gamma}. \tag{8.20}$$

Here, γ is a pseudocritical index

$$\theta = \frac{T - T_{\min}}{T_{\max} - T_{\min}}.$$

The dependence of the form (8.20) was also confirmed in [18]. The theory of [17] is strictly applicable in the vicinity of the spinodal (point B, Fig. 8.1), but it says nothing about the behavior of c_p on the isobar throughout the metastable region (interval AB, Fig. 8.1). The question about the exact value of the pseudocritical index within the interval given in [17] $1/3 \le \gamma \le 1/2$ also remains to the answered. A possible answer to these questions is contained in the theses [19, 20] published in Germany. For instance, in [19], on the basis of calculations made for different equations of state, the value of $\gamma = 1/2$ is recommended. Now, the scaling law (8.20) takes on the form

$$\frac{c_{p\min}}{c_p(\theta)} = \sqrt{1 - \theta}. \tag{8.21}$$

Figure 8.2 compares dependences (8.21) with the results of the numerical solution carried out in [20] on the basis of the Berthelot equation of state for the reduced pressure $p/p_{\rm cr} = 0.6$. Substituting (8.20) into (8.17) gives the relation for the Stefan number averaged over the entire metastable region

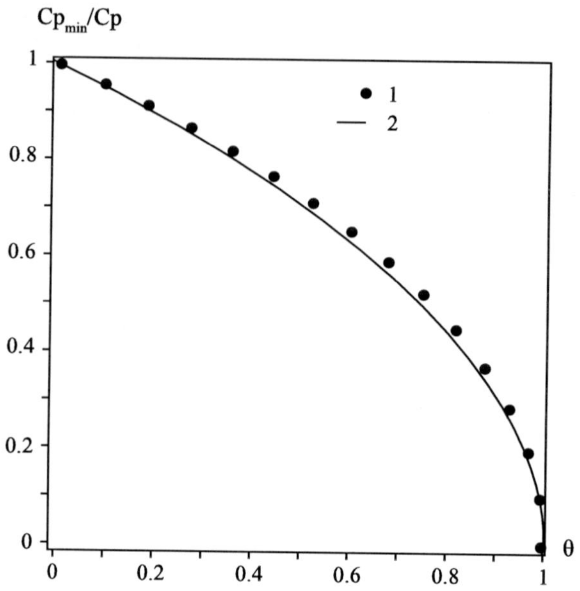

Fig. 8.2 Scaling law of change in the isobar heat capacity in the metastable region for the case of $p/p_c = 0.6$. 1 Calculation by the Soave–Redlich–Kwong equation, 2 calculation by formula (8.21)

$$S_{max} = \frac{1}{L}\int_{T_{min}}^{T_{max}} c_p dt = \frac{c_{pmax}(T_{max} - T_{min})}{(1-\gamma)L_{min}}. \tag{8.22}$$

8.5 Effervescence of the Butane Drop

In [21], with the help of high-speed filming with a time resolution of 10^{-3}, the process of effervescence of the butane drop in glycol at atmospheric pressure was investigated. When the surface of the drop reached a temperature close to the spinodal temperature, a vapor bubble began to grow in its volume, and in -100 the surface of this bubble reached the drop boundary. Let us analyze the laws of bubble growth as applied to the experimental conditions of [21]. Using the scaling law (8.21), after a number of elementary transformations we obtain a dependence of the form (8.16). The necessary thermal properties given in the form of a table were approximated by means of quadratic splines. For our analysis, it is important to emphasize that according to the calculation the initial value of the Stefan number $S_{max} = 1.26$. From this, it follows that from the point of view of the present model the experiment in [21] was performed in the region of pressure blocking. The calculated time dependence of the bubble growth rate is given in Fig. 8.3.

Here, we can distinguish three main growth stages. In the very short initial period ($t < 10^{-6}$ μs), the bubble grows by the Rayleigh law (8.2). The prolonged intermediate stage (10^{-6} $\mu s < t < 10^4$ μs) proceeds under the mutual influence of the

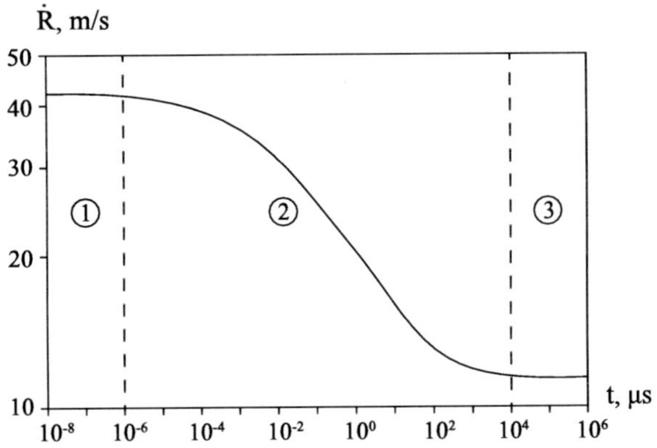

Fig. 8.3 Calculated time dependence of the growth rate of the bubble as applied to the experimental conditions of [21]. 1 The initial Rayleigh law, 2 the intermediate inertial-thermal law, 3 the asymptotic Rayleigh law

inertial and the thermal mechanisms. Finally, at $t > 10^4$ μs the pressure blocking effect shows up: the bubble grows in accordance with the asymptotic Rayleigh law (8.18). The vapor pressure in it does not decrease to below $p_* \approx 2.1$ bar.

Figure 8.4 shows the change in the local Stefan number according to the modified growth modulus m_* (Fig. 8.4a) and the time (Fig. 8.4b). Numerical integration of a relation of the form (8.16) with the use of the program package of the system of computer algebra Maple leads to the sought curve of the bubble growth $R(t)$. This curve in the range of 5 μs $< t <$ 100 μs is approximated fairly exactly by the simple dimensional relation

$$R \approx 0.0242t^{0.9}. \tag{8.23}$$

Here, R is given in mm, and t is given in μs.

The results of numerical calculations are presented in Fig. 8.5. The straight lines 1 and 2 describe the corresponding Rayleigh growth laws—the initial (8.2) and asymptotic (8.18) one. As is seen from Fig. 8.5, the inertial-thermal branch of the growth curve (region 2 in Fig. 8.3) is in good agreement with the experimental data of [21]. The above results can be regarded as a visual illustration of the pressure blocking effect. Thus, the pressure blocking effect is not a certain abstraction and can be realized in a concrete experiment. A very good agreement with the experimental data of [21] was also attained in the numerical investigation in [12]. The author of [12] does not indicate the method of calculating the isobar heat capacity in the metastable region. However, here the use of standard method of temperature approximations can be supposed. This is supported by the range (2.34 – 3.20) kJ/(kg K) of c_p values, as given on page 454 of [12] for the (273 – 378) K temperature interval. This exactly corresponds to the range of variation of the isobar heat capacity of butane along the binodal between points B and C at its effervescence at atmospheric pressure (Fig. 8.1).

From here for the conditions of numerical calculations [12], we obtain $S_{max} =$ 0.868. Consequently from the viewpoint of the approach used in [12], the pressure

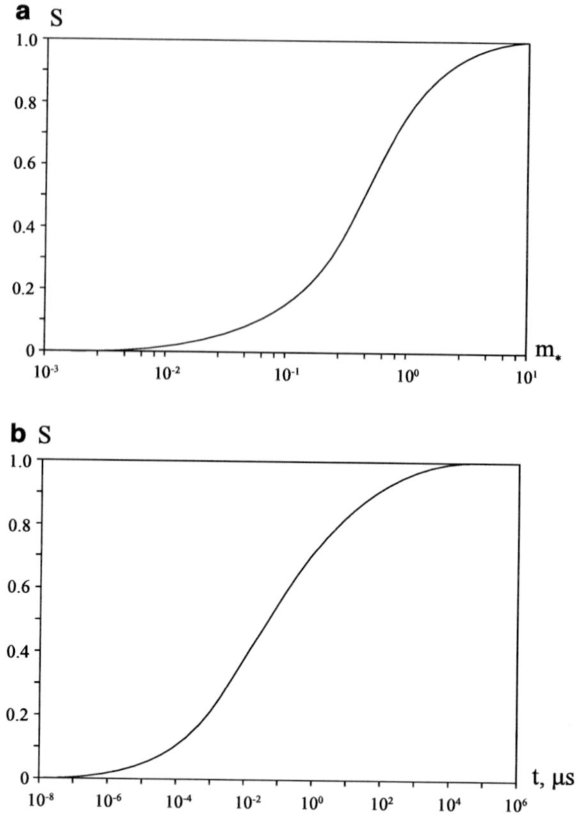

Fig. 8.4 Local Stefan number as a function of the modified growth modulus (**a**) and time (**b**) as applied to the experimental conditions of [21]

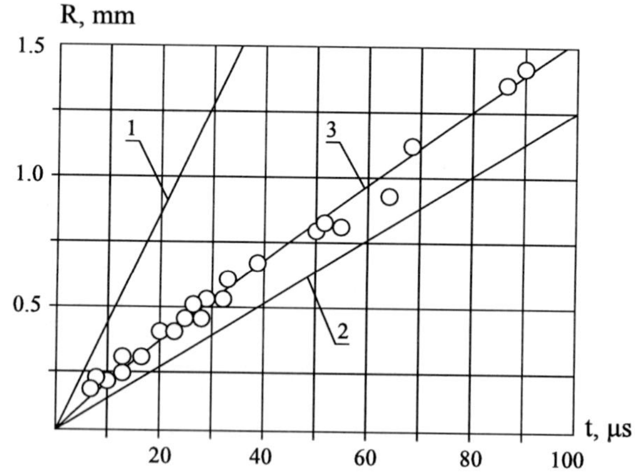

Fig. 8.5 Curve of the bubble growth in effervescence of the butane drop. Dots show the experimental data of [21]. 1 The initial Rayleigh law, 2 the asymptotic Rayleigh law, and 3 calculated dependence

blocking could not be revealed in principle. Thus, we have two radically different qualitative interpretations of the same experimental data [21], and in both cases there is good quantitative agreement between the model and the experiment. The author of [12] believes that "... the high growth rates observed in experiments can be explained by the long inertial stage during which a substantial pressure difference between the liquid and the bubble is sustained ...".

Consequently, according to [12], in the course of time the inertial-thermal law of growth must "fall down" to the asymptotic thermal law. The present model is based on the following mechanism of pressure blocking (Fig. 8.3). During the very short initial period (stage 1), the bubble grows according to the inertial Rayleigh law $R \sim t$. During the second (inertial-thermal) stage the growth curve is approximated by the power law $R \sim t^n$, and for the conditions of the experiment of [12] the exponent varies over the range of $0.9 < n < 1$. Finally, in the course of time the inertial-thermal law of growth will smoothly go over into the asymptotic Rayleigh law (stage 3).

The author of [11] managed to explain the points obtained in [21] only for the region of short times: $t < 40$ μs. At the same time, the difference between the calculated values of the bubble radius obtained for $t < 40$ μs and the experimental values widens [21], at $-t100$ μs the difference being more than two times as great. As it seems, such a wide discrepancy is a consequence of the essential shortcomings of the computational procedure [11], which in essence is an extrapolation to the region $S \sim 1$ of the original model of [10] with its all rigid assumptions. For instance, the authors of [10, 11] took into account the change with time in the bubble pressure but ignored the corresponding change in the density which changes meanwhile by a factor of about.

17 (!) as the bubble moves from the point $C\{p_{\max}, T_{\max}\}$ to the point $D\{p_*, T_*\}$ (Fig. 8.1).

Then, the approximation of a portion of the saturation curve by a straight line segment can strongly distort the real picture of the interaction between the inertial and the thermal growth mechanisms [9]. Finally, in calculating the growth modulus on the right side of (8.15) in [11] the Scriven integral approximation [7], whose error in the region of S -1 exceeds 10%, was used. With such "synthesis" of contradictory initial positions used without substantiating their reliability, it is practically impossible to determine the range of applicability of the computational model [11]. It should also be recognized that the criticism in [11] of the universally recognized Plesset–Zwick formula, which is the basis for calculating the bubble growth [8–10], is groundless. For our analysis, it is important that the calculated maximal Stefan number in [11] obtained by the method of temperature approximations did not exceed the value of $S = 0.7$, which initially excludes the pressure blocking effect.

8.6 Seeking an Analytical Solution

8.6.1 Qualitative Analysis

Relation (8.23) represents an approximation of the results of a single numerical experiment, in which at given values of the temperatures ($T_{\min}$, $T_{\max}$) and pressures ($p_{\min}$, $p_{\max}$) for a given liquid (butane) the dependence $R(t)$ was obtained. The laws of growth of vapor bubbles are usually not the immediate goal of analysis in applied problems. As a rule, it is necessary to know the growth law for constructing physical models of the heat transfer at nucleate boiling [6, 9]. Therefore, seeking approximate analytical equations describing the bubble growth in a wide range of diagnostic variables is topical. A broad spectrum of analytical solutions of problems of single-phase thermo-hydrodynamics is given in Weigand's monograph [22]. Following [22], let us list the advantages of the classical analytical approach over numerical methods.

(1) The importance of the analytical approach is that it provides the possibility of closed qualitative description of the considered process, revealing the complete list of dimensional diagnostic variables, and hierarchical classification of the above variables according to the degree of their importance.
(2) Analytical solutions feature the necessary generality, therefore, varying their boundary and initial conditions permits parametric investigation of a wide class of problems.
(3) To test numerical solutions of input exact equations, it is necessary to have basic analytical solutions of simplified equations. The latter is obtained as a result of physical estimations of the significance of individual terms and rejection of secondary effects.
(4) The necessary condition for putting results of numerical calculations into practice is their validation of known classical solutions. Consequently, direct check of the correctness of setting numerical investigations can only be made on the basis of available analytical solutions.

In the light of the foregoing, it may be stated that the above advantages of the analytical approach were realized in investigating the limiting schemes of bubble growth in Labuntsov's paper [6] which is still topical.

On the basis of the above considerations, we can propose the following algorithm for constructing an analytical solution of the problem on the vapor bubble growth in the region of strong superheating

- the dependence $R(t)$ in the inertial-thermal region (region 2 in Fig. 8.3) can approximately be described by the power law $R = At^n$, where A is some dimensional complex of thermal properties (generally speaking, hitherto unknown),
- according to (8.24), (8.27), the exponent is bounded from below: $n \geq 3/4$,
- according to the physical content of the inertial-thermal growth scheme, the exponent is bounded from above: $n \leq 1$,
- according to (8.23), the exponent is a function of the Stefan number: $n = n(\mathrm{S})$.

8.6.2 Asymptotically Analytical Solution

As is known [23], infinitive divergent series are frequently useful in calculating required values, which are sought in the form of partial sums of series. In general, the desired function is expanded in a function series with respect to some variable. In this case, the approximation given by the first few members of the series is the better, the closer the independent variable is to a certain limit value. In many cases, the members of a series first rapidly decrease, and then begin to increase again. With such formally divergent series, one can generally work in the same way as with the convergent function series (performing addition, multiplication, integration, and differentiation operations). Historically, the first source of divergent asymptotic series was the Euler–MacLaurin summation formula [24]. In this case, the divergence is associated with the properties of Bernoulli numbers, which are coefficients of the divergent series under consideration. The concept of an "asymptotic series" was introduced in 1886 by Poincaré [25] when studying the divergent series

$$\sum\nolimits_{n=0}^{\infty} \frac{A_n}{x^n}, 0 < A_n < 1.$$

The earliest and most significant book on the applied theory of singular perturbations was the famous book by Van Dyck [26]. It was the first to present the method of consistent asymptotic (or internal and external) expansions. The main idea of the method is to search for a region of overlap in which the internal and the external expansions agree.

In this case, the internal expansion is based on the asymptotic formula (as $t \to 0$) in which the first term is described by the Rayleigh formula (8.2). The thermal balance equation is written as the Plesset–Zwick formula

$$\rho_{\mathrm{v}} L\dot{R} = \frac{k\Delta T}{\sqrt{\alpha t}}. \tag{8.24}$$

However, since $k \to \infty$ on the spinodal line, it is impossible to get the second term from the internal expansion. This singularity creates fundamental difficulties in the studies of vapor bubble growth as $t \to 0$.

To find the external expansion, we consider the asymptotics as $t \to \infty$, in which the first term is described by formula (8.18). The thermal balance equation in this asymptotics assumes the form

$$\rho_v L\dot{R} = \sqrt{\frac{3}{2}} \frac{\kappa \Delta T}{\sqrt{(1-S)\alpha t}}. \tag{8.25}$$

Let us now express the temperature difference in terms of the pressure difference using the Clapeyron–Clausius equation near the blocking point $D\{p_*, T_*\}$ (Fig. 8.1)

$$\frac{T_{\rm v}-T_*}{T_*}=\frac{p_{\rm v}-p_*}{L\rho_{\rm v}}. \tag{8.26}$$

Using the Rayleigh formula

$$\dot{R}=\sqrt{\frac{2}{3}\frac{(p_{\rm v}-p_{\min})}{\rho}}, \tag{8.27}$$

from the solution of the biquadratic equation, we get

$$u=\left[\frac{1}{2}+\left(\frac{1}{4}+\frac{1}{\tau}\right)^{1/2}\right]^{1/2}. \tag{8.28}$$

Here, $u=\dot{R}/\dot{R}_*$ is the dimensionless velocity of bubble growth, $\tau=t/t_0$ is the dimensionless time, $t_0=\frac{\alpha L^2}{\varepsilon T_*\langle c_p\rangle \dot{R}_*^4}$ is the time scale, $\langle c_p\rangle$ is the isobaric heat capacity as averaged over the entire metastable region using Eq. (8.21).

Integrating (8.28) with respect to time, we get the expression for the dimensionless bubble radius $r=R/R_0$, where $R_0=\dot{R}_*t_0$ is the scale length. This relation is extremely cumbersome and contains a linear function, a logarithmic function, and a hypergeometric function. However, the asymptotic of the solutions are quite simple

$$\left.\begin{array}{c} t\to 0: r_0=\frac{4}{3}\tau^{3/4}, \\ t\to\infty: r_\infty=\tau. \end{array}\right\} \tag{8.29}$$

An approximation of the analytical solution, which is accurate in asymptotics (8.29) and whose error is at most 1.5%, can be represented as

$$r=\left(r_0^n+r_\infty^n\right)^{1/n}, \tag{8.30}$$

where $n=2.68$.

Calculation via (8.30) gives a good agreement with the experimental data of [21]. Moreover, the analytical curve (Eq. (8.30)) passes very close to the numerical curve (Eq. (8.23)) and deviates from it by no more than ±10%. This is why we do not show the dependence (8.30) in Fig. 8.5.

Strictly speaking, the analytical solution thus obtained can be applied to the case $\tau >> 1$, and therefore, the noted agreement is somewhat unexpected. However, the above result supports the physical rule stated by Labuntsov [6]: "The correct description of the "late stage" of vapor bubble growth allows one, as a rule, to correctly extrapolate it to the entire growth curve". In this sense, the asymptotically analytical solution (8.30) can be recommended for describing the pressure blocking effect in a growing vapor bubble.

8.7 Conclusions

The problem on the vapor bubble growth in a liquid whose superheating enthalpy exceeds the phase transition heat was considered. It was that when the Stefan number exceeds 1 there arises a feature of the mechanism of heat input from the liquid to the vapor leading to the effect of pressure blocking in the vapor phase. The known theoretical Plesset–Zwick formula was extended to the region of strong superheating. To calculate the Stefan number in the metastable region, we used the scaling law of change in the isobar heat capacity. The problem for the conditions of the experiment on the effervescence of the butane drop was solved numerically. An algorithm was proposed for constructing an approximate analytical solution for the range of Stefan numbers greater than unity.

References

1. Prosperetti A (2004) Bubbles. Phys Fluids 16(6):1852–1865
2. Lohse D (2006) Bubble Puzzles. Nonlinear Phenom Complex Syst 9(2):125–132
3. Straub J (2001) Boiling heat transfer and bubble dynamics in microgravity. Adv Heat Transf 35:157–172
4. Picker G (1998) Nicht-Gleichgewichts-Effekte beim Wachsen und Kondensieren von Dampfblasen, Dissertation, Technische Universitat München, München
5. Zudin YB (2015) Binary schemes of vapor bubble growth. J Eng Phys Thermophys 88(3):575–586
6. Labuntsov DA (1974) Current views on the bubble boiling mechanism. In: Heat transfer and physical hydrodynamics. Nauka, Moscow, pp 98–115 (in Russian)
7. Scriven LE (1959) On the dynamics of phase growth. Chem Eng Sci 10(1/2):1–14
8. Plesset MS, Zwick SA (1954) The growth of vapor bubbles in superheated liquids. J Appl Phys 25:493–500
9. Labuntsov DA, Yagov VV (1978) Mechanics of simple gas-liquid structures. Moscow Power Engineering Institute, Moscow (in Russian)
10. Mikic BB, Rosenow WM, Griffith P (1970) On bubble growth rates. Int J Heat Mass Transf 13:657–666
11. Avdeev AA (2014) Laws of vapor bubble growth in the superheated liquid volume (thermal growth scheme). High Temp 40(2):588–602
12. Aktershev SP (2004) Growth of a vapor bubble in an extremely superheated liquid. Thermophys Aeromech 12(3):445–457
13. Skripov VP (1974) Metastable liquid. Wiley, New York
14. Korabel'nikov AV, Nakoryakov VE, Shraiber IR (1981) Taking account of non-equilibrium evaporation in the problems of the vapor bubble dynamics. High Temp 19(4):586–590
15. Vukalovich MP, Novikov II (1948) Equation of state of real gases. Gosenergoizdat, Moscow (in Russian)
16. Reid RC, Prausnitz JM, Poling BE (1988) The properties of gases and liquids, 4th edn. McGraw-Hill Education, Singapore
17. Novikov II (2000) Thermodynamics of spinodal and phase transitions. Nauka, Moscow (in Russian)
18. Boiko VG, Mogel KJ, Sysoev VM, Chalyi AV (1991) Characteristic features of the metastable states in liquid-vapor phase transitions. Usp Fiz Nauk 161(2):77–111 (in Russian)

19. Thormahlen I (1985) Grenze der Überhitzbarkeit von Flüssigkeiten: Keimbildung und Keimaktivierung, Fortschritt-Berichte VDI. Verfahrenstechnik. VDI-Verlag, Düsseldorf, Reihe 3, Nr. 104
20. Wiesche S (2000) Modellbildung und Simulation thermofluidischer Mikroaktoren zur Mikrodosierung, Fortschritt-Berichte VDI. Wärmetechnik/Kältetechnik. VDI-Verlag, Düsseldorf, Reihe 19, Nr. 131
21. Shepherd JE, Sturtevant B (1982) Rapid evaporation at the superheat limit. J Fluid Mech 121:379–402
22. Weigand B (2015) Analytical methods for heat transfer and fluid flow problems, 2nd edn. Springer, Berlin
23. Steinrück H (2010) Asymptotic methods in fluid mechanics: survey and recent. Springer, Wien
24. DeVries PL, Hasbun JE (2011) A first course in computational physics. Jones and Bartlett Publishers
25. Poincare H (1886) Sur les integrals irregulieres des equations lineaires. Acta Mathematica 8:295–344
26. Van Dyke M (1964) Perturbation methods in fluid mechanics. (Applied mathematics and mechanics, Vol 8). Academic Press

Chapter 9
Evaporating Meniscus on the Interface of Three Phases

Abbreviations

BC Boundary condition

Symbols

B	Kinetic parameter
k	Thermal conductivity
K	Curvature
L	Heat of phase transition
P	Pressure
q	Heat flux
T	Temperature

Greek letter symbols

β	Evaporation–Condensation coefficient
ε	Heat flux parameter
δ	Film thickness
σ	Surface tension

Subscripts

0	State at $x = 0$
l	State at $x = l$
m	Meniscus
v	Vapor

Y. B. Zudin, *Non-equilibrium Evaporation and Condensation Processes*, Mathematical Engineering, https://doi.org/10.1007/978-3-030-67553-0_9

9.1 Introduction

Modern progress in the nanotechnology, micro- and nano-electronics depends on a detailed analysis of the behavior of the interphase boundary in microscopic objects, and in particular, on the "liquid–gas" interphase boundary. Of special importance here are the manifestations of the effect of the intermolecular and superficial forces, which control the motion of macroscopically thin films. Super-thin (nanoscale) films occur in practice only in crystal growth processes, treating of printed-circuit boards, in biological microreactors, etc. Nanotechnology is concerned with polar fluids, of which water is the most common one. On the interface of two media, polar fluids may form a double electric layer, which has an effect on the behavior of this interface. This effect is manifested in the form of specific intermolecular and superficial forces as the rigid and liquid surfaces become in contact.

Systematic experimental and numerical studies of flows in evaporating thin films on a heated surface were begun by Wayner with collaborators in a series of papers started in [1]. Without mentioning all studies in the series of papers, we note one of the last ones [2]. According to [1, 2], a film consists of the following three regions (Fig. 9.1).

(1) the adsorbed microfilm of thickness of the molecular size $\delta_0 = \left(10^{-10} - 10^{-9}\right)$ nm (the nanoscale film),
(2) the evaporating film of variable thickness $\delta(x)$ (viz., the meniscus of the liquid film),
(3) the macrofilm of thickness δ_l.

In region "1", the effect of the dispersion of Van der Waals forces from the side of solid boundary is predominant, these forces are known to impede the evaporation. In region "2", the intensity of evaporation increases as the film becomes more thick and the dispersion forces become weaker. In region "3", the evaporation intensity again decreases as a consequence of an increase in the thermal resistivity δ/k of the

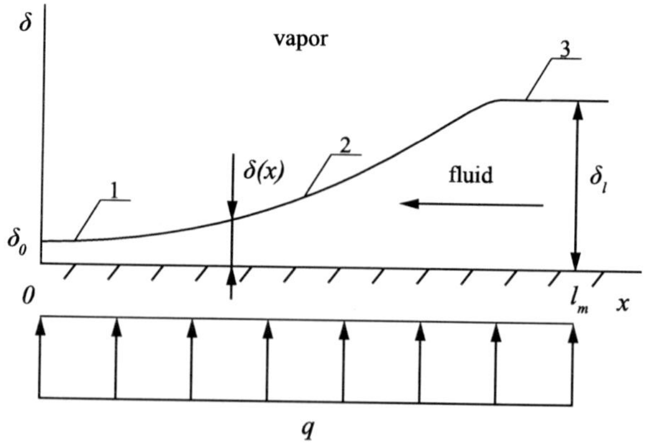

Fig. 9.1 The scheme of the flow in the meniscus of an evaporating film. 1 The adsorbed microfilm, 2 the evaporating film of variable thickness (viz., the meniscus of the liquid film), and 3 the macrofilm

liquid film. Thus, the main heat extraction of the evaporation film from the heated surface corresponds to the meniscus region ($\delta_0 < \delta < \delta_l$).

A theoretical investigation of thermo-hydrodynamics of a thin liquid film wetting the groove of a heat pipe was made in the papers [3, 4], which extended the theoretical studies of [1, 2] to an important application area. On the whole, the results of [3, 4] were supported by the numerical investigations [5, 6]. Numerical modeling of the problem of evaporation of the meniscus in circular capillary tubes was made in [7–9]. The authors [10] studied this problem with the help of the molecular dynamics method.

The papers [11, 12] (see also the book [13]) put forward a physical model of nucleate boiling heat transfer, which takes into account powerful heat sinks near the "dry path" on the heating surface. The model of [11, 12] is based on the analysis of thermo-hydrodynamic properties of the evaporating meniscus in the region of contact of three phases along the periphery of the dry path. It was shown that, despite the small area of the dry region, their role in the total heat balance of nucleate boiling can be fairly significant (even decisive in the high-pressure region).

The model of [11–13] supplements the Labuntsov classical theory of nucleate boiling [14] by extending it to the high-pressure region. This approach is of integral type: it is oriented toward the calculation of the averaged heat transfer by considering the total contribution of coupling various effects (the oscillation of the thin liquid film under large size vapor agglomerations, the flow near the dry path, nucleation sites on the surface).

The local approach to the problem of nucleate boiling [15–17], which is an alternative to that of [11–13], depends on the analysis of the thermo-hydrodynamic properties of the evaporating meniscus under a single vapor bubble growth on the heating surface. The principal assumption of the local approach was formulated in [15]: the growth of vapor bubbles is completely governed by the evaporation process on the unsteady region of interface of three phases which is attached to the bubble.

Experimental and numerical studies of the bubble growth problem on the basis of the model of evaporating meniscus were made in [18], where the effect of the travel velocity of the three-phase boundary on the local heat removal from the heat surface was discovered. We note that vertical chains of vapor bubbles that separate in succession from relatively stable nucleation sites on the wall can be observed only for relatively small heat fluxes [13]. This is precisely the regime of individual bubbles in which one usually performs cinematographic recordings and obtains experimental information on the dynamics of vapor bubble growth and detachment. As the heat flux density increases, bubbles become to merge and transform into vapor conglomerates which grow and separate from the wall.

It seems that the direct numerical simulation of the multiple-factor process of nucleate boiling is a matter of remote future. This shows that the calculation scheme of [15–18] is oversimplified and can serve as a basis for the analysis of boiling thermo-hydrodynamics only for very small thermal loadings. The study of the flow laws in thin films stimulated the appearance of various analytical methods of the solution of nonlinear equations based on asymptotic expansions. Reynolds [19] with his analysis of the lubrication flow was the first to study the theory of flows of thin layers on a hard surface. At present, the hydrodynamic lubrication theory, which is an individual branch of mathematical physics [20], is widely useful in modeling flows in thin films. With the help of the asymptotic approach, this theory is capable of reducing the Navier–Stokes equations to far more simple partial differential equations. These equations, which conserve the principal physical regularities of the initial problem, are known to be highly nonlinear.

In the present chapter, we shall be concerned with the hydrodynamic of the evaporating meniscus of a thin liquid film on a heated surface. We put forward an approximate method of the solution, which is capable of finding the effect of the kinetic molecular phenomena on the geometric meniscus parameters and on the intensity of heat removal from the hard wall. The methods depend on a substantial simplification of the real flow pattern in the evaporating meniscus. The purpose of this method is to obtain an approximate analytical solution describing the flow thermo-hydrodynamics of an evaporating thin film on the interface of three phases.

The scheme of the flow in the meniscus of an evaporating film is depicted in Fig. 9.1. The flow of liquid in the negative direction of the longitudinal coordinate x is caused by the drop of pressure controlled by the curvature gradient of phase interphase. Liquid evaporates as the flow progresses in the meniscus toward its thinning. It is assumed that the stationary of the process is secured by liquid makeup from the side of the macrofilm. The flow in the meniscus becomes gradually more slow and terminates at the conventional boundary with the adsorbed thin film (the nanoscale film) of thickness δ_0, where the process of liquid evaporation terminates subject to the Van der Waals forces. A nanoscale film is an intriguing physical object, which simultaneously manifests the action of the viscos and intermolecular forces, as well as the kinetic intermolecular effects.

The study of the phenomena occurring in adsorbed thin films was initiated by Van der Waals [21], who called the intermolecular forces the “inner pressure”. Van der Waals explained the appearance of the inner pressure by the difference of liquid properties in the interfacial transitory layers. In the theory of intermolecular forces [22], it is assumed that the Van der Waals forces are long-range forces of molecular attraction of magnetoelectric nature. It is also assumed that in the bulk phase the inner pressure is governed by the coupling of liquid molecules. In the interfacial layers, this is superimposed by the exposure of molecules from the phases in contact with the liquid, which results in the appearance of the pressure difference in the nanofilm.

9.2 Evaporating Meniscus

Let us consider the fluid flux in comparison with the viscous force in an evaporating meniscus in the direction from the macrofilm toward the adsorbed film, which is affected by the pressure gradient in the liquid phase along the x-axis (Fig. 9.1). Wayner and Coccio [1] were the first to show that, under a constant surface tension coefficient on the surface interface phase interface and under a constant pressure in the vapor phase ($\sigma = \text{const}$, $p_v = \text{const}$), the only driving force in the liquid film is the curvature of the phase interface K

$$K = \frac{d^2\delta/dx^2}{\left[1 + (d\delta/dx)^2\right]^{3/2}} \approx \frac{d^2\delta}{dx^2}. \tag{9.1}$$

In Eq. (9.1), it was taken into account that the estimate $(d\delta/dx)^2 << 1$ holds for any point on the phase interface. Correspondingly, the evaporation of liquid in the meniscus results in the curvature gradient of the phase interface, which in turn, according to the Laplace formula, is responsible for the motion of liquid in the direction where the film becomes more thin

$$\frac{dp}{dx} = -\sigma\frac{dK}{dx} \approx -\sigma\frac{d^3\delta}{dx^3}. \tag{9.2}$$

For the scheme of Fig. 9.1, this means that $dK/dx < 0$, that is, the curvature phase interface decreases in the positive direction of the x-coordinate. The process of evaporation of the meniscus in the thin layer approximation [20] is described by the equation of motion

$$\frac{1}{3}\frac{\sigma\delta^3}{\nu}\frac{d^3\delta}{dx^3} + \Gamma = 0, \tag{9.3}$$

and the heat balance equation

$$\frac{d\Gamma}{dx} = \frac{q}{L}. \tag{9.4}$$

Here, Γ is the liquid mass flux.

The heat flux q, as transferred from the solid surface through the liquid film, is related to the temperature difference ΔT by the addition procedure of thermal resistance of film δ/k and kinetic thermal resistance $1/h_k$, which is responsible for the non-equilibrium of the evaporating process

$$q = \Delta T\left(\frac{\delta}{k} + \frac{1}{h_k}\right)^{-1}. \tag{9.5}$$

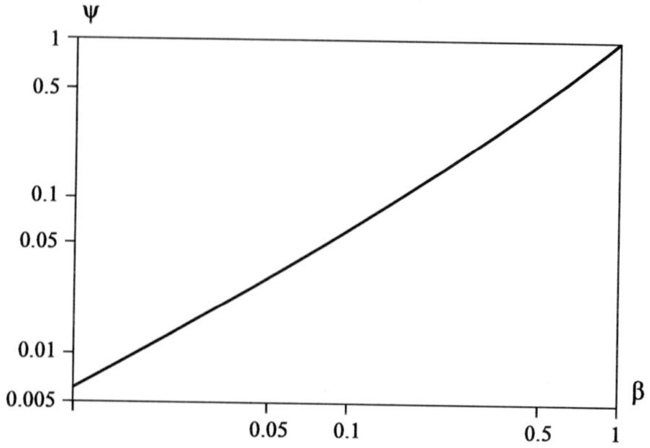

Fig. 9.2 The function $\psi(\beta)$

The kinetic heat transfer coefficient h_k is an important characteristic of the linear kinetic theory [23], which is defined by the relation

$$h_k = \frac{2}{3} \frac{C^{1/2} \psi \rho_v L^{3/2}}{T_s}. \tag{9.6}$$

Here, $C = L/R_g T_s \approx 10$, L is the heat of phase transition, R_g is the individual gas constant, and ρ_v is the vapor density

$$\psi(\beta) = \frac{0.6\beta}{1 - 0.4\beta} \tag{9.7}$$

is the function of the evaporation–condensation coefficient, so that $\psi(\beta) = 1$ for $\beta = 1$. The dependence $\psi(\beta)$ is shown in Fig. 9.2.

In the framework of the kinetic theory, the mass flux of evaporation appears as the disbalance of the two oppositely directed molecular fluxes: the one emitted from the phase interface and the one incident to it from the vapor content. In the general case, a portion of the incident flux (of quantity, corresponding to $\beta \leq 1$) is captured by the surface, the remaining part, which proportional to $1 - \beta$, is reflected from the surface without capturing. The quantity β is the averaged characteristic of the surface and of physical nature of condensate phase [23]. According to Eq. (9.5), the entire heat flux dissipated from the solid surface is spent for the evaporation of liquid. Differentiating both sides of Eq. (9.3) in x and inserting the result into Eq. (9.4), it follows by Eq. (9.5) that

$$\left(\frac{\delta}{k} + \frac{1}{h_k}\right)\left(\delta^3 \frac{d^4\delta}{dx^4} + 3\delta^2 \frac{d^3\delta}{dx^3}\right) + 3\frac{\nu \Delta T}{\sigma L} = 0. \tag{9.8}$$

Equations (9.3)–(9.8) use the following nomenclature: ν, k are, respectively, the kinematic viscosity and thermal conductivity of fluid, Γ is the mass flux of liquid in the meniscus (per unit length in the direction perpendicular to the figure). The nonlinear fourth-order ordinary differential Eq. (9.8) has no analytical solutions. Numerical investigations [7–9] reveal very interesting thermo-hygrometric properties of the evaporating meniscus. However, these results lie uncomfortably with each other and are not capable of giving a complete process pattern.

It seems that the problem here is not exhausted by the difficulties which could be circumvented by numerical methods, but has deeper physical reason. The hydrodynamic theory of lubrication [20] deals with flows of a liquid in a thin gap between solid bodies of known geometry, which enables one to uniquely specify the required number of Boundary Conditions (BCs). Hence, the procedure of the solution involves as a rule certain technical difficulties. The statement of the flow problem in the meniscus is, however, not that transparent. For example, the presence of the interfacial surface, whose form, strictly speaking, should be ascertained in the process of the solution (the free surface) calls for the averaging in the transverse section of the meniscus (the thin layer approximation [20]). Otherwise, in the study of the flow in the meniscus, one would have recourse to a nonlinear integro-differential equation.

Substantial uncertainty also comes from the procedure of conjugation of the meniscus itself with the external flow. As a result, it becomes difficult to formulate four boundary value conditions for Eq. (9.8) from fairly plausible physical considerations. In the above four numerical investigations, one seeks a solution to the fourth-order Eq. (9.8) (or of a more general system of equations) with the following boundary value conditions: the four BCs with $x = 0$ [1, 2, 7], two BCs with $x = 0$ and two BCs with $x = l$ [8], three BCs with $x = 0$ and one BC with $x = l$ (see [9]). Complete conclusions cannot also be made from the results of the interesting study of a flow in an evaporating meniscus [10], which depends on the methods of molecular dynamics. The above arguments justify to our opinion the search of an approximate analytical solution in the problem of meniscus hydrodynamics.

The author of the present book proposed a method of the solution depending on the change of the boundary value problem for the fourth-order equation by the Cauchy problem for the second-order equation (with BCs with $x = 0$) [24]. To this aim, a reduction of order of the initial equation was made in [2] using simplifying assumptions and plausible physical estimates. In the present chapter, which continues the studies of [24], we examine the kinetic molecular effects on the hydrodynamic of meniscus.

9.3 Approximate Analytical Solution

According to the physical analysis of flows in the meniscus, the process of evaporation in the meniscus decreases as the thickness decreases and stops in the region of the adsorbed nanoscale film $x = 0$ (Fig. 9.1). We now estimate the characteristics of the meniscus near this point. For $x = 0$, Eq. (9.5) assumes the form

$$q_0 = \Delta T \left(\frac{\delta_0}{k} + \frac{1}{h_k} \right)^{-1}. \tag{9.9}$$

Taking approximately $q \approx q_0$ in Eq. (9.4) and integrating in x with the BC

$$x = 0 : \Gamma = 0. \tag{9.10}$$

This gives

$$\Gamma = -\frac{q_0}{L} x. \tag{9.11}$$

From Eqs. (9.3), (9.11), we have the third-order equation

$$\delta^3 \frac{d^3 \delta}{dx^3} = -3\varepsilon x. \tag{9.12}$$

Here, $\varepsilon = \frac{q_0 \nu}{\sigma L}$ is the parameter, which has the sense of the dimensionless heat transfer with $x = 0$ (the "heat flux parameter"). A remarkable feature of Eq. (9.12) is that it preserves its form when changing to the dimensionless coordinates (that is, with an arbitrary length scale). Let us formulate three BCs for this equation. The first BC follows from the condition of conjugation of the meniscus and the microfilm (Fig. 9.1)

$$x = 0 : \delta = \delta_0. \tag{9.13}$$

The second BC is the smoothing requirement for such a conjugating

$$x = 0 : \frac{d\delta}{dx} = 0. \tag{9.14}$$

As the third BC, we take the natural physical condition of degeneration of the curvature meniscus when it changes to a macrofilm

$$x \to \infty : \frac{d^2 \delta}{dx^2} \to 0. \tag{9.15}$$

Assuming that $\varepsilon \ll 1$ we shall use the method of small parameter. For $\varepsilon = 0$ using Eq. (9.12), we find that

$$\frac{d^3 \delta}{dx^3} = 0, \frac{d^2 \delta}{dx^2} = K_0, \frac{d\delta}{dx} = K_0 x,$$

where K_0 is the curvature of the meniscus with $x = 0$. This gives us the meniscus profile in the zero approximation

$$\delta = \delta_0 + \frac{3}{2} K_0 x^2. \tag{9.16}$$

Substituting δ from Eq. (9.16) on the left of Eq. (9.12), this gives

$$\frac{d^3\delta}{dx^3} = -\frac{3\varepsilon x}{\left(\delta_0 + K_0 x^2/2\right)^3}. \tag{9.17}$$

Integrating Eq. (9.17) with the BC (9.15), we find that

$$\frac{d^2\delta}{dx^2} = \frac{6\varepsilon}{K_0\left(2\delta_0 + K_0 x^2\right)^2}. \tag{9.18}$$

Equating the values of the curvature with $x = 0$ in the first and zero approximations

$$\left.\frac{d^2\delta}{dx^2}\right|_{x=0} = \frac{3}{2}\frac{\varepsilon}{K_0 \delta_0^2} \approx K_0.$$

This gives us the following approximate relation between the nanoscale film thickness δ_0, the initial curvature of the meniscus K_0, and the parameter ε

$$K_0 = \left(\frac{3}{2}\varepsilon\right)^{1/2} \frac{1}{\delta_0}. \tag{9.19}$$

Introducing the dimensionless nomenclature

$$\tilde{\delta} = \frac{\delta}{\delta_0},\ \tilde{x} = \kappa \frac{x}{\delta_0},\ \kappa \equiv \left(\frac{3\varepsilon}{8}\right)^{1/4},$$

we write the expression for the curvature of the meniscus surface in the dimensionless form

$$\frac{d^2\tilde{\delta}}{d\tilde{x}^2} = \frac{2}{\left(1 + \tilde{x}^2\right)^2}. \tag{9.20}$$

Now, the BCs (9.13), (9.14) for Eq. (9.20) assumes the form

$$\tilde{x} = 0 : \tilde{\delta} = 1,\ \frac{d\tilde{\delta}}{d\tilde{x}} = 0. \tag{9.21}$$

Integrating in succession Eq. (9.20) with the BC (9.21), we find the relations for the slope of the meniscus contour

$$\frac{d\tilde{\delta}}{d\tilde{x}} = \frac{\tilde{x}}{1+\tilde{x}^2} + \arctan(\tilde{x}) \tag{9.22}$$

and the meniscus profile

$$\tilde{\delta} = 1 + \tilde{x}\arctan(\tilde{x}). \tag{9.23}$$

Equations (9.20), (9.22), (9.23) constitute the required analytical solution for the evaporating meniscus parameters. Figure 9.3 shows the curves calculated from these relations. As a result, we have the following asymptotics for the macrofilm

$$\tilde{x} \to \infty : \frac{d^2\tilde{\delta}}{d\tilde{x}^2} \approx \frac{1}{\tilde{x}^4} \to 0, \frac{d\tilde{\delta}}{d\tilde{x}} \approx \frac{\pi}{2}, \tilde{\delta} \approx \frac{\pi}{2}\tilde{x} \to \infty.$$

From Fig. 9.3, it is seen that an increase in $\tilde{x}$ (in the actual fact, for $\tilde{x} \geq 3$) results in an abrupt decrease of the meniscus curvature, its slope angle assuming a constant value, the thickness growing according to a linear law. Let us now write in more detail the asymptotics of the solution (9.23) in the dimensional form

$$x \to 0 : \delta = \delta_0 + 0.245\left(\frac{\beta}{1-0.4\beta}\frac{\rho_v L \nu \Delta T}{\sigma}\right)^{1/2} \frac{x^2}{\delta_0 R_g^{1/4} T_s^{3/4}}, \tag{9.24}$$

$$x \to \infty : \delta = \frac{\pi}{2}x. \tag{9.25}$$

The solution (9.23), which was obtained with a number of simplifying assumptions, cannot be that accurate as the numerical solutions [3–10] for specific regions of parameters, for which they were carried out. At the same time, the asymptotics (9.24), (9.25) demonstrate the effect of fairly complicated complexes of parameters

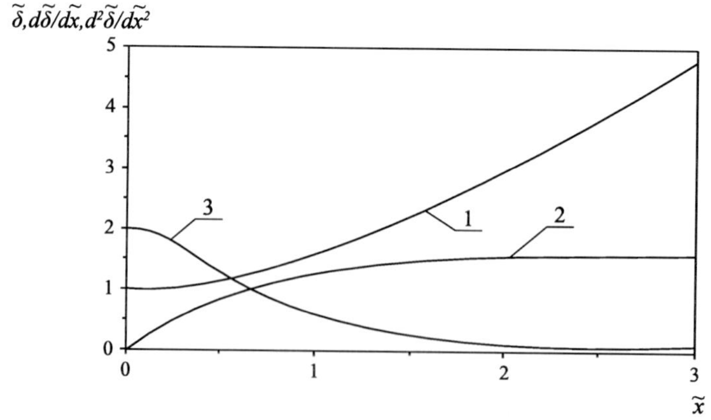

Fig. 9.3 Analytical solution for the evaporating meniscus parameters. 1 $\tilde{\delta}$, 2 $d\tilde{\delta}/d\tilde{x}$, and 3 $d^2\tilde{\delta}/d\tilde{x}^2$

and thermophysical properties, which cannot be achieved in principle in numerical solutions. It seems that the analytical approach has a number of advantages in comparison with the numerical methods. According to [25], such advantages are as follows

- the importance of the analytical approach is in the possibility of a closed qualitative description of the process and in the detection of the list of the dimensionless key parameters,
- analytical solutions are fairly general. Hence, by varying the boundary value conditions one may perform the parametric analysis for a wide class of problems,
- test of numerical solutions of the initial exact equations should depend on the basic analytical solutions of the simplified equations, which could be obtained by assessing the physical values of the separate terms.

9.4 Nanoscale Film

Detailed experimental and theoretical studies of the structure of the superficial forces acting on the nanoscale film were carried out by Israelachvili [26] in a number of studies. A modern account of this independent branch of chemical physics may be found in [27]. Below we shall use some results from the fundamental studies [26, 27]. From Eq. (9.9), it follows that the relative contribution of the thermal conductivity and kinetic molecular resistance depends on the dimensionless parameter $B = \frac{h_k \delta_0}{k}$.

Finding B requires an additional relation, which characterizes the conjugation of the meniscus and the adsorbed film. For this purpose, we shall use the dynamic condition of force balance, which controls the flow in the meniscus. As was already pointed out, the flow of liquid in the meniscus is controlled by the curvature gradient of the phase interface. Calculation by Eq. (9.20) shows that for $\tilde{x} \geq 1$ the curvature abruptly decreases (for $\tilde{x} = 3$, it amounts to less than 2% of its initial value, see Fig. 9.3). So, given the meniscus length $\tilde{l}_m \geq 3$, the total pressure difference through the meniscus Δp_m will in fact be equal to the absolute pressure at the point of conjugation of the meniscus and the macrofilm

$$\Delta p_m \approx p_0 = \sigma K_0 = \sqrt{\frac{3}{2}} \frac{\varepsilon^{1/2} \sigma}{\delta_0}.$$

A nonevaporating nanoscale film is subject to the dispersion Van der Waals forces $p_d \approx E_0/\delta_0^3$ [21]. Here, $E_0 \approx \left(10^{-22} - 10^{-21}\right)$ J, is the Hamaker constant [22], which depends on the values of the dielectric constants of media in contact. The Hamaker constant has important physical meaning of the interaction energy between molecules from various surfaces. Let us now formulate the conjugation condition of the meniscus and the nanoscale film. To this aim, we shall employ the method of [28, 29] for the analysis of unsteady impingement of film to the surface. Equating the differential of the pressures through the meniscus and the dispersion Van der Waals forces, we have $p_0 \approx c_2 p_d$. As a result, we arrive at the equation

$$\frac{B^4}{1+B} = \frac{2}{3}\frac{Lh_k^3 E_0^2}{\nu\sigma k^4 \Delta T}, \tag{9.26}$$

from which we may estimate the parameter B. For $B \ll 1$ from Eq. (9.24), we have

$$\delta_0 \approx c_1 E_0^{1/2}\left(\frac{2}{3}\frac{L}{h_k \nu\sigma \Delta T}\right)^{1/4}. \tag{9.27}$$

The method of conjugation being approximate [28, 29], Eq. (9.27) is written up to an unknown number constant c_1, which was estimated by a comparison with the results of the numerical investigation [7] for the case of an evaporating meniscus of saturated octane with $\beta = 1$. Using Eq. (9.6) to expand the relation for h_k, we find that $c_1 \approx 0.41$. Taking this into account, we rewrite Eq. (9.27) in the form

$$\delta_0 \approx 0.41\frac{E_0^{1/2}}{(\psi \Delta T \rho_v \nu\sigma)^{1/4}}\left(\frac{T_s}{R_g}\right)^{1/8}. \tag{9.28}$$

This simultaneously simplifies Eq. (9.9) for the heat flux through the microfilm to read $q_0 \approx h_k \Delta T$, the parameter ε from the right of Eq. (9.12) will be determined by the relation $\varepsilon \approx h_k \Delta T \nu/\sigma L$. The limit case $B \equiv h_k \delta_0/k \ll 1$ means physically that the thermal resistance of the nanoscale film is much smaller than that of the kinetic thermal resistance. It is worth pointing out that case when the heat transfer is controlled by the kinetic molecular effects is very rarely found in applications. This stresses one more time the unique character of the above problem of evaporation of the meniscus on the interface of three phases.

9.5 The Averaged Heat Transfer Coefficient

From the available meniscus profile using Eq. (9.23) one may find the heat flux density, as averaged over the meniscus length, which is transmitted from the wall, and hence, find the averaged heat transfer coefficient: $h = h_k J/\tilde{l}_m$. Here,

$$\tilde{l}_m \equiv (3\varepsilon/8)^{1/4} x/\delta_0 = \kappa x/\delta_0,$$

$$J = \int_0^{\tilde{l}} \frac{d\tilde{x}}{1 + B\left[1 + \tilde{x}\arctan(\tilde{x})\right]} \approx \int_0^{\tilde{l}} \frac{d\tilde{x}}{1 + B\tilde{x}\arctan(\tilde{x})},$$

l is the conventional length of the meniscus. The denominator of the integrand can be adequately approximated by the relation

$$1 + B(1 + \tilde{x}\arctan(\tilde{x})) \approx 1 + B\tilde{x}\arctan(\tilde{x}) \approx 1 + (\pi/2)B\tilde{x}^2.$$

Calculating the integral, we find that

$$h \approx \frac{2}{\pi}\frac{k}{\delta_m}\ln\left(1 + \frac{\pi}{2}\frac{h_k\delta_m}{k}\right). \tag{9.29}$$

Here,

$$\delta_m \approx \delta_0 + \kappa l_m \arctan\left(\frac{\kappa l_m}{\delta_0}\right) \tag{9.30}$$

is the thickness of the macrofilm: $\delta_m = \delta|_{x=l_m}$. Let us find the asymptotic dependences $h(h_k)$ as $\beta \to 0$. From Eq. (9.6) of the kinetic heat transfer coefficient, it is found Eq. (9.7) that

$$\beta \to 0 \Rightarrow h_k \to 0 \Rightarrow h \to h_k. \tag{9.31}$$

From Eq. (9.31) it follows, already for anomalously small coefficients of evaporation/condensation, that not only the local ($q_0 \approx h_k\Delta T$), but also the heat transfer intensity ($q \approx h_k\Delta T$), as averaged over the meniscus length, is governed by the kinetic molecular mechanism. Equations (9.28)–(9.30) give the minimum and maximum thickness of the meniscus, as well as the heat flux through it. Under usual conditions, the thickness of the macrofilm exceeds by several order that of the adsorbed film (Fig. 9.1, $\delta_m \gg \delta_0$). Hence, from Eq. (9.30), one may find the slope angle of the meniscus

$$\left.\frac{d\delta}{dx}\right|_{x\to\infty} = \frac{\delta_m}{l_m} = \frac{\pi}{2}\kappa \approx 1.23\varepsilon^{1/4}.$$

To estimate the argument of the logarithm in Eq. (9.29), we write its second term in the form $\frac{h_k\delta_m}{k} = \frac{h_k\delta_m}{k} = B\frac{\delta_m}{\delta_0}$.

Under usual conditions, inequality $\delta_m/\delta_0 \gg 1$ is stronger than the inequality $B \ll 1$ and so $h_k\delta_m/k \gg 1$. In view of this, Eq. (9.29) can be rewritten in the form

$$h \approx \frac{2}{\pi}\frac{k}{\delta_m}\ln\left(\frac{h_k\delta_m}{k}\right). \tag{9.32}$$

9.6 The Kinetic Molecular Effects

The thickness of the nanoscale film δ_0 can be determined from Eq. (9.28) regardless of the macroscopic effects. However, the thickness of the macrofilm δ_m (Eq. 9.30) and the averaged heat transfer coefficient h (Eq. 9.29) will depend on the coupling conditions of the meniscus with the external flow. The following external objects of such a conjugation are considered in the literature.

- The liquid droplet [10]. It seems that this fairly exotic problem was realized only in the paper [10], which was carried out using the molecular dynamics method.
- The growth of vapor bubble [17, 18]. Here, considerable role is played by the unsteady effects and by the adopted model of vapor bubble growth.
- The round capillary. The considerable effect of [7, 8] of the axial-symmetricity and of the capillary radius is beyond the scope of our approximate approach.
- Groove of a heat pipe [3–6]. It seems that this case reflects most adequately the two-dimensional problem considered above. Besides, the absence of the above circumstantial factors enables one to study the problem in the pure form.

The schemes of conjugation of the meniscus with the external flow in the cases of round capillary (the axisymmetric effects) and groove of a heat pipe (the two-dimensional problem) are practically identical (Fig. 9.4). It is worth noting that such a conjugation, which involves intermediate regions, is fairly complicated. Kinetic molecular effects in the pure form can be estimated using the planar geometry. In order to find the range of possible variation of β we consider the paper [30], which investigated the kinetic molecular effects versus the vapor bubble growth laws. The paper [30] was carried out in the framework of a unique experiment on the study of refrigerant—11/refrigerant—113 boiling) on a platform falling down from a tower of 110 m height. The so-obtained measured data and available relations of the kinetic molecular theory were used in [30] to calculate the coefficients of evaporation/condensation: $10^{-2} \leq \beta \leq 0.7$ for R11, $8.1 * 10^{-3} \leq \beta \leq 1.0$ for R113. Following [30], we shall consider the range of possible variation $10^{-2} \leq \beta \leq 1$.

Figure 9.5 shows the dependences $\delta(x)$ with various values of β under the conditions considered in the numerical investigations [4, 6]. The reference dependence

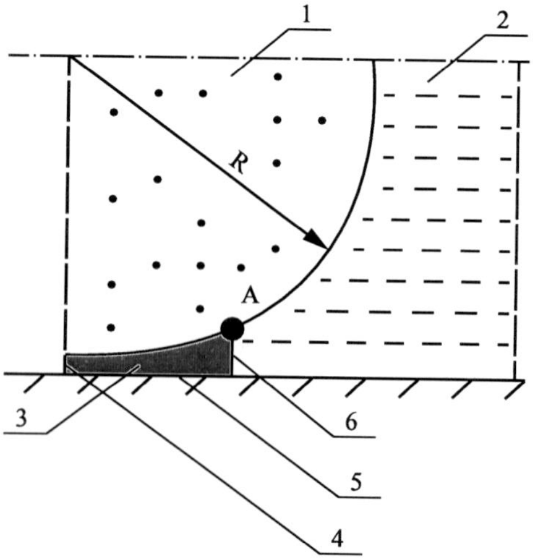

Fig. 9.4 The schemes of conjugation of the meniscus with the external flow. 1 The vapor phase, 2 the liquid phase, 3 the meniscus, 4 the adsorbed film, 5 the solid boundary, and 6 the microfilm, point A marks the border of the meniscus

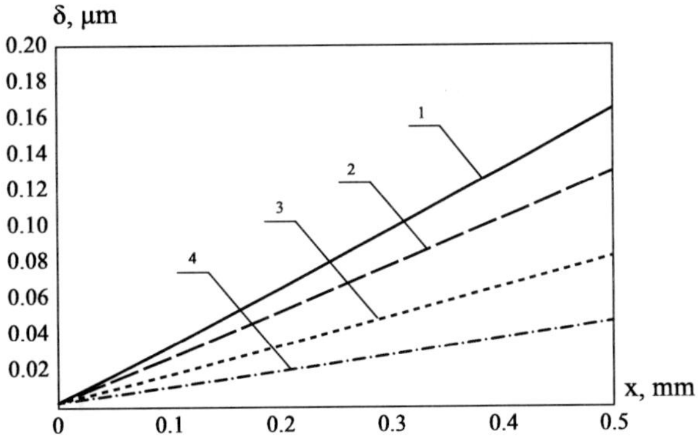

Fig. 9.5 The shape of a meniscus in macroscale with various values of the evaporation/condensation coefficient ($0 \le x \le 0.5$ mm). 1 $\beta = 1$, 2 $\beta = 0.5$, 3 $\beta = 0.1$, and 4 $\beta = 0.01$

with $\beta = 1$ corresponds to Eq. (9.23) with the empirical constant $c_2 = 0.71$

$$\tilde{x} \approx 1 + 0.71\tilde{x} \arctan(\tilde{x}). \tag{9.33}$$

Figure 9.5 shows that in the macroscale the adsorbed film is practically invisible and the meniscus is nearly a one-parameter family of straight lines of the form $\delta \approx kx$, where $k = k(\beta)$. The general trend here is that the meniscus slope angle decreases with the evaporation–condensation coefficient. Besides, curve 1 almost exactly coincides with the corresponding curve of the paper [4]. We note that the results of the numerical solutions from [4, 6], which were obtained under the same conditions, differ from each other by $\approx 8\%$. This quantity can serve as a measure of admissible error of calculations.

Figure 9.6 shows the dependence $\delta(x)$ near the conjugation of the meniscus with the nanoscale film (in the microscale). Here, two opposite trends of the effect of a decrease in β are manifested: decrease of the slope angle and thickening of the nanoscale film, hence, the curves $\delta(x)$ must intersect.

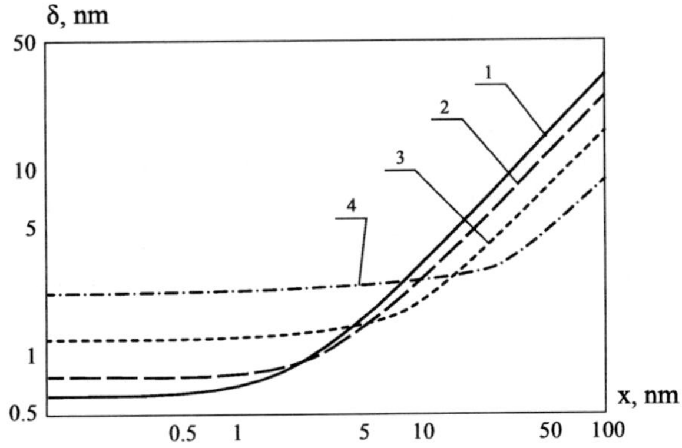

Fig. 9.6 The shape of a meniscus in microscale with various values of the evaporation/condensation coefficient ($0 \le x \le 100$ nm). 1 $\beta = 1$, 2 $\beta = 0.5$, 3 $\beta = 0.1$, and 4 $\beta = 0.01$

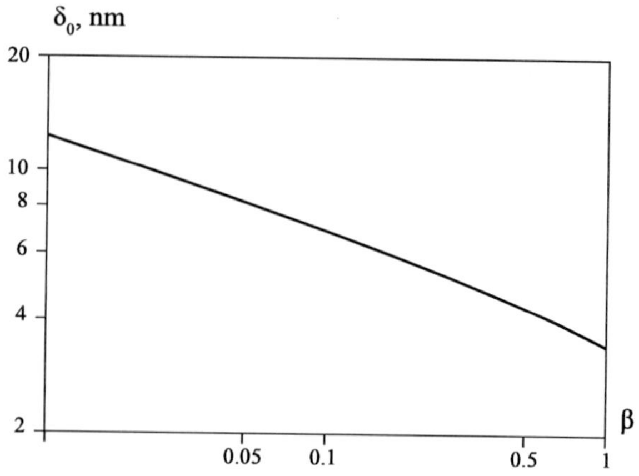

Fig. 9.7 The nanoscale film thickness versus the β

Figure 9.7 depicts the nanoscale film thickness versus the β under the conditions of the numerical investigation [7]. A decrease in the evaporation–condensation coefficient by two orders is seen to approximately quadruple δ_0.

9.7 Conclusions

The thermo-hydrodynamic problem of evaporating meniscus of a thin liquid film on a heated surface is considered. An approximate solving method is presented capable of finding the influence of the kinetic molecular effects on the geometric parameter of the meniscus and on the heat transfer intensity. The method depends substantially on the change of the boundary value problem for the fourth-order differential equation (describing the thermo-hydrodynamics of the meniscus) by the Cauchy problem for a second-order equation. This is achieved by reducing the order of the initial equation by introducing simplifying assumptions and plausible physical estimates. The meniscus is conjugated with the adjacent adsorbed thin film using an approach used in the analysis of unsteady impingement of a film on a solid surface. Analytical expressions for the evaporating meniscus parameters are obtained from the analysis of interaction of the intermolecular, capillary and viscous forces, and the study of the kinetic molecular effects (up to two empirical constants). The latter effects are shown to depend substantially on the evaporation–condensation coefficient. This unexplored effect should be taken into account in the study of problems involving thermo-hydrodynamics of flows of microfilms under the conditions of phase change.

References

1. Wayner PC Jr, Coccio CL (1971) Heat and mass transfer in the vicinity of the triple interline of a meniscus. AIChE J 17:569–575
2. Panchamgam SS, Chatterjee A, Plawsky JL, Wayner PC Jr (2008) Comprehensive experimental and theoretical study of fluid flow and heat transfer in a microscopic evaporating meniscus in a miniature heat exchanger. Int J Heat Mass Transf 51:5368–5379
3. Stephan P (1992) Wärmedurchgang bei Verdampfung aus Kapillarrillen in Wärmerohren. Ph.D. thesis, Univdersität Stuttgart
4. Stephan P, Busse CA (1992) Analysis of the heat transfer coefficient of grooved heat pipe evaporator walls. Int J Heat Mass Transf 35:383–339
5. Do KH, Kim SJ, Garimella SV (2008) A mathematical model for analyzing the thermal characteristics of a flat micro heat pipe with a grooved wick. Int J Heat Mass Transf 51:4637–4650
6. Akkuş Y, Dursunkaya Z (2016) A new approach to thin film evaporation modeling. Int J Heat Mass Transf 101:742–748
7. Wang H, Garimella SV, Murthy JY (2007) Characteristics of an evaporating thin film in a microchannel. Int J Heat Mass Transf 50:3933–3942
8. Dhavaleswarapu HK, Murthy JY, Garimella SV (2012) Numerical investigation of an evaporating meniscus in a channel. Int J Heat Mass Transf 55:915–924
9. Janeček V, Doumenc F, Guerrier B, Nikolayev VS (2015) Can hydrodynamic contact line paradox be solved by evaporation–condensation? J Colloid Interface Sci 460:329–338
10. Van den Akker EAT, Frijns AJH, Kunkelmann C, Hilbers PAJ, Stephan PC, van Steenhoven AA (2012) Molecular simulations of the microregion. Int J Thermal Sci 59:21–28
11. Yagov VV (1988) Heat transfer with developed nucleate boiling of liquids. Therm Eng 2:65–70
12. Yagov VV (1988) A physical model and calculation formula for critical heat fluxes with nucleate pool boiling of liquids. Therm Eng 6:333–339
13. Labuntsov DA, Yagov VV (2007) Mechanics of two-phase systems. Moscow Power Energetic Univ. (Publ.), Moscow (in Russian)
14. Labuntsov DA (2000) Physical foundations of power engineering. Selected works, Moscow Power Energetic Univ (Publ.). Moscow (in Russian)
15. Stephan P, Hammer J (1994) A new model for nucleate boiling heat transfer. Wärme- und Stoffübertragung 30:119–125
16. Stephan P, Kern J (2004) Evaluation of heat and mass transfer phenomena in nucleate boiling. Int J Heat Fluid Flow 25:140–148
17. Kunkelmann C (2011) Numerical modeling and investigation of boiling phenomena. Ph.D. thesis. Technische Universität Darmstadt
18. Ibrahem K, Schweizer N, Herbert S, Stephan P, Gambaryan-Roisman P (2012) The effect of three-phase contact line speed on local evaporative heat transfer: experimental and numerical investigations. Int J Heat Mass Transf 55:1896–1904
19. Craster RV, Matar OK (2009) Dynamics and stability of thin liquid films. Rev Mod Phys 81:1131–1198
20. Loitsyanskii LG (1988) Mechanics of liquids and gases. Pergamon Press, Oxford
21. Parsegian A (2006) Van der Waals forces: a handbook for biologists, chemists, engineering and physicists. Cambridge University Press
22. Dzyaloshinskii IE, Lifshitz EM, Pitaevskii LP (1961) General theory of van der Waals' forces. Sov Phys Usp 4:153–176 (in Russian)
23. Muratova TM, Labuntsov DA (1969) Kinetic analysis of the processes of evaporation and condensation. High Temp 7(5):959–967
24. Zudin YB (1993) The calculation of parameters of the evaporating meniscus of a thin liquid film. High Temp 31(5):777–779
25. Weigand B (2015) Analytical methods for heat transfer and fluid flow problems, 2nd edn. Springer, Berlin, Heidelberg

26. Israelachvili JN (1992) Intermolecular and surface forces. Academic Press, London
27. Plawsky JL, Fedorov AG, Garimella SV, Ma HB, Maroo SC, Li C, Nam Y (2014) Nano- and microstructures for thin film evaporation: a review. Nano Microscale Thermophys Eng 18:251–269
28. Iliev SD, Pesheva NC (2011) Dynamic meniscus profile method for determination of the dynamic contact angle in the Wilhelmy geometry. Colloids Surf A 385(1–3):144–151
29. Snoeijer JH, Andreotti B (2013) Moving contact lines: scales, regimes, and dynamical transitions. Annu Rev Fluid Mech 45:269–292
30. Picker G (1998) Nicht-Gleichgewichts-Effekte beim Wachsen und Kondensieren von Dampfblasen. Dissertation. Technische Universität München

Chapter 10
Kinetic Molecular Effects with Spheroidal State

Symbols

b	Capillary constant
j	Mass flux
L	Heat of phase transition
k	Thermal conductivity
p	Pressure
r	Radial coordinate
Re	Reynolds number
T	Temperature
u	Axial velocity
w	Radial velocity
x	Axial coordinate

Greek Letter Symbols

β	Evaporation-Condensation coefficient
μ	Dynamic viscosity
ρ	Density

Subscripts

l	Liquid
s	Saturation state
v	Vapor
w	Surface

Y. B. Zudin, *Non-equilibrium Evaporation and Condensation Processes*,
Mathematical Engineering, https://doi.org/10.1007/978-3-030-67553-0_10

10.1 Introduction

Cooling of a hot surface by dropwise jets is widely useful in various engineering problems: power systems, metallurgy, cryogenic systems, space, and fire-fighting engineering. Progress in this field is retarded by the lack of full comprehension of the entire realm of phenomena occurring for an incident flow of fluid jet to a surface. The principal question governing the entire process of jet cooling pertains to the study of the coupling of dynamic and thermal drops with the surface. The problem of coupling drops with a hard hot surface has a long history. In 1966, a fragment of a treatise by Leidenfrost (a German physician and theologist) of 1756 was published (Leidenfrost JG, De Aquae Communis Nonnullis Qualitatibus Tractatus. Duisburg, 1756) [1]. This insightful manuscript appeared before the energy conservation law was discovered and before the real nature of heat was unveiled, in particular, the concept "heat of evaporation" was not formulated. The principal result, which made Leidenfrost's name immortal, was his discovery of a new physical fact: at a certain temperature, a metal surface ceases to be wettable by water (and other liquids).

Earlier, in 1732, Boerhaave, a Dutch physician, botanist, and chemist, also reported that the alcohol spilled above a hot surface does not "touch the fire", rather forming "bright droplets looking like mercury". However, it was Leidenfrost, who examined in detail this phenomenon, which now bears his name. In the modern view, Leidenfrost's phenomenon consists of the appearing of a vapor cushion between a liquid drop and the hard surface superheated over some limiting temperature. Once this limiting temperature (the spinodal temperature [2]) is reached, the existence of the liquid phase becomes impossible by the laws of thermodynamics.

In 1836, Boutigny published the book [3] with the results of a detailed study of Leidenfrost's phenomenon. Boutigny introduced the term "spheroidal state", which he considered as the "fourth state of matter". The achievement of Boutigny is that he again paid attention to Leidenfrost's phenomenon. The term "spheroidal state" was also proved highly successful and is now widely adopted in science.

The spheroidal state is the state of a liquid, as water, when, on being thrown on a surface of highly heated metal, it rolls about in spheroidal drops or masses, at a temperature several degrees below ebullition, and without actual contact with the heated surface, a phenomenon due to the repulsive force of heat, the intervention of a cushion of nonconducting vapor, and the cooling effect of evaporation. In 1876, Gesechus [4] took his master's degree in physics for the work entitled "Application of electric current to the study of the spheroidal state of fluids". The spheroidal state was also experimentally studied by Kristensen (1888) and others. In these experiments (which are described in the books by Rosenberger [5]) the following conclusions were made:

- the fluid spheroid, as soon as certain temperature of the surface is reached, starts to rest on the vapor cushion,
- the vapor due to the supply of heat energy to the spheroid from the hot surface through the vapor cushion at the expense of thermal conductivity egresses from the drop with constant,

- the thickness of the vapor cushion is $l \approx (50–250)\ \mu\text{m}$, it increases with the temperature,
- higher temperatures lead to instability, a drop lifts up, moves, oscillates, assuming sometimes a star form, sometimes the drop sporadically contacts with the hot surface,
- the smallest surface temperature, for which the liquid is in the spheroidal state, is much higher than the boiling temperature.

As low as practicable surface temperature, at which a portion of the liquid is in the spheroidal state (that is, it is separated from the surface by a thin vapor cushion) is called the "Leidenfrost temperature". At present, a large number of experiments were carried out on the measurement of the Leidenfrost temperature. We note, in particular, the recent papers [6, 7]. It was shown that the results of measurements depend substantially on a great number of additional factors. For example, for water, various researchers propose the Leidenfrost temperature to range in the wide interval from 150 to 455 °C. In [8], the behavior of water drops on a smooth aluminum surface was studied with a gradually increasing temperature. The Leidenfrost temperature was assumed to be 165 °C, which corresponds to the appearance of a stable vapor film. The authors of [9] supplemented their studies by considering falling drops and jets. The Leidenfrost phenomenon was also experimentally studied in [10].

10.2 Assumptions in the Problem Description

The spheroidal state is characteristic of the continuous evaporation of drops levitating above the vapor cushion. As the process develops, the volume of the drop monotone decreases and changes in succession through a series of forms: from a large drop to a small one, until the complete vaporization. The experiments of [6–10] show, however, that the process of evaporation of a drop is small due to the low intensity of the heat transfer. With the aim of exposing the principal provisions of the theory, we introduce the following assumptions:

- we consider the quasi-stationary process of evaporation of a fixed drop pressed against the hot surface by the force of gravity,
- the radiation heat flux to the drop is neglected,
- the evaporation from the lateral and upper parts of the drop surface is neglected,
- a drop is a disk of radius R and height H (Fig. 10.1),
- the thickness of the vapor cushion under a drop is assumed to be constant and is independent of the radial coordinate,
- the thermophysical properties of vapor are assumed to be homogeneous and corresponding to the mean temperature in the vapor cushion.

Let us consider a drop of volume $4/3\pi R_0^3 = \pi R^2 H$ separated from the hot surface by vapor cushion of thickness l and which is in equilibrium with its own vapor. Here, R_0 is the radius of a sphere of the same volume with the droplet (the equivalent radius

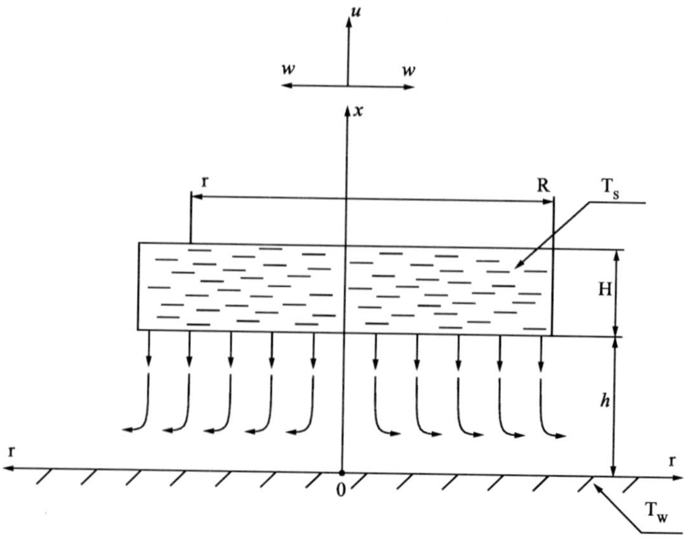

Fig. 10.1 Drop evaporation

of drop). The pressure p_s on the outer boundary $r = R$ of this cushion corresponds to the saturation temperature T_s. We shall consider a large drop satisfying the condition $R_0 \gg b$. Here, $b = \sqrt{\frac{\sigma}{(\rho_l - \rho)g}}$ is the capillary constant, σ is the surface tension, ρ_l is the density of liquid, ρ is the density of gas, g is the gravitational acceleration. According to [6], the height and radius of the disk are estimated as follows

$$H = 2b,\ R = \sqrt{\frac{2}{3}\frac{R_0^{3/2}}{b^{1/2}}}. \tag{10.1}$$

10.3 Hydrodynamics of Flow

The flow of gas in the cushion between two flat surfaces (the hot surface and the bottom of the drop) is consequent on the continuous mass injection through the bottom at the expense of evaporation (Fig. 10.1). The final aim of the analysis is the evaluation of the pressure field in the cushion. Let the origin of the coordinate system be at the center of the lower surface. We introduce the following dimensionless variables

$$\tilde{x} = \frac{x}{l},\ \tilde{r} = \frac{r}{l},\ \tilde{u} = \frac{u}{u_0},\ \tilde{w} = \frac{w}{u_0},\ \tilde{p} = \frac{p}{\rho u_0^2}.$$

Here, x, r are the axial and radial coordinates, u is the axial and radial velocities, l is the width of the cushion, $u_0 = \text{const}$ is the constant velocity for the injection from the bottom of the drop inside the vapor cushion, p is the pressure in the vapor cushion. The Navier–Stokes equations for the given system of equations in the dimensionless form read as

$$\tilde{w}\frac{\partial \tilde{w}}{\partial \tilde{r}} + \tilde{u}\frac{\partial \tilde{w}}{\partial \tilde{x}} = -\frac{\partial \tilde{p}}{\partial \tilde{r}} + \frac{1}{\text{Re}}\left[\frac{\partial^2 \tilde{w}}{\partial \tilde{r}^2} + \frac{\partial}{\partial \tilde{r}}\left(\frac{1}{\tilde{r}}\frac{\partial}{\partial \tilde{r}}(\tilde{r}\tilde{w})\right)\right], \tag{10.2}$$

$$\tilde{u}\frac{\partial \tilde{u}}{\partial \tilde{x}} + \tilde{w}\frac{\partial \tilde{u}}{\partial \tilde{r}} = -\frac{\partial \tilde{p}}{\partial \tilde{x}} + \frac{1}{\text{Re}}\left[\frac{1}{\tilde{r}}\frac{\partial}{\partial \tilde{r}}\left(\tilde{r}\frac{\partial \tilde{u}}{\partial \tilde{r}}\right) + \frac{\partial^2 \tilde{u}}{\partial \tilde{r}^2}\right], \tag{10.3}$$

$$\frac{1}{\tilde{r}}\frac{\partial}{\partial \tilde{r}}(\tilde{r}\tilde{w}) + \frac{\partial \tilde{u}}{\partial \tilde{x}} = 0. \tag{10.4}$$

Here, $\text{Re} = \frac{\rho u_0 l}{\mu}$ is the Reynolds number constructed from the injection velocity, μ is the dynamic viscosities of gas. We set the following boundary conditions

$$\tilde{x} = 0, \tilde{u} = -1, \tilde{w} = 0, \tag{10.5}$$

$$\tilde{x} = 0, \tilde{u} = 0, \tilde{w} = 0. \tag{10.6}$$

We shall search for a self-similar solution of the system of Eqs. (10.2)–(10.4). Their principal property is that the axial velocity in the vapor cushion depends only on the axial coordinate: $\tilde{u} = \tilde{u}(\tilde{x})$. From Eq. (10.4), one can express the radial velocity as follows

$$\tilde{w} = -\frac{1}{2}\tilde{r}\tilde{u}'. \tag{10.7}$$

Here, the prime means the derivative in $\tilde{x}$. Substituting $\tilde{u}, \tilde{w}$ into Eqs. (10.2), (10.3), we have

$$\frac{1}{2}\left(\tilde{u}'\right)^2 - \tilde{u}\tilde{u}'' + \frac{1}{\text{Re}}\tilde{u}''' = -\frac{2}{\tilde{r}}\frac{\partial \tilde{p}}{\partial \tilde{r}}, \tag{10.8}$$

$$\tilde{u}\tilde{u}' - \frac{1}{Re}\tilde{u}'' = -\frac{\partial \tilde{p}}{\partial \tilde{x}}. \tag{10.9}$$

Differentiating both sides of Eq. (10.8) in $\tilde{r}$, we find

$$\frac{\partial^2 \tilde{p}}{\partial \tilde{x}\partial \tilde{r}} = 0. \tag{10.10}$$

Using this equality and differentiating both sides of Eq. (10.9) in $\tilde{x}$, this gives

$$-\tilde{u}\tilde{u}''' + \frac{1}{\text{Re}}\tilde{u}^{IV} = 0. \tag{10.11}$$

In view of Eq. (10.7), the boundary conditions (10.5), (10.6) can be rewritten as

$$\left.\begin{aligned} \tilde{x} = 1 : \tilde{u} = -1, \tilde{u}' = 0, \\ \tilde{x} = 0 : \tilde{u} = 0, \tilde{u}' = 0. \end{aligned}\right\} \tag{10.12}$$

The fourth-order nonlinear differential Eq. (10.11) and the four boundary conditions (10.12) determine the flow of an incompressible fluid in a planar vapor cushion between two surfaces with a constant injection velocity u_0 through the upper surface. Equations (10.11) and (10.12) show that the flow character depends solely on Re. In the general case, the solution of Eq. (10.11) is not expressible in elementary functions.

Let us consider the limit case of the viscous stream: Re $\to$ 0. The solution to Eq. (10.11) will be searched up to the first term from the small-parameter expansion Re $\ll$ 1

$$\tilde{u} = \tilde{u}_0 + \tilde{u}_1\text{Re}. \tag{10.13}$$

The equation for $\tilde{u}_0$ reads as

$$\tilde{u}^{IV} = 0. \tag{10.14}$$

From Eqs. (10.11)–(10.14), we get the following equation for $\tilde{u}_1$

$$\tilde{u}_1^{IV} + 12\left(2\tilde{x}^2 - 3\tilde{x}^3\right) = 0. \tag{10.15}$$

The boundary conditions for Eq. (10.15) are as follows

$$\left.\begin{aligned} \tilde{x} = 1 : \tilde{u} = 0, \tilde{u}' = 0, \\ \tilde{x} = 0 : \tilde{u} = 0, \tilde{u}' = 0. \end{aligned}\right\} \tag{10.16}$$

The solution to Eq. (10.15) satisfying the boundary conditions (10.16) reads as

$$\tilde{u}_1 = -\frac{13}{70}\tilde{x}^2 + \frac{9}{35}\tilde{x}^3 - \frac{1}{10}\tilde{x}^6 + \frac{1}{35}\tilde{x}^7. \tag{10.17}$$

Using Eqs. (10.15) and (10.17) in Eq. (10.13), we have the following equation for the axial velocity with Re $\ll$ 1

$$\tilde{u} = -2\tilde{x}^2 + 3\tilde{x}^3 + \left(-\frac{13}{70}\tilde{x}^2 + \frac{9}{35}\tilde{x}^3 - \frac{1}{10}\tilde{x}^6 + \frac{1}{35}\tilde{x}^7\right)\text{Re}. \tag{10.18}$$

Using Eq. (10.7) and employing Eq. (10.17), we find the relation for the radial velocity with Re $\ll$ 1

$$\tilde{w} = \tilde{r}\tilde{x}\left[3(1-\tilde{x}) + \left(-\frac{13}{35} + \frac{27}{35}\tilde{x} - \frac{3}{5}\tilde{x}^4 + \frac{1}{5}\tilde{x}^5\right)\mathrm{Re}\right]. \tag{10.19}$$

Integrating Eqs. (10.8) and (10.9) and taking into account Eq. (10.18), we find the pressure field in the vapor cushion with Re $\ll 1$

$$\tilde{p}(\tilde{x},\tilde{r}) = \tilde{p}_0 + \left(-3\tilde{r}^2 - 6\tilde{x} + 6\tilde{x}^2\right)\mathrm{Re}^{-1} \\ - \frac{27}{70}\tilde{r}^2 - \frac{13}{35}\tilde{x} + \frac{27}{35}\tilde{x}^2 - \frac{9}{2}\tilde{x}^4 + \frac{27}{5}\tilde{x}^5 - \frac{9}{5}\tilde{x}^6. \tag{10.20}$$

Here, $\tilde{p}_0 = \tilde{p}(0,0)$ is the dimensionless pressure at the origin ($\tilde{x} = 0, \tilde{r} = 0$). Let us find from Eq. (10.20) the pressure distribution on the upper surface of the vapor cushion ($\tilde{x} = 1$)

$$\tilde{p}(1,\tilde{r}) = \tilde{p}_0 - \frac{3}{\mathrm{Re}}\left(1 + \frac{9}{70}\mathrm{Re}\right)\tilde{r}^2 - \frac{1}{2}. \tag{10.21}$$

Making Re $\to 0$, we have the limiting case of viscous flow

$$\tilde{p}(1,\tilde{r}) = \tilde{p}_0 - \frac{3}{\mathrm{Re}}\tilde{r}^2. \tag{10.22}$$

10.4 Equilibrium of Drop

The heat energy is transferred from the hot surface through the vapor film by heat conduction and is completely spent for evaporation

$$q = \frac{k\Delta T}{l}, \tag{10.23}$$

where $\Delta T = T_w - T_s$ is the temperature difference. The heat balance condition for vapor reads as

$$u_0 = \frac{q}{L\rho}, \tag{10.24}$$

where L is the heat of phase transition. Using Eqs. (10.23) and (10.24), we write down the Reynolds number for the flow in the vapor cushion

$$\mathrm{Re} \equiv \frac{\rho u_0 l}{\mu} = \frac{k\Delta T}{L\mu}. \tag{10.25}$$

It can be shown that in this problem, the second term in the round brackets on the right of Eq. (10.21) is always negligible. Changing Eq. (10.21) to the dimensional nomenclature and using Eq. (10.25), we find the average value of the hydrodynamic pressure difference Δp_h (with respect to the bottom of drop)

$$\Delta p_h = p_h - p_s = \frac{3}{2} \frac{\mu k \Delta T R^2}{L \rho l^4}. \tag{10.26}$$

In the general case, for a description of evaporation, one should take into account the non-equilibrium effects in the vapor space associated with the formation of the Knudsen layer of thickness of the order of mean free path of molecules near the condensed-phase surface. In the Knudsen layer, the standard gas dynamic descriptions become improper—the phenomenological gas parameters (temperature, pressure, density, velocity), as given according to the conventional laws of statistical averaging, lose their macroscopic sense.

The linear kinetic theory of evaporation was first formulated in the papers by Labuntsov and Muratova [11, 12], who showed that the boundary condition with evaporation on the phase interface becomes much more involved than it is assumed in the equilibrium condition. In particular, it was shown that the pressure within the Knudsen layer is constant and is equal to $p_{v\infty}$, so that the condensed phase is under the same pressure as the vapor. This means that the actual vapor pressure near the phase interface (beyond the Knudsen layer) is not equal to the saturated pressure p_{ws}, which corresponds to the temperature of the liquid surface: $p_{ws} = p_{ws}(T_s)$. The difference $p_{ws} - p_{v\infty}$ (the kinetic pressure difference) is determined from the relation

$$\Delta \tilde{p} = -2\sqrt{\pi}\frac{1-0.4\beta}{\beta}\tilde{j} - 0.44\tilde{q}. \tag{10.27}$$

Here, $\Delta \tilde{p} = (p_{vs} - p_{v\infty})/p_{vs}$, $\tilde{j} = j/\rho_{vs}U = U/v_s$, $\tilde{q} = q/p_{v\infty}v_s$, $v_s = \sqrt{2R_g T_s}$ is the thermal velocity of a molecule, j, q is the mass and heat flux crossing the phase interface, respectively, β is the evaporation/condensation coefficient.

According to the linear theory [11, 12], positive values of j and q correspond to flows transferred to the vapor phase. In our problem, the heat flux is directed from vapor towards the drop, while the mass flux is directed from the drop towards the vapor. Hence, in the case under consideration, we have $\tilde{q} < 0$, $\tilde{j} > 0$ (Fig. 10.1). Accordingly, Eq. (10.27) can be rewritten in the form

$$\Delta \tilde{p} = -2\sqrt{\pi}\frac{1-0.4\beta}{\beta}\tilde{j} + 0.44|\tilde{q}|. \tag{10.28}$$

Changing to the dimensional nomenclatures and taking into account relations (10.23) and (10.24), we rewrite Eq. (10.28) to read

$$\Delta p_k = p_k - p_s = \frac{k\Delta T}{\sqrt{Ll}} f. \tag{10.29}$$

Here

$$f = 0.313\sqrt{A} + \frac{1}{\sqrt{A}} - \frac{2.505}{\sqrt{A}\beta}, \tag{10.30}$$

p_k is the kinetic pressure, $A = L/R_g T_s$, R_g is the individual gas constant. From formula (10.30), we see that the function f in the general case is sign-variable.

In order to find the range of possible variation of β, we consider the papers [13–15], which investigated the kinetic molecular effects versus the vapor bubble growth laws. The papers [13–15] was carried out in the framework of a unique experiment on the study of refrigerant -11/refrigerant -113 boiling) on a platform falling down from a tower of 110 m height. The so-obtained measured data and available relations of the kinetic molecular theory were used in [13–15], to calculate the coefficients of evaporation/condensation: $10^{-2} \le \beta \le 0.7$ (for R11), $8.1 * 10^{-3} \le \beta \le 1.0$ (for R113). Following [13–15], we shall consider the range of possible variation $10^{-2} \le \beta \le 1$. Setting $f = 0$, from Eq. (10.30), we find the limiting value of the evaporation–condensation coefficient

$$\beta_* = \frac{2.505}{1 + 0.313A}. \tag{10.31}$$

The following cases are possible

$$\left.\begin{array}{r} \beta_*\langle\beta \le 1 : \Delta p_k\rangle 0, \\ \beta = \beta_* : \Delta p_k = 0, \\ 0 < \beta < \beta_* : \Delta p_k < 0. \end{array}\right\} \tag{10.32}$$

So, the sign of the kinetic pressure difference depends on the value of β. Using the thermophysical properties sheet, one may approximate the dependence of the value A on the pressure, and then use Eq. (10.31) to evaluate the function $\beta_*(p)$. In particular, for water in the pressure range $10^{-2}\text{bar} \le p \le 10^2\text{bar}$ the following approximation (with error not exceeding 1%) was obtained

$$\beta_* = 0.263\left(1 + 1.1p^{1/5} + 7.5 * 10^{-3} p^{6/5}\right)\left(1 + 0.115p^{1/5} + 7.82 * 10^{-4} p^{6/5}\right)^{-1}, \tag{10.33}$$

where $[p] = \text{bar}$.

From Fig. 10.2, it follows that for $p = 10^2\text{bar}$, we have[1]: $\beta_* \approx 1$. In view of relations (10.32) this means the absence of the domain $\Delta p_k > 0$, that is, the kinetic pressure with respect to the droplet may be only "attracting" ($\Delta p_k < 0$). As

---

[1] Of course, such an excellent agreement (up to 2%) is only a happy accident, which enables one to clearly visualize the phenomenon.

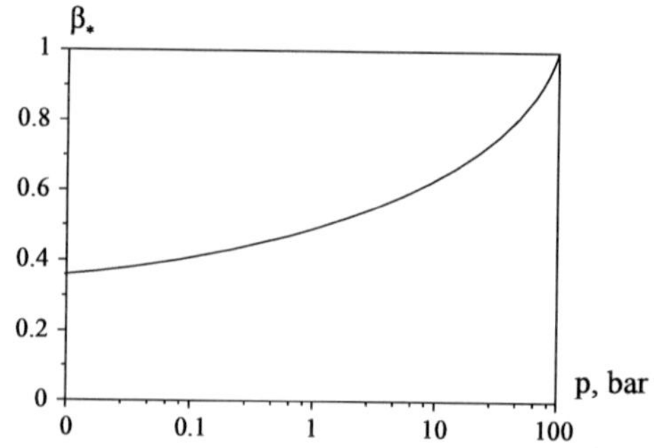

Fig. 10.2 The limiting value of the evaporation/condensation coefficient versus the pressure calculated for water

the pressure reduces, there appears and monotonically increases the range of β, in which the kinetic pressure is "repulsing"($\Delta p_k > 0$). However, for $\beta < \beta_*$, we again have $\Delta p_k < 0$. Moreover, from Eq. (10.30), it follows that, for any pressure, there is always a sufficiently small value of β_*, which secures the existence of the domain of repulsing kinetic pressure. In particular, for $p = 10^{-2}$ bar, we have $\beta_* \approx 0.359$. We shall be concerned with a large disk-shaped drop of radius $R_0 = 5b$. Closing relation of analysis is the equilibrium equation for drop

$$\Delta p_h + \Delta p_k = \rho_l g H. \tag{10.34}$$

Here ρ_l is the density of a liquid.

From Eqs. (10.26), (10.29), and (10.30), we have the fourth-order equation for the thickness of the vapor cushion, which separates the drop from the hot surface

$$\tilde{l}^4 - A\tilde{l}^3 - B. \tag{10.35}$$

Here

$$\tilde{l} \equiv \frac{l}{b}, \; A = \frac{1}{2}\frac{fk\Delta T}{\sigma L^{1/2}}, \; B = \frac{(g\rho_l)^{1/2}\nu k\Delta T}{\sigma^{3/2}L}.$$

It is worth pointing out that under the standard approach, the kinetic pressure difference is not taken into account. Putting $\Delta p_k = 0$ in Eq. (10.34), we get the standard value of the vapor film

$$l_* = 0.76\left(\frac{\nu k\Delta T}{\sigma L}\right)^{1/4} R_0^{3/4}. \tag{10.36}$$

From Fig. 10.3, it is seen that for $\beta = 1$, the thickness of the vapor film, as calculated from (10.35), exceeds its standard value. This is a consequence of the

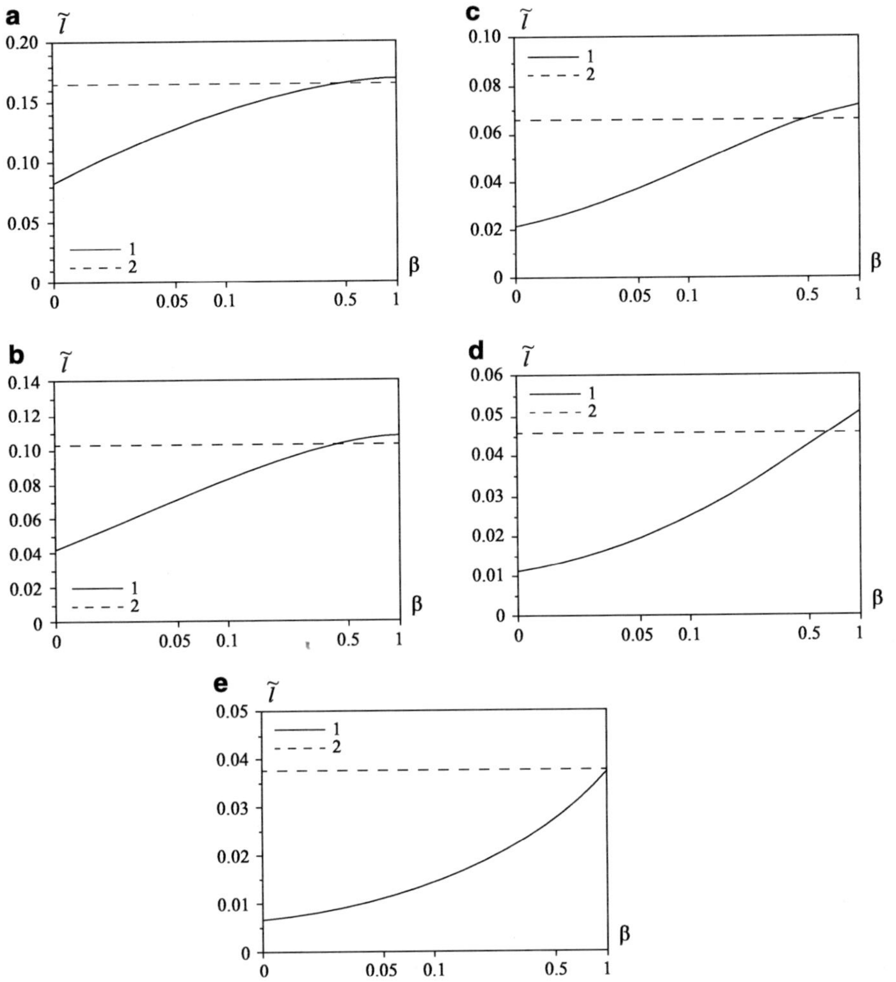

Fig. 10.3 Vapor film thickness versus the evaporation/condensation coefficient for various values of the pressure calculated for water. 1 Obtained by solving the Eq. (10.35), 2 standard value with $A = 0$, **a** $p = 0.01$ bar, **b** $p = 0.1$ bar, **c** $p = 1$ bar, **d** $p = 10$ bar, **e** $p = 100$ bar

repulsing character of the kinetic pressure, which monotonically relaxes as the pressure increases. As a corollary, during this process, the thickness of the vapor film somehow exceeds the standard value, which was calculated from relation (10.36) without consideration of the non-equilibrium effects. As β decreases, the attracting effect due to the non-equilibrium of evaporation manifests itself more and more. This effect increases with increasing pressure. For example, with $\beta = 0.01$, $p = 100$ bar in Fig. 10.3e, the thickness of the vapor film decreases by more than four times with respect to the standard value (Fig. 10.3e).

Figure 10.4 depicts the family of functions $l(\beta)$ calculated for water in the pressure

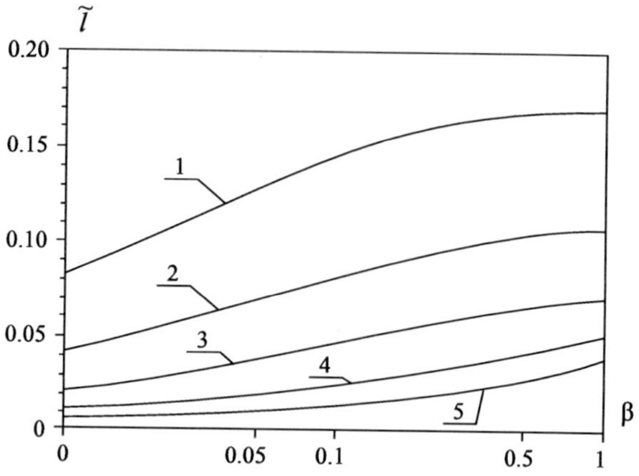

Fig. 10.4 The family of functions $\tilde{l}(\beta)$ calculated for water. 1 $p = 0.01$ bar, 2 $p = 0.1$ bar, 3 $p = 1$ bar, 4 $p = 10$ bar, 5 $p = 100$ bar

range $10^{-2}\text{bar} \le p \le 10^{2}\text{bar}$. Figure 10.4 clearly demonstrates the general trend that the non-equilibrium influence increases with increasing pressure. It is worth noting that such a trend is nontrivial. The thing is that according to the linear kinetic theory of evaporation [11, 12], the thermal effect of non-equilibrium is manifested prima facie for low pressures, that is, for small vapor densities. This is indicated by the consideration of the kinetic heat transfer coefficient h_k

$$h_k = \frac{0.4\beta}{1 - 0.4\beta}\frac{\rho L^2}{R_g^{1/2}T_s^{3/2}}. \tag{10.37}$$

From Eq. (10.37), it is seen that a decrease in ρ results in a decrease of h_k (that is, in an increase of the kinetic thermal resistivity due to non-equilibrium). This being so, the application of the kinetic molecular analysis to the Leidenfrost phenomena leads us to nontrivial qualitative and profound quantitative phenomena.

Let us now consider a hypothetical case of abnormally small values of the evaporation–condensation coefficient. Following Eqs. (10.26) and (10.29), we write the drop equilibrium conditions for a drop as $\beta \to 0$ in the general form

$$1.5\frac{\mu k \Delta T R^2}{L\rho l^4} + \frac{k\Delta T}{\sqrt{L}l} f = \rho_l g H. \tag{10.38}$$

Making $\beta \to 0$, we get from Eq. (10.30) that

$$f \to -\frac{2.505}{\beta}\left(\frac{R_g T}{L}\right)^{1/2}. \tag{10.39}$$

Now Eq. (10.38) assumes the form

$$1.5\frac{\mu k\Delta T R^2}{L\rho l^4} - \frac{2.505}{\beta}\frac{\left(R_g T\right)^{1/2} k\Delta T}{Ll} = \rho_l g H. \tag{10.40}$$

With $\beta \to 0$ the second (kinetic attracting) term on the right of Eq. (10.40) unboundedly increases, and hence to guarantee the drop equilibrium, the first (hydrodynamical repulsing) term also goes to infinity. Besides, the hydrostatic pressure on the right of Eq. (10.40) is a small difference of two infinitely large quantities, so that from Eq. (10.40), it follows that

$$1.5\frac{\mu k\Delta T R^2}{L\rho l^4} \approx \frac{2.505}{\beta}\frac{\left(R_g T\right)^{1/2} k\Delta T}{Ll}, \tag{10.41}$$

and hence

$$l \approx 0.843\beta^{1/3}\frac{\nu^{1/3} R^{2/3}}{\left(R_g T_s\right)^{1/6}}. \tag{10.42}$$

From Eq. (10.42), it follows that as $\beta \to 0$ the width of the vapor cushion ceases to depend on the superheat of the hard surface and the drop weight. From the physical point of view, such an exotic situation is explained by an unboundedly increasing kinetic attracting effect. To offset this effect, the drop should go down at a so small distance from the hot surface in order to create the required repulsing effect. This distance secures the required drop equilibrium, independent of the surface superheat and the drop height. For the case of "large drop" under consideration, we shall have $R_0 = 5b$. Now relation (10.42) assumes the form

$$l = \frac{2.46}{\left(R_g T_s\right)^{1/6}}\left(\frac{\beta\nu\sigma}{g\rho_l}\right)^{1/3}. \tag{10.43}$$

10.5 Conclusions

A theoretical analysis of the well-known problem of evaporation of a drop levitating over the vapor cushion is performed under the generally adopted assumptions. The question on the influence of the kinetic molecular effects on the drop equilibrium conditions was considered for the first time. The hydrodynamic pressure difference in a vapor film is calculated using the well-known problem of the motion of gas in the cushion between two flat surfaces (a hot hard surface and an upper drop bottom), which is consequent on the mass injection by evaporation. Results of the linear kinetic theory of evaporation are used to evaluate the kinetic pressure difference due to non-equilibrium conditions of the evaporation process. It is shown that, depending on the value of the evaporation/condensation coefficient, the kinetic pressure with respect to a drop may have either repulsing or attracting character. The analytical dependence for the thickness of the vapor film for a wide range of β is found. A fairly exotic asymptotics formula for the solution with $\beta \to 0$ is obtained, describing the balance between the repulsing and attracting phenomena.

References

1. Leidenfrost JG (1966) On the fixation of water in diverse fire. Int J Heat Mass Transf 9:1153–1166
2. Debenedetti PG (1996) Metastable liquids: concepts and principles. Princeton University Press, Princeton
3. Boutigny PH (1857) Études sur les corps a l'État sphéroïdal. Nouvelle branche de hysique. 3rd éd. Victor Masson, Paris
4. Gesechus N (1876) Electric current in the study of the spherodial state of liquids. (St Petersbourg), (in Russian
5. Rosenberger F (2007) Geschichte der Physik I. ThUL
6. Kruse C, Anderson T, Wilson C, Zuhlke C, Alexander D, Gogos G, Ndao S (2013) Extraordinary shift of the leidenfrost temperature from multiscale micro/nanostructured surfaces. Langmuir 29:9798–9806
7. Quere D (2013) Leidenfrost dynamics. Annu Rev Fluid Mech 45:197–215
8. Bernardin JD, Mudawar I (2002) A cavity activation and bubble growth model of the Leidenfrost point. J Heat Trans-T ASME 124:864–874
9. Bernardin JD, Mudawar I (2004) A Leidenfrost point model for impinging droplets and sprays. J Heat Trans-T ASME 126:272–278
10. Biance A-L, Clanet C, Quér D (2003) Leidenfrost drops. Phys Fluids 5(6):1632–1637
11. Labuntsov DA (1967) An analysis of the processes of evaporation and condensation. High Temp 5(4):579–647
12. Muratova TM, Labuntsov DA (1969) Kinetic analysis of the processes of evaporation and condensation. High Temp 7(5):959–967
13. Picker G (1998) Nicht-gleichgewichts-effekte beim wachsen und kondensieren von dampfblasen. Dissertation, Technische Universität München
14. Winter J (1997) Kinetik des blasenwachstums. dissertation, Technische Universität München
15. Straub J (2001) Boiling heat transfer and bubble dynamics in microgravity. Adv Heat Transfer 35:57–172

Chapter 11
Solution of Special Problems of Film Condensation

Abbreviations

HTC	Heat transfer coefficient
NCG	Noncondensable gas
MM	Mechanistic model

Symbols

A	Relative heat transfer coefficient
c_p	Isobaric heat capacity
D	Molecular diffusivity
h_{fg}	Heat of phase transition
h	Heat transfer coefficient (HTC)
K	Phase transformation number
k	Thermal conduction
Le	Lewis number
Pr	Prandtl number
p	Pressure
Re	Reynolds number
St	Stanton number
St_m	Mass Stanton number
Sc	Schmidt number
T	Temperature

Greek Letter Symbols

α	Thermal diffusivity
β	Mass transfer coefficient
δ	Film thickness
ξ	Darcy friction factor
ϕ	Angular coordinate
ν	Kinematic viscosity
ρ	Density

Y. B. Zudin, *Non-equilibrium Evaporation and Condensation Processes*, Mathematical Engineering, https://doi.org/10.1007/978-3-030-67553-0_11

ψ NCG influence coefficient

Subscripts

0 External
b Flow core
δ State a liquid film surface
m Vapor-Gas mixture
v Vapor
w State at wall surface

11.1 Condensation of Vapor Flowing around a Cylinder

11.1.1 Introduction

The problem of vapor condensation on a solid surface has been traditionally referred to as a classical problem of two-phase thermo-hydrodynamics. The best-studied case is the condensation of a steady-state vapor on a vertical plate [1, 2] when the hydrodynamics of the laminar flow of a condensate film is determined by the interaction between the gravity forces (the driving force) and the viscous friction on the wall.

The analytical solution to this problem was obtained in 1916, in fundamental studies by Nusselt [3, 4]. A significant landmark was the works by Kutateladze [5] and Labuntsov [6]. In these studies, the effects of nonisothermicity of the wall on the heat transfer and wave formation on the surface of a liquid film were taken into account and the turbulent mode of its flow was investigated. When vapor moves concurrently with the film, a new driving force emerges, that is, a tangential stress on the interfacial surface that induces film acceleration due to the difference in the rates of vapor and liquid phases (the pure friction effect). The mass flow conditioned by vapor condensation results in an increase in the rate gradient in the vapor boundary layer (the suction effect). In turn, it leads to the enhancement of interfacial friction.

Kholpanov and Shkadov [7] cover the condensation of vapor on a vertical wall under conditions of concurrent and reverse-current phase motion with account for the wave formation on the surface of condensate film and the influence of capillary forces on it, as well as upon the turbulent flow mode, in great detail.

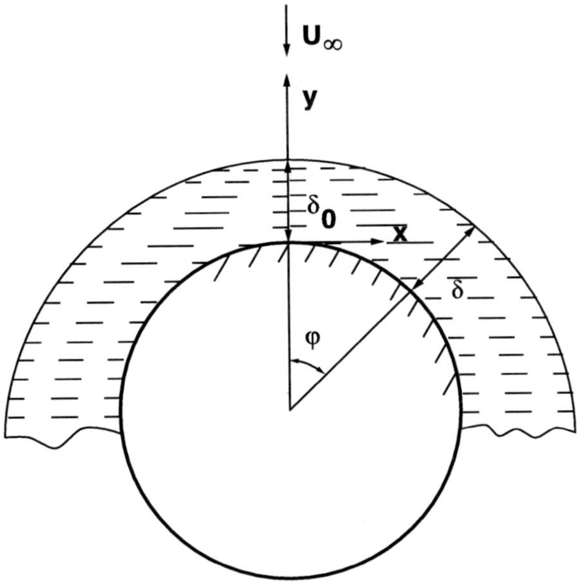

Fig. 11.1 Vapor condensation

The case of vapor condensation upon the transversal flow of a horizontal cylinder (Fig. 11.1) has been studied to a considerably lesser extent. There have been only a few computational-theoretical studies devoted to this field [8–10]. The feature they have in common is that, along with the forces of gravity and friction on the wall, certain additional driving factors were selectively taken into account in the force balance for a flowing film. Meanwhile, other substantial factors were not taken into consideration, frequently without explicit physical grounding.

Thus, in the groundbreaking study [8], only the component associated with the transversal mass flow (the suction friction) is taken into account during the calculation of the tangential stresses on the interfacial boundary. However, it is known from the boundary layer theory [11] that it is an asymptotic case that is implemented only at high rates of mass suction through a permeable surface. In their analysis, the authors of [8, 9] neglected the changes in the pressure along the cylinder circumference. However, they always take place upon the transversal flow around the cylinder [11]. The influence of the pressure gradient was first taken into account in [10]. However, the tangential stresses at the interfacial boundary, just as in [8], were determined on the basis of the suction friction.

Thus, the calculation and theoretical studies [8–10] have a common methodology: the numerical study was carried out on taking into account appreciably strict assumptions that eliminate the effect of a particular factor. As a result, the final analytical approximations constructed [8–10] are incomplete from the very beginning. The empirical formulas proposed in [12] are not a reasonable solution either, since they are adequate only within a certain range of parameters that were covered by experiments.

We should take note of the work of Gaddis [13], where the study discussed was investigated in its complete mathematical formulation. The equations of continuity, motion, and energy for a liquid film and the equations of continuity and motion for the interfacial layer of vapor were written down. The continuity conditions were set at the interphase boundary: the equality of tangential velocities and tangential stresses of contacting phases. Unfortunately, the author of [13] illustrates the results of his numerical calculations only with individual tables for limited parameter ranges. That neither physical generalization nor analytical approximation of the calculated data have been performed makes the use of the results of [13] almost impossible.

From what was said, it can be claimed that the numerical partial solutions to the problem of vapor condensation upon flowing around a horizontal cylinder that have been found so far shed little light on the physical picture of this sophisticated process. Therefore, we adopted an approach that consisted of sequential analysis of asymptotic variants of the problem, controlled by only one parameter each time, under the assumption that there is no influence of the remaining parameters. For each of these asymptotic variants, the strict analytical solution to the problem can be performed. Then, the limiting heat transfer laws obtained can be successively complicated by taking into account additional factors that have an effect on the process.

11.1.2 Limiting Heat Transfer Laws

As follows from [8–10], the thickness of a condensate film is always much lower than the cylinder diameter ($\delta \ll D$), while the fluid velocity on the interfacial surface is much lower than the velocity of the incident flow ($u_\delta \ll U_\infty$). It means that the fluid film has virtually no observable effect on vapor hydrodynamics. Therefore, the interfacial friction will be almost equal to the surface friction, which emerges upon gas (vapor) flowing around a solid cylinder with a permeable surface. The withdrawal of vapor masses through the interface boundary will be controlled by the intensity of the condensation process. Within the framework of the boundary layer model [11], the distribution of the tangential velocity U in the vapor phase near the cylinder surface is described by the relationship for the potential flow of an ideal fluid, $U = 2U_\infty \sin\phi$, while the change in pressure p is given by the following equation

$$p = p_0 - 2\rho_v U_\infty^2 \sin^2\phi. \tag{11.1}$$

Here, p_0 is the pressure in any critical point (Fig. 11.1).

The laminar flow in the liquid phase is described by the impulse balance equation

$$\mu \frac{d^2u}{dy^2} + \rho g \sin\phi + \frac{dp}{dx} = 0,$$

and the boundary conditions

$$y = 0 : u = 0,\ y = \delta : \mu \frac{du}{dy} = \tau_\delta.$$

Here, g is the acceleration due to gravity, u is the fluid velocity in the film, x and y are longitudinal and transversal coordinates, respectively, $\phi = 2x/D$ is the angular coordinate counted off the critical point (Fig. 11.1), D is the cylinder diameter, τ_δ are tangential stresses at the interface boundary, ρ and ρ_v are the fluid and vapor densities, respectively, and μ is the dynamic viscosity of the fluid.

As follows from Eq. (11.1)

$$\frac{dp}{dx} = -4 \frac{\rho_v U_\infty^2}{D} \sin 2\phi.$$

The integral of the impulse balance equation gives the expression for the specific (per unit of length of the cylinder) volume flow rate of the fluid j [2, 10]

$$j = \frac{1}{3} \frac{g\delta^3}{\nu} \sin\phi + \frac{4}{3} \frac{\rho_v U_\infty^2 \delta^3}{\mu D} \sin 2\phi + \frac{1}{2} \frac{\tau_\delta \delta^2}{\mu}. \tag{11.2}$$

Here, δ is the film thickness and ν is the kinematic viscosity of the fluid.

The right-hand part of Eq. (11.2) involves the driving forces that ensure the flow of a fluid film against viscosity forces: the first term is the force of gravity, the second term is the force conditioned by changes in the static pressure at the cylinder circumference (for brevity, referred to as the pressure force), and the third term is the interfacial friction force. The heat transfer through a laminar fluid film is performed by thermal conduction. Therefore, in the absence of temperature gradient in saturated vapor, the Heat Transfer Coefficient (HTC) will be determined by the relation $h = k/\delta$, where k is the thermal conduction of a fluid. Thus, the local Nusselt number will be as follows

$$\text{Nu}(\phi) \equiv \frac{h(\phi)D}{k} = \frac{D}{\delta(\phi)}. \tag{11.3}$$

Equation (11.3) can be used to calculate

$$\langle \text{Nu} \rangle \equiv \frac{\langle h \rangle D}{k} = \frac{D}{\pi} \int_0^{\phi_*} \frac{d\phi}{\delta(\phi)}. \tag{11.4}$$

Here, $0 \le \phi_* \le \pi$ is the limit of integration (values ϕ are counted in radians), $\langle h \rangle$ is the HTC, averaged over the corresponding range of angular coordinate.

Taking into account (11.3), the heat balance equation for the liquid phase will be written as

$$\frac{dj}{d\phi} = \frac{1}{2}\frac{Dk\Delta T}{\delta\rho L}, \tag{11.5}$$

where L is the heat of phase transition and ΔT is the "surface—vapor" temperature drop. Relations (11.2)–(11.5) allow carrying out the analysis of the limiting heat transfer laws. To do so, only one driving force has to be taken into account in the right-hand part of Eq. (11.2).

11.1.3 Asymptotics of Immobile Vapor

Expression for the specific (per unit of length of the cylinder) volume flow rate of fluid (11.2) in the case in which the film is flowing under gravity forces performing work against viscosity forces on the wall acquires the following form

$$j = \frac{1}{3}\frac{g\delta^3}{\nu}\sin\phi. \tag{11.6}$$

We obtain the equation for the film thickness from Eqs. (11.5) and (11.6)

$$\frac{d(\delta^4)}{d\phi} + \frac{4}{3}(\cot\phi)\delta^4 = \frac{2\nu k\Delta T D}{\rho g L}\frac{1}{\sin\phi},$$

its solution will appear as

$$\left(\frac{\delta}{D}\right)^4 = 2\frac{\mu k\Delta T}{\rho^2 g L D^3}\frac{\int_0^\phi \sin^{1/3}\phi d\phi}{\sin^{4/3}\phi}. \tag{11.7}$$

As can be seen in Eq. (11.7), at $\phi = 0$, the film has finite thickness $\delta_0 > 0$, which is cardinally different from the case of condensation on a flat plate, where the initial thickness of the film is equal to zero $\delta_0 \equiv 0$ [1–4]. Thus, value δ monotonically increases and is directed toward infinity with $\phi \equiv \phi_* = \pi$. The use of Eq. (11.7) in integral (11.4) provides the classical Nusselt solution [3, 4]

$$\langle \mathrm{Nu}_g \rangle \equiv \frac{\langle h_g \rangle D}{k} = 0.728\left(\frac{D^3\rho^2 g L}{k\mu\Delta T}\right)^{1/4}. \tag{11.8}$$

We note that according to Eqs. (11.3) and (11.7), the expression for the Nusselt number in the critical point will appear as Eq. (11.8), but the value of the numerical constant will be larger by a factor of 1.24.

11.1.4 Pressure Asymptotics

As follows from Eq. (11.2), at $\phi = \pi/2$, the force conditioned by the changes in pressure along the circumference (the pressure force) changes its sign and becomes decelerating instead of accelerating. The emerging positive pressure gradient (at appreciably high values of dynamic pressure of the vapor flow $\rho_v U_\infty^2$ may result in a complete stop ("flooding") of the liquid film even before it reaches the rear critical point of the cylinder $\phi = \pi$. Let us consider the limiting case in which only the pressure force is retained in Eq. (11.2)

$$j = \frac{4}{3}\frac{\rho_v U_\infty^2 \delta^3}{\mu D}\sin 2\phi. \tag{11.9}$$

By substituting formula (11.9) into Eq. (11.5), we obtain the following equation for the film thickness

$$\frac{d(\delta^4)}{d\phi} + \left(\frac{8}{3}\cot 2\phi\right)\delta^4 = \frac{1}{2}\frac{\nu k \Delta T D^2}{\rho_v L U_\infty^2}\frac{1}{\sin 2\phi}.$$

with its solution in the following form

$$\left(\frac{\delta}{D}\right)^4 = \frac{1}{2}\frac{1}{\mathrm{Re}}\frac{k\Delta T}{\rho_v \nu L}\frac{\int_0^\phi \sin^{1/3}2\phi d\phi}{\sin^{4/3}2\phi}. \tag{11.10}$$

Here, $\mathrm{Re} = \frac{U_\infty D}{\nu}$ is the Reynolds number constructed using the vapor flow velocity at infinity U_∞, the kinematic viscosity of the fluid ν, and the cylinder diameter D. As follows from Eq. (11.10), the film thickness in the midsection is directed toward infinity, since the pressure gradient here becomes zero (the flooding effect). Therefore, the HTC defined by the relationship $h = k/\delta$ will be equal to zero with $(\phi = \phi_* = \pi)/2$. Assuming that the rear surface $\pi/2 \le \phi \le \pi$ does not participate in the heat transfer, we obtain the HTC averaged over the entire cylinder surface from Eqs. (11.10) and (11.4)

$$\langle \mathrm{Nu}_p \rangle \equiv \frac{\langle h_p \rangle D}{k} = 0.612\sqrt{\mathrm{Re}}\left(\frac{\rho_v \nu L}{k\Delta T}\right)^{1/4}. \tag{11.11}$$

According to Eqs. (11.4) and (11.11), the Nusselt number in the critical point will be higher than its own mean value by a factor of 2.48. Numerical investigation of this condensation taking into account the effect of the longitudinal pressure gradient was first carried out in [10]. Unfortunately, Rose [10] did not study the asymptotics with respect to pressure. The final calculation formula was selected based on the approximation of the numerical results. This formula provides

$$\langle \mathrm{Nu}_p \rangle = 1.13\sqrt{\mathrm{Re}}\left(\frac{\rho_v \nu L}{k\Delta T}\right)^{0.209}.$$

which coincides with correct asymptotics (11.11) in terms of neither the exponent of power nor numerical constant.

11.1.5 Tangential Stresses at the Interface Boundary

The equation for the specific volume flow rate of fluid (11.2) in the case in which film flows under the action of tangential stresses τ_δ at the interface boundary (which perform work against the viscosity forces on the wall) acquires the following form

$$j = \frac{1}{2}\frac{\tau_\delta \delta^2}{\mu}. \tag{11.12}$$

As already mentioned, the value τ_δ can be adopted with good approximation from the solution to the single phase problem of flow in the boundary layer near a solid cylinder. The normal component of the vapor velocity at the interface boundary will be equal to the mass suction rate "w" conditioned by the phase transition. As far as is known, there has been no solution to the hydrodynamic problem of the flow in the boundary layer with suction at the cylinder surface yet. Therefore, we consider the limiting cases $w \to \infty$ (suction friction) and $w = 0$ (pure friction) separately. However, let us emphasize that the specified variants are not asymptotics in the strict sense. They are two limiting branches with unified asymptotics of interphase friction. As $w \to \infty$, the tangential stresses at the interphase boundary will be determined by the impulse flow transferred through the boundary $\tau_\delta = 2\rho_v w U_\infty \sin\phi$ [11]. By determining the vapor suction rate from the heat balance $w = k\Delta T/\delta L \rho_v$, we obtain the expression for the suction friction

$$\tau_\delta = 2\frac{k\Delta T U_\infty}{L\delta}\sin\phi. \tag{11.13}$$

Substitution of Eq. (11.13) into Eq. (11.5) gives the equation for the film thickness

$$\left(\frac{\delta}{D}\right)^2 = \frac{1}{\mathrm{Re}(1+\cos\phi)}. \tag{11.14}$$

As follows from Eq. (11.14), in the asymptotics of suction friction, the condensate film (similarly to the case of asymptotics of immobile vapor) covers the entire cylinder surface. The averaging of the Nusselt number over the entire cylinder surface gives

$$\langle \mathrm{Nu}_s \rangle \equiv \frac{\langle h_s \rangle D}{k} = \frac{2\sqrt{2}}{\pi}\sqrt{\mathrm{Re}}. \tag{11.15}$$

According to Eqs. (11.4) and (11.15), the Nusselt number in the critical point will exceed the averaged number by $\pi/2 \approx 1.57$. We note that the approximation of the numerical solution in [8] provides the asymptotics of the suction friction

$$\langle \mathrm{Nu}_s \rangle \approx 0.905\sqrt{\mathrm{Re}},$$

which virtually coincides with the accurate solution (11.15).

Now let us consider the case of pure friction when the suction rate is negligibly small ($w \ll U_\infty$). The flow rate of fluid in the film will still be determined by Eq. (11.12). The interphase friction is calculated as the surface friction upon the laminar flow around an impermeable cylinder [11]

$$\tau_\delta = \frac{\rho_v U_\infty^2}{\sqrt{\mathrm{Re}_v}} f(\phi). \tag{11.16}$$

In the above formula, $\mathrm{Re}_v = U_\infty D/\nu_v$ is the Reynolds number constructed using the vapor flow velocity at the infinity U_∞, the kinematic viscosity of vapor ν_v, and the cylinder diameter D,

$$f(\phi) = 4.93\phi - 1.93\phi^3 + 0.206\phi^5 - 0.0129\phi^7 + 0.304 * 10^{-4}\phi^9$$
$$- 0.814 * 10^{-4}\phi^{11}$$

is a function of angular coordinates with the maximum at $\phi \approx 1.02$ becoming zero at $\phi = \phi_* \approx 1.9$. It corresponds to the angle of separation of the boundary layer $\approx$109°, which is considerably higher than the value of the angle $\approx$85° obtained in the experiments on airflow around the cylinder [14]. This discrepancy is likely to be conditioned by the fact that the theory of a steady laminar boundary layer is inadequate to describe the regularities of separation. Nevertheless, in order to illustrate the limiting heat transfer laws, we will use the "classical" value of the angle of separation $\approx$109°.

The equation for film thickness follows from Eqs. (11.5), (11.9), and (11.13)

$$\frac{d(\delta^3)}{d\phi} + \frac{3}{2}\frac{d(\ln f)}{d\phi}\delta^3 = \frac{3}{2}\frac{D^3}{\sqrt{\mathrm{Re}}}\left(\frac{\mu L}{k\Delta T}\sqrt{\frac{\mu_v \rho_v}{\mu\rho}}\right)^{1/3}\frac{1}{f},$$

the solution for it will appear as

$$\left(\frac{\delta}{D}\right)^3 = \frac{3}{2}\frac{1}{\mathrm{Re}^{3/2}}\frac{k\Delta T}{\mu L}\sqrt{\frac{\mu\rho}{\mu_v\rho_v}}\frac{\int_0^\phi f^{1/2}d\phi}{f^{3/2}}. \tag{11.17}$$

Since with $\phi = \phi_* \approx 109°$ the friction is zero, $f = 0$, we obtain from Eq. (11.17) in the point of separation of the vapor flow that $\delta_* \to \infty$, $h_* = 0$. Assuming again that there is no heat transfer at the rear side of the cylinder $\phi_* \leq \phi \leq \pi$, we obtain the Nusselt number averaged over the entire surface

$$\langle \mathrm{Nu}_f \rangle \equiv \frac{\langle h_f \rangle D}{k} = 0.793\sqrt{\mathrm{Re}}\left(\frac{\mu L}{k\Delta T}\sqrt{\frac{\mu_v\rho_v}{\mu\rho}}\right)^{1/3}. \tag{11.18}$$

According to Eqs. (11.4) and (11.18), the Nusselt number at the critical point will be higher than the average value by a factor of 2.15. We note that a similar formula that gives asymptotics of form Eq. (11.18) was proposed in [9] based on approximation of the numerical solution. However, the constant used there is 0.9. This discrepancy is likely to be accounted for by the fact that, in the model used in [9], the flow around the cylinder was without separation. Whereas our solution was obtained taking into account the effect of separation of the boundary layer. Unfortunately, the authors of work [9] provided no comments on this circumstance that contradicts the conclusions of the theory of laminar boundary layer [11, 15].

11.1.6 Analysis of the Results

The monotonous rise in the thickness of a liquid film from δ_0 at $\phi = 0$ to infinity at a certain limiting value of the angular coordinate $\phi = \phi_*$ is common for all limiting cases considered above (Fig. 11.1). Hence, it follows that the HTC will drop from the maximum value $h = k/\delta_0$ in the critical value to zero.

Figure 11.2 shows the distributions of the relative HTC $h(\phi)/h_0$ over the angular coordinate.

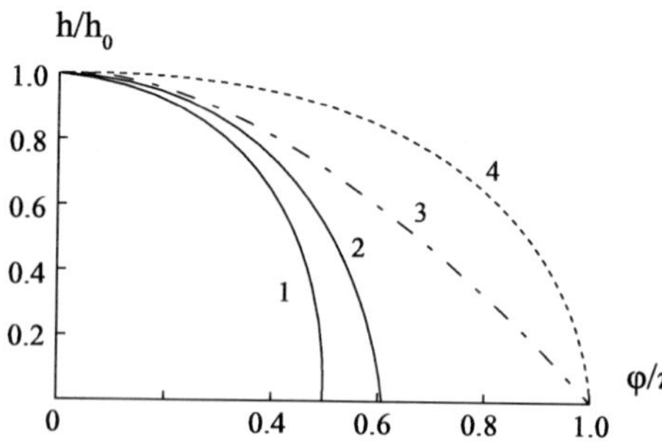

Fig. 11.2 The relative heat transfer coefficient for asymptotics. 1. Pressure, 2. pure friction, 3. suction friction, 4. gravity

- For the case of steady-state vapor, the liquid film exists over the entire cylinder surface and $\phi_* = \pi$
- In the asymptotics of suction friction, the flow of a cylinder without separation is also implemented up to $\phi = \pi$. However, the distribution of the HTC over the circumference is more gently sloping here
- The limiting case of pure friction is governed by the external flow in the boundary layer. Therefore, in the point of separation of the vapor flow $\phi \approx 1.9$, the HTC becomes zero
- In the case in which the pressure gradient is the driving force, its turning into zero in the midsection results in flooding of the liquid film: $h = 0$ at $\phi = \pi/2$.

Thus, we have studied all four limiting laws: (11.8), (11.11), (11.15), and (11.18). Since in the actual process, the gravity force is a permanent factor, it is reasonable to introduce the relative heat transfer laws $\psi = h/h_g$, where h_g is determined using the Nusselt formula (11.8) for an immobile vapor. Then, the limiting laws obtained above will acquire the following form: asymptotics of pressure, index p (pressure)

$$\psi_p \equiv \frac{\langle h_p \rangle}{\langle h_g \rangle} = 0.841 \left(\frac{\rho_v U_\infty^2}{\rho g D} \right)^{1/4}, \tag{11.19}$$

asymptotics of suction friction, index s (suction)

$$\psi_s \equiv \frac{\langle h_s \rangle}{\langle h_g \rangle} = 1.236 \left(\frac{U_\infty^2}{gD} \frac{k \Delta T}{\mu L} \right)^{1/4}, \tag{11.20}$$

asymptotics of pure friction, index f (friction)

$$\psi_f \equiv \frac{\langle h_f \rangle}{\langle h_g \rangle} = 1.09 \left(\frac{U_\infty^2}{gD} \right)^{1/4} \left(\frac{\mu L}{k \Delta T} \right)^{1/12} \left(\frac{\mu_v \rho_v}{\mu \rho} \right)^{1/6}. \tag{11.21}$$

The relative heat transfer laws determined by formulas (11.19)–(11.21) can be used to give comparative estimations of the effects of various factors on heat transfer upon condensation. Let us note that, regardless of the great practical significance and the lengthy period of investigation of this problem, a relatively small amount of experimental data on the condensation of water vapor [12, 16, 17], R21 [2], and R113 [17, 18] upon the horizontal flow of vapor around the horizontal cylinder have been accumulated thus far. The corresponding ranges of variation of mode parameters and the calculated values ψ are listed in Table 11.1.

In practice, the process of condensation on the cylinder surface is widely used and is an important component of the thermo-hydrodynamics of apparatuses at thermal and nuclear power plants, including condensing units for turbines, condensers for desalination plants, high-pressure heaters for turbo units at nuclear power plants, and turbines at thermal power plants. The range of characteristics of vapor parameters for the given applications is listed in Table 11.2.

Table 11.1 The ranges of parameters of the experimental studies of heat exchange upon vapor condensation on the surface of a horizontal cylinder

Author	Medium	Pressure p (bar)	Vapor flow rate U_∞ (m/s)	Cylinder diameter D (mm)	ψ_s	ψ_f	ψ_p
Berman and Tumanov [12]	Vapor	0.032–0.48	1–12	19	0.3–2.6	0.22–0.64	0.089–0.77
Fujii et al. [9]	Vapor	0.026	22–73	14	1.2–2.6	1.2–2.3	0.44–0.77
Michael et al. [16]	Vapor	1.0	5–81	14	1.3–7.4	0.93–4.3	0.48–2.0
Lee and Rose [17]	Vapor	0.05–1.0	0.3–26	12.5; 25	0.33–1.9	0.15–2.1	0.05–0.53
Lee and Rose [17]	R_113	0.4–1.05	0.3–26	12.5	1.0–1.4	0.27–1.3	0.19–0.77
Honda et al. [18]	R_113	1.0–1.2	1–16	8; 19	0.39–4.0	0.47–3.0	0.33–1.8
Gogonin et al. [2]	R_21	3–5.2	1–5	2.5; 16	0.53–2.8	0.75–2.4	0.51–1.3

Table 11.2 Characteristic ranges of mode parameters of vapor in condensation units at thermal and nuclear power plants

Apparatus	Medium	Pressure p (bar)	Vapor flow rate U_∞ (m/s)	Cylinder diameter D (mm)	ψ_s	ψ_f	ψ_p
Turbine condensers	Vapor	0.03–0.5	1–100	≈25	0.3–8.4	0.2–3.8	0.08–1.6
Condensers for desalination plants	Vapor	0.3–1	≈10	≈25	1.2–3	1–1.4	0.45–0.6
Heaters at thermal and nuclear plant	Vapor	10–80	10–70	≈25	1.6–16	1.3–5.9	0.83–4.9

It clearly flows from comparing Tables 11.1 and 11.2 that the experimental investigations performed are far from complete embracing the range of high velocities and pressures, which for practical needs is considerable. Thus, for the water-vapor system, all experimental data were obtained for pressures $p \leq 1$ bar, whereas the pressure range for the actually operating power units extends up to $p \approx 80$ bar. The greatest study of relatively high pressures ($p \approx 5.2$ bar) was made in [2] while studying the condensation of R21 vapors. When calculated for the conditions of water (with account for the pressure in the thermodynamic critical point), these experimental data correspond to pressure $p \approx 220$ bar.

Analysis of Tables 11.1 and 11.2 demonstrates that intensification of heat transfer, as compared with the case of steady-state vapor, may be as high as eightfold. The absence of experimental data in the ranges of mode parameters that are significant for practical applications results in considerable difficulties when elaborating calculation procedures. At a fixed pressure, the thermophysical properties of vapor and liquid phases remain virtually constant. Two major mode parameters can be varied in the experiments: the velocity of the incident flow U_∞ and wall-vapor temperature differential ΔT. As can be seen from relationships (11.11), (11.15), and (11.18), for moving vapor, all three limiting laws acquire the form $\langle \mathrm{Nu} \rangle \sim \sqrt{\mathrm{Re}}$. The effect of temperature difference on heat transfer in pressure asymptotics (11.11) is the same as that in the case of steady-state vapor $\langle h_p \rangle \sim \langle h_g \rangle \sim \Delta T^{-1/4}$.

In the limiting case of pure friction (11.18), the HTC also drops with increasing temperature difference $\langle h_f \rangle \sim \Delta T^{-1/3}$. Asymptotics of suction friction (11.15) is characterized by a purely hydrodynamic dependence $\langle h_s \rangle \sim \sqrt{U_\infty}$, which does not comprise ΔT: $\langle h_s \rangle =$ idem at a constant velocity and pressure values. Therefore, it should be expected that, in the case of appreciably high velocities and under low-temperature differences, the effect of pure friction $(\psi_f \gg \psi_s)$, is dominating, whereas the effect of suction friction is dominating at high $(\psi_s \gg \psi_f)$. This hypothesis is attested to by the experimental data [2] obtained upon condensation of R21 vapor on a cylinder with diameter $D = 2.5$ mm and at velocities $U_\infty = (2.9–3.8)$ m/s.

Figures 11.3 and 11.4 show that, for the region of small ΔT, the experimental points are fairly well described by pure friction asymptotics $\langle h \rangle \approx \langle h_f \rangle$. As the temperature difference increases, the experimental curve $\langle h \rangle(\Delta T)$ gradually descends to a horizontal shelf that is described by the suction asymptotics $\langle h \rangle = \langle h_s \rangle$. As follows from Figs. 11.3 and 11.4, under the conditions studied, the intensification of heat transfer due to the effect of the velocity factor was $\psi \approx 2.5$. In physical terms, this means that it is likely that, in this case, there is a simultaneous impact of gravity and friction factors, each of these making contributions of approximately the same order of magnitude.

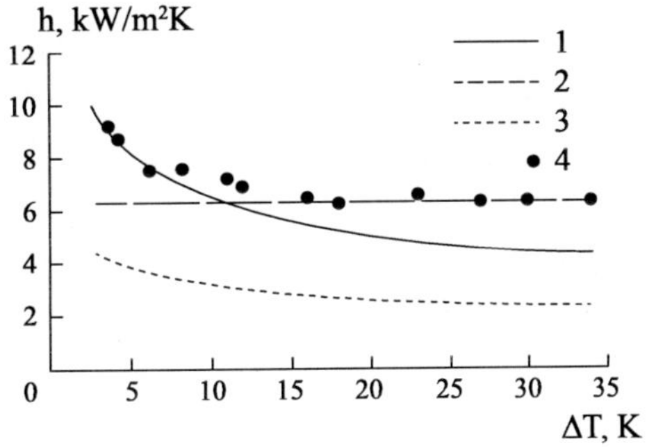

Fig. 11.3 Condensation heat transfer of R21, $D =$ 2.5 mm, $p = 5.2$ bar, $U_\infty =$ 2.9 m/s, dots represent the experimental data [2]. The limiting heat transfer laws, 1. steady-state vapor, 2. pure friction, 3. suction friction

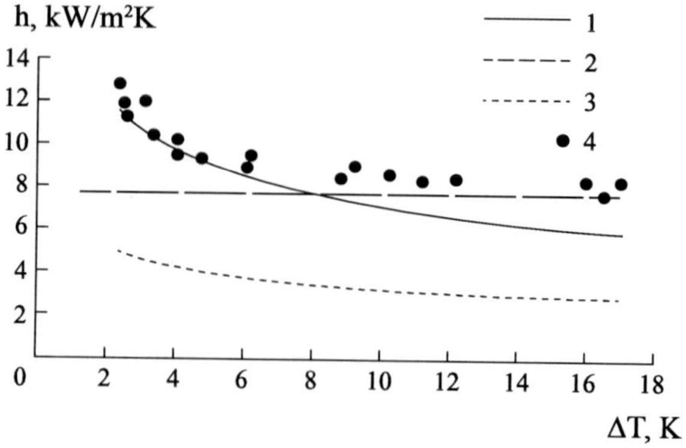

Fig. 11.4 Condensation heat transfer of R21, $D =$ 2.5 mm, $p = 3$ bar, $U_\infty =$ 3.8 m/s, dots represent the experimental data [2]. The limiting heat transfer laws. 1. steady-state vapor, 2. pure friction, 3. suction friction

Let us note that Tables 11.1 and 11.2 demonstrate a comparable quantitative effect of values ψ_f and ψ_s for broad ranges of variation of mode parameters. Therefore, the experimental data shown in Figs. 11.3 and 11.4 [2], which are explicitly described by the corresponding asymptotic laws, are unique to a certain extent. The impact of pressure on the heat transfer manifests itself mainly through the effect of the ratio between the densities of the vapor and liquid phases. Here, we obtain

$$\langle h_s \rangle = \text{idem}, \langle h_f \rangle \sim \left(\frac{\rho_v}{\rho} \right)^{1/6}, \langle h_p \rangle \sim \left(\frac{\rho_v}{\rho} \right)^{1/4}.$$

Thus, the effect of pressure for asymptotics (11.11) is to a certain extent higher than that for asymptotics (11.18). The trends of this effect of pressure coincide, which additionally attests to the mutual dependence of the limiting heat transfer laws. In the general case, a simultaneous effect of all three factors (gravity, pressure gradient, and suction and pure friction) is likely to be manifested. With the increasing velocity of vapor in leakage, the effect of gravity is eliminated. All three limiting laws ψ_p, ψ_s, and ψ_f are similar with respect to the vapor flow velocity (and to the cylinder diameter). Only stratification over the temperature differential is observed here at fixed thermophysical properties. The simultaneous action of all influencing factors results in the fact that in the actual problem (with the exception of the boundary variant of steady-state vapor), asymptotic variants can be isolated infrequently. Therefore, strictly speaking, the present problem should be always considered in its complete formulation.

The analysis of limiting heat transfer laws given in this chapter is the first step in this direction. The next stage should comprise the consideration of the mixed heat-exchange laws when several influencing factors are taken into account in the force balance. The analytical solutions for the limiting heat transfer laws were developed in the paper of the author of the present book [19].

11.2 Condensation from a Vapor-Gas Mixture

11.2.1 Survey of the Literature

It is known that even a small content of Noncondensable Gas (NCG) can appreciably reduce the intensity of heat transfer during condensation of the vapor. The corresponding negative effect can be quite substantial even if the original NCG content is small. Because of the vapor reduction by condensation, the velocity of the vapor–air mixture flow decreases downstream, while the concentration of inert gases is increasing, in turn, this results in a decrease in the mass-transfer intensity. The presence of NCG makes much more complications in the calculation of condensation in channels. In contrast to the case of pure vapor condensation, in this setting, one should evaluate the local characteristics of the heat-and-mass transfer. In this way, the concentration and temperature in the flow and on the interface surface should be determined, as well as the friction and the transverse mass flow on the phase interface. Theoretical investigations of the effect of the presence of NCG on the heat transfer by condensation have begun in [20] (published in 1934) and continue until today.

The condensation in the presence of NCG is a very complex phenomenon. A rigorous approach to the study of condensation in the presence of NCG is based on the solution of the fully coupled, liquid and mixture, conservation equations. Within this approach, different solution methods can be applied and in some cases, an analytical solution can be obtained. Application of numerical methods allows one to formally solve two three-dimensional fully coupled systems of conservation equations for both liquid and gaseous phases. This, however, may involve difficulties associated with the physical picture of the process.

The method of heat and mass transfer analogy has proved quite potent. This approach, which is based on the similarity between the heat and mass fraction transport equations, involves two important dimensionless relations, namely, the Nusselt number and the Sherwood number. The use of the similarity between heat, mass fraction, and momentum equations underlies the well-known Chilton–Colburn analogy [21, 22]. A more rigorous approach is based on the solution of the system of conservation equations, for both the liquid film and gaseous mixture phases. These equations describe the conservation of mass, momentum, and energy, which can be written in the form of boundary layer equations. Together with interface conditions, such a system of equations can be solved analytically in a number of cases. This approach is general and formally does not rely on any empirical correlations. Fast developments achieved in numerical methods (for example, computational fluid dynamics (CFD) methods) and tools give even more general capabilities for detailed investigation. In connection with rational physical estimates, numerical results can be properly analyzed.

For the physical background of the effect of NCG on the heat transfer, see the books [22–24].

Sparrow et al. [25] investigated the condensation of a mixture with NCG past a horizontal surface from. The system of boundary layer equations is solved by the similarity method.

Felicione and Seban [26] solved the boundary layer equations for the vapor–gas mixture using the integral approach. Both the similarity solution and the integral method were found to give very similar results.

Prosperetti [27] studied boundary conditions at the liquid–gas interface. A generic equation, which takes into account the interface balances of mass, momentum, and energy, has been presented with a correct mathematical formalism.

One should also mention Siddique et al. [28], which theoretically investigated vapor–gas mixtures during condensation inside a vertical pipe.

Hassan and Banerjee [29] implemented the six-equation model for condensation from a binary mixture of gases based on the heat and mass transfer analogy.

Yao et al. [30] discussed the implementation of the one-dimensional model in the RELAP5/MODE3 code. The system of equations was closed by the correlations developed for the calculation of the condensation HTC, as well as by the heat and mass transfer analogy.

Using the diffusion layer model, Peterson [31] showed that some correlations can produce significant errors. As a result, mechanistic modeling was recommended.

No and Park [32] investigated the condensation of vapor from the vapor–air mixture. The resulting model was based on the heat and mass transfer analogy through empirical correlations.

Siow et al. [33] showed numerical investigations of the condensation from binary mixtures, on a horizontal surface. A numerical approach was applied to the film and mixture boundary layer equations.

Maheshwari et al. [34] studied theoretically the heat and mass transfer during condensation from a vapor–gas mixture (VGM) for a wide range of the film and gas Reynolds numbers in a vertical pipe. The analogy between heat and mass transfer was applied.

Slow et al. [35] presented numerical investigations of the heat transfer process during condensation from the binary mixture flowing between two parallel vertical surfaces.

Martin et al. [36] investigated the film condensation model implemented in a CFD code. These models rely on the Chilton–Colburn heat-mass transfer analogy and the diffusion layer model.

Stephan [37] studied the condensation from a binary mixture of condensable and noncondensable gases. Based on the solution of boundary layer equations, a correlation for the HTC was been proposed.

Oh and Revankar [38] investigated the condensation of vapor from a vapor–gas mixture inside a vertical pipe. Relying on the heat and mass transfer analogy, a correlation for the HTC was developed.

Stevanovic et al. [39] numerically investigated the condensation from the VGM in a tube. A mechanistic CFD approach was recommended for future investigations.

Siow et al. [40] applied the finite volume method in order to solve the system of boundary layer equations for condensation from binary mixtures of gases in a channel.

Li [41] simulated the condensation of water vapor in the presence of NCG using computational fluid dynamics (CFD) for turbulent flow in a vertical cylindrical condenser tube. The simulation accounts for the turbulent flow of the gas mixture, the condenser wall, and the turbulent flow of the coolant in the annular channel with no assumptions of constant wall temperature or heat flux. The CFD results also show, at least for flows involving high water vapor content, that the axial velocity of the gas mixture at the interface between the gas mixture and the condensate film is in general not small and cannot be neglected.

Gorpinyak and Solodov [42] considered a one-dimensional differential model describing the condensation of a vapor–gas mixture in tubes and its computer implementation if the visual basic integrated development environment is presented. The compute kernel is based on mathematical models that take into account the main significant effects during condensation such as gravitation, friction at the phase interface boundary, availability of noncondensing, admixtures, different external cooling methods, and the possibility of dangerous operation conditions to occur. A mathematical formulation of the problem involves the conservation equations (of momentum, energy, and mass of mixture components) for averaged flows of coolants.

A detailed survey of theoretical investigations of the effect of gases on vapor condensation known as of 1988 is given in [43].

11.2.2 The Mechanistic Model

Traditionally, a "Mechanistic Model" (MM), refers to a model based on a mathematical description of a phenomenon or process. Such models should not be confused with strictly deterministic MM. A classical example of this model is Newtonian mechanics. The term "empirical models" is used for models based on relatively simple approximations. It should be noted that there is no clear line between the "mechanistic" and "empirical" models. Therefore, there are various intermediate types of models.

The problem of heat transfer with film vapor condensation from a VGM is a conjugate problem of heat-and-mass transfer on the discontinuity surface for the "condensate film–VGM" system. In accordance with the reference case of condensation of pure vapor, the mathematical description of the problem involves in addition the equation for turbulent diffusion for the vapor component. The diffusion equation describes the penetration of vapor molecules through the gas mass at rest in a gas cluster. Such a description is certainly legitimate in the case when the vapor mass fraction is small relative to the gas mass fraction. However, the mathematical model becomes less and less representative of the physical picture of the process as the gas content in the VGM is reduced. What is more, with a vanishingly small concentration of the inert gas in the VGM, we arrive at an abnormal situation: the diffusion equation should describe the penetration of large vapor clusters through the "lattice" of separate gas molecules. To eliminate this paradox for numerical investigation, one has to "unplug" the diffusion equation starting from some allowed minimum value (a "threshold") of the gas content. This implies the self-contradictoriness of the mathematical description of condensation from a VGM involving the diffusion equation.

Below, we shall be concerned with a variant of MM free from this contradiction. An algebraic MM is based on the well-known theory of mass transfer of the film (or the film theory), which was formulated by Colburn and Drew in 1937 [44]. The film theory is widely used to describe the heat and mass transfer at the contact of liquid and gaseous turbulent flows separated by an interface. The film theory assumes that the resistance to turbulent heat and mass transport is concentrated in thin layers (films) adjoining on both sides to the interface. The works [45–47] show that the methods based on this calculation model are quite accurate and can be applied in a wide range of operating parameters.

In condensation, as rule, the circulation circuit of coolant serves as a heat-removing external environment from the outer wall of the channel. The heat flow q, which is removed from the VGM to the external environment, goes in succession through a chain of thermal resistances: of the liquid film $1/h_\delta$, of the hard wall δ_w/k_w, and of the outer environment $1/h_0$. The total external thermal resistance is given by their sum

$$\frac{1}{h_\Sigma} = \frac{1}{h_0} + \frac{\delta_w}{k_w} + \frac{1}{h_\delta}.$$

We shall consider in the first approximation the "internal" part of the problem from the gas/vapor core to the wall internal surface (Fig. 11.5). Here the "external" thermal resistances are not taken into account: $\delta_w/k_w \approx 1/h_0 \approx 0$, $h_\Sigma \approx h_\delta$. For the distribution of the temperatures over the channel section, this means that $T_0 \approx T_{w0} \approx T_w$. This approximation, which substantially simplifies our calculations,

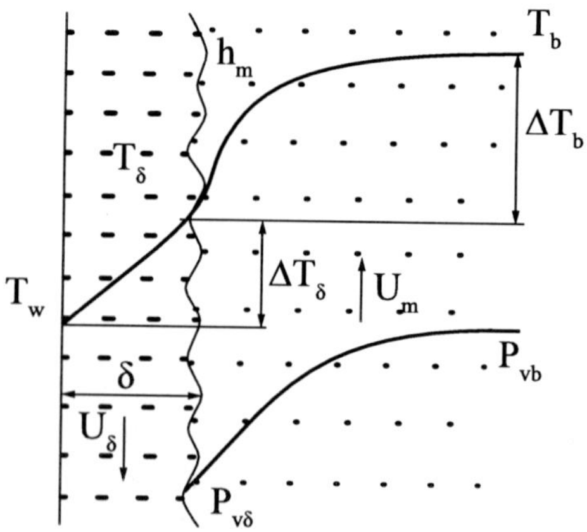

Fig. 11.5 Vapor film condensation from the vapor–air mixture

is capable of elucidating the effect of the principal physical factors on the problem under consideration. Next, having the result of the analytical solution at our disposal, we can easily include in it the external thermal resistances and thereby extend it to the general case. In view of the above assumptions, the heat flow carried to the wall can be written as

$$q = h_\delta (T_\delta - T_w). \tag{11.22}$$

Here T is the temperature, the subscripts δ, w, and 0 correspond to the surface of condensate liquid film, the wall, and the external coolant, respectively.

Within the algebraic MM, it is assumed that the heat transfer from the VGM to the phase interface is affected by two independent mechanisms. The diffusive heat flow q_1 is related to the mass flow j_v of the vapor condensed on the wall by the thermal balance condition

$$q_1 = j_\text{v} h_{fg}, \tag{11.23}$$

where h_{fg} is the latent heat of vaporization. The convective heat flow q_2 is carried away via the turbulent heat transfer mechanism from the superheated VGM to the film surface

$$q_2 = F h_m (T_b - T_\delta), \tag{11.24}$$

where h_m is the HTC. The right-hand side of (11.24) involves the Akkerman correction factor F, which takes into account the effect of the transverse mass flow of condensation vapor on the heat transfer [48, 49]

$$F = \frac{\sigma}{1 - \exp(-\sigma)}. \tag{11.25}$$

Here

$$\sigma = \frac{j_\text{v} c_{p\text{v}}}{h_m} \tag{11.26}$$

is the dimensionless vapor flow rate, c_p is the specific heat capacity at constant pressure, the subscript "v" corresponds to vapor properties. By the principle of superposition [50], the resulting heat flow arrived in the liquid film consists of the above two parts

$$q = q_1 + q_2. \tag{11.27}$$

The quantity σ assumes its maximum value with smallest possible vapor partial pressure on the condensate film surface determined by the coordinates of the triple point. In thermodynamics, the triple point of a substance is the temperature and pressure at which the three phases (gas, liquid, and solid) of that substance coexist in thermodynamic equilibrium [1]. It is that temperature and pressure at which meet the sublimation curve, the fusion curve, and the vaporization curve. For condensation of water vapor at atmospheric pressure, $\sigma_{\max} = 0.901$.

We introduce the diffusion factor

$$\Lambda = \ln\left(\frac{1 - X_{v\delta}}{1 - X_{vb}}\right), \tag{11.28}$$

where

$$X_{vb} = \frac{p_s(T_b)}{p_\Sigma}, X_{v\delta} = \frac{p_s(T_\delta)}{p_\Sigma} \tag{11.29}$$

is the dimensionless partial pressure of vapor, respectively, in the gas/vapor core and on the phase interface.

During condensation, the film surface is impermeable for inert gas. Hence, using the continuity condition for VGM, we get the expression for the mass flow of the vapor going through the phase interface

$$j_v = \rho_{vb}\beta\Lambda, \tag{11.30}$$

where β is the mass transfer coefficient. Relations (11.28), (11.30) determine the dependence of the vapor flow (and hence, the condensation intensity) on the mass transfer coefficient and the differentials of the vapor phase concentration. From (11.22)–(11.30) we get the heat balance equation for condensation from a VGM

$$h_\delta(T_\delta - T_w) = \beta\rho_{vb}h_{fg}\ln\left(\frac{1 - X_{v\delta}}{1 - X_{vb}}\right) + Fh_m(T_b - T_\delta). \tag{11.31}$$

11.2.3 The Chilton–Colburn Analogy

The Chilton–Colburn J-factor analogy [21] is a widely used analogy between friction, heat transfer, and mass transfer. This analogy comes from the approximate similarity of the physical mechanisms of turbulent transport. Let us derive the Chilton–Colburn analogy based on our present knowledge [22]. We first write down the well-known Petukhov–Kirillov correlations for heat and mass transfer [48], which hold in the range $10^4 < Re < 5 * 10^6$ for the Reynolds number

$$\left.\begin{aligned} \mathrm{St} &= \frac{\xi/8}{1.07+12.7\sqrt{\xi/8}(\mathrm{Pr}^{2/3}-1)}, \\ \mathrm{St}_m &= \frac{\xi/8}{1.07+12.7\sqrt{\xi/8}(\mathrm{Sc}^{2/3}-1)}. \end{aligned}\right\} \tag{11.32}$$

Here $\mathrm{St} = \frac{h_m}{\rho_m c_{pm} U}$ is the Stanton number, $\mathrm{St}_m = \frac{\beta}{U}$ is the mass Stanton number, $\mathrm{Pr} = \frac{\nu_m}{\alpha_m}$ is the Prandtl number, $\mathrm{Sc} = \frac{\nu_m}{D_m}$ is the Schmidt number, ν_m, α_m is the kinematic viscosity and thermal diffusivity of VGM, respectively, D_m is the molecular diffusivity, U is the stream velocity of VGM, the subscript m corresponds to the properties of VGM.

The Stanton number and the mass Stanton number are related as

$$\frac{\mathrm{St}}{\mathrm{St}_m} = \frac{1.07 + 12.7\sqrt{\xi/8}\left(\mathrm{Sc}^{2/3} - 1\right)}{1.07 + 12.7\sqrt{\xi/8}\left(\mathrm{Pr}^{2/3} - 1\right)}.$$

Making $\mathrm{Pr} \to \infty$, $\mathrm{Sc} \to \infty$, from (11.32) we have $\mathrm{St}/\mathrm{St}_m \approx (\mathrm{Sc}/\mathrm{Pr})^{2/3}$, or, what is the same, $\mathrm{StPr}^{2/3} \approx \mathrm{St}_m\mathrm{Sc}^{2/3}$. Changing from the approximate equality to the exact equality, we get the main postulate of the Chilton–Colburn analogy

$$\mathrm{StPr}^{2/3} = \mathrm{St}_m\mathrm{Sc}^{2/3}. \tag{11.33}$$

Despite its semi-empirical nature, the Chilton–Colburn analogy is capable of quite accurately describing the relationship between the coefficients of friction, heat transfer, and mass transfer for the range of parameters [22]

$$0.6 < \mathrm{Pr} < 60,\ 0.6 < \mathrm{Sc} < 3000.$$

Consequently, the heat and mass transfer coefficients are related by

$$\frac{h_m}{\beta} = \rho_m c_{pm}\left(\frac{h_m}{D_m}\right)^{2/3} = \rho_m c_{pm}\mathrm{Le}^{2/3},$$

where $\mathrm{Le} = \frac{\mathrm{Sc}}{\mathrm{Pr}} = \frac{\alpha_m}{D_m}$ is the Lewis number.

11.2.4 The Principal Equation of the Mechanistic Model

In view of (11.33), the principal equation of the MM reads as

$$h_\delta(T_\delta - T_w) = \frac{h_m h_{fg}}{c_{pm}\mathrm{Le}^{2/3}}\ln\left(\frac{1 - X_{v\delta}}{1 - X_{vb}}\right) + F h_m(T_b - T_\delta). \tag{11.34}$$

To analyze the effect of inert gases on the heat transfer intensity during condensation, we introduce the "NCG influence coefficient"

$$\psi = \frac{h_\Sigma}{h_\delta}. \tag{11.35}$$

Here

$$h_{\Sigma} = \frac{q}{T_{\delta} - T_w} \tag{11.36}$$

is the required HTC for vapor condensation from VGM,

$$h_{\delta} = \frac{q}{T_{\delta} - T_w} \tag{11.37}$$

is the reference HTC for pure vapor condensation. In view of (11.36), (11.37), the NCG influence coefficient can be written as the ratio of the corresponding temperature differences

$$\psi = \frac{T_{\delta} - T_w}{T_b - T_w}. \tag{11.38}$$

From (11.38), it follows that the transfer from the case of VGM understudy to the reference case of condensation of a pure vapor amounts to the following change: $T_b \to T_{\delta}$. This happens because of the shift of the partial pressure in the flow core due to the presence of the superheated gas/vapor core

$$p_b \equiv p_s(T_b) \to p_{\delta} \equiv p_s(T_{\delta}). \tag{11.39}$$

Note that physically it is more legitimate to write

$$T_b \equiv T_s(p_b) \to T_{\delta} \equiv T_s(p_{\delta}) \tag{11.40}$$

in place of (11.39), which vividly reflects the cause and effect relationship of the NCG effect (Fig. 11.5).

An introduction of inert gas in the vapor flow results in the appearance of a superheated gas/vapor core in the flow, which in turn, decreases the partial pressure of the vapor. As a result, the saturation temperature of the condensing vapor follows the pressure in accordance with the Antoine equation [22]. This, in turn, reduces the driving force of the condensation process (the "saturated vapor–the condensate film" temperature difference). Consequently, this reduces the heat flow removed to the wall, and therefore, the mass flow of the vapor. So, the NCG influence coefficient, as defined by (11.35) or by the equivalent relation (11.38), is an "omnibus test" of the process of condensation from a VGM. This quantity has the following limit values. For the pure vapor condensation (with no inert gas present, $X_{vb} \to 1$), we have $T_{\delta} = T_b$, $\psi = 1$. For an extremely small content of vapor ($X_{vb} \to 0$), we have $T_{\delta} = T_w$, $\psi = 0$.

Dividing both parts of (11.34) by $h_m(T_b - T_w)$, we get the dimensionless form of the principal equation of MM

$$A\psi = B\Lambda K + F(1 - \psi). \tag{11.41}$$

Equation (11.41) contains the following three dimensionless numbers, which control the condensation from a VGM.

- the dimensionless vapor mass flow

$$\phi = B\Lambda, \tag{11.42}$$

where $B = \frac{c_{pv}}{c_{pm}} \frac{1}{\mathrm{Le}^{2/3}}$ is the complex of thermophysical properties,
- the phase transformation number (Kutateladze [51]), which characterizes the ratio of the heat amount spent for the phase transition to the heat amount spent for condensate cooling

$$K = \frac{h_{fg}}{c_{pv}(T_b - T_w)}, \tag{11.43}$$

- the relative HTC

$$A = \frac{h_\delta}{h_m}. \tag{11.44}$$

In calculations, the thermophysical properties of the mixture are evaluated by the additivity rule

$$\rho_m = \rho_g(1 - W_{vb}) + \rho_v W_{vb}, \, c_{pm} = c_{pg}(1 - W_{vb}) + c_{pv} W_{vb}.$$

Here W_{vb} is the vapor mass fraction, the subscript g corresponds to gas properties. The unknowns X_{vb}, $X_{v\delta}$ were evaluated from the relation between the partial pressure of the vapor and the temperature via the Antoine equation $p_v(T)$

$$\ln p_v(bar) = C_1 - \frac{C_2}{C_3 + T(\mathrm{C^o})}.$$

The empirical constants C_1, C_2, C_3 are specified for each liquid in a particular interval of temperatures. In particular, for water in the range 0–300 °C of temperatures we have $C_1 = 11.9648$, $C_2 = 3984.923$, $C_3 = 233.426$ (see [52]). The Antoine equation is a class of semi-empirical correlations describing the relation between vapor pressure and temperature for pure substances. The Antoine equation is derived from the Clausius–Clapeyron relation.

A general approach to the solution of the MM is based on the use of an equation of type (11.41) for the iteration search of the phase interface temperature T_δ and $X_{v\delta}$ from the known values of the VGM temperature T_b, the coolant temperature T_0 (or the wall temperature T_w), and the mass content of vapor in VGM X_{svb}. Hence, we can find the NCG influence coefficient, and therefore, using (11.35), we can determine the required HTC to the wall: $h_\Sigma = \psi h_\delta$. However, the iteration procedure hinders the influence analysis of the defining parameters to the NCG influence coefficient.

Hence, it is more reasonable to find an adequate iteration-free procedure for finding the dependence $\psi(X_{vb})$.

11.2.5 Iteration-Free Solution Procedure

We write down the principal equation of MM (11.41) in the form

$$A\psi = \sigma K + F(1 - \psi). \tag{11.45}$$

The Akkerman correction factor (11.25) with maximal error 3% can be approximated by the relation

$$F \approx \frac{1 + 0.8\sigma + 0.3\sigma^2}{1 + 0.3\sigma}. \tag{11.46}$$

The iteration-free procedure runs as follows

- substituting (11.46) into (11.45) and solving it for ϕ, we get the dependence $\phi(\psi)$,
- from (11.41), we find $X_{v\delta}(\psi)$,
- substituting the dependences $\phi(\psi, A)$, $X_{v\delta}(\psi)$ into (11.41), we get the dependence $X_{vb}(\psi, A)$.

Thus, the iteration-free procedure of solution provides a complete description of condensation from a VGM in the form of a similarity equation relating six parameters

$$f(X_{vb}, X_{v\delta}, \psi, A, K, B) = 0.$$

In contrast to direct evaluation, in which the dependence $\psi(X_{vb})$ is found iteratively, the iteration-free calculation proceeds in the opposite way. Given ψ, which is a function of the unknown temperature T_δ, we determine the mass content of vapor in the volume X_{vb} (which in fact is known ab initio). Strictly speaking, the iteration-free procedure of the solution also involves iterations. The fact is that VGM properties should be known for calculations. For this purpose, one should know, in turn, the mass content of vapor X_{vb}. But X_{vb} is unknown and should be determined via an iteration-free procedure. In fact, one can easily use approximate values of VGM thermophysical properties, which can be specified in advance. Calculations show that, as a rule, one preliminary iteration is sufficient. At the same time, the advantage of the iteration-free procedure is the possibility of parametric analysis of the problem under consideration.

11.2.6 Evaluation by the Mechanistic Model

The iteration-free approach was used for MM calculations under the following conditions: VGM pressure, $p = 1$ bar, wall temperature $T_w = 20\,°\mathrm{C}$, VGM temperature $T_b = 100\,°\mathrm{C}$. Special attention was paid to the region with high gas content. In calculations, the relative HTC A, as defined by (11.44), was varied, the parameters B, K were assumed to be constant: $K = 15.64$, $B = 1.3$. Approximation (11.46) was also used in calculations.

The results of calculations are presented in Fig. 11.6 as the dependences of ψ and of the difference $X_{vb} - X_{v\delta}$ on the mass content of vapor VGM. Figure 11.6a shows that a decrease in X_{vb} reduces ψ, the rate of decrease becoming faster as the value of the parameter A increases. So, for $A = 50$, the presence of 20% of NCG reduces ψ by a factor of 2.

For further analysis of the results, we write down in an expanded form the expression for the diffusion heat flow q_1, which is given by (11.23), (11.28)

$$q_1 = \frac{h_m h_{fg}}{c_{pm}\mathrm{Le}^{2/3}}\Lambda.$$

According to the physical picture of condensation, the mass flow of vapor (and hence, the heat flow q_1) is always directed from the VGM core towards the phase interface. This means that the following evident inequalities hold: $X_{vb} - X_{v\delta} > 0$, $q_1 > 0$. Here, to the equality $X_{vb} = X_{v\delta} = 1$, there corresponds the reference case of condensation of pure vapor: $\psi = 1$.

Figure 11.6b shows that if the vapor mass content decreases, then the difference of the concentrations $X_{vb} - X_{v\delta}$ passes through its maximum and, for low partial pressures of vapor, passes through the zero and changes its sign. Hence, the diffusive heat flow q_1 also changes its sign. At the same time, due to the presence of NCG, the flow core always remains superheated with respect to the liquid phase: $\Delta T_b = T_b - T_\delta > 0$. According to (11.24), the convective flow of heat q_2 is always directed from the flow core to the wall, $q_2 > 0$. It follows that the abnormal situation

$$q_2 > 0, q_1 < 0, j_v < 0 \tag{11.47}$$

will occur under certain conditions. Moreover, we simultaneously have

$$q = q_1 + q_2 > 0. \tag{11.48}$$

From (11.47) and (11.48), we have the following two paradoxes of the MM.

(1) The diffusion paradox. The heat flow arrives from the VGM to the wall, while the mass flow is directed from the film surface to the flow core. This leads to the physically impossible inversion of the vapor mass flow.

(2) "The condensation lock" paradox. A further decrease in the mass content of vapor in VGM will result in an increase in the absolute value of the negative

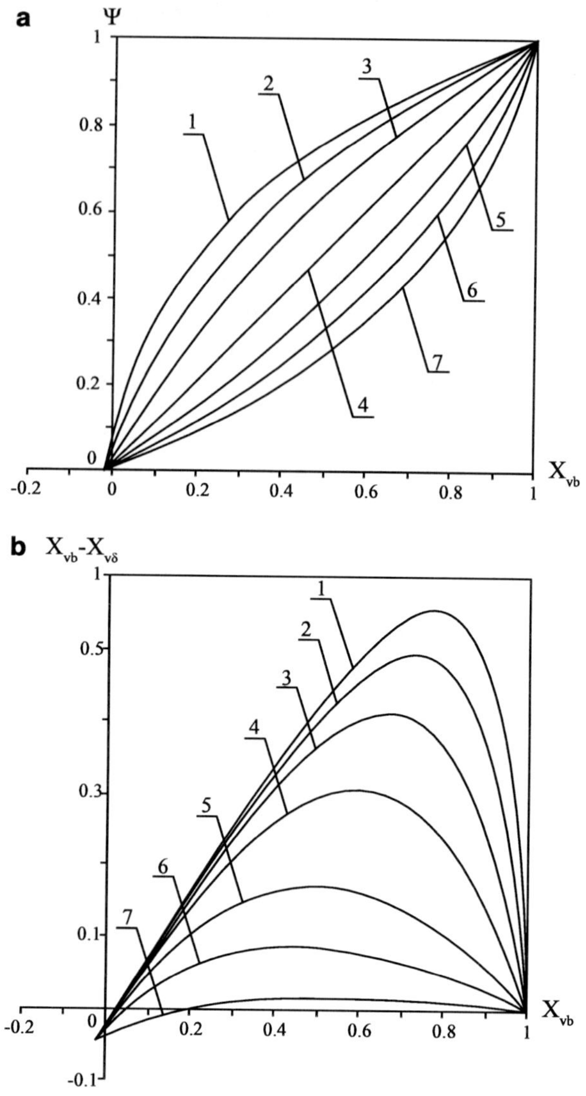

Fig. 11.6 Standard mechanistic model. Dependences ψ and $X_{\mathrm{vb}} - X_{\mathrm{v}\delta}$ on X_{vb} for various values of the relative heat transfer coefficient

quantity q_1. Finally, for some value of X_{vb}, the total heat flow vanishes and the process of condensation terminates.

Figure 11.6 shows that both paradoxes of the stereotyped model are interrelated and are manifested sooner the lower is the parameter A. The anomalous effects are manifested most vividly for very low mass contents of vapor and commensurable values of the HTC ($h_\delta \le h_m$), i.e., for $A \sim 1$.

The above analysis shows that the stereotyped variant of MM cannot be applied for a high concentration of inert gases in VGM. A possible reason for such paradoxes is the postulation of the rule of superposition. Formula (11.27) closely relates two physically different heat flows for the entire range of variation of NCG concentration. As a presumable corollary, this might imply the above paradoxical behavior of differentials of concentrations and temperatures. Calculations show that the above anomalies are manifested most vividly for the range of parameters: $K < 5$, $B < 0.4$. It is worth pointing out that, for the traditional field of practical application of use MM, there corresponds strong excess of the heat transfer intensity through the condensate film over the heat surface of the turbulent core of the vapor-air flow: $A >> 1$. It seems that the above fact is the reason that such obvious unphysicality has remained unnoticed.

11.2.7 The Modified Mechanistic Model

The above analysis shows that the main reason for decreasing the condensation intensity is the decrease of the vapor partial pressure in VGM due to inert gases. This, in turn, implies the thermal non-equilibrium of the gas/vapor core, which is manifested in the appearance of the core superheat with respect to the temperature of the condensate film surface (Fig. 11.5). We shall try to find an MM modification, which will enable us to rectify the above anomalies of the method of calculation.

In the mid-1960s, Kutateladze [51], when studying boiling problems, proposed to use the effective latent heat of the phase transition

$$h_{fg*} = h_{fg} + c_{pm}\Delta T_b. \tag{11.49}$$

In our setting, (11.49) means that, for vapor condensation in the presence of NCG, it is necessary not only overcome the energy barrier of the phase transition h_{fg}, but also to remove from VGM the amount of heat equal to its superheat enthalpy $c_{pm}\Delta T_b$. In view of (11.49), the Kutateladze number (11.43) can be replaced by its effective value

$$K_* = \frac{h_{fg} + c_{pm}(T_b - T_\delta)}{c_{pm}(T_b - T_w)} = K + 1 - \psi. \tag{11.50}$$

So, within the framework of the modified MM, we can abandon the excessive hard method of addition of heat flows, and replace it by the introduction of an equivalent supplement to the phase transition heat. This will enable us to omit the expression for the convective flow q_2 (11.24). So, from the chain of thermal resistances, we exclude the link responsible for the gas/vapor core (Fig. 11.5). In this case, the principal relation of the modified model (11.45) assumes the following form

$$A\psi = (K + 1 - \psi)B\Lambda. \tag{11.51}$$

A comparison of Eqs. (11.41) and (11.51) shows that the transition from the stereotyped to the modified model results in a radical simplification of the mathematical description. Below, both methods (the stereotyped and modified ones) will be applied for the solution of the test problem of condensation from a VGM ($p = 1$ bar, $T_b = 100$ °C, $T_w = 20$ °C). The results of the calculations are given in Fig. 11.7.

A comparison of Figs. 11.6 and 11.7 shows that the general character of the dependence $X_{vb} - X_{v\delta}$ is preserved for both variants of the MM, including the presence of the maximum and the trend that parameter A is influential. In the range of variation of the parameters $A \gg 1$, $X_{vb} \geq 0.5$, the results of calculations via both

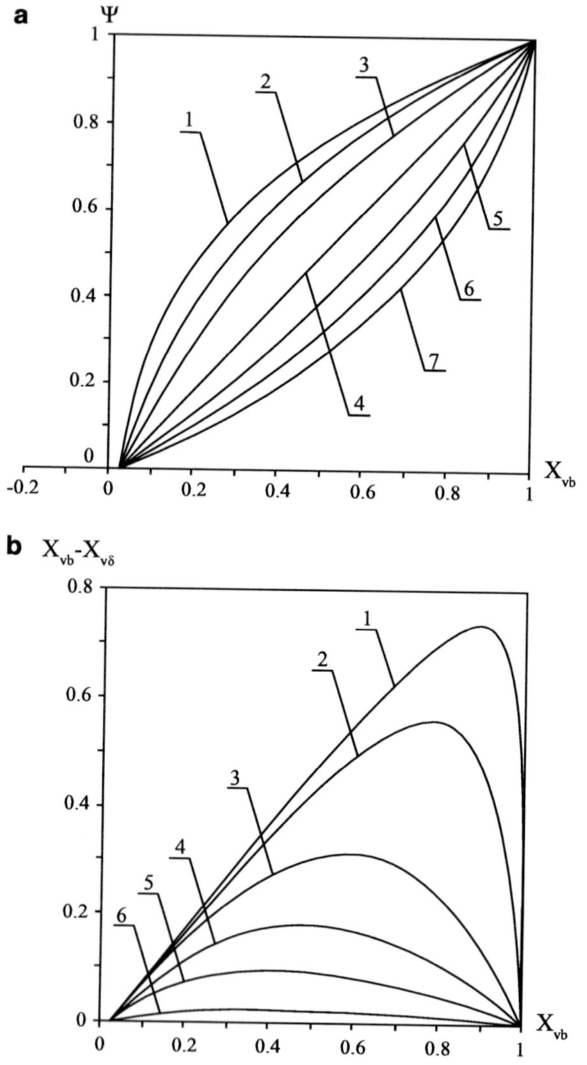

Fig. 11.7 A modified mechanistic model. The dependences of ψ (**a**) and $X_{vb} - X_{v\delta}$ (**b**) on X_{vb} with various values of the relative heat transfer coefficient

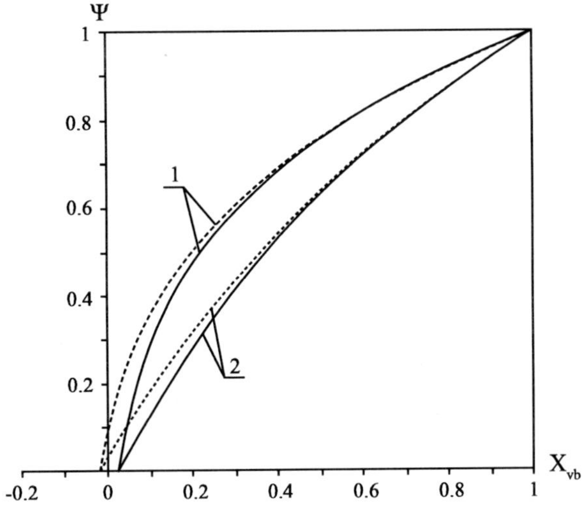

Fig. 11.8 The coefficient ψ versus the X_{vb}, – – the stereotyped model, — the modified model

models are practically the same. However, Fig. 11.7 shows that in the framework of the modified model, the previously existing abnormal region for small values of X_{vb} is absent. Moreover, the calculation curves do not change their sign as $X_{vb} \to 0$. The difference in the concentrations $X_{vb} - X_{v\delta}$ vanishes for all values of the parameter A for the smallest possible value of X_{vb}.

The above results suggest that the introduction of the term $1/h_m$ in the chain of thermal resistances (Fig. 11.5) in the region $X_{vb} \to 0$ is disputable. On the one hand, in the range $A \gg 1$, $X_{vb} \geq 0.5$, this term is negligible (which explains the actual agreement of the calculation results via the two methods). On the other hand, in a different parameter range, the presence of the term $1/h_m$ leads to the paradoxes: the first one is the inversion of the heat flow and the second one is the condensation lock.

Figure 11.8 depicts the calculated dependences of the influence coefficient of NCG for three various values of the parameter A. It is seen that the curves calculated via the modified model originate from the point with coordinates (0.0234, 0). This agrees with the physical picture of the process: condensation begins when the partial pressure of vapor exceeds the saturation pressure at a temperature $T_w = 20$ °C (0.0234 bar). In turn, for the stereotyped model, to the zero value of ψ there correspond negative values of X_{vb}. Moreover, to the actual initial vapor content $X_{vb} = 0.0234$, there correspond some positive values of ψ, which in addition depend on the parameter A.

11.2.8 Asymptotic Analysis of Mechanistic Model

Let us now give the asymptotic analysis of the modified MM. Consider the asymptotic behavior of the dependence $\psi(X_{gb})$ with the parameter A. In the limit case of a vanishingly small content of inert gas in VGM ($X_{gb} \to 0, X_{vb} \to 1$), we have $\psi \to 1$. Here (11.51) simplifies to read: $A = KB\Lambda$. Taking approximately $\frac{c_{pv}}{c_{pm}} = 1$, we get

$$\frac{1 - X_{v\delta}}{1 - X_{vb}} = \exp\left(\frac{\mathrm{ALe}^{0.6}}{K}\right).$$

For the partial pressure of vapor, we will use the linear Clapeyron–Clausius equation

$$\frac{\Delta p_s}{\Delta T_s} = \frac{r\rho_{vb}}{T_b}.$$

Expressing the vapor density from the equation of state for an ideal gas $p = \rho R_g T$, we get the following asymptotics for the influence coefficient of NCG

$$\psi = 1 - \frac{R_g T_b^2}{h_{fg}(T_b - T_w)}\left[\exp\left(\frac{\mathrm{ALe}^{0.6}}{K}\right)\right](1 - X_{vb}),$$

where R_g is the individual gas constant. This asymptotics shows the strong effect of NCG even in the initial stage $X_{gb} \ll 1$ for large values of the parameter A.

Let us now consider the asymptotics of extremely large content of gases ($X_{vb} \to 0, X_{gb} \to 1$). The following estimates hold: $X_{v\delta} \ll 1$, $X_{vb} \ll 1$. In this case, the diffusion factor (11.28) assumes the form

$$\Lambda \approx X_{vb} - X_{v\delta}.$$

As a result, we get the asymptotics of the high gas contents

$$\psi = \frac{X_{vb} - (X_{v\delta})_{\min}}{E + F}.$$

Here $(X_{v\delta})_{\min}$ is the minimal value of the vapor phase concentration, $E = \frac{A}{(K+1)}$, $F = \frac{h_{fg}\Delta T_0}{R_g T_b^2}$ are the dimensionless parameters. Note that the case of limit NCG content is important for estimation of the full condensation length, when the vapor content decreases to the smallest possible value. The basic facts about the MM were first published in the paper [52].

11.3 Conclusions

The problem of vapor condensation upon transversal flow around a horizontal cylinder was considered. The analytical solutions for the limiting heat transfer laws, which correspond to the effect of only one factor (gravity, longitudinal pressure gradient, or interfacial friction) was obtained under the assumption that there is no effect of the remaining factors. The results of the solution were presented as relative (with respect to the case of steady-state vapor) heat-exchange laws. The qualitative analysis of the effect of model parameters on heat transfer upon condensation was carried out. The conclusion was drawn that the earlier performed experimental studies did not embrace the range of parameters, which may be of interest for practical purposes. The analysis of the limiting heat transfer laws demonstrated their mutual interdependence, which impedes the isolation of simple asymptotics of the problem under consideration with respect to individual parameters.

A mechanistic model of condensation from a vapor–gas mixture is considered. We propose an iteration-free procedure of the solution of the main equation of the mechanistic model, from which parametrical analysis with arbitrary different mass contents of the inert gas can be carried out. For high gas contents, the mechanistic model is shown to be unphysical, which is manifested in the inversion of the diffusion component of the heat flow (the diffusion paradox) and vanishing of the total heat flow (the condensation lock paradox). We develop a modified mechanistic model of condensation from a vapor–gas mixture based on the introduction of the effective heat of phase transition. It is shown that the proposed modified model is capable of correctly describing the condensation process in the entire range of variation of the inert gas mass content.

References

1. Faghri A, Zhang Y (2006) Transport phenomena in multiphase systems. Academic Press
2. Fujii T (1991) Theory of laminar film condensation. Springer, New York
3. Nußelt W (1916) Die Oberflächenkondensation des Wasserdampfes. VDI-Zeitschrift 60:541–546
4. Nußelt W (1916) Die Oberflächenkondensation des Wasserdampfes. VDI-Zeitschrift 60:569–575
5. Kutateladze SS (1963) Fundamentals of heat transfer. Edward Arnold, London
6. Labuntsov DA (2000) Physical foundations of power engineering. Selected works, Moscow Power Energetic Univ. (Publ.), Moscow (in Russian)
7. Kholpanov LP, Shkadov VY (1990) Hydrodynamics of heat and mass exchange with interfaces. Nauka, Moscow (in Russian)
8. Shekriladze IG, Gomelauri VI (1966) Theoretical study of laminar film condensation of flowing vapor. Int J Heat Mass Transf 9:581–591
9. Fujii T, Uehara H, Kurata C (1972) Laminar filmwise condensation of flowing vapour on a horizontal cylinder. Int J Heat Mass Transf 15:235–246
10. Rose JW (1984) Effect of pressure gradient in forced convection film condensation on a horizontal tube. Int J Heat Mass Transf 27(1):39–47
11. Schlichting H (1974) Boundary layer theory. McGraw-Hill, New York

12. Berman LD, Tumanov YA (1962) Flow over a horizontal tube. Therm Eng 10:77–84 (In Russian)
13. Gaddis ES (1979) Solution of the two-phase boundary-layer equations for laminar film condensation of vapor flowing perpendicular to a horizontal cylinder. Int J Heat Mass Transf 22:371–382
14. Fransson JHM, Konieczny P, Alfredsson PH (2004) Flow around porous cylinder subject to continuous suction or blowing. J Fluids Struct 19:1031–1048
15. Loitsyanskii LG (1966) Mechanics of liquids and gases. Pergamon Press, New York
16. Michael AG, Rose JW, Daniels LC (1989) Forced convection condensation on a horizontal tube—experiments with vertical downflow of steam. J Heat Transf 111:792–797
17. Lee WC, Rose JW (1984) Forced convection film condensation on a horizontal tube with and without non-condensing gases. Int J Heat Mass Transf 27:519–528
18. Honda H, Zozu S, Uchima B, Fujii T (1986) Effect of vapor velocity on film condensation of R-113 on horizontal tubes in across flow. Int J Heat Mass Transf 29:429–438
19. Avdeev AA, Zudin YB (2011) Vapor condensation upon transversal flow around a cylinder (limiting heat transfer laws). High Temp 49(4):558–565
20. Colburn AP, Hougen OA (1934) Design of cooler condensers for mixtures of vapors with noncondensing gases. Ind Eng Chem 26:1178–1182
21. Chilton TH, Colburn AP (1934) Mass transfer coefficients: predictions from data in heat transfer and fluid friction. Ind Eng Chem 26:1183–1187
22. Bird RB, Stewart WE, Lightfoot EN (2002) Transport phenomena. Wiley
23. Whalley PB (1996) Two-phase flow and heat transfer. Oxford University Press
24. Baehr HD, Stephan K (2006) Heat and mass transfer, 2nd ed. Springer
25. Sparrow EM, Minkowycz WJ, Saddy M (1967) Forced convection condensation in presence of noncondensables and interfacial resistance. Int J Heat Mass Transf 10:1829–1845
26. Felicione FS, Seban RA (1973) Laminar film condensation of a vapour containing a soluble, noncondensing gas. Int J Heat Mass Transf 16:1601–1610
27. Prosperetti A (1979) Boundary conditions at a liquid-vapour interface. Meccanica 34–47
28. Siddique M, Golay MW, Kazimi MS (1993) Local heat transfer coefficients for forced-convection condensation of steam in a vertical tube in the presence of a noncondensable gas. Nucl Technol 102:386–402
29. Hassan YA, Banerjee S (1996) Implementation of a non-condensable model in RELAP5/MOD3. Nucl Eng Des 162:281–300
30. Yao GF, Ghiaasiaan SM, Eghbali DA (1996) Semi-implicit modelling of condensation in the presence of non-condensables in the RELAP5/MOD3 computer code. Nucl Eng Des 166:277–291
31. Peterson PF (1996) Theoretical basis for the Uchida correlation for condensation in reactor containments. Nucl Eng Des 162:301–306
32. No HC, Park HS (2002) Non-iterative condensation modelling for steam condensation with non-condensable gas in a vertical tube. Int J Heat Mass Transf 45:845–854
33. Siow EC, Ormiston SJ, Soliman HM (2002) Fully coupled solution of a twophase model for laminar film conensation of vapour-gas mixtures in horizontal tubes. Int J Heat Mass Transf 45:3689–3702
34. Maheshwari NK, Saha D, Sinha RK, Aritomi M (2004) Investigation on condensation in presence of a noncondensable gas for a wide range of Reynolds number. Nucl Eng Des 227:219–238
35. Siow EC, Ormiston SJ, Solliman HM (2004) A two-phase model for laminar film condensation from steam-air mixtures in vertical parallel-plate channels. Heat Mass Transf 40:365–375
36. Martin-Valdepenas JM, Jimenez MA, Martin-Fuertes F, Fernandez-Benitez JA (2005) Comparison of film condensation models in presence of noncondensable gases implemented in CFD code. Heat Mass Transfer 41:961–976
37. Stephan K (2006) Interface temperature and heat transfer in forced convection laminar film condensation of binary mixtures. Int J Heat Mass Transf 49:805–809

38. Oh S, Revankar ST (2006) Experimental and theoretical investigation of film condensation with noncondensable gas. Int J Heat Mass Transf 49:2523–2534
39. Stevanovic VD, Stosic ZV, Stoll U (2006) Three-dimensional numerical simulation of non-condensables accumulation induced by steam condensation in a non-vented pipeline. Int J Heat Mass Transf 49:2420–2436
40. Siow EC, Ormiston SJ, Soliman HM (2007) Two-phase modelling of laminar film condensation from vapour-gas mixtures in declining parallel-plate channels. Int J Therm Sci 46:458–466
41. Li JD (2013) CFD simulation of water vapour condensation in the presence of non-condensable gas in vertical cylindrical condensers. Int J Heat Mass Transf 57(2):708–721
42. Gorpinyak MS, Solodov AP (2019) Vapor–gas mixture condensation in tubes. Therm Eng 66:388–396
43. Jensen MK (1988) Condensation with noncondensables and in multicomponent mixtures. In: Shah RK, Subbarao EC, Mashelkar RA (eds) Heat transfer equipment design. Hemisphere Publishing Corp., pp 497–512
44. Colburn AP, Drew TB (1937) The condensation of mixed vapors. Trans AIChE 33:197–215
45. Mitrovic J, Gneiting R (1996) Kondensation von Dampfgemischen - Teil 1. Forsch Ingenieurwes 62:1–10
46. Mitrovic J, Gneiting R (1996) Kondensation von Dampfgemischen - Teil 2. Forsch Ingenieurwes 62:33–42
47. Fullarton D, Schlünder E-U (2006) Jb: Filmkondensation von binären Gemischen mit und ohne Inertgas. VDI Wärmeatlas. 10. Springer, Berlin, Heidelberg, Kap. Jb: 1029–1040
48. Ackermann G (1937) Heat transfer and molecular mass transfer in the same field of high temperatures and large partial pressure differences, Ver. Deutsch, Ing. Forschungsheft 8(382):1–10
49. Krishna R, Panchal CB, Webb DR, Coward I (1976) An Ackermann-Colburn and Drew type analysis for condensation of multicomponent mixtures. Lett Heat Mass Transf 3:163–172
50. Cengel YA, Turner RH (2004) Fundamentals of thermal-fluid sciences. McGraw-Hill, Boston
51. Kutateladze SS (1961) Boiling heat transfer. Int J Heat Mass Transf 4(1):31–45
52. Karnaukhov E, Ustinov VS, Ganzhinov AM, Zudin YB, Lukashenko ML (2016) Comparative analysis of computational techniques taking into account the influence of noncondensing gases on vapor condensation. High Temp 54:731–736

Chapter 12
Nucleate Pool Boiling

Abbreviations

ATHTC	Averaged true heat transfer coefficient
EHTC	Experimental heat transfer coefficient
HTC	Heat transfer coefficient
THTC	True heat transfer coefficient

Symbols

b	capillary constant
c_p	Isobaric heat capacity
Fo	Fourier number
h	Heat transfer coefficient (HTC)
h_{fg}	Heat of phase transition
h_e	Experimental heat transfer coefficient (EHTC)
K	Curvature
k	Thermal conductivity
m	Growth modulus
n_F	Nucleation site density
p	Pressure
R	Bubble radius
q	Heat flux
T	Temperature
t	Time

Greek Letter Symbols

α	Thermal diffusivity
δ	Thickness
ε	Conjugation parameter
ρ	Density
ν	Kinematic viscosity
ϑ	Temperature difference

Y. B. Zudin, *Non-equilibrium Evaporation and Condensation Processes*,
Mathematical Engineering, https://doi.org/10.1007/978-3-030-67553-0_12

$\hat{\vartheta}$	Oscillating temperature
σ	Surface tension

Subscripts

s	Saturation state
spin	State at spinodal
cr	State at critical point
v	Vapor
w	State at wall

12.1 Metastable Liquid

The first documented mention of superheated liquid occurred in 1777 when the London Royal Society issued a recommendation to place a thermometer bulb not in boiling water itself, but rather in its vapors. In 1873, a century later, Gibbs [1] was first to carry out a physical analysis of specific features of the superheated ("metastable") state of liquid. Interesting facts about early observations of superheated liquid can be found in "Course in Physics" by Khvolson (1923).

The dependence of the temperature of liquid on the pressure in the state of thermodynamic equilibrium is described by the equation of the saturation curve (which is also called the "binodal"). If T is increased with p = const or if p is reduced with T = const, then the liquid transits (without phase change) into a metastable state, which in theory is unstable relative to external disturbances. Since the probability of spontaneous evaporation of any significant mass of liquid is very small, it follows that the liquid in a metastable state can exist arbitrarily long. However, for some "threshold" level of fluctuations of the parameters (pressure, temperature), the unstable equilibrium of a metastable liquid breaks down, consequently, an initially weak perturbation will increase in the course of time due to the reaction of the thermodynamic system itself.

The instability increases the more the deeper is the penetration of the liquid in the metastable domain (for example, the isobaric superheating of liquid). As the perturbation develops, the liquid vaporizes (a "phase transition"). The maximally possible temperature, under which a substance can still exists in the form of a liquid phase ("the limit superheating temperature") is described by the spinodal equation $T_{spin}(p)$, which is a direct corollary to the equation of state for real gas conditions.[1] In the domain $T > T_{spin}$, the liquid is in the state of absolute thermodynamic instability and should inadvertently vaporize.

More than a hundred state equations based on the classical van der Waals equation had been already known at the time the book [2] was written. Since this time, their number has been significantly increased, however, the "old" equations (the Dieterici,

[1] In a more rigorous form this equation is known as the "thermal equation of state".

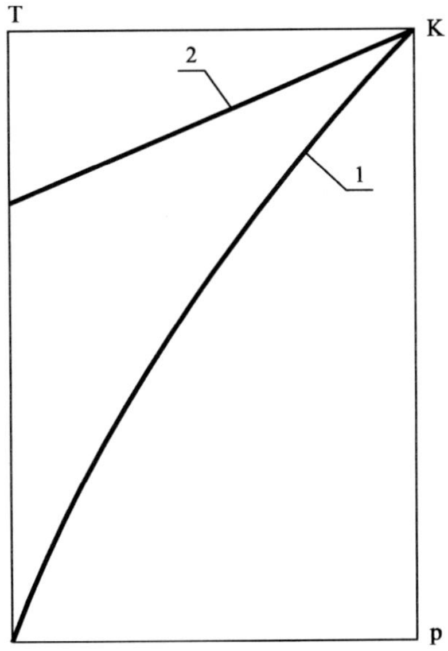

Fig. 12.1 Binodal (1) and spinodal (2) in the *p–T* diagram

Berthelot, Redlich–Kwong equations, etc.) are the most popular in engineering analysis [3]. For example, from Soave–Redlich–Kwon's equation [4], one can derive the following approximation of the dependence of the spinodal temperature $T_{\rm spin}$ on the pressure

$$\frac{T_{\rm spin}}{T_{\rm cr}} = 0.89 + 0.11\frac{p}{p_{\rm cr}}. \tag{12.1}$$

Here, $T_{\rm cr}$, $p_{\rm cr}$ are, respectively, the temperature and pressure at the critical point K (Fig. 12.1).

Following Gibbs [1], one can single out two principally different ways of theoretical description of the state of limit superheating of a metastable liquid.

- The study of the thermodynamic stability of liquids on the basis of the state equation. To this aim, one employs the mathematical tools involving the classical Maxwell and Gibbs equations
- Reaching the spinodal by monotone motion in the domain of metastable state of liquid. Here, the analysis is based on the theory of homogeneous nucleation, which describes the generation of spherical vapor nuclei in the mother phase.

Experiments on reaching the limit superheating temperature for various liquids show that the experimental points fall on the curves which are close to the theoretical spinodal curves. The agreement between the theoretical and experimental limit curves, which were obtained by completely different methods, supports the confidence of theoretical concepts about metastable liquid [5, 6].

12.2 Conditions for the Onset of Boiling

The formation of a vapor bubble in a bulk of liquid which is associated with the formation of a new interface surface is connected with overcoming the energetic barrier. Hence a necessary condition for the beginning of boiling is the attainment of a certain superheat of liquid with respect to the saturation temperature T_s at a given pressure p in the system. Consider a spherical vapor bubble of radius R_* in the state of a thermodynamic equilibrium in liquid in which a constant pressure is maintained. The pressure of vapor p_v in the bubble can be found by Laplace's formula, which describes the jump of the normal component of the stress tensor in terms of the curvilinear interface surface

$$p_v = p + \frac{2\sigma}{R_*}. \tag{12.2}$$

The temperature of saturated vapor in the bubble $T_v = T_s(p_v)$ is greater than the saturation temperature at pressure in liquid $T_v > T_s(p)$. As a corollary, the liquid under the conditions of an isothermal system becomes superheated. Below by $T_s \equiv T_s(p)$, we denote the saturation temperature at the pressure of the liquid surrounding the bubble. The superheat value of liquid (the temperature difference) $\vartheta \equiv T_v - T_s$ is related to the Laplace jump of pressures $\Delta p \equiv p_v - p$ by the equation of the saturation curve $T_s(p_s)$. In linear approximation (i.e., for small differences of temperatures and pressures) we have the Clapeyron–Clausius equation

$$\Delta p \approx \frac{h_{fg}\vartheta\rho_v}{T_s}, \tag{12.3}$$

where ρ_v is the vapor density. In view of (12.2) and using (12.3), we get

$$R_* = \frac{2\sigma T_s}{\rho_v h_{fg}\vartheta}. \tag{12.4}$$

The quantity R_*, which is called the "critical radius of a vapor nucleus", corresponds to an absolutely unstable equilibrium. A decrease in the bubble size results in an increase in the curvature of the phase interface, thanks to which the pressure of vapor in the bubble exceeds the saturation pressure. This will result in condensation of vapor and collapse of the bubble. As the bubble size is further increasing ($R > R_*$) the pressure of vapor in the bubble becomes smaller than the saturation pressure, the superheated liquid on the bubble surface will evaporate, and the bubble will grow. So, the (purely theoretical) quantity R_* means the boundary of two oppositely directed processes: collapse and growth of the bubble. From formula (12.4), it follows that in a liquid in thermodynamic equilibrium a vapor nucleus (which is in a dynamical equilibrium with the mother phase) should have infinite radius: $R_* \to \infty$ as $\vartheta = 0$. For $\vartheta > 0$, the nucleus has a finite radius, which decreases as the depth of entering in the metastable state increases. As the liquid reaches the spinodal temperature, the critical nucleus has the smallest possible (for a given pressure in the system) radius

$$R_{*\text{spin}} = \frac{2\sigma T_s}{\rho_v h_{fg} \vartheta_{\text{spin}}},$$

where $\vartheta_{\text{spin}} \equiv T_{\text{spin}} - T_s$ is the temperature difference at the spinodal.

A fluctuation formation of vapor nuclei in the mother phase in the absence of "weak points" (admixtures, spatial inhomogeneities, rigid surfaces) is called a "homogeneous nucleation". The process of nucleate boiling at weak spots (concavities, hollows, cracks) on a hard heated surface is a process of "heterogeneous nucleation". The theory of homogeneous nucleation predicts very strong superheating of liquid (up to the spinodal temperature, see [5, 6]). An experimental implementation of homogeneous nucleation involves the difficult problem of preventing the appearance of vapor nuclei in weak spots. In boiling under normal conditions, vapor bubbles are formed practically always on the heated surface. Here, elements of microroughness of the wall featuring degraded local wettability serve as "nucleation sites" [7–9].

In the majority of practical situations, the superheat of the surface necessary for evaporation is small: for water at atmospheric pressure, it is about (5–7°) K, and at high pressures, it is about several fractions of a degree. Substitution of these values into formula (12.4) gives an estimate for the linear scale of elements of the surface roughness, which serves as nucleation sites. At atmospheric pressure for water, this size is about several micrometers. A growth in the reduced pressure results in a decrease of R_*. So, for liquid helium at atmospheric pressure (the reduced pressure $p/p_{cr} \approx 0.45$) and for feasible superheats of the wall $\vartheta \approx 0.04$ K, we have $R_* \approx 7 \times 10^{-8}$ m.

12.3 Nucleation Sites

Nucleation sites with linear size of order R_* start to act in a stable way with a certain temperature difference $\vartheta = T_w - T_s$. A detachment from the surface of developed vapor bubbles takes place, as a rule, without penetration of the liquid inside concavities. Hence, a small concavity filled with vapor serves as a generator of a chain of bubbles. The periodicity of separation of bubbles is quite different for concavities of different sizes. The nucleation sites with linear size exceeding R_* are "protected" to a greater extent from penetration of liquid inside concavities, and hence they operate in a more stable way as compared with sites of smaller size. For small concavities, there are periods of shutdown of a boiling site, when the phase interface penetrates inside the concavity. The division into "large" and "small" concavities is only conventional, because the scale size R_* varies with the degree of superheat ϑ (and also depends on the type of a boiling liquid).

The number of active nucleation sites with fixed temperature depends on local adhesive characteristics (local wettability) and the surface relief. Physically, it is clear that the surface density of nucleation sites n_F should be proportional to the number of hollows and concavities per unit area of the surface and whose sizes are commensurable with R_*. Besides, the proportionality coefficient depends on the dynamical wetting angle θ. The character of the distribution of the number of irregularities in sizes depends on the number of factors: the structure and composition of the surface material, the way of treatment of the surface, etc. Unfortunately, in the literature, there are no systematic data on such dependences. Hence, based on the analysis of dimensions one can adopt the most simple law

$$n_F = \frac{\gamma_1}{R_*^2} \tag{12.5}$$

for the distribution function of irregularities.

Physically, conjecture (12.5) means that in the range of R_* which is relevant for boiling of liquids, microconcavities on the rigid surface have no preferable sizes. Hence, a reduction of R_* will result in an increase in the number of surface concavities which can develop in nucleation sites. It is this law that is reflected by formula (12.5), which was proposed [10]. The number constant γ_1 in (12.5), which is estimated by comparison with experimental data, is as follows: $\gamma_1 \approx 10^{-8}–10^{-7}$. Hence, for uniform distribution of nucleation sites over the surface, the characteristic distance l_* between them is given by the relation

$$l_* \equiv \frac{1}{\sqrt{n_F}} = \gamma_1^{-1/2} R_*. \tag{12.6}$$

Since $\gamma_1^{-1/2} \sim 10^3–10^4$ and since R_* is quite small, the total area of dry spots on the surface should be negligibly small, which agrees well with experimental results.

12.4 Boiling Regimes

12.4.1 Boiling Curve

For boiling of liquid on a rigid surface, an increase in the wall superheat $\vartheta = T_w - T_s$ results in an increase in the number of simultaneously acting nucleation sites, which in turn, increases the heat transfer intensity. A striking feature of the boiling process, which makes it drastically different from the single-phase convection and condensation, is the strong dependence of the heat flow density q removed from the wall on the degree of its superheating with respect to the saturation temperature. The dependence $q(\vartheta)$ is known as the "boiling curve" or the "Nukiyama curve" (after the Japanese researcher, who first described it in 1935).

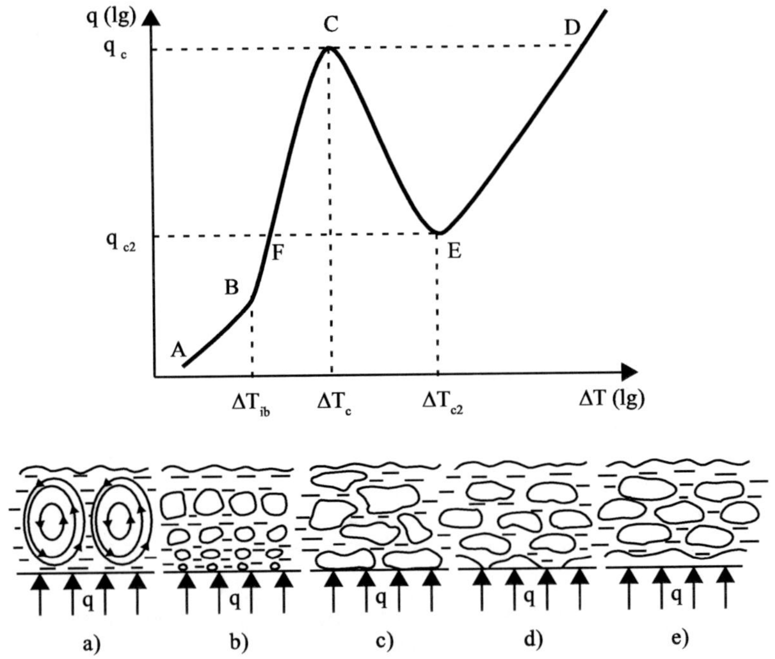

Fig. 12.2 Boiling curve and scheme of different heat transfer mechanisms

Figure 12.2 shows a typical boiling curve with a schematic heat transfer mechanism with various combinations of the heat flow q and the temperature differences. Until the wall superheating with respect to T_s fails to reach the value ϑ_*, which is sufficient for the formation of vapor bubbles ("onset of boiling"), the heat from the heated surface is removed by free convection (Fig. 12.2a), while from the liquid it is removed due to evaporation from its free surface. In the case of free-convection motion, the dependence $q(\vartheta)$ is as follows: $q \sim \vartheta^{4/3}$ (the interval AB in Fig. 12.2). After the onset of boiling, the $q(\vartheta)$ abruptly becomes steeper, and on the interval BC, we have $q \sim \vartheta^3$.

12.4.2 Nucleate Boiling

A regime of individual bubbles is realized for relatively small densities of the heat flow. Each bubble nucleates in a nucleation site, attains the detachment size, and ascends through the action of the buoyant force. The bubble is replaced by the new portion of liquid, which is warmed up near the wall (up to the beginning of the new bubble cycle), see Fig. 12.2b. It is the regime of individual bubbles in which one usually carries out cinematic studies of boiling and gets experimental information about the dynamics of growth and detachment of vapor bubbles. The structure of the two-phase mixture changes substantially with increasing heat flow density.

(a) neighboring vapor bubbles start to coalesce (both in the vertical and lateral directions),
(b) the regime of coalescent bubbles ("vapor conglomerates") that periodically originate and detach from the wall is formed (Fig. 12.2),
(c) the phase interface undergoes quick-changing irregular fluctuations.

On the one hand, in the regime of vapor conglomerates, it is difficult to identify the domains filled by vapor and liquid (both in visual observation and in cinematic studies). On the other hand, experiments strongly suggest that under vapor conglomerates there is a liquid film covering an absolutely predominant part of the rigid surface. From the available direct measurements (for boiling of water, ethanol, and methanol at atmospheric pressure), the film thickness ranges between the interval $(2 \times 10^{-5}–3 \times 10^{-4})$m and has many disruptions that form "dry spots[2] of a very small area on the surface. Such spots are nucleation sites, which were preserved after the coalescence of vapor bubbles that developed on them. The total area of the dry spots surrounding the nucleation sites is at most 10% even for small values of the heat flow".

Despite the fact that the boiling picture is drastically different in the regimes of individual and coalescent bubbles, the dependence $q \sim \vartheta^3$ on the interval BC remains practically unchanged. It seems that this remarkable dependence is indicative of the existence of an integrated interior mechanism of the process. From the viewpoint of applications, it is convenient to call them simply "nucleate boiling".

12.4.3 *Film Boiling*

The continuous curve $q(\Delta T)$ in Fig. 12.2, in fact, is an approximation of a family of discrete points that were obtained in experiments when the corresponding stationary state was attained after a stepwise change in the heat flow. With electrical heating of the wall (i.e., when the process is "controlled" by the density of the heat flow) the nucleate boiling regime terminates abruptly as some limiting value $q = q_c$ is reached (the point C in Fig. 12.2). Even a little (smaller than by 3%) increase of q in the vicinity of q_c results in a snowballing growth of the area of dry spots and the formation of a continuous vapor film on the heated surface.

The new stationary state (the point D in Fig. 12.2) is now settled in the "film boiling" regime (Fig. 12.2d), the process of transition from nucleate boiling to film boiling is called the "boiling crisis". On the other hand, in the film regime of boiling, the wall temperature exceeds the spinodal temperature, which excludes the possibility of direct contact between the liquid and the wall. The heat from the wall to the phase interface is transmitted through the vapor film by means of thermal conductivity

[2] In Fig. 12.2c this film is depicted as a blackened strip.

and single-phase convection in vapor and also due to radiation. According to the available theoretical concepts [9], two types of hydrodynamic instability develop simultaneously in the vapor film: the Taylor instability and the Helmholtz instability. As a result, vapor bubbles are periodically formed on the phase interface and ascend under the action of gravitation (Fig. 12.2e).

For film boiling, the heat transfer intensity is smaller by one to two orders of magnitude than that for nucleate boiling with the same values of the heat flow. For liquids with low temperature of normal boiling (cryogenic liquids, light hydrocarbons, many freons), in a transition to film boiling the wall temperature does not reach the level under which the mechanical integrity can be impaired. For water and other high-boiling liquids, such a transition usually results in the thermal breakdown of the surface. For this reason, the "crisis of nucleate boiling" is sometimes called "burnout".

Stationary film boiling is observed at various heat loads which either exceed or are substantially much smaller than the heat flow at point *D*. Besides, here the surface temperature undergoes pulsations due to oscillation of the thickness of the vapor film. Such a regime is preserved with decreasing q, until the wall temperature drops down to the limit superheating temperature of the liquid. In this case, the vapor film breaks rapidly and the nucleate boiling regime again occurs (the transition *EF*). This transition, whose rate is controlled by the heat capacity of the heater, is called the "crisis of film boiling". The corresponding quantity q is called the "second critical" heat flow: $q \equiv q_{c2}$.

12.4.4 Transition Boiling

Assume that the outer surface of the wall is heated by circulating liquid or condensing vapor or that electrical heating is used in combination with convective cooling (using a fairly involved system of automatic regulation). Then in the experiment, the wall temperature is "controlled", which enables one study in stationary regime a fairly specific process of "transition boiling". This branch of the boiling curve is characterized by the "abnormal" dependence $q(\vartheta)$, when the heat flow decreases (the segment *CE* in Fig. 12.2) with increasing the wall superheat.

Since during the transition boiling, the wall temperature is majorized by the spinodal temperature, it follows that the contact of liquid with the heating surface is thermodynamically admissible. For such a contact, due to high overheat, liquid instantaneously boils and is pushed aside from the wall by the resulting vapor. The presence of contact points of liquid with the superheated surface is shown in Fig. 12.2d. It is worth noting that for the transition boiling of cryogenic liquids (helium, nitrogen, oxygen) the heating surface is relatively "cold" (for nitrogen boiling its temperature is at most 130 K). With increasing ϑ along the segment *CE*, the characteristic time of periodic wetting of the wall by the liquid decreases monotonically. As the surface temperature reaches the value T_{spin}, the liquid ceases to be in contact with it and the film boiling regime onsets.

The above stationary process is realized with a monotonic increase of the wall temperature when the boiling curve is traversed "from left to right". In a number of unsteady processes (quenching of metal products, simulation of post-failure cooling of a nuclear fuel element in nuclear reactors), the boiling curve is traversed with decreased wall temperatures in time, i.e., "from right to left". Under this approach, all the above characteristic areas *DE, EC, CB, BA* are adequately replicated. Quantitative differences from the results of stationary studies are mainly manifested near the nucleate boiling crisis (the point *C*): smaller values of q_c are registered under unsteady cooling conditions.

12.5 Vapor Bubble Growth Laws

12.5.1 Bubble Growth in a Bulk Liquid

We first consider the classical growth problem of vapor bubble in a bulk of uniformly superheated liquid. The growth rate of a bubble with $R \gg R_*$ depends, in general, on the resistance of the liquid that is pushed away (dynamical effects) and on the intensity of evaporation on the phase interface (energy effects). The dynamical effects are due to the inertia of liquid and its viscosity, while the energy effects depend on the heat supply conditions to the phase interface and the kinetics of the evaporation process. For an increasing vapor bubble, all the above physical effects act simultaneously, however, in each specific case one some of them are play a defining role. Labuntsov [10] proposed a classification, in accordance with which four principal "limit growth schemes" are singled out. Each of such schemes corresponds to the manifestation of one of the principal effects (under the assumption that the remaining effects are immaterial). In [10], he showed that the real growth rate of a vapor bubble is always smaller (or is equal in the limit) than the smallest of the velocities determined by the limit schemes.

The dynamic viscous scheme corresponds to the case when at any time the difference of vapor pressures in the bubble and in the liquid is offset by the normal component of the viscous stress tensor in liquid on the bubble boundary. This limit scheme determines the growth rate of a vapor bubble in very viscous liquid and for very small bubble radii. Estimates show that in a real process of bubble growth the viscosity effects are immaterial.

The dynamic inertial scheme assumes that the pressure differential between vapor in the bubble and in the surrounded liquid are maintained to be constant. Besides, here one postulates the conditions of homogeneity of temperature in the entire domain under consideration, including the vapor bubble. Physically this means that the liquid is characterized by infinitely large thermal conductivity. It is quite clear that in the real condition such a restrictive condition cannot be fulfilled.

The energy molecular-kinetic scheme of bubble growth can play a crucial role only for very low values of the accommodation coefficients of molecular flows from the side of vapor phase.

The energy thermal scheme is the most important from the practical point of view. In this limit scheme, it is assumed that the bubble growth rate depends on the intensity of the heat supply from the superheated liquid to the phase interface. Analysis of the energy thermal scheme is based on the following simplifying assumptions

- the growth of a bubble is constrained by the mechanism of unsteady thermal conductivity in the superheated liquid surrounding the bubble,
- the heat income on the phase interface is spent only on the evaporation of liquid,
- during the entire growth process, the pressure of vapor in the bubble p_v is constant and is equal to the pressure in the liquid,
- the vapor temperature in the bubble is equal to the saturation temperature with a given pressure $T_v = T_s(p_v)$.

The above limit schemes have different significance in practice. For example, the dynamic viscous scheme and the energy molecular-kinetic scheme are exotic to a large extent. The dynamic inertial scheme can be manifested only in the initial period of bubble growth. The energy thermal scheme is dominating over practically the entire time of bubble growth, this scheme is based on the heat balance equation

$$4\pi h_{fg}\rho_v R^2 \frac{dR}{dt} = Q_f. \tag{12.7}$$

Here, Q_f is the heat flow supplied to the growing bubble from the surrounding liquid. According to [11], under these conditions, the bubble radius changes in time according to the diffusion law

$$R = m\sqrt{\alpha t}. \tag{12.8}$$

Here, $m \equiv R/\sqrt{\alpha t}$ is the "growth modulus". Consider the "Jakob number"

$$\mathrm{Ja} = \frac{\rho c_p \vartheta}{h_{fg}\rho_v}, \tag{12.9}$$

defined as the ratio of the enthalpy of liquid superheating to the phase transition enthalpy (both quantities are related to a unit volume). For Ja $\to \infty$ in [11], the following relation for the growth modulus was obtained

$$m = 2\sqrt{\frac{3}{\pi}}\mathrm{Ja}. \tag{12.10}$$

The asymptotic formula (12.10) was obtained earlier by Plesset and Zwick [12] derived from the interplay of a thin thermal boundary layer in liquid on the bubble surface for the domain Ja $\gg$ 1. Scriven [13] was the first to obtain an exact analytical solution in the bubble growth problem in the framework of the energy thermal scheme. Scriven's results were given in a tabular form for the dependence of the growth rate module on the Jakob number and the ration of the densities of phases: $m(\mathrm{Ja}, \rho_v/\rho)$. In the domain $p/p_{cr} \ll 1$, the following approximation of Scriven's solution was put forward

$$m = \sqrt{\frac{3}{\pi}}\mathrm{Ja} + \sqrt{\frac{3}{\pi}\mathrm{Ja}^2 + 2\mathrm{Ja}}. \tag{12.11}$$

As Ja $\to \infty$ relation (12.11) becomes the "unsteady" asymptotic formula (12.10), while as Ja $\to$ 0 it is transformed to the "quasi-stationary" limit case

$$m = \sqrt{2\mathrm{Ja}}. \tag{12.12}$$

Expression (12.12) is obtained if the heat flow corresponding to the stationary heating problem of a rigid ball in an infinite medium $Q_f = 4\pi k R \vartheta$ is substituted into the heat balance Eq. (12.7). Approximation (12.11) describes exactly the limit cases and compares favorably with the solution of [13] in the entire domain of variation of the Jakob numbers ($0 < \mathrm{Ja} < \infty$) with an error smaller than 2%.

12.5.2 *Bubble Growth on a Rigid Surface*

For liquids boiling on a rigid surface, the growth of vapor bubbles takes place in the conditions of a highly inhomogeneous temperature field [14]. Besides, the vapor space has a boundary not only with the liquid phase, but also with the wall. The contact line of three media (solid, liquid, vapor ones) along the bubble boundary is also called the "boundary line". The knowledge of the bubble growth mechanism and their subsequent detachment from the surface has great importance for understanding the boiling process. This explains considerable interest in theoretical and experimental studies of the dynamics of vapor bubbles on the wall.

The presently available experiments cover a wide range of pressures (from 0.01 to 100 bars, i.e., from 1 kPa to 10 MPa) and deal with a great variety of liquids (water, alcohols, hydrocarbons, halocarbons, cryogenic liquids). Analysis of experimental findings enables one to make qualitative conclusions and serve as a basis for the development of semi-empirical models. It is worth pointing out that in reality one can only speak about statistically averaged growth laws of vapor bubbles. Cinematic studies show that even in a single experiment with fixed values of the pressure and temperature of the wall, the growth rates of bubbles may differ by several times. Hence, of practical importance are only approximate models, which reflect the effect of the principal factors and describe quantitative interrelations for some "average" conditions.

A scheme of heat supply to a bubble is shown in Fig. 12.3. Since vapor concavities

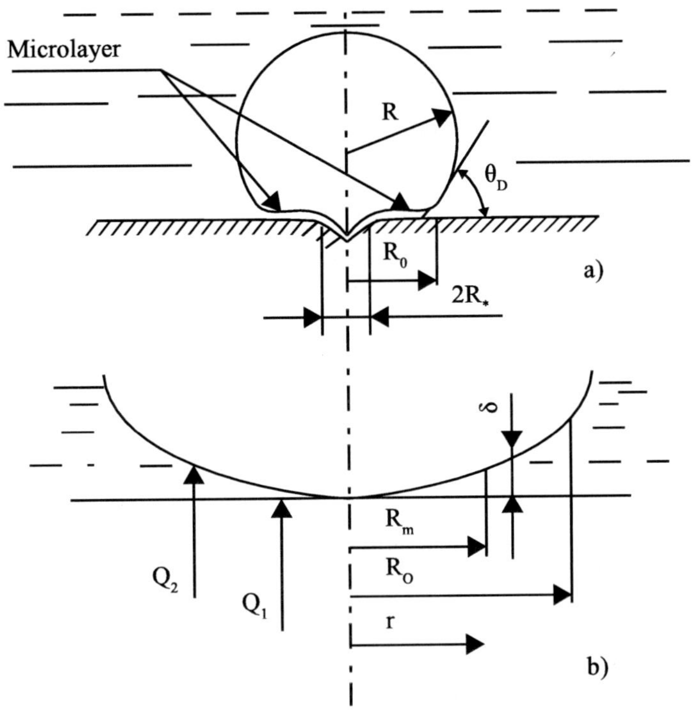

Fig. 12.3 Scheme of a vapor bubble growing on a hard wall (**a**) and the theoretical heat supply model to the bubble surface through the microlayer from the wall and from the superheated liquid (**b**). R the of the bubble spherical part, R_1 the radius of the "dry spot", R_m the conditional outer boundary of the microlayer heat-conducting part , R_0 the bubble base radius, δ the liquid microlayer thickness, θ the dynamical boundary angle, Q_w, Q_f the heat inputs to the phase interface, respectively, from the hard wall and the superheated liquid

serve as nucleation sites, it follows that in the core part of the bubble base there is a domain of direct contact of vapor with the surface (a dry spot). The characteristic size of active concavities is close to the equilibrium radius R_* of vapor nucleus, as defined by (12.4). In typical conditions, R_* is commensurable to a micrometer or (at high reduced pressures) in the range of tenths of a micrometer. Hence, save only for the very short initial growth period, a dry spot is only a fraction of a per cent of the area of the bubble projection to the heated surface. The space between the bubble surface and the rigid wall is filled with a thin layer of liquid (a "microlayer").

In constructing theoretical models of bubble growth, it is assumed that the thickness of the microlayer increases from zero (on the boundary line) to some finite value near the outer boundary of the bubble base (Fig. 12.3a). Near the boundary line, there is a layer of adsorbed molecules (the nonevaporable part of the meniscus liquid film of a thickness of the order of intermolecular distances in liquid ~10^{-9} m). As the layer of the liquid becomes more thick away from the boundary line, the effect of adsorption forces (hindering evaporation) decreases monotonically, and there begins evaporation from the film surface. The radius R_1 of a dry spot and the length of the adsorbed film are both negligibly small in the radial direction in comparison with the radius R_0 of the bubble base. Hence, it can be assumed that the microlayer of liquid in the vapor base forms a conical surface. Figure 12.3b depicts the section of the microlayer by an arbitrary plane passing through the bubble symmetry axis.

If the heat reserve in the superheated liquid in a think microlayer is relatively small, then the heat for evaporation of liquid from its surface is taken directly from the rigid wall. This conclusion is supported by experimental measurements of a rapid decrease of temperature of a heated surface at the initial moment of bubble growth (usually, within 1–3 ms). Such measurements, in which thin plates (with relatively small thermal conductivity) were used as the heating surface, served as a basis for the conjecture about the evaporating microlayer of liquid under a growing bubble. Later, the existence of a microlayer under a bubble growing on a transparent heat surface was supported by direct optical studies with the help of laser interferometry.

A survey of modern state of the art in the theory of microlayer and a detailed numerical investigation of the problem can be found in the recent paper [15]. The physical mechanisms associated with the evolution of a microlayer beneath a bubble and the transition between the contact line and microlayer regimes are investigated with fully resolved numerical simulations in the framework of nucleate pool boiling. Capturing the transition between these two regimes has been possible for the first time using very refined grids and parallel computations. The simulations are used to analyze the physical processes involved in the formation and depletion of a microlayer.

From these results, the limit conditions between nucleate boiling in microlayer and contact line regimes are deduced. Neglecting the microlayer would lead to erroneous results, because it has a strong influence on the overall bubble growth. Therefore, the results of [15] could be of major interest in designing models of nucleate pool boiling in larger scales computations when the microlayer cannot be resolved. Temperature measurement in a bulk of liquid showed that superheated liquid covers the part of the spherical surface (dome) of a growing bubble which is nearest to the wall. Hence, under certain conditions, the excessive enthalpy of liquid pushed out by the bubble from the temperature boundary layer on the wall should also have an effect on the growth process.

The model of homogenous boiling of uniformly superheated liquid proves incapable of describing the growth of a vapor bubble on a rigid surface. This is manifested most vividly in the range $p/p_{\mathrm{cr}} > 0.1$, when the superheating enthalpy is much smaller than the phase transition heat, i.e., for a small value of the Jakob number (Ja < 1). Here the growth of a bubble is completely determined by the heat flow supplied from the heating surface to the bubble (near the base of its surface). The experimental findings described above allow one to use the bubble model from Fig. 12.3. We shall assume that during its growth, the bubble retains the form of a truncated sphere and that the bubble shape is similar at any time. This means that all geometrical characteristics are assumed to be proportional to the "equivalent radius" R_e, i.e., to the radius of the sphere, whose volume is equal to that of the bubble

$$V = \frac{4\pi}{3} R_e^3. \tag{12.13}$$

The thickness of the liquid microlayer at the bubble base is assumed to be proportional to the distance to the symmetry axis, i.e., $\delta = \gamma_2 r$, where $\gamma_2 \ll 1$ is a number constant. The rigid surface is assumed to be isothermal—this assumption is legitimate in the case of high thermal conductivity of its material. The local density of the heat flow supplied from the wall reads as

$$q_w = \frac{k\vartheta}{\delta} = \frac{k\vartheta}{\gamma_2 r}. \tag{12.14}$$

We shall assume that on the interval $0 < r < R_m$ (Fig. 12.3b), the excessive enthalpy of the liquid has no effect on the intensity of evaporation from the microlayer surface, as a result, the heat flow from the heated surface to the phase interface is as follows

$$Q_1 = \int_0^{R_m} \frac{k\vartheta}{\gamma_2 r} 2\pi r dr = \frac{2\pi}{\gamma_2} k\vartheta R_m. \tag{12.15}$$

Using the assumption about the preservation of bubble form during its growth, we can write $R_m = \gamma_3 R$, where $\gamma_3 < 1$ is a number constant. Now (12.15) can be put in the form

$$Q_1 = \gamma_4 k\vartheta. \tag{12.16}$$

Here, γ_4 is a number constant reflecting the "degree of the lack of knowledge" of the real form of the microlayer and the growing bubble.

When the microlayer becomes sufficiently "thick", the excessive enthalpy of the superheated liquid starts to play an important role in the heat balance. As a conventional border between the "thin" and "thick" parts of the microlayer, one can take the film thickness $\delta_m \approx \sqrt{\alpha t}$, where t is the time elapsed from the beginning of the bubble growth, α is the thermal diffusion coefficient of liquid. For nonmetallic liquids, we have the estimate $\delta_m \approx 10^{-5}$ m. In our approximate model, such a boundary can be found from the ratio $\delta_m = \beta_2 R_m$. We note that the partition of the microlayer segments into the thin (heat conducting) and thick (thermal capacity) parts is merely notional. The expression for the heat flow from the superheated liquid to the phase interface corresponds to the asymptote of the Scriven solution for Ja $\gg 1$, i.e., to Plesset–Zwick's formula

$$q = 2\sqrt{\frac{3}{\pi}} \frac{k\vartheta}{\sqrt{\alpha t}}. \tag{12.17}$$

For a bubble growing in the wall, the heat flow from the superheated liquid can be written as

$$Q_2 = \gamma_5 \frac{k\vartheta}{\sqrt{\alpha t}} R^2. \tag{12.18}$$

Here, the number constant γ_5 reflects the following effects

- the superheated liquid covers only a part of the bubble surface (more precisely, the surface of area $4\pi R^2$ which has the same volume as the bubble),
- the liquid superheat is in the average smaller than the superheat ϑ of the wall, which was used in (12.14).

For a bubble in the form of a truncated sphere (Fig. 12.3a), it is assumed that the superheated liquid covers the area of the bubble surface, which is approximately equal to the area of its base πR_0^2. In this case, the coefficient β_5 is expressed as a function of the wetting angle θ. At any time, the bubble satisfies the energy balance equation

$$h_{fg}\rho_v \frac{dV}{dt} = Q_1 + Q_2,$$

which in view of (12.16), (12.18) can be written as

$$R^2 \frac{dR}{dt} = \gamma_6 \frac{k\vartheta R}{h_{fg}\rho_v} + \gamma_7 \frac{k\vartheta R^2}{\sqrt{\alpha t}}, \tag{12.19}$$

where $\gamma_6 = \gamma_4/4\pi$, $\gamma_7 = \gamma_5/4\pi$ are number constants.

Introducing the Jakob number by formula (12.9), we have

$$R \frac{dR}{dt} = \gamma_6 \mathrm{Ja}\alpha + \gamma_7 \mathrm{Ja} R \sqrt{\frac{\alpha}{t}}. \tag{12.20}$$

Using in (12.19) the diffusion law (12.8) and the initial condition $t = 0$, $R = 0$, we arrive at the quadratic equation with respect to the growth modulus. As a result, we get the following dependence of the bubble growth rate modulus on the Jakob number (Fig. 12.4):

$$m \equiv \frac{R}{\sqrt{\alpha t}} = \gamma_7 \mathrm{Ja} + \sqrt{\gamma_7^2 \mathrm{Ja}^2 + 2\gamma_6 \mathrm{Ja}}. \tag{12.21}$$

With constant coefficients γ_6, γ_7 the equation derived above has the following asymptotes:

- For Ja $\ll 1$, the supply of heat from the superheated liquid is negligibly small. This means that the heat capacity of the liquid has no effect on the growth law. Hence, using (12.21), we have

$$m = \sqrt{2\gamma_6 \mathrm{Ja}}. \tag{12.22}$$

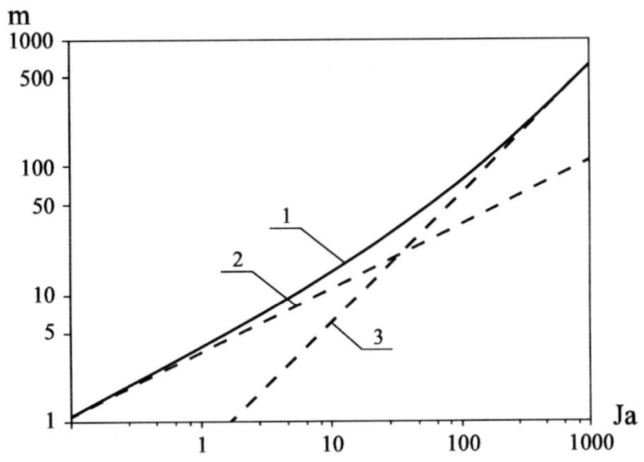

Fig. 12.4 Dependence of the bubble growth modulus on the Jacob number. 1 Calculation by Eq. (12.21), 2 calculation by Eq. (12.22), 3 calculation by Eq. (12.23)

This formula, which was obtained by Labuntsov [10] in 1963, holds for the region of high reduced pressures with $\gamma_6 = 6$.

- For Ja $\gg$ 1, the supply of heat from the superheated liquid becomes predominant

$$m = 2\gamma_7 \mathrm{Ja}. \tag{12.23}$$

Formula (12.23) differs from Plesset–Zwick's formula only by the number coefficient β_7, which in the context of the analysis should be smaller than the factor $\sqrt{3/\pi}$ in Plesset–Zwick's formula. Relation (12.23) holds for the range of low reduced pressures with $\gamma_7 = 0.3$. Relation (12.21) with $\gamma_6 = 6, \gamma_7 = 0.3$ enables one to satisfactorily describe the experimental growth curves of bubble growth for boiling of various liquids in a wide range of pressures (in the range 0.1 < Ja < 500 of the Jakob numbers). It is interesting to compare the asymptotic growth rates of a bubble in a large bulk (12.11) and in a layer of nonuniformly superheated liquid on the heated surface (12.21). In the "stationary" asymptotics for Ja $\ll$ 1, the bubble on the wall (formula 12.22) grows much slower than in the volume (formula 12.12). The opposite picture is observed in the "nonstationary asymptotics" Ja $\gg$ 1, where the bubble growth rate in the volume (formula 12.10) substantially exceeds the rate of its growth on the wall (formula 12.23).

12.6 Mechanisms of Nucleate Boiling Heat Transfer

12.6.1 Applied Significance of Nucleate Boiling

Nucleate boiling regime is certainly most important for practical applications. Being an integral part of many widely differing technologies, nucleate boiling has no rival in the problem of cooling of rigid surfaces subject to high-intensity thermal effects (structure elements of nuclear fusion installations, high-power lasers, physical targets, etc.). The very strong dependence of the heat flow density on the wall superheat enables one to take away flows of energy of huge density with relatively small temperature differences ($\vartheta = T_w - T_s$). The crisis of nucleate boiling serves here as a constrain, which in turn, can be moved away in the domain of particularly high densities of heat flows by increasing the velocity of forced motion and underheating of liquid to the saturation temperature. It is worth pointing out that the dependence $q \sim \vartheta^3$, which is characteristic of nucleate boiling and which is shown in Fig. 12.2 for the conditions of natural convection in the volume, is also preserved for the forced motion in the channel. The heat transfer mechanism with forced convection starts to prevail only in the domain of very high velocities of the liquid.

Moreover, it was found that the heat transfer with nucleate boiling is practically independent of mass forces of various natures (in particular, gravitational forces) and of electric and magnetic fields. This suggests the existence of a wide domain of regime parameters in which nucleate boiling is "independently controlled" by its own internal mechanisms. This domain is called the regime of developed nucleate boiling. At the same time, the heat transfer with nucleate boiling can depend significantly on the factors which play a little role in other forms of heat transfer: wettability and microrelief of the surface, the presence of admixtures in liquid and on the surface, the material of the wall. These "weak factors" can result (with the same value of q) in twofold (or even greater) differences in ϑ. Such a unique and exotic combination of the immunity to "strong" external actions and instability to "weak" external actions causes some pessimism regarding the feasibility of a closed description of the process of developed nucleate boiling.

12.6.2 *Classification of Utilized Liquids*

Various boiling heat transfer media are used depending on the requirements on the specifics of heat transfer equipment and temperature level. These liquids can be subdivided into the following groups:

Ordinary liquids (water and organic liquids: benzene, heptane, pentane, alcohols, freons) feature no peculiarities and anomalies. They are widely useful in engineering, for them the temperatures range from cryogenic ones (below 100 K) to several hundred degrees Kelvin (for water $T_{cr} = 647.3$ K).

Cryogenic liquids include condensed gases: helium, hydrogen, neon, nitrogen, oxygen, argon. Cryogenic liquids feature high wettability and their temperatures range from several degrees to 100 K. In comparison with ordinary liquids, they have enhanced cleanliness (in particular, liquid helium, in which all other gases and liquids are frozen out). Cryogenic liquids are widely useful in space and aero hardware for the purposes super conductivity, for simulation of the outer space under in ground conditions,and in hydrogen energy.

Liquid metals (mercury, sodium, potassium, rubidium, cesium) feature high thermal conductivity, thoroughly wet rigid metal surfaces, and are chemically active. They are useful in power engineering industry, nuclear engineering, and chemical engineering.

Quantum liquids helium isotopes like He^4 (or He-II) and He^3. The vanishingly small viscosity of quantum liquids is responsible for the amazing phenomenon of "superfluidity". Due to their extremely high thermal conductivity, the heat is directly removed from the liquid film to vapor without the formation of vapor bubbles—this excludes the possibility of the regime of nucleate boiling.

Liquid mixtures The use of binary and multicomponent mixtures in the petrochemical industry and refrigerating engineering enables one to choose a heat carrier with required thermophysical properties in the given range of temperatures.

12.6.3 Heat Transfer Modeling (Dynamics of Bubbles)

The author would like to point out in advance that he does not pretend on giving a survey or a classification of the presently available models of nucleate boiling heat transfer. During many years of exhaustive investigation of this extremely complicated, multifactorial, and intriguing process, there appeared several survey papers and books on this topic (see, for example, [8, 10, 16]). The purpose of the author of the present book is to give a coherent treatment of the semi-empirical theory developed in the scientific school of Prof. Labuntsov and of its later modifications.

Experimental studies [16] show that both the number of nucleation sites and the growth rates of bubbles depend on the local wettability and the character of roughness of the boiling surface. Hence, in applications of these factors, the mechanism and quantitative law of heat transfer vary considerably. This means that for nucleate boiling one cannot single out some universal dependence $q(\vartheta)$. The latter considerably transforms depending on the superficial conditions.

With the developed regime of nucleate boiling, the liquid near the heating surface is subject to intensive pulsations induced from nucleation, growth, and separation of vapor bubbles. Such a picture suggested, at the early phase of research, various conjectures about the "quasiturbulent" character of nucleate boiling heat transfer. However, it soon developed that the principal thermal resistance is concentrated in the near-wall ("viscosity-constrained") layer of liquid between single bubbles growing on the wall. The thickness of this specific layer, which is locally ruptured by bubbles appearing on the wall, is determined from the balance of viscous and inertial forces. Outside this near-wall layer, the key role is played by the fluctuating flow and "turbulent" mixture, whose intensity is characterized by some average fluctuation velocity w. This quantity can be estimated from the balance of the kinetic energy of the fluctuating flow ρw^2 and the energy transferred to the liquid from growing vapor bubbles. In a similar manner, the average thickness of the thermal layer can be taken to be equal to the thickness of the viscous sublayer in Prandtl's turbulence model: $\delta \approx 10\nu/w$.

Following Labuntsov, we consider two main ways of nucleate boiling heat transfer.

- Heat transfer due to thermal conductivity through the boundary layer with the thermal resistance δ/k towards the principal mass of liquid. The corresponding density of the heat flow is estimated as $q_1 \sim k\vartheta/\delta$.
- Maintenance of boiling on the surface, when the heat is supplied to bubbles near their bases and is spent directly for evaporation. For developed boiling, the average mass rate of removal of the vapor phase from the surface is of order ρw. The corresponding density of the heat flow is of order $q_2 \sim h_{fg}\rho w$.

From the comparison of the two components of the general heat flow, we find that

$$\frac{q_2}{q_1} \approx 10\frac{h_{fg}\rho\nu}{k\vartheta}. \tag{12.24}$$

From (12.24), it is seen that the ratio of the heat flow components changes drastically with pressure. The heat conduction mechanism is prevailing for low pressures. As the pressure increases, the intensities of both mechanisms become first commensurable and then the direct evaporation on the surface becomes prevailing. The value of the fluctuation velocity can be found from the balance relation

$$\rho\left(\frac{dR}{dt}\right)^2 R^2 n_F \sim \rho w^2. \tag{12.25}$$

Substituting into (12.20) the bubble growth rate dR/dt and the density of nucleation sites n_F and solving the equation for w, we get $w \approx 10\frac{k\vartheta^2}{\sigma T_s}$. Since both heat transfer mechanisms act simultaneously, it follows that the total heat flow is determined by the sum of two components

$$q = q_1 + q_2 \approx 10^{-3}\frac{k\vartheta^3}{\sigma T_s}\left(1 + 5\frac{h_{fg}\rho\nu}{k\vartheta}\right). \tag{12.26}$$

Formula (12.26), which was obtained by Labuntsov [10], with number constants chosen from comparison with experiments, is the first known analytical solution for nucleate boiling heat transfer.

12.6.4 Heat Transfer Modeling (Integral Characteristics)

Relation (12.26) was derived before the appearance of the first cinematic studies of the process of the developed nucleate boiling. Later, it became clear that for the larger part of the range of variation of heat flows in the near-wall zone, there exist unstable vapor conglomerates formed as a result of the consolidation of separate bubbles. Below we present the theoretical model, which was developed in the paper [10] and which is based on integral characteristics of the process. In [10], first, a detailed analysis of micro characteristics of nucleate boiling (the growth rate of a vapor bubble, the thickness of the near-wall liquid film, the distance between nucleation sites, etc.) was carried out. As a result, relations involving motivated estimates of the number of constants involved in them were written down. We shall be interested first of all in the physical content and in the summarizing recommendations of the theoretical model [10]. So in what follows we shall drop the aforementioned number constants and write instead of the symbol "∼" (which means the order equality).

The starting point of the model of [10] is the assumption about the simultaneous effect of the following two principal factors:

- On the prevailing part of the area of the heated surface (i.e., in the space between nucleation sites), the heat to liquid is transferred by virtue of convection due to the formation and growth of vapor bubbles on the wall
- In the contact area of three phases along the periphery of the dry spot, there exist heavy heat sinks. Despite the small area of dry spots, their role in the total thermal balance can be pretty significant.

Denoting the convective part of the flow by q_1 and denoting by q_2 the heat flow due to evaporation on the boundary of dry spots, we have, for the total heat flow for boiling

$$q = q_1 + q_2. \tag{12.27}$$

The component q_1 can be written in the form

$$q_1 = \frac{k\vartheta}{\delta}, \tag{12.28}$$

which is traditional for a single-phase convection. Here, δ is the effective thickness of the heat-conducting layer of liquid on the wall.

In the regime of coalescent bubbles, the quantity δ is a liquid film of thickness δ_0 under vapor conglomerates. This quantity is frequently called a "macrofilm", as distinct from the "microfilm" (or microlayer) in the base of a single bubble. In the regime of a single bubble, the heat-conducting layer occupies (for nonmetallic liquids) some part of the depth of the specific dynamic boundary layer on the wall. For the characteristic rate of the process, we take the average rate of vapor generation $w = q\left(h_{fg}\rho_{\mathrm{v}}\right)^{-1}$, by the characteristic linear scale, we shall mean the linear scale, which is the distance l_* between nucleation sites. Now δ can be estimated by analogy to the thickness of the dynamic boundary layer as $\delta \sim \sqrt{l_* \nu / w}$. Using this result and taking into account (12.24), we find q_1

$$q_1 \sim k\vartheta \sqrt{\frac{q}{R_* \nu h_{fg} \rho_{\mathrm{v}}}}. \tag{12.29}$$

The heat flux q_2 can be written as $q_2 = Qn_F$, where Q is the heat sink intensity per one dry spot. Figure 12.5 shows the calculation scheme of the flow of liquid near the dry sports typical for the regime of coalescent bubbles. A process is assumed to be stationary, in which all liquid supplied to the zone of intensive evaporation is evaporated due to the heat flow from the wall. At some distance R_m from the center of the dry spot, there is a zone of the most intensive evaporation. In linear scales, the thickness of the film δ_m occupies an intermediate position. On the one hand, this thickness is much smaller than the average thickness δ_0 of the "macrofilm", and so its thermal resistance δ_m/k due to thermal conductivity is negligibly small. On the other hand, δ_m is much larger than intermolecular distances in liquid, and hence, in this case, the adsorption forces do not impede evaporation anymore. The evaporation zone near the circumference $r = R_m$ has some finite width.

From physical considerations, it follows that the flow of liquid in a thin plane film (meniscus) is due to the curvature gradient of its surface. In model experiments [17, 18], it was shown by direct measurements that the surface curvature K in the area of the most intensive evaporation increases during evaporation from the meniscus surface of a liquid film. For a scheme of Fig. 12.5, this means that $dK/dr < 0$, i.e., the film surface curvature decreases away from the symmetry axis. Since the pressure in the vapor is homogeneous (p_v = const), from Laplace's formula

$$p_v - p_f(r) = 2\sigma K(r)$$

we find that

$$\frac{dp_v}{dr} = -\sigma \frac{dK}{dr}. \tag{12.30}$$

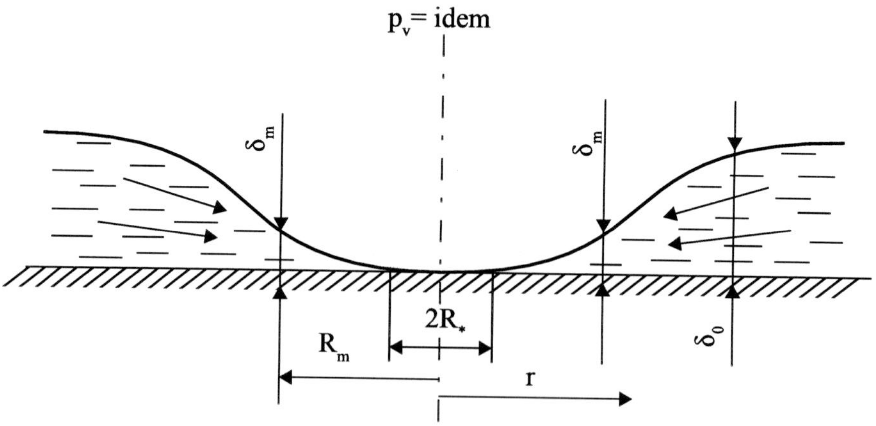

Fig. 12.5 Flow calculation scheme of liquid near the boundary of a dry spot in the regime of vapor conglomerates

Hence, in liquid, there exists a pressure gradient generating its flow to each boiling center encircled by a circular liquid film. From the solution of the problem of flow of a plane film in the radial direction, one can determine the mass-flow rate of liquid G_R per unit length of the perimeter in the zone of intensive evaporation: $G_R = \frac{\sigma\delta_m^2}{\nu}\frac{dK}{dr}$. Detailed numerical studies of the problem of evaporation of liquid near a boiling site were carried out in [19–21]. The solutions, which were obtained in these papers for specific conditions, shed some light on the very interesting laws of the boiling process. However, the applied relevance of these results is very limited due to the very strong idealization of the flow scheme under consideration. The thing is that the film thickness near the intensive evaporation zone is very small $\left(10^{-8}–10^{-7}\mathrm{m}\right)$, and hence any mechanical surface cannot be considered smooth with respect to the thin film. Besides, the choice in [19–21] of the boundary, wetting angle involved in numerical experiments is quite controversial.

The author of the present book proposed a method of the solution depending on the change of the boundary value problem for the fourth-order equation by the Cauchy problem for the second-order equation. Consider an approximate method of physical estimates for the determination of the curvature gradient of the meniscus in (12.26), which was developed in [22, 23].

We introduce the following assumptions

- thickness of the liquid film is much larger than the radius of the vapor critical nucleus: $\delta_m \gg R_*$,
- the maximal curvature of the meniscus is of order $1/R_*$,
- the minimal curvature of the meniscus is of order $1/\delta_m$,
- the meniscus curvature varies in the interval $R_m \approx \delta_m$.

Thus, we get the following estimate for the curvature gradient of the meniscus

$$\frac{dK}{dr} \sim \frac{\delta_m}{R_*}.$$

As a result, we get $G_R \sim \frac{\sigma\delta_m^2}{\nu R_*}$. If the entire liquid evaporates near the zone of radius R_m, then the linear density heat flow near the boundary of this zone is $q_R = G_R h_{fg}$, and the heat sink intensity per one dry spot is

$$Q \equiv 2\pi R_m q_R \sim \frac{\sigma\delta_m^2 R_m h_{fg}}{\nu R_*}.$$

On the other hand, one can estimate Q from a linear approximation of the meniscus profile on the interval of intensive evaporation: $\delta \sim r$. Then, we get the following expression for the film thickness in the intermediate zone of interest $\delta_m \sim \sqrt{k\vartheta\nu R_*/\sigma h_{fg}}$. Recalling the above assumption $\delta_m \approx R_m$, we get

$$q_2 \sim \left(\frac{\nu}{\sigma h_{fg}}\right)^{1/2}. \tag{12.31}$$

If we substitute this relation into (12.24), we get a quadratic equation with respect to the heat flow density q. As a result, we get the following resulting relation for the heat transfer during nucleate boiling

$$q = 3.43 \times 10^{-4} \frac{k^2 \Delta T^3}{\nu \sigma T_s} \left(1 + \sqrt{1 + 800B} + 400B\right). \qquad (12.32)$$

Here, $B \equiv \frac{h_{fg}(\rho_v \nu)^{3/2}}{\sigma (kT_s)^{1/2}}$ is the "pressure parameter".

It should be noted that the two resulting numerical coefficients involved in Eq. (12.32) were picked in [10, 22] after matching with a large amount of data on boiling of more than twenty different liquids in a wide range of reduced pressures. The above model of nucleate boiling heat transfer, which is based on integral characteristics of the process, describes experimental data for boiling of various liquids (except cryogenic ones) with scatter $\pm 35\%$ of points from the theoretical curve $h(q)$. As of today, formula (12.32) can be looked upon as the most reliable relation for heat transfer calculation during nucleate boiling of ordinary liquids. We note, however, that it does not contain such important factors for nucleate boiling like the roughness of the heat transfer surface, wettability, and thermal-inertial properties of the wall material. Note that in a number of cases (mostly for boiling of helium and nitrogen), thermophysical properties and thickness of the wall have a substantial effect on the experimentally measured AHTC. This new physical effect will be considered in detail later.

12.6.5 Difficulties of Theoretical Descriptions

As far as it is known to the author, in spite of the long-term theoretical and experimental investigations of nucleate boiling, nobody has created so far a closed theory of this extremely complex process. This conclusion is supported by extensive survey works [24–33], which just reproduced or modified the equations for the Heat Transfer Coefficient (HTC) calculation developed during 1970s. A rather interesting direction in the field of the boiling theory based on numerical modeling of this process [16] uses a series of initial assumptions, contains a significant number of numerical constants, and is consequently still quite far from its final completion.

At last, attempts to bypass the basic difficulties connected to insufficient knowledge of the internal structure of the boiling process with the help of the formal mathematical methods borrowed from other areas of physics are believed to be unpromising. This relates, in particular, to the "fractal model" of heat transfer at nucleate boiling suggested in the work [17]. Concepts of a "fractal" and "fractal geometry" introduced in 1975, by Mandelbrot [27], relate to the irregular (chaotic) structures possessing a property of self-similarity. In simplified terms, it means that some small part of a fractal already contains in a compressed form the information on the entire fractal as a whole. Fractals indeed play an important role in the theory of nonlinear dynamic systems, where they allow, with the help of simple algorithms, investigating complex and nontrivial structures [19].

In view of the above said, it would be possible to expect from the work [17] some qualitatively new results in the modeling of nucleate boiling, like it, for example, did happen at the fractal analysis of chaotic structures [19]. However, in fact, the authors of the work [17] restricted themselves with a search of the formulas for the parameters of nucleate boiling (characteristic time of growth of a vapor bubble, thickness of a temperature boundary layer, nucleation site density, HTC, etc.) already available in the literature, and then the authors simply put certain concepts from the theory of fractals in correspondence with these equations.

To summarize, it is possible to agree with the capacious definition stated by the author of the work [10]: "...Heat transfer at boiling is always determined by simultaneous influence of numerous mechanisms controlling transfer of a substance...". These words can be understood as an expression of constrained pessimism concerning an opportunity for the solution of the problem of boiling in the foreseeable future. Here an analogy arises to the known problem of the theory of turbulence, which is known to be also rather far from the final completion. Such a stand-point gets even the greater weight in view of the fact that the author of the work [10] was one of those who originated the development of the semi-empirical formulas for heat transfer at nucleate boiling (see [21]), which then have been brought (and remain there at the time being) into the standard handbooks (see, for example, [22]).

12.7 Periodic Model of Nucleate Boiling

12.7.1 Oscillations of the Thickness of a Liquid Film

One of the possible models of nucleate boiling is considered below. This model has incorporated a minimal number of numerical constants. The basic emphasis is done on an independent validation of the separate components in the model, which in a narrow sense can be understood as a verification of the values of these constants. In order to undertake a more detailed analysis of the thermal effect of a body on heat transfer at nucleate boiling, it is necessary first of all to analyze the spatial and temporal periodicity of the process. In accordance with the quasi-stationary character of the process of boiling, it is natural to believe that the thickness of the film will undergo periodic oscillations in time with a certain period t_0. On the other hand, the presence of the fixed sites of boiling assumes unavoidable spatial non-uniformity (waviness) of the film with a certain lengthscale (wavelength) L. It is possible with a good degree of accuracy to reflect the mentioned spatiotemporal periodicity via setting harmonic oscillations of the film thickness (Fig. 12.6)

$$\delta = \langle\delta\rangle\left\{1 + b\cos\left[2\pi\left(\frac{z}{L} - \frac{t}{t_0}\right)\right]\right\}. \tag{12.33}$$

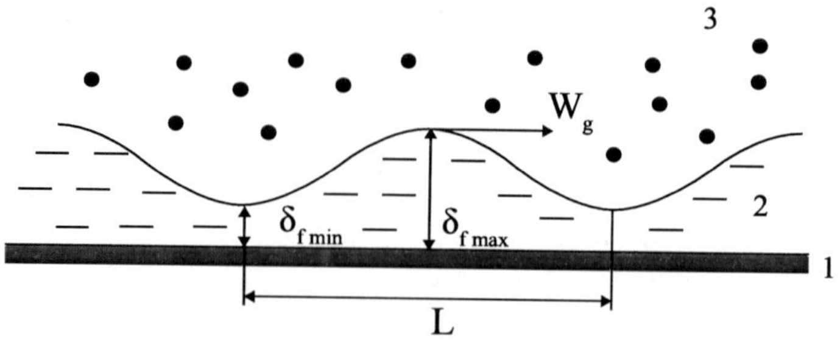

Fig. 12.6 Periodic model of nucleate boiling. 1 Heated surface, 2 oscillating liquid film, 3 vapor conglomerates

As the results of visual investigations show, nucleate boiling is characterized by some microroughness with a linear lengthscale of the order of magnitude comparable with the diameter of a critical vapor nucleus, which represents a lengthscale of some bubble microroughness on the heated surface. Basing on this fact, let us assume that the minimal film thickness for the period of oscillations

$$\delta_{\min} = \langle \delta \rangle (1 - b), \tag{12.34}$$

becomes equal to

$$\delta_{\min} = 2R_*. \tag{12.35}$$

Based on the analogy with a near-wall turbulent flow, one can assume that the maximal film thickness

$$\delta_{\max} = \langle \delta \rangle (1 + b) \tag{12.36}$$

is proportional to the thickness of the viscous sublayer based on the nucleation rate

$$\delta_{\max} = \gamma_1 \frac{\nu}{w}. \tag{12.37}$$

According to Eq. (12.33), the averaged HTC can be determined in this case from the following relation

$$\langle h \rangle = \frac{k}{\sqrt{\delta_{\min} \delta_{\max}}}. \tag{12.38}$$

Finally, the law of heat transfer at nucleate boiling can be derived from Eqs. (7.11–7.16)

$$q = \frac{1}{4\gamma_1} \frac{k^2 \vartheta^3}{\nu \sigma T_s}. \tag{12.39}$$

12.7.2 Nucleation Site Density

Let us consider now in more detail the problem connected with the determination of the nucleation site density

$$n_F = \frac{1}{L^2}. \tag{12.40}$$

According to Labuntsov's theory [10], this value should be determined as

$$n_F \approx \frac{10^{-8}}{R_*^2}. \tag{12.41}$$

Equation (12.41) predicts a square-law dependence of the nucleation site density on the temperature difference

$$n_F \sim \vartheta^2. \tag{12.42}$$

However, an analysis of the experimental investigations into the near-wall structure of nucleate boiling carried out up to the present time reveals that the power exponent should be given much larger numerical values [26, 33–35]

$$n_F \sim \vartheta^{3\ldots5}. \tag{12.43}$$

The estimate (12.42) was obtained by Labuntsov based on the assumption that the radius of a vapor nucleus (micro-lengthscale), is the unique characteristic lengthscale of the entire process. Therefore, the experimental proof of the higher power exponent in this dependence indirectly points out the existence of the second (macro-)lengthscale. Another indirect evidence of the insufficiency of Labuntsov's model also consists of the abnormally small value of the numerical constant in Eq. (12.41). A simple physical model of a flow in a near-wall liquid film on a heated surface between the boiling nucleation sites proposed below allows determining the aforementioned macro-lengthscale.

Let us consider a stationary flow over the length of a liquid film that directly adjoins to its thinnest part (Fig. 12.7). Considering the thickness of the layer constant and which is equal to $\delta_{\min}$, one can thus receive a case of a viscous flow of a liquid in a layer with a constant suction rate on its top boundary equal to the rate of evaporation of the liquid

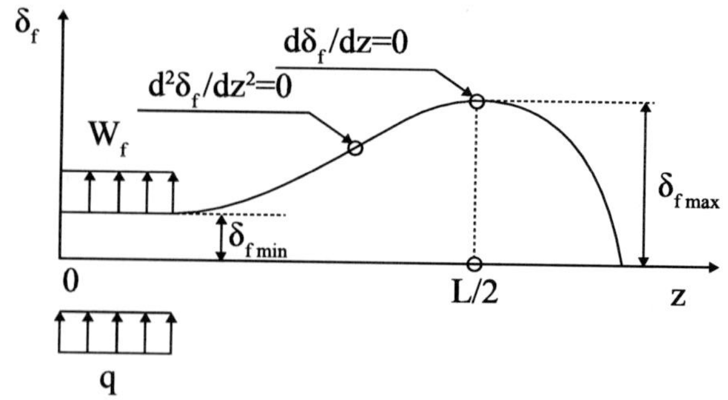

Fig. 12.7 Determination of the effective length of the liquid film

$$w = \frac{k\vartheta}{\delta_{\min} h_{fg} \rho}. \tag{12.44}$$

The pressure gradient in the liquid in the z-direction can be expressed as [36]

$$\frac{dp}{dz} = 3\frac{\mu w}{\delta_{\min}^3} z. \tag{12.45}$$

As shown in the work [37] with the reference to a problem of film condensation, the flow of a liquid against viscose forces for the considered case of a very small (microscopic) film thickness can effectively exist mainly at the expense of capillary forces

$$\frac{dp}{dz} = -\sigma\frac{dK}{dz}. \tag{12.46}$$

Here, K is the curvature of the film surface, which for small values of the derivative $d\delta/dz \ll 1$ can be approximately determined as

$$K \approx \frac{d^2\delta}{dz^2}. \tag{12.47}$$

From Eqs. (12.44)–(12.47), one can derive the following differential equation

$$\frac{d^3\delta}{dz^2} = -Az, \tag{12.48}$$

where

$$A = 3\frac{\nu k\vartheta}{h_{fg}\sigma\delta_{\min}^4}. \tag{12.49}$$

From the conditions of the conjugation of a film with a layer of the bubble microroughness, two boundary conditions for Eq. (12.48) physically follow

$$z = 0 : \delta = \delta_{\min}, \frac{d\delta_{\min}}{dz} = 0. \tag{12.50}$$

Then, a simple integration of Eq. (12.48) leads to the following equation for the dependence of the liquid film thickness on the longitudinal coordinate

$$\delta = \delta_{\min} + \frac{1}{2}Cz^2 - \frac{1}{24}Az^4. \tag{12.51}$$

It follows from Eq. (12.51) that the dependence $\delta(z)$ exhibits consecutively a growing branch ($d\delta/dz > 0$), an inflection point $\left(d^2\delta/dz^2 = 0\right)$, a point of maximum $d\delta/dz = 0$ and a descending branch ($d\delta/dz < 0$) (Fig. 12.7). Since the descending branch is physically unjustifiable, it is necessary to trim the dependence $\delta(z)$ at a certain point, i.e., to determine in doing so, both the constant C and the effective length of the film L (or, in other words, the spatial lengthscale of periodicity $z_0 = L$. From the reasons of symmetry of the film profile (or, in other words, smooth interface between two adjacent boiling nucleation sites), let us accept that the condition of trimming is fulfilled at the point of the maximum of the dependence $\delta(z)$

$$z = \frac{1}{2}L, \frac{d\delta}{dz} = 0. \tag{12.52}$$

Thus, we have obtained a picture of the wave

$$\delta = \langle\delta\rangle\left\{1 + b\cos\left(2\pi\frac{z}{L}\right)\right\}. \tag{12.53}$$

One can further find out from Eqs. (12.50)–(12.52) that

$$\frac{L}{\delta_{\min}} = \beta_2\left(\frac{h_{fg}\sigma\delta_{\max}}{\nu k\vartheta}\right)^{1/4}. \tag{12.54}$$

One should point out that the estimate of the relation between the maximal and minimal thickness of the evaporating liquid film ($\delta_{\min} \ll \delta_{\max}$) suggested in the works [38, 39] was used at the derivation of Eq. (12.54). An interrelation between the macroscale L and microscale $\delta_{\min}$ of the process of nucleate boiling can be deduced from Eqs. (12.37), (12.54)

$$\frac{L}{\delta_{\min}} = \beta_3\frac{(\nu\rho_v T_s)^{1/4}\left(h_{fg}\sigma\right)^{1/2}}{k^{3/4}\vartheta}. \tag{12.55}$$

For a transition from the frozen (Eq. 12.53) to the running (Eq. 12.33) progressive wave of oscillations of the film thickness, it is necessary to find out the period of temporal oscillations. It is natural to believe that oscillations of the heat transfer intensity extend along the surface of a body with a phase speed of the order of magnitude comparable with the vapor nucleation rate. Then the time scale of periodicity can be determined from the relation

$$t_0 = \frac{L}{w}. \tag{12.56}$$

Knowing the macroscale L, it is possible to determine the required nucleation site density

$$n_F = \beta_4 \frac{h_{fg}(k\rho_v)^{3/2}\vartheta^4}{\nu^{1/2}T_s^{5/2}\sigma^3}. \tag{12.57}$$

From Eq. (12.57), one can conclude that the theoretical model represented here provides a qualitatively true dependence of the nucleation site density on the temperature difference and agrees well with the correspondent tendencies documented in works [26, 33–35].

12.8 Effect of Thermophysical Characteristics of Wall Materials

According to experiments, thermophysical characteristics of a heat-transmitting wall were found to have a considerable effect on the heat transfer intensity in boiling of cryogenic liquids. In studies [40, 41], which were concerned with the interpretation of the above experimental data, a physical model of such an effect was put forward. Below, we give a critical analysis of this model. Experiments of boiling of helium and nitrogen on surfaces of various metals (copper, nickel, bronze, stainless steel) show that the intensity of the heat transfer (the other conditions being equal) is proportional to the coefficient of the thermal activity of a wall $\sqrt{k_w c_w \rho_w}$ [16].

The authors of [40, 41] give the following physical interpretation of the effect of thermal influence of a wall on the characteristics of nucleate boiling. Bubbles that grow on the surface can be looked upon as periodically engaged "thermal micropumps". After taking some heat from the heat transferring wall and from the superheated liquid, they grow to a detachment diameter and ascend in the gravity field in the liquid volume. During the time of growth of a bubble to its detachment diameter, the temperature of the surface of the liquid drops down. Following the detachment of the bubble from the wall, the liquid is again heated to the initial temperature during some "exposure time", and a new bubble appears in the boiling center. In turn, this bubble grows and again engages the "thermal micropump", the bubbling cycle repeating again and again.

Consequent on this chain of arguments, the authors of [40, 41] include the coefficient of the thermal activity of a wall $\sqrt{k_w c_w \rho_w}$ in the expression (12.20) for the bubble growth rate. Next, the use is made of the more early Labuntsov's model (see Sect. 6.3), which is based on the interpretation of boiling as efficient near-wall turbulence induced by bubbles that grow on the wall. As a result, in [40, 41], some dependences $h(q)$ were obtained with due account of the effect of the coefficient of the thermal activity of a wall. However, the above interpretation of the wall material on heat transfer for nucleate boiling appears to be incorrect for the following reasons.

The model [40, 41] uses the solution of the heat conduction problem in a semi-infinite body with a step jump in surface temperature. It is known that in solving this problem, one specifies a homogeneous initial distribution of temperatures in the body [42]. So, in this setting, one ignores the important fact that the statement of the classical with step jump excludes a heat flux at infinity of the body. But this means that the authors of [40, 41] replace the required consideration of the periodic structure of the temperature field in a heat-transmitting wall by a speculative explanation about "thermal micropumb" on the rigid surface. As an alternative to the method of [40, 41], the author of the present book proposed a correct physical model of the process of conjugate heat transfer with periodic intensity [14]. The principal features of this model are briefly outlined below.

12.9 Effect of Oscillating Interface on Heat Transfer

12.9.1 Oscillation Structure of Convective Heat Transfer

As a rule, real stationary processes of heat transfer can be looked upon as stationary only on average. Actually (except for the purely laminar cases), flows are always subjected to various periodic, quasi-periodic and other casual oscillations of velocities, pressure, temperatures, momentum and energy fluxes, vapor content, and interphase boundaries about their average values. Such oscillations can be smooth and periodic (wave flow of a liquid film or vapor, a flow of a fluctuating coolant over a body), sharp and periodic (hydrodynamics and heat transfer at slug flow of a two-phase media in a vertical pipe, nucleate and film boiling process), or can have complex stochastic character (turbulent flow). Oscillations of parameters have in some cases spatial nature, in others they are temporal, and generally one can say that the oscillations have mixed spatiotemporal character.

The theoretical base for studying instantly the oscillations and at the same time stationary on the average heat transfer processes are the unsteady differential equations of momentum and energy transfer, which in the case of two-phase systems can be rotated for each of the phases separately and be supplemented by transmission conditions (transmission conditions). An exhaustive solution to the problem could be a comprehensive analysis with the purpose of a full description of any particular fluid flow and heat transfer pattern with all its detailed characteristics, including various fields of oscillations of its parameters.

However, at the time being such an approach cannot be realized in practice. The problem of modeling turbulent flow [43] can serve as a vivid example. As a rule at its theoretical analysis, Reynolds-averaged Navier–Stokes equations are considered, which describe time-averaged quantities of fluctuating parameters, or in other words turbulent fluxes of the momentum and energy. To provide a closed description of the process, these correlations by means of various semi-empirical hypotheses are interrelated with time-averaged fields of velocities and enthalpies. Such schematization results in the statement of a stationary problem with spatially variable coefficients of viscosity and thermal conductivity. Therefore, as boundary conditions here, it is possible to set only respective stationary conditions on the heat transfer surface of such a type as, for example, "constant temperature", "constant heat flux".

It is necessary to specially note, that the replacement of the full "instant" model description with the time-averaged one inevitably results in a loss of information on the oscillations of fluid flow and heat transfer parameters (velocities, temperatures, heat fluxes, pressure, friction) on a boundary surface. Thus, the theoretical basis for an analysis of the interrelation between the temperature oscillations in the flowing ambient medium and in the body is omitted from the consideration. And generally saying, the problem of an account for the possible influence of thermophysical and geometrical parameters of a body on the heat transfer at such on approach becomes physically senseless. For this reason, such a "laminarized" form of the turbulent flow description is basically not capable of predicting and explaining the wall effects on the heat transfer characteristics, even if these effects are observed in practice. The problem becomes, especially complicated by imposing external oscillations on the periodic turbulent structure that takes place, in particular, flows over aircraft and spacecraft. Unresolved problems of closing the Navier–Stokes equations in combination with difficulties of numerical modeling make a problem of detailed prediction of a temperature field in the flowing fluid very complicated. In some cases, differences between the predicted and measured local HTC exceeds 100%.

In this connection, the direction in the simulation of turbulent flow based on the use of the primary transient equations [43] represents significant interest. The present book represents results of numerical modeling of the turbulent flow in channels subjected to external fields of oscillations (due to vortical generators, etc.). It is shown that in this case, an essentially anisotropic and three-dimensional flow pattern emerges strongly different from that described by the early theories of turbulence. In the near-wall zone, secondary flows in the form of rotating "vortical streaks" are induced that interact with the main flow. As a result, oscillations of the thermal boundary layer thickness set on, leading to periodic enhancement or deterioration of heat transfer. Strong anisotropy of the fluid flow pattern results in the necessity of a radical revision of the existing theoretical methods of modeling the turbulent flow. So, for example, the turbulent Prandtl number being in early theories of turbulence a constant of the order of unity (or, at the best, an indefinite scalar quantity), becomes a tensor. It is necessary to emphasize that all the mentioned difficulties are related to the nonconjugated problem when the role of a wall is reduced only to the maintenance of a boundary condition on the surface between the flowing fluid and the solid wall.

12.9.2 Correct Averaging of the Heat Transfer Coefficient

The basic applied task of the book is the investigation of the effects of a body (its thermophysical properties, linear dimensions, and geometrical configuration) on the traditional HTC, which is measured in experiments and used in engineering calculations. Processes of heat transfer are considered stationary on average and fluctuating instantly. A new method fox investigating the conjugate problem "fluid flow—body" is presented. The method is based on a replacement of the complex mechanism of oscillations of parameters in the flowing coolant by a simplified model employing a varying "True Heat Transfer Coefficient" (THTC) specified on a heat transfer surface.

The essence of the developed method can be explained rather simply. Let us assume that we have perfect devices measuring the instant local values of temperature and heat fluxes at any point of the fluid and heated solid body. Then the hypothetical experiment will allow finding the fields of temperatures and heat fluxes and their oscillations in space and in time, as well as their average values and all other characteristics. In particular, it is possible to present the values of temperatures (exact saying, temperature heads or loads, i.e., the temperatures counted from a preset reference level) and heat fluxes on a heat transfer surface in the following form

$$\vartheta = \langle\vartheta\rangle + \widehat{\vartheta}, \tag{12.58}$$

$$q = \langle q\rangle + \widehat{q}, \tag{12.59}$$

i.e., to write them as the sum of the averaged values $\langle\vartheta\rangle$, $\langle q\rangle$ and their temporal oscillations $\widehat{\vartheta}$, $\widehat{q}$. For the general case of spatiotemporal oscillations of characteristics of the process, the operation of averaging is understood here as a determination of an average with respect to time t and along the heat transferring surface (with respect to the coordinate z).

The THTC is determined on the basis of Eqs. (12.45), (12.59) according to Newton's law of heat transfer [44]

$$h = \frac{q}{\vartheta}. \tag{12.60}$$

This parameter can always be presented as a sum of an averaged part and a fluctuating additive

$$h = \langle h\rangle + \widehat{h}. \tag{12.61}$$

It follows from here that the correct averaging of the HTC is as follows

$$\langle h\rangle = \langle\frac{q}{\vartheta}\rangle. \tag{12.62}$$

Therefore, we shall call parameter $\langle h\rangle$ an "Averaged True Heat Transfer Coefficient" (ATHTC). The problem consists in the fact that the parameter $\langle h\rangle$ cannot be directly used for applied calculations, since it contains initially the unknown information of oscillations $\widehat{\vartheta}$, $\widehat{q}$. This fact becomes evident if Eq. (12.62) is rewritten with the help of Eqs. (12.58)–(12.59)

$$\langle h\rangle = \langle \frac{\langle q\rangle + \widehat{q}}{\langle \vartheta\rangle + \widehat{\vartheta}}\rangle. \tag{12.63}$$

The purpose of the heat transfer experiment is the measurement of averaged values on the temperature difference $\langle \vartheta\rangle$ and a heat flux $\langle q\rangle$ on the surfaces of a body and determination of the traditional HTC

$$h_e = \frac{\langle q\rangle}{\langle \vartheta\rangle}. \tag{12.64}$$

The parameter h_e is fundamental for carrying out engineering calculations, designing heat transfer equipment, composing thermal balances, etc. However, it is necessary to point out that transition from the initial Newton's law of heat transfer (12.60) to the restricted Eq. (12.64) results in the loss of the information of the oscillations of the temperature $\widehat{\vartheta}$ and the heat fluxes $\widehat{q}$ on the wall. So, it is logical to assume that the influence of the material and the wall thickness of the body taking part in the heat transfer process on HTC uncovered in experiments is caused by the non-invariance of the value of HTC with respect to the Newton's law of heat transfer. For this reason, we shall refer further to the parameter h_e as to an "Experimental Heat Transfer Coefficient (EHTC). Thus, we have two alternative procedures of averaging the HTC: true Eq. (12.62) and experimental Eq. (12.64).

The physical reason for the distinction between $\langle h\rangle$ and the h_e can be clarified with the help of the following considerations

- local values ϑ and q on a surface where heat transfer takes place are formed as a result of the thermal contact of the flowing fluid and the body,
- under conditions of oscillations of the characteristics of the coolant, temperature oscillations will penetrate inside the body,
- owing to the conjugate nature of the heat transfer in the considered system, both fluctuating $\widehat{\vartheta}$, $\widehat{q}$ and averaged $\langle \vartheta\rangle$, $\langle q\rangle$ parameters on the heat transfer surface depend on the thermophysical and geometrical characteristics of the body,
- the ATHTC $\langle h\rangle$ directly follows from Newton's law of heat transfer (12.60) (which is valid also for the unsteady processes) and consequently, it is determined by hydrodynamic conditions in the fluid flowing over the body,
- the EHTC h_e by definition does not contain the information on oscillations $\widehat{\vartheta}$, $\widehat{q}$, and consequently, it is in the general case a function of parameters of the interface between fluid and solid wall,

- aprioristic denying of the dependence of the EHTC on material properties and wall thickness is wrong, though under certain conditions quantitative effects of this influence might be insignificant.

From the formal point of view, the aforementioned differences between the true (12.62) and experimental (12.64) laws of averaging of the actual HTC is reduced to a rearrangement of the procedures of division and averaging. This situation is illustrated evidently in Fig. 12.8.

Using the concepts introduced above, the essence of a suggested method can be explained rather simply. We shall assume that for the case under investigation the HTC is known: $h = h(z, t)$, where z and t are the coordinate along a surface where heat transfer takes place and the time, respectively. According to the internal structure of the considered processes, this parameter should have periodic, quasi-periodic or generally fluctuating nature, varying about its average value $\langle h \rangle$: $h = \langle h \rangle + \hat{h}(z, t)$. This information is basically sufficient for the definition of actual driving temperature difference $\vartheta(z, t)$ and heat fluxes $q(z, t)$ in a massive heat transferring body, and hence, on the heat transfer surface. Thus, the calculation is reduced to a solution of a boundary value problem of the unsteady heat conduction equation [45, 46]

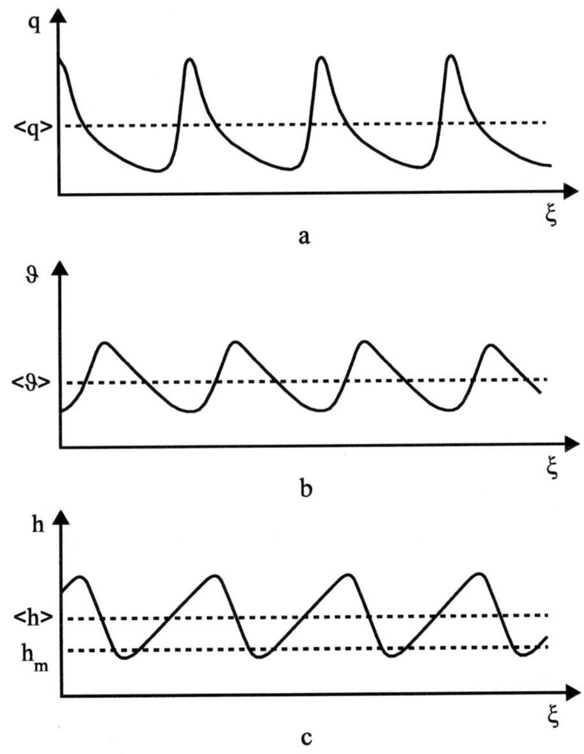

Fig. 12.8 True and experimental laws of the averaging of the heat transfer coefficient. **a** Heat flux density on the heat transfer surface, **b** temperature difference "wall–ambiance", **c** heat transfer coefficient

$$\frac{\partial \vartheta}{\partial t} = \alpha \left(\frac{\partial^2 \vartheta}{\partial x^2} + \frac{\partial^2 \vartheta}{\partial z^2} \right) \quad (12.65)$$

with the boundary condition of the third kind on the heat transfer surface

$$-k \frac{\partial \vartheta}{\partial x} = h\vartheta, \quad (12.66)$$

and suitable BC on the external surfaces of the body.

It is essential for our analysis that up to the same extent in which the information about the function $h = h(z, t)$ is trustworthy, the computed parameters $\vartheta(z, t)$ and $q(z, t)$ are determined also authentically. The basis for such a statement is the fundamental theorem of uniqueness of the solution of a boundary value problem for the heat conduction equation. In other words, the temperature field ϑ and heat flux q found in the calculation should appear identical to the actual parameters, which could be in principle measured in a hypothetical experiment. Further, based on the known distributions ϑ and q, it is possible to determine corresponding average values $\langle \vartheta \rangle$ and $\langle q \rangle$ and finally (from Eq. 12.64) the parameter h_e, which appears to be a function of the parameters of conjugation. It follows from the basic distinction of procedures of averaging of Eqs. (12.62) and (12.64) that an experimental value of the actual HTC is not equal to its averaged true value

$$h_e \neq \langle h \rangle. \quad (12.67)$$

The analytical method schematically stated above, in which "from the hydrodynamic reasons" the following relation is stated

$$h(z, t) = \langle h \rangle + \widehat{h}(z, t), \quad (12.68)$$

and further from the solution of the heat conduction equation in a body, the parameter h_e is determined, which outlines the basic essence of the approach developed in the present analysis.

12.9.3 Model of Periodical Contacts

A simple evident model of the conjugate problem "fluid flow—body" is a scheme of periodic collisions with a surface of a solid body (conductive supply of heat into the system) of the volumes of fluid constantly replacing each other (convective removal of heat)—Fig. 12.9.

Since a constant heat flux is supplied from the depth of a solid body, the distribution of the average temperature in the body should look like linear functions. On this linear

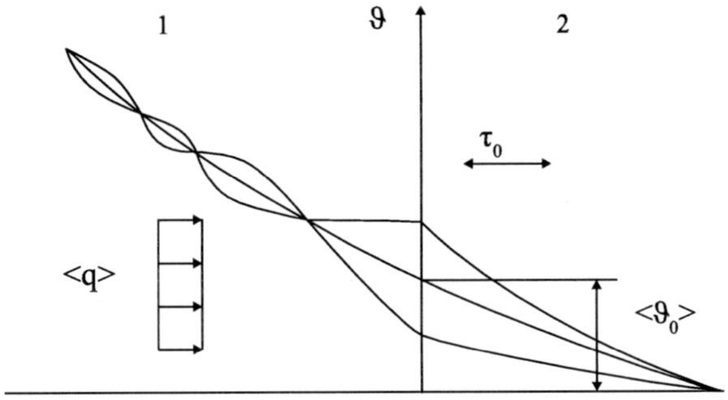

Fig. 12.9 Schematic of the periodical contacts of two media. 1 Body, 2 ambient fluid

distribution, temperature oscillations with increasing amplitude (as approaching the surface) will be imposed. In doing so, the "conductive condition of periodicity" should be fulfilled: temperature distribution in the solid body at time $t = t_0$ should exactly repeat the respective distribution at time $t = 0$. The temperature of the surface of the next cold fluid volume will always grow in time (stepwise at the initial moment of time, and then as a monotonic function during the entire period of interaction).

"The convective condition of periodicity" will be expressed in the replacement of a heated volume after the end of the interaction with a wall with a new cold volume. The mathematical description of the problem includes the unsteady one-dimensional equations of heat conduction for the solid body and the volume of fluid completed with the conditions of conjugation at the interface (equality of temperatures and heat fluxes). The described model of periodic contacts contains a unique dimensionless parameter, which is the ratio of coefficients of the thermal activity of the contacting media

$$N = \sqrt{\frac{k_w c_w \rho_w}{k c \rho}}.$$

Nevertheless, the apparent simplicity of the problem is deceptive. Its solution with the help of the Green's function [45] results in obtaining a complex integro-differential equation. Let us consider the heat conduction equation for a volume of fluid for the limiting cases allowing an analytical solution.

ϑ = const. The limiting case for ϑ = const will be reached for $N \to \infty$. In this case, oscillations of temperatures and averaged temperature gradient in a body will be negligibly small. The known solution [45] for a case ϑ = const gives: $q = k\vartheta/\sqrt{\pi \alpha t}$. It follows from here that the heat flux averaged over the period of contact t_0 will be equal to $\langle q \rangle = 2k\vartheta/\sqrt{\pi \alpha t_0}$. Under these conditions, the EHTC and ATHTC will be equal to each other $h_e = \langle h \rangle = 2k/\sqrt{\pi \alpha t_0}$. One should note that in the general case, final values of the complex N under conditions of conjugation of a flowing fluid and body temperature oscillations will penetrate into the body, and the isothermal wall condition thus will be broken.

q = const. In the other limiting case $\to 0$, temperature oscillations in a body will reach their maximum. It follows from the Fourier law that at $k \to 0$ an infinitely large average temperature gradient corresponds to a final average heat flux in a body. This means from a physical point of view an unlimited increase in the heat flux rate, in relation to which any finite oscillations will be considered negligibly small. This corresponds to a limiting case $q = \text{const}$. The known solution [45] for this case gives the law of monotonic increase of temperature in time: $\vartheta = 2/\sqrt{\pi}q\sqrt{\alpha t}/k$. It means that in the limiting case $N \to 0$, the surface temperature at the change of the volumes of fluid falls down abruptly to zero value, and then starts to increase monotonically.

Let us obtain relations for the following quantities

- averaged temperature difference $\langle\vartheta\rangle = \frac{4}{3\sqrt{\pi}}\frac{q\sqrt{\alpha t}}{k}$,
- ATHTC $\langle h\rangle = \sqrt{\pi}\frac{k}{\sqrt{\alpha t}}$,
- EHTC $h_e = \frac{3\sqrt{\pi}}{4}\frac{k}{\sqrt{\alpha t}}$.
  An analysis of the transition from the case $N \to \infty$ to $N \to 0$ results in the following conclusions.
- Despite the radical reorganization of the temperature field, oscillations in a body, the EHTC, and the ATHTC differ from each other insignificantly (no more than by 25%). Even though this fact is unexpected, it agrees with the physically natural (in other words, physically expected) way of the thermal effect of a wall.
- The EHTC not only does not decrease, but, on the contrary, increases by ≈18%. This result is completely unexpected. The reason for this metamorphosis consists, apparently, in reorganization actually of the ATHTC: for the case of $q = \text{const}$, it appears to be $\pi/2$ times higher, than for the case of $\vartheta = \text{const}$.
- An uncontroversial conclusion follows from the above-mentioned limiting estimations that there is practically no effect of the thermal conjugation within the framework of the model of periodic contacts. More precisely: this effect is so weak that it is not visible on the background of the changes in the character of oscillations of the THTC. This discouraging circumstance can induce quite a critical analysis of applications of the model of periodic contacts available in the literature.

12.10 Conjugate Heat Transfer Problem in Boiling

12.10.1 Nucleate Boiling

On the basis of the above analysis, it can be definitely asserted that the problem of the effect of thermophysical characteristics of a heat-transmitting wall on the heat transfer in nucleate pool boiling is still in essence unexplored. At the same time, a noticeable influence of the complex (coefficient of the thermal activity of a wall) on the measured HTC at nucleate boiling of a liquid is an experimentally established fact. So, it was found in experiments [40, 41] that replacing the heater's material from copper to stainless steel results in a decrease in the heat transfer intensity at boiling cryogenic liquids by an order of magnitude by 12 times at boiling of nitrogen and by 40 times at boiling of helium.

Therefore, there is an open question in front of the theory of nucleate boiling to search for the correct model describing the thermal influence of a wall on the average intensity heat transfer. The conjugate heat transfer model based on periodic collisions of a bulk of liquid with a heat transfer body leads to the following striking result: EHTC as $N \to 0$ is found to be bigger by 11% than as $N \to \infty$. However, this directly contradicts the principal conception of the theory of periodic conjugate heat transfer (which is outlined in [14]), according to which as $N \to 0$ EHTC should tend to its minimal value.

This negative result leads to the conclusion on the possible effect of other physical factors on heat transfer characteristics during nucleate boiling. Let us consider in this respect the problem of the growth of an individual bubble. According to experiments, after the detachment of a bubble from the wall, a new bubble appears at a boiling site not at once, but rather after some "exposure time". During this "passive period", the local wall temperature increases monotonically up to some threshold value necessary for the formation of a new vapor nucleus in the boiling site. As was already pointed out, the regimes of individual and coalescent bubbles have a common internal structure, which is manifested on the macroscopic level by the common dependence $q \sim \vartheta^3$. On this basis, it is possible to extend the laws related to alternation of the "active" and "passive" heat transfer regimes also to the regime of vapor conglomerates. We shall assume that the main cause of the strong effect of a body on the characteristics of a periodic heat transfer is the shielding of the rigid surface by dry spots (this phenomenon is time alternating). In order to model this effect, we shall use a simple approximate model, which describes spatiotemporal variations of the heat transfer intensity. According to the method of [14], we define the pulsations of the THTC as an asymmetric step function

$$\left.\begin{aligned} 0 \le \xi \le \xi_* &: h = h_*, \\ \xi_* \le \xi \le 1 &: h = 0. \end{aligned}\right\} \tag{12.69}$$

Here, $\xi = t/t_0 \mp z/z_0$ is the generalized coordinate of the progressive wave developing from left to right with the coordinate z along the heat transfer surface, z_0, t_0 are the spatial and time periods of oscillations, respectively.

Conditions (12.69) describe the propagation of stepwise (discontinuous) perturbations of THTC along the surface. The Fourier number

$$\mathrm{Fo} = \frac{\alpha_w t_0}{z_0^2}, \tag{12.70}$$

which defines the ("spatial" or "temporal") type of pulsations, serves as an important example of a periodic heat transfer. The limit case $\mathrm{Fo} \to \infty$ corresponds to unbounded "tensioning" of the time period:$t_0 \to \infty$. Besides, the progressive wave of heat transfer oscillations becomes "frozen", and THTC varies along the heat transfer surface according to the space-periodic law $h(\xi) \rightarrow h(z)$. The Biot number, which is the principal parameter of the conjugate problem, is as follows

$$\langle \bar{h} \rangle = \langle h \rangle z_0 / k_w. \quad (12.71)$$

The limit case Fo $\rightarrow 0$ can be obtained if the propagation velocity of the progressive heat transfer wave goes off to infinity along the body surface. This will correspond to unbounded "stretching" of the spatial period of pulsations, $z_0 \rightarrow \infty$. In this case, THTC varies synchronously over the entire heat transfer surface according to the temporal periodic law $h(\xi) \rightarrow h(\tau)$. Besides, the linear scale of the problem under consideration is redefined:

$z_0 \underset{Fo\rightarrow 0}{\Rightarrow} \sqrt{\alpha_w t_0}$, moreover, the Biot number, which describes the "temporal problem", assumes the form

$$\langle \tilde{h} \rangle = \frac{\langle h \rangle \sqrt{\alpha_w t_0}}{k}. \quad (12.72)$$

The quantity $\sqrt{\alpha_w t_0}$ is also known as the "thermal wavelength". In order to find the type of periodicity of heat transfer, we use the physical characteristics of nucleate boiling: the distance between nucleation sites (the length scale) $z_0 = l_*$ and the rate of vapor generation $w = q/h_{fg}\rho_v$ (the velocity scale). From these, one can find the time scale from the clear relation $t_0 = z_0/w$.

Moreover, in accordance with the available ideas, we shall consider separately the case of low and high pressures. To simplify the analysis, we shall define symmetric stepwise oscillations of the heat transfer intensity, when the active and passive (adiabatic) periods are equal: $\xi_* = 1/2$. Below we give theoretical relations with numerical constants, which were obtained from detailed estimates on the basis of the theory developed in the book [14].

Nitrogen boiling. For the case of nitrogen boiling at atmospheric pressure, we get the estimate: Fo $\gg 1$. This means that here we have a spatial law of oscillations. The Biot number assumes the form

$$\langle \bar{h} \rangle = 4.96 \frac{k}{k_w} \frac{\left(1 + \sqrt{1 + 800B} + B\right)^{1/3}}{B^{1/2}}. \quad (12.73)$$

The "conjugation parameter"

$$\varepsilon \equiv \frac{h_e}{\langle h \rangle} = \frac{1 + 0.218\langle \bar{h} \rangle}{1 + 1.07\langle \bar{h} \rangle + 0.103\langle \bar{h} \rangle^2}. \quad (12.74)$$

Expresses the quantitative degree of the thermal effect of the wall. Figure 12.10 shows that, for nitrogen boiling on copper, the conjugation parameter significantly exceeds the corresponding quantity for boiling on stainless steel.

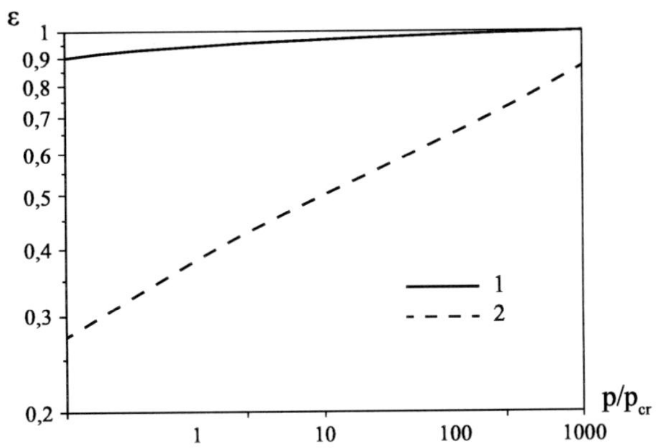

Fig. 12.10 Nucleate boiling of nitrogen. Dependence of the conjugation parameter on the dimensionless pressure. 1 Boiling on copper, 2 boiling on stainless steel

Water boiling. We now estimate the time scale in the form of a concrete number constant

$$t_0 = 214\frac{bh_{fg}\rho_v}{q}$$

is the period of pulsations, $b = \sqrt{\frac{\sigma}{\rho g}}$ is the capillary constant. Here from the estimates of the Fourier number, we get: Fo $\ll$ 1. Hence we have here the temporary type of THTC oscillations. The Biot number is as follows

$$\widetilde{\langle h\rangle} = \frac{1}{\sqrt{2\pi}}\frac{\langle h\rangle\sqrt{\alpha_w t_0}}{k_w}. \quad (12.75)$$

The conjugation parameter reads as

$$\varepsilon \equiv \frac{h_e}{\langle h\rangle} = \left(1 + 0.0178\widetilde{\langle h\rangle} + 1.32\widetilde{\langle h\rangle}^2\right)\left(1 + 0.87\widetilde{\langle h\rangle} + \widetilde{\langle h\rangle} + 1.36\widetilde{\langle h\rangle}^3\right)^{-1}. \quad (12.76)$$

It is seen from Fig. 12.11 that here the picture is similar to the case of nitrogen boiling.

The (substantial) difference is in the transition from the spatial to the temporary character of variation of the heat transfer intensity.

So, on the basis of a simple physical model, we constructed a correct model of periodic conjugate heat transfer in relation to nucleate boiling of liquid. As a result, it proved possible to identify the principal influencing factor—the exposure time of the surface—and obtain an explicit form of the thermal effect of the wall material on the standard averaged ("experimental") HTC. It is worth pointing out

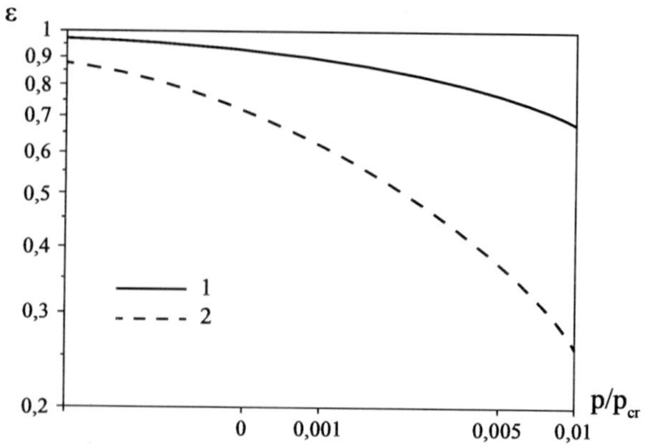

Fig. 12.11 Nucleate boiling of water. Dependence of the conjugation parameter on the dimensionless pressure. 1 Boiling on copper, 2 boiling on stainless steel

that this influence is not exhausted only by the effect of the principal conjugation parameter (the Fourier and Biot numbers). In the general case, one should also take into account the effect of thickness of the heat transfer body (plate). Moreover, one should differentiate between two limit possible cases of external heating of a plate: a constant temperature and a constant heat flow on the outer surface. According to the book [14], the latter case also describes the type of heating with a constant density of bulk heat sources (i.e., electrical heating).

12.10.2 Transition Boiling

Transition boiling of liquids serves as a vivid manifestation of purely temporal pulsations in heat transfer intensity. Whereas there is an entire series of semi-empirical heat transfer models for the nucleate boiling, in the case of the transition boiling the theory is unexplored. It is worth specially mentioning that the author of the present book has no pretensions on creating the model of transition boiling heat transfer. The purpose of the analysis that follows is the specific narrow problem: find the conjugate parameters for this "abnormal" boiling regime.

In constructing the model of thermal effect, we shall be based on the experimentally revealed picture of transition boiling. The active period of heat transfer corresponds to a short-time contact of the layer of liquid with a highly superheated surface. Then follows the passive period, when the surface is shielded by the vapor of negligible density and the heat transfer proceeds in the regime of film boiling. The total time of the characteristic cycle of the transition boiling exceeds substantially that of the active period. Hence, to calculate ε one can use the asymmetric time function THTC: $\xi_* \ll 1$. We shall assume that we have at our disposal the original "nonconjugated" dependence of transition boiling heat transfer for the case of an isothermal rigid surface of the wall.

This dependence, which describes the segment CE in Fig. 12.2, can be directly obtained from an experiment in accordance with the adopted methods.

- The active period of heat transfer: evaporation of thin film of liquid of thickness $\delta \sim R_*$. At the phase when the liquid is in contact with the wall the heat flow

$$q_* \equiv \frac{k\vartheta_*}{\delta} = \frac{\sigma T_s}{\vartheta_* \tau_0} \tag{12.77}$$

  is removed from it, where ϑ_* is the ϑ as$N \equiv \sqrt{k_w c_w \rho_w / kc\rho} \to \infty$. The evaporation rate of liquid is $w = \delta h_{fg} \rho / q_*$. The time of complete evaporation is $t_* \sim \delta / w$.
- The passive period of heat transfer: growth of a vapor bubble during film boiling. The total time of the cycle of transition boiling is determined from the dimensional analysis as a time scale of capillary-gravitational waves, $t_0 \sim \sqrt{b/g}$.
- The averaged heat flow over the total heat transfer cycle is as follows

$$\langle q \rangle \equiv q_* \frac{t_*}{t_0} = \frac{\sigma T_s}{\vartheta_* t_0}. \tag{12.78}$$

- Due to the sharp difference of intensities in the active and passive periods of heat transfer, the film boiling may to a good approximation be treated as an "adiabatic process". It is assumed that when the growing vapor bubble reaches some limit size, it is detached from the surface. Then the entire cycle is repeated. This completely describes the physical picture of transition boiling.

The above analysis reveals the following remarkable result: the conjugation parameter in the context of the asymmetric stepwise function as $\xi_* \to 0$ depends on the only parameter: the Biot number, as constructed from the characteristic of the active heat transfer period. This means that the universal parameter

$$\langle \tilde{h}_* \rangle = \sqrt{\frac{h_{fg} k \rho}{\Delta T_* k_w c_w \rho_w}} \tag{12.79}$$

is responsible for the thermal effect of the wall.

Here, it is worth again pointing out some remarkable features of the "Nukiyama curve" (Fig. 12.2). The presence of the ascending (nucleate boiling) and descending (transition boiling) compromises the standard concept of "heat transfer intensification". Indeed, the physical effect of conjugation is manifested in the increased average difference of temperatures with fixed (averaged over the entire period) heat flux: $\langle q \rangle = \text{idem}$. Let us now see how the conjugation effect is manifested with fixed averaged difference of temperatures: $\langle \vartheta \rangle = \text{idem}$. Figure 12.1 shows that in this setting the heat flow increases for the "normal" bubble regime, but for the "abnormal" transition boiling regime it is decreasing. So here it better to speak about a "shift to the right" along the temperature scale over the entire boiling curve.

12.11 Conclusions

The principal constituents of the general problem are considered: conditions for the inception of boiling, formation of nucleation sites, boiling regimes (bubble, film, and transition regimes). Growth laws of a vapor bubble in a bulk of liquid and on a rigid surface are described. A microlayer of liquid under a vapor bubble, a macrolayer under vapor conglomerates, and dry spots on the heat surfaces are studied. A brief description of heat transfer models for nucleate boiling is given. These models are based on the bubble dynamics and integral characteristics of the process. A special attention is given to a debating problem on the effect of thermophysical characteristics of a heat-transmitting wall. An approximate model for periodic conjugate heat transfer problem for boiling is given. Calculation results of the conjugation factor for boiling and transition boiling regimes are given.

References

1. Gibbs JW (1906) The scientific papers of J. Willard Gibbs (Longmans Green, London), vol 1, pp 252–258 (Reprinted by Dover, New York, 1961)
2. Vukalovich MP, Novikov II (1948) Equation of state of real gases. Gosenergoizdat, Moscow (In Russian)
3. Bjerre A, Bak TA (1969) Two-parameter equations of state. Acta Chem Scand 23:1733–1744
4. Soave G (1972) Equilibrium constants from a modified Redlich-Kwong equation of state. Chem Eng Sci 27(6):1197–1203
5. Skripov VP (1974) Metastable liquid. Wiley, New York
6. Debenedetti PG (1996) Metastable liquids: concepts and principles. Princeton University Press, Princeton
7. Lahey RT (1992) Boiling heat transfer: modern developments and advances New York Elsevier
8. Stephan K (1992) Heat transfer in condensation and boiling. Springer, Berlin, Heidelberg, New York
9. Zuber N (1958) On the stability of boiling heat transfer. ASME J Heat Transfer 80(2):711–772
10. Labuntsov DA (2000) Physical foundations of power engineering. Moscow Power Energetic Univ, Moscow (In Russian), Selected works
11. Birkhoff G, Margulis R, Horning W (1958) Spherical bubble growth. Phys. Fluids 1:201–204
12. Plesset MS, Zwick SA (1954) The growth of vapor bubbles in superheated liquids. J Appl Phys 25(4):493–500
13. Scriven LE (1959) On the dynamics of phase growth. Chem Eng Sci 10(1/2):1–14
14. Zudin YB (2017) Theory of periodic conjugate heat transfer, 3rd ed. Springer
15. Urbano A, Tanguy S, Huber G, Colin C (2018) A direct numerical simulation of nucleate boiling in micro-layer regime. Int J Heat Mass Transf 123:1128–2113
16. Gorenflo D (2002) Behältersieden (Sieden bei freier Konvektion). Springer, Berlin, VDI—Wärmeatlas, Hab

17. Wayner PC Jr, Coccio CL (1971) Heat and mass transfer in the vicinity of the triple interline of a meniscus. AIChE J 17:569–575
18. Panchamgam SS, Chatterjee A, Plawsky JL, Wayner PC Jr (2008) Comprehensive experimental and theoretical study of fluid flow and heat transfer in a microscopic evaporating meniscus in a miniature heat exchanger. Int J Heat Mass Transf 51:5368–5379
19. Stephan P, Kern J (2004) Evaluation of heat and mass transfer phenomena in nucleate boiling. Int J Heat Fluid Flow 25:140–148
20. Ibrahem K, Schweizer N, Herbert S, Stephan P, Gambaryan-Roisman P (2012) The effect of three-phase contact line speed on local evaporative heat transfer: experimental and numerical investigations. Int J Heat Mass Transf 55:1896–1904
21. Kunkelmann C (2011) Numerical modeling and investigation of boiling phenomena. PhD thesis. Technische Universität Darmstadt
22. Yagov V (2009) Nucleate boiling heat transfer: possibilities and limitations of theoretical analysis. Heat Mass Transf 45:881–892
23. Zudin YB (1993) The calculation of parameters of the evaporating meniscus of a thin liquid film. High Temp 31(5):777–779
24. Pioro IL, Rohsenow W, Doerffer SS (2004) Nucleate pool-boiling heat transfer. I: review of parametric effects of boiling surface. Int J Heat Mass Transf 47: 5033–5044
25. Kenning D, Golobic I, Xing H et al (2006) Mechanistic models for pool nucleate boiling heat transfer: input and validation. Int J Heat Mass Transf 42: 511–527
26. Dhir VK (2006) Mechanistic prediction of nucleate boiling heat transfer–achievable or a hopeless task? ASME J Heat Transfer 123:1–12
27. Dhir VK (2001) Numerical simulations of pool-boiling heat transfer. AIChE J 47:813–834
28. Yu B, Cheng P (2002) A fractal model for nucleate pool boiling heat transfer. ASME J Heat Transfer 124:1117–1124
29. Mandelbrot BB (1982) The fractal geometry of nature. Freeman WH (ed). New York
30. Eanshaw RA (ed) (1993) Application of fractals and chaos. Springer, Berlin
31. Stephan K (1963) Mechanismus und Modellgesetz des Wärmeübergangs bei der Blasenverdampfung. Chem-Ing-Tech 35(11):775–784
32. Stephan K, Abdelsalam M (1980) Heat-transfer correlations for natural convection boiling. Int J Heat Mass Transfer 23:73–87
33. Kenning D, Golobic I, Xing H et al (2006) Mechanistic models for pool nucleate boiling heat transfer: input and validation. Heat Mass Transf 42:511–527
34. Qi Y, Klausner JF (2006) Comparison of nucleation site density for pool boiling and gas nucleation. ASME J Heat Transfer 128:13–20
35. Benjamin RJ, Balakrishnan AR (1997) Nucleation site density in pool boiling of saturated pure liquids: effect of surface microroughness and surface and liquid physical properties. Exp Thermal Fluid Sci 15:32–42
36. Sherman FS (1990) Viscous flow. McGraw-Hill
37. Rose JW (2004) Surface tension effects and enhancement of condensation heat transfer. Trans IChemE, Part a Chem Eng Res Des 82:419–429
38. Wayner PC Jr, Kao YK, LaCroix LV (1976) The Interline heat transfer coefficient on an evaporating wetting film. Int J Heat Mass Transf 19:487–492
39. Straub J (2001) Boiling heat transfer and bubble dynamics in microgravity. Adv. Heat Transfer 35:57–172
40. Ametistov EV, Grigoriev VA, Pavlov YM (1972) Effect of thermophysical properties of heating surface material on heat transfer during boiling of water and ethanol. High Temp 10:821–823
41. Grigoriev VA, Pavlov YM, Ametisov EV, Klimenko AV, Klimenko VV (1977) Concerning the influence of thermal properties of heating surface material on heat transfer intensity of nucleate pool boiling of liquids including cryogenic ones. Cryogenics 2:94–96
42. Labuntsov DA, Zudin YB (1984) Heat-transfer processes of periodic intensity. Energoatomizdat, Moscow (in Russian)
43. Piquet J (2002) Turbulent flows. Models Phys. Springer
44. Cebeci T (2002) Convective heat transfer. Springer, Berlin, Heidelberg, New York

45. Carslaw HS, Jaeger JC (1992) Conduction of heat in solids. Clarendon Press, London, Oxford
46. Dietz C, Henze M, Neumann SO, von Wolfersdorf J, Weigand B (2005) Numerical and experimental investigation of heat transfer and fluid flow around a vortex generator using explicit algebraic models for the turbulent heat flux. In: Proceedings of the 17th international symposium on airbreathing engines, Munich, Germany, Paper ISABE-2005–1197

Chapter 13
Heat Transfer in Superfluid Helium

Symbols

T	Temperature
c_p	Isobaric heat capacity
k	Thermal conductivity
L	Heat of phase transition
p	Pressure

Greek Letter Symbols

ρ	Density
μ	Dynamic viscosity
ζ	Chemical potential

Subscripts

eq	Equilibrium
f	Liquid
n	Normal component
s	Superfluid component
sat	Saturation state
v	Vapor
w	State at surface
λ	State at λ-point

13.1 Introduction

The superfluidity phenomenon of helium, which was discovered in the 1930s, by Kapitza [1], is about 100 years old. The superfluidity, which a macroscopic quantum phenomenon, is related to the formation ("condensation") of a finite number of particles in one quantum state. This condensate of particles features some properties of

Y. B. Zudin, *Non-equilibrium Evaporation and Condensation Processes*,
Mathematical Engineering, https://doi.org/10.1007/978-3-030-67553-0_13

the dissipation-free motion, which are responsible for the superfluidity phenomenon [2]. Below, we shall be concerned only with the superfluidity of Helium-4, which is the most common of the two isotopes of helium in nature. The superfluidity of the second isotope (Helium-3), which is a much more involved insufficiently known problem [3], is beyond the scope of this book. So, below by “helium” we shall mean Helium-4.

Helium is an extremely unusual system. This inert gas condenses only at a few degrees Kelvin and only helium remains a fluid down to absolute zero $T = 0$. Solidification of helium requires a pressure of about 30 atm. Because of this, the phase diagram for helium does not contain the triple point, which is standard for all other substances. At normal pressure, helium boils at 4.2 K, the thermodynamic critical point corresponds to 5.19 K at a pressure of 2.24 atm. The “reluctance” of helium to form crystals can be explained by quantum effects—because of the small mass of atoms and weakness of their interactions, their deflections from the equilibrium position in helium crystal are comparable with the interatomic distance, which leads to the “delocalization” of atoms in the crystal. To a certain extent, the smallness of the amplitude of the zero-point vibrations of atoms in the crystal lattice is analogous to the behavior of electrons in an ordinary metal. A remarkable property of quantum crystals of helium is their ability to generate crystallization waves, which can be looked upon as a dissipationless recrystallization of the surface.

The structure of liquid helium, which was first produced by Kamerlingh Onnes in 1908 (see [2]), is even more startling. A phase transition of the second kind occurs in helium if the temperature drops below ~2.2 K. Kapitza [1] was the first to observe that helium loses its viscosity below this temperature. The name of “superfluidity” for this highly unusual state was also given by Kapitza. It turned out that helium flows through thin capillaries with velocity characteristic for a flow of a perfect liquid, and hence, it does not follow Poiseuille’ law [4]. A phase transition in helium was discovered by Schmitt [2], later in the 1920s, from a distinctive anomaly of the heat capacity (resembling the Greek letter). Because of this peculiarity, this transition is now called the λ-point. To indicate the states of helium, Keesom introduced “He-I” (for $T > T_\lambda$) and “He-II” (for $T < T_\lambda$).

The dependence of the helium heat capacity on the temperature in the range below 0.5 K is similar to Debye’s law $C \sim T^3$, which is very surprising for a liquid medium. This means that in the range below the λ-point, in liquid helium there is an additional transport of heat, whose intensity exceeds that of the process of normal boiling. The experimentally measured thermal conductivity of helium below the λ-point is extremely high and has the following unusual peculiarity: it depends on linear sizes of an experimental setup. Such physical picture is characteristic of a high-purity rigid body in the range of low temperatures when the thermal conductivity is controlled by the scattering of quasiparticles (photons) on the boundaries of the rigid body.

The coincident (cubical) form of the temperature dependences on the heat capacity and thermal conductivity ($C \sim k\sim T^3$) is also indicative of the specific nature of the scattering mechanism for “heat carriers” on the body boundaries. According to [2], at absolute zero any thermodynamic system is at its principal quantum–mechanical state. Properties of such systems can be described by the application of the method of small perturbations to the distribution of particles over the states. This enables one to describe thermodynamic systems consisting of many particles using "quasiparticles" (elementary excitations of quantum liquid [5]). Physically, a quasiparticle reflects some collective motion of all real particles in liquid. Liquid helium is a classical example of realization of the conceptions of quasiparticles.

13.2 Practical Applications of Superfluidity

The extremely low-temperature level of liquid helium allows one to use it for efficient cooling of electronic devices. Unique cooling properties of liquid helium have been used in space infrared telescope and in ESA's Infrared Space Observatory. The use of He-II as a cryogenic agent has the following important advantages: high values of critical heat flux, independence of heat emission and boiling crisis on surface orientation, the absence of vapor phase up to high thermal loads. Featuring high intensity of heat transfer, He-II provides a reliable contact between elements of cooling systems. Due to its superfluidity, He-II is capable of penetrating in very small channels, thereby increasing the efficiency of cooling.

Heat-exchange applications of superfluidity call for the creation of principally new theoretical models. At the same time, the available at present bank of experimental data is insufficient to formulate a theory of heat transfer in He-II. Another unique property of He-II is that the proper thermal resistance of this liquid is negligible due to the high efficiency of heat transfer. In these conditions, a definite role in the two-phase heat transfer is played by molecular-kinetic phenomena on the phase interface.

13.3 Two-Fluid Model of Superfluid Helium

According to the classical experiment by Andronikashvili with a rotating pendulum [6], the fraction of mass of liquid helium not taken by the pendulum increases monotonically if the temperature decreases below the λ-point. The entire liquid becomes superfluid only at ultralow temperatures. This means that the fraction of the medium entrained by perturbations behaves like a normal fluid, which decelerates because of friction with the wall. In turn, the superfluid fraction of liquid should exist not only at zero, but also at nonzero temperature.

These physical considerations underlie Landau's two-fluid model of He-II flow [7]. Landau's theory is based on the conjecture that in He-II, two kinds of motion exist: the dissipation-free (superfluid) motion with velocity v_s and the normal (viscous) motion with velocity v_n. The fraction of fluid entrained by perturbations (the normal component) has density ρ_n. The remaining part of the liquid, which is not involved in normal motion, has density ρ_s of the superfluid component. It is assumed that both the components depend monotonically on the temperature, and moreover, $\rho_s = 0$ at the λ–point and $\rho_n = 0$ if the temperature is zero.

Landau [7] emphasized that the interpretation of a liquid as a mixture of two components is only a convenient way for the description of the superfluidity phenomenon and does not mean that the flow of liquid indeed splits into two separate parts. One can assert that such interpretation stems from the traditional problem of insufficiency of standard stereotypes for the description of quantum phenomena. Since the flow of the superfluid component is not subject to external perturbations, it follows from the definition that it cannot transport thermal energy. As a result, the heat flux (transport of entropy) in liquid helium is related to the motion of the normal component only. This important conclusion enables one to find out to which reference frame there corresponds the observed spectrum of energy excitations of superfluid helium. By the setup of the experiment, the perturbations are generated in the superfluid component, and hence the spectrum of excitations is measured in the frame in which the superfluid component is at rest.

13.4 Peculiarities of "Boiling" of Superfluid Helium

13.4.1 Some Properties of Superfluid Helium

Consider the phase diagram of He-II (Fig. 13.1). There are two liquid phases in helium. For temperatures ranging from the critical point ($T_{cr} = 5.2$ K, $p_{cr} = 2.27 \cdot 10^5$ Pa) to the so-called λ-point on the saturation line ($T_\lambda = 2.17$ K, $p_\lambda = 0.05 \cdot 10^5$ Pa), liquid helium (He-I) features all properties of ordinary liquids. For temperatures below the λ-point, helium experiences a second-order phase transition without generation or absorption of heat and changes to the He-II state. Due to the pressure

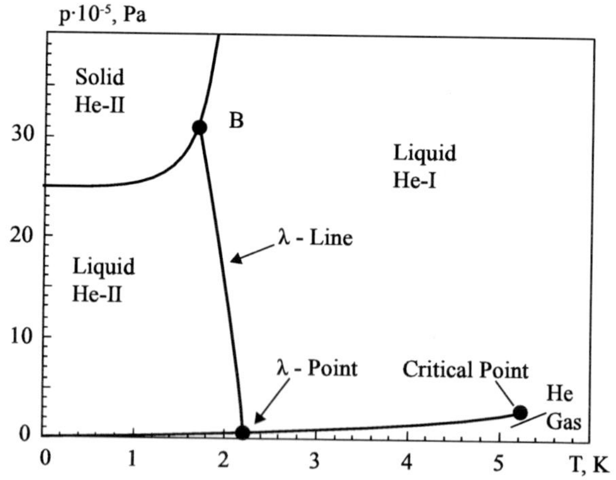

Fig. 13.1 Helium phase diagram

of saturated vapors at normal pressure, He-II does not solidify at any arbitrarily low temperature (up to the absolute zero). It proves possible to obtain solid helium only at pressures exceeding $25 \cdot 10^5$ Pa.

The point B of intersection of the line λ (which separates two phases of helium, He-I and He-II) with the curve bounding the rigid states, is characterized by the parameters$T_B = 1.176$ K, $p_B = 30.13 * 10^5$ Pa. As the temperature becomes smaller than the λ-point, liquid helium (He-II) becomes a quantum liquid, i.e., a liquid, whose macroscopic volume manifests quantum properties of its atoms. In this state, liquid helium features extremely high "thermal conductivity" (which is by many million times greater than that of He-I and exceeds by almost three orders than that of pure silver), considerable heat capacity, and vanishingly small viscosity (superfluidity).

13.4.2 Heat Transfer in Superfluid Heilum

According to experimental data, the intensity of heat transfer from He-II is commensurable with the nucleate boiling heat transfer in He-I. Because of exceedingly high thermal conductivity, a heat transfer in He-II is not accompanied by the formation of vapor bubbles. Hence, the dependence of the heat flux q on the difference of temperatures ΔT is linear, $q \sim \Delta T$ (this recalls the single-phase convection). This dependence features no principal differences from the standard (for bubble boiling) cubic dependence of the form $q \sim \Delta T^3$, which is also characteristic of He-I. However, the mechanism of heat removal from the wall for He-II is cardinally different from that in the He-I setting.

Consider the curve $q(\Delta T)$ from Fig. 13.2. The position of this curve depends, in general, on the size and configuration of the heater (a flat surface, wires, channels), the depth of its immersion into the volume, and the saturation temperature T_{sat}. Two zones

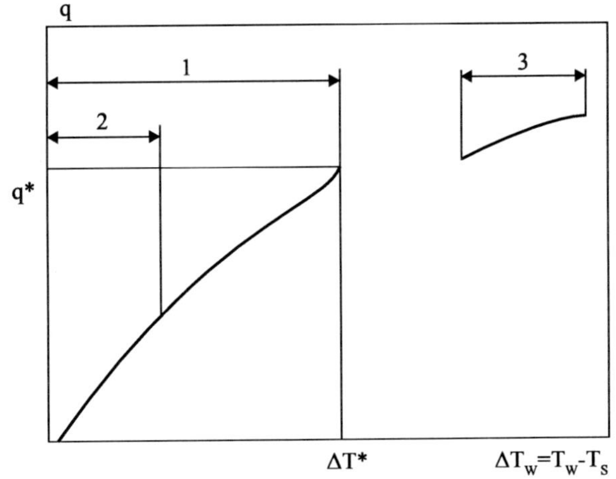

Fig. 13.2 Typical curve of He-II heat transfer. 1 Filmless boiling, 2 Kapitza resistance regime, 3 film boiling

can be singled out in Fig. 13.2. The first zone (1) describes the "filmless boiling". Some part of it is occupied by zone (2) (the "Kapitza resistance regime"). In this regime, the density of the heat flux is proportional to the difference of temperatures of the heat surface T_w and of the liquid T_{sat}, $\Delta T = T_w - T_{sat}$. Such a dependence is characteristic of small densities of heat flux and is bounded by the difference of temperatures $\Delta T \leq 2$ K. The dependence $q(\Delta T)$ ceases to be linear as ΔT increases. As the heat flux attains some maximal value ($q = q^*$) in the "filmless boiling" region, there appears a different stage of heat transfer (zone (3)). This stage, in analogy with the boiling of conventional liquids, is known as "film boiling". Here, the character of heat transfer is similar (by external characteristics) to the conventional film boiling. However, the intensity of heat transfer, as well as the quantity q^*, depends on the depth of immersion of the heater in liquid and its geometry.

The extremely high thermal conductivity of He-II can be explained from the Landau's [7] phenomenological two-fluid model. According to this theory, for any temperature, the density of liquid He-II can be looked upon as consisting of the density of the normal component ρ_n and the density of the superfluid component ρ_s, the total density for any temperature remaining constant (Fig. 13.3)

$$\rho = \rho_n + \rho_s = \text{const.} \tag{13.1}$$

It is assumed that the properties of the normal component are analogous to those of He-I "diluted" by the second component, whereas there are no thermal disturbances in the superfluid component. The mechanism of heat transfer to He-II can be represented as an oppositely directed motion of two components: from the heat surface, there is a flow of normal component, and toward it comes the flow of the superfluid component, which is equal to it in absolute value. Besides, there is no total macroscopic flow in He-II.

So, the presence of the temperature gradient in He-II triggers a thermal counterflow (an internal convection of two interpenetrating components) in it. Hence, the heat

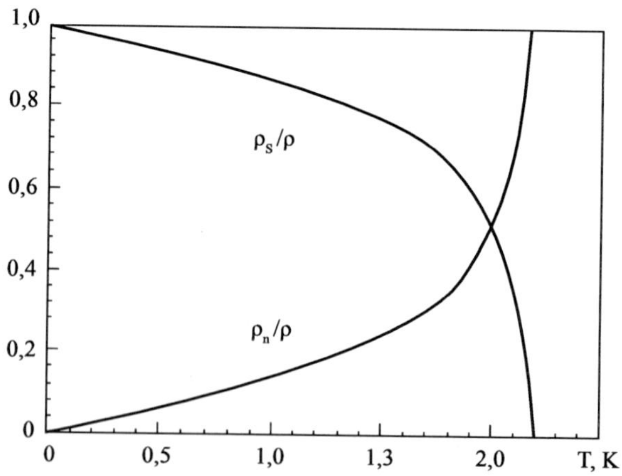

Fig. 13.3 Dependence of the specific normal and superfluid densities on the temperature

transfer of He-II "at rest" (i.e., with zero total mass rate) is attributable not only to the diffusion process. In view of the presence of the superfluid component of liquid, the difference of temperatures results in the appearance in the mass of He-II of powerful convective flows. We note that He-II is not subject to Fourier's law. Hence, in the context of his He-II, the concept of "thermal conductivity" is used only to compare it with conventional media. So, if the "thermal conductivity" is formally computed, then it turns out that this coefficient highly depends on the temperature gradient.

13.4.3 Kapitza Resistance

Because of the exceptionally high thermal conductivity of He-II, the thermal resistance of a helium layer is extremely small and there is practically no temperature gradient over the liquid column height on the heat surface. Under these conditions, the process of heat transfer is controlled solely by the thermal resistance on the rigid body–He-II boundary. The existence of contact thermal resistance on the boundary of He-II with a rigid body was first experimentally discovered in 1941, by Kapitza [1]. This effect is now known as the "Kapitza resistance". The Kapitza resistance, which appears in a thin (~0.01 mm) layer in the phase interface, is realized in the form of a temperature jump between the rigid body and He-II. Kapitza explained this jump by the presence of some hypothesized temperature layer, which is characterized by the quantity $h_0 = q/\Delta T$. In essence, h_0 is the heat transfer coefficient from the rigid wall to He-II (the "Kapitza resistance"). It is easily seen that this quantity is the inverse of the Kapitza resistance

$$R_0 \equiv h_0^{-1} = \frac{\Delta T}{q}. \tag{13.2}$$

Later, it was found that the Kapitza resistance is caused by molecular-kinetic phenomena on the phase interface.[1] We note that the heat transfer to He-II from a heat surface for small heat fluxes and temperature differences (not exceeding 0.05 K) is characterized by a constant value of R_0. Theoretical justification of the jump of temperatures on the rigid body–He-II boundary was proposed in 1952, by Khalatnikov [8]. It was assumed that, for all temperatures below the λ-point, the heat transfer on the interphase boundary proceeds in two ways. On the one hand, the energy is transferred from the hot surface to liquid helium by radiation of quasiparticles (phonons), which are quanta of the oscillatory motion of atoms (of the rigid body or liquid). On the other hand, the reverse process of absorption by the rigid body surface of photons from He-II takes place.

The energy flow appearing on the interphase boundary is controlled by the difference between these two flows: the one directed from the rigid body towards liquid and the one from liquid to the rigid body. So, according to [8], the Kapitza resistance

[1] In usual conditions, the phase thermal resistance is always negligible.

is the resistance to the passage of heat photons moving from both sides of the interphase boundary. It can be shown that the resulting density of heat flux is proportional to the difference of the fourth powers of temperatures of the rigid body and liquid helium

$$q \sim T_w^4 - T_{sat}^4. \tag{13.3}$$

For small differences in temperatures ($\Delta T \ll T_{sat}$), we have

$$q = h_0 \Delta T. \tag{13.4}$$

The quantity h_0 is proportional to T_{sat}^3 and depends on the acoustic and elastic properties of a rigid body and liquid. The following relation was obtained in [8] for the heat transfer coefficient

$$h_0 = 9.7 * 10^5 \frac{\rho_f w_f F T_{sat}^3}{M \Theta_D^3}. \tag{13.5}$$

Here, ρ_f is the density of liquid helium, w_f is the velocity of the first sound in liquid helium, $F(w_t/w_e)$ is the function of elastic constants of the rigid body, w_t, w_e are, respectively, the velocities of the longitudinal and transverse propagation of sound in a rigid body, M is the molecular mass of the rigid body, Θ_D is the Debye temperature for which all oscillation modes are excited (from zero to the maximal frequency of oscillations).

Assume that the mechanism of heat transfer which is based on the concept of Kapitza resistance is extended to arbitrary values of the difference of temperatures. Then the heat transfer coefficient in the region $\Delta T \approx T_s$ can be written as

$$h = h_0 \left(1 + \frac{3}{2}\theta + \theta^2 + \frac{1}{4}\theta^3 \right), \tag{13.6}$$

where $\theta = \Delta T / T_{sat}$.

Dependence (13.6) can be used to find the heat transfer coefficient for He-II in the range of high fluxes (up to critical ones). To this end, we need to know the values of h_0 that were obtained experimentally for small values of ΔT or by the theoretical formula (13.5). Experimental data for h give slightly underestimated values relative to (13.6) (up to 40% with critical density of the heat flux). A probable cause of such a departure is the appearance of the layer of normal component between the heat surface and the superfluid helium (if the wall temperature exceeds T_λ). This layer can bring additional thermal resistance and hence decrease the rate of heat transfer. According to experimental data, in this temperature range, there is an effect of the size of a heating element on the process of transfer of heat from He-II. This is the evidence of the considerable role of buoyant forces (and moreover, for high-temperature differences, of the effect of the depth of immersion of a specimen). We note that the last effect is possible only if there is a vapor phase on the heat surface.

13.5 Theory of Laminar Film Boiling of Superfluid Helium

13.5.1 Laminar Film Boiling

The process of film boiling of He-II also has very specific features. According to experimental studies, as distinct from the case of film boiling of conventional liquids, in the case of He-II, the thickness of the vapor film does not practically vary in time. Because of this, no characteristic features of film boiling are manifested in this setting: accumulation of vapor in the film and departure of vapor bubbles.

At present, there are two different physical interpretations of this anomaly. In [9], it is assumed that film boiling of He-II is accompanied by heat flux q_b removed from the vapor–liquid interface inside the volume of He-II. The quantity q_b is interpreted in [9] as a specific characteristic of He-II, which depends on the thermodynamic parameters of the system (the pressure and saturation temperature T_s in a two-phase system) and on the depth of immersion H of a heater in the liquid. At the same time, q_b is independent of the form and size of the heater, the heat flux density on its surface, and the surface temperature T_w. According to the model of [9], the heat flux q_v supplied from vapor to the interphase boundary has some limit value q_b. If this value is exceeded, then the excess heat $|q_v| - |q_b|$ is spent on the formation of vapor j on the interphase boundary

$$|q_v| - |q_b| = Lj. \tag{13.7}$$

Here, L is the heat of phase transition. For $|q_v| - |q_b| < 0$, relation (13.7) controls the condensation rate.

An alternative physical interpretation of the heat flux q_b was proposed by Labuntsov [10] based on the analysis of molecular-kinetic effects on the interphase boundary. The approach of [10] is based on relations of the linear kinetic theory of evaporation and condensation [11], which enable one to establish a link between flows of mass j, momentum Δp, and heat q_v on the interphase boundary. These relations in the dimensionless form can be written as

$$\Delta \tilde{p} + 2\sqrt{\pi}\frac{1-0.4\beta}{\beta}\tilde{j} + 0.44\tilde{q} = 0. \tag{13.8}$$

Here, $\tilde{j} = j/\rho_v\sqrt{2R_g T_{sat}}$, $\Delta\,\tilde{p} = (p_v - p_{sat})/p_v$, β is the evaporation–condensation coefficient, the values of the flows of mass $\tilde{j}$ and heat $\tilde{q}$ in (13.8) are positive if they are directed inside the vapor phase.

The value $\Delta\,\tilde{p}$ is equal to the dimensionless difference of the physical pressure of vapor in the film and the theoretical saturation pressure corresponding to the temperature of liquid (condensed) phase. As there is no accumulation of vapor in the film, one should put $\tilde{j} = 0$ in Eq. (13.8). The difference of pressures in height of the film is due to the weight of the liquid column of height H: $p_v - p_{sat} = \rho_l g H$. Now using (13.8), we get

$$q^* = 2.27\rho_f g H\sqrt{2R_g T_{sat}}. \tag{13.9}$$

If the heat flux supplied from vapor to the boundary exceeds q^* in absolute value, then from (13.8), it follows that evaporation occurs ($j > 0$). For $|q_v| < q^*$ the vapor condenses ($j > 0$). Thanks to such self-regulation (i.e., the negative feedback) the thickness of the film is set automatically so that the heat flux q_v be equal to q^*.

Let us consider the case of laminar film boiling of superfluid helium on heated surfaces of two types: a cylinder and a vertical plate (Fig. 13.4). The thickness of the vapor film is assumed to be constant in time, but it can in general vary over the cylinder circumference and the plate height. The thickness of the vapor layer is minimal near the lower generator of the cylinder. It is assumed that the maximal thickness of the vapor layer is much smaller than the cylinder diameter. The vapor temperature in the film varies from T_w (the heating surface temperature) to T_s' (the interphase temperature close to the temperature T_{sat}), see Fig. 13.5.

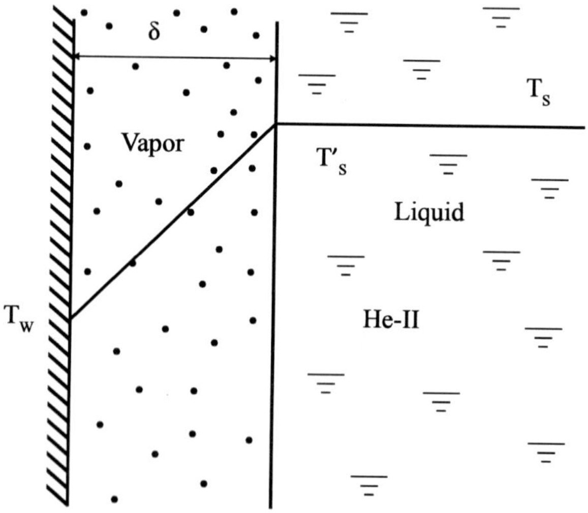

Fig. 13.4 Distribution of temperatures in the boundary layer for film boiling of He-II

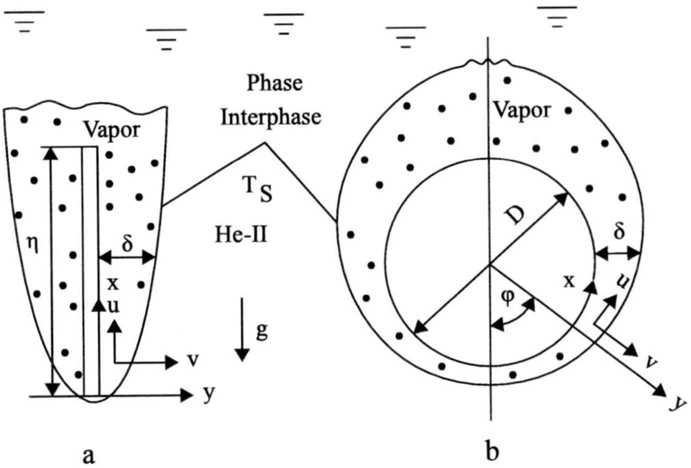

Fig. 13.5 Film boiling of He-II vertical plate (**a**) horizontal cylinder (**b**)

The temperature difference $\Delta T^{'} = T^{'}_{\text{sat}} - T_{\text{sat}}$, which is proportional to the heat flux density q at the interphase boundary, is negligible in comparison with the total temperature difference $\Delta T = T_w - T_{\text{sat}}$. Hence, it can be assumed with good approximation that the temperature in the film varies continuously from T_w to T_{sat}. The effect on q^* of the thermal resistivity to heat transfer on the interphase boundary from the side of liquid (i.e., the inequalities $T^{'}_{\text{sat}} \neq T_{\text{sat}}$) is achieved by introducing into expression (13.9) an empirical constant of order approximately 1.

We write the conservation equations for energy, mass, and momentum for an element of the vapor boundary layer (Fig. 13.5) as follows

$$q_w - q_{\text{v}} = c_{p\text{v}} \frac{d}{dx}\left(\int\limits_0^{\delta} \rho_{\text{v}} u (T - T_{\text{sat}}) dy \right), \tag{13.10}$$

$$(\rho_{\text{v}} \upsilon)_{\text{sat}} = -\frac{d}{dx}\left(\int\limits_0^{\delta} \rho_{\text{v}} u dy \right), \tag{13.11}$$

$$g_x \rho_f + \mu_{\text{v}} \frac{\partial u}{\partial y} = 0. \tag{13.12}$$

Here x, y are, respectively, the longitudinal (along the surface) and lateral coordinates, u, υ are the velocity components in the directions of x, y, respectively, g_x is the projection of the acceleration due to gravity to the x-axis, μ_{v} is the vapor viscosity, $c_{p\text{v}}$ is the isobaric heat capacity of vapor, ρ_{v}, ρ_f are the densities, respectively, of the vapor and liquid phases, the index "w" corresponds to the heating surface, index "s" means the direction to the phase interface.

The boundary condition on the He-II—vapor phase interface can be obtained from Eq. (13.8) by taking into account that $\Delta\,\tilde{p} + 0.44\,\tilde{q} = 0$ for $\tilde{j} = 0$

$$q_{\mathrm{v}} = q^* - 8.05\frac{1-0.4\beta}{\beta}p_{\mathrm{v}}\left(\frac{j}{\rho_{\mathrm{v}}}\right)_{\mathrm{sat}}, \tag{13.13}$$

or

$$q_{\mathrm{v}} = q^* - \mathrm{const}(p_{\mathrm{v}}\upsilon)_{\mathrm{sat}}. \tag{13.14}$$

From expression (13.11), we get

$$\rho_{\mathrm{vsat}}\upsilon_{\mathrm{sat}} = -\frac{d}{dx}\left(\int_0^\delta \rho_{\mathrm{v}}u\,dy\right), \tag{13.15}$$

where $\rho_{\mathrm{vsat}} \equiv \rho_{\mathrm{v}}(T_s, p_{\mathrm{v}})$ is the density of the saturated vapor at temperature T_s. Substituting (13.14) and (13.15) into expression (13.10), gives

$$q_w - q^* = \frac{d}{dx}\left\{\int_0^\delta\left[\mathrm{const}\frac{p_{\mathrm{v}}\rho_{\mathrm{v}}}{\rho_{\mathrm{vsat}}}u + c_{p\mathrm{v}}\rho_{\mathrm{v}}u(T - T_{\mathrm{sat}})\right]dy\right\}. \tag{13.16}$$

Assuming that the phase interface temperature is much larger than the saturation temperature ($T_w \gg T_{\mathrm{sat}}$), we get $T - T_{\mathrm{sat}} \approx T$. As a result

$$c_{p\mathrm{v}}\rho_{\mathrm{v}}u(T - T_{\mathrm{sat}}) \approx \frac{5}{2}p_{\mathrm{v}}u,\quad \frac{p_{\mathrm{v}}\rho_{\mathrm{v}}}{\rho_{\mathrm{vsat}}}u \ll \frac{5}{2}p_{\mathrm{v}}u.$$

Taking into account the above transformations, expression (13.16) can be written in the form

$$q_w - q^* = \frac{5}{2}p_{\mathrm{v}}\frac{d}{dx}\langle u\delta\rangle. \tag{13.17}$$

Here, $\langle u\delta\rangle \equiv \int_0^\delta u\,dy$ is the average volumetric flow rate of vapor in the film. Integrating (13.17), gives

$$\langle u\delta\rangle = \frac{2}{5}\frac{q_w - q^*}{p}x. \tag{13.18}$$

To find $\langle u\delta\rangle$, we employ the motion equation of vapor in the film (13.12). On this equation, one does not take into account the inertial forces, because such forces do

not introduce considerable error for the process of film boiling. The dependence of the vapor thermal conductivity and viscosity in the film on the temperature can be expanded into power functions

$$k_v = aT^n, \mu_v = bT^n. \tag{13.19}$$

Here the exponent n is not constant in a wide range of temperatures and varies from approximately 1 (with very low temperatures corresponding to He-II temperatures) to 0.8–0.6 (at the wall temperature). It can be shown that the variation of the exponent "n" in these limits has no substantial effect on the final result. With this approximation, we get $k_v/\mu_v \approx$ const. Using the Fourier law

$$q = -k_v \frac{dT}{dy} \tag{13.20}$$

in (13.12), we change from the independent variable y to the variable T. We also assume that for the regimes of film boiling for which q_w is not too far from q^*, the value of q in (13.20) is approximately constant and is equal to the mean value

$$q \approx q_m = \frac{1}{2}(q_w + q^*). \tag{13.21}$$

Transforming Eq. (13.12) with due account of the above assumptions, we get a relation for u and T

$$g_x \rho_f + \frac{q_m^2}{k_v} \frac{\partial}{\partial T}\left(\frac{\mu_v}{k_v} \frac{\partial u}{\partial T}\right) = 0. \tag{13.22}$$

The boundary-value conditions for Eq. (13.22) assume the form

$$y = 0, T = T_w : u = 0,$$
$$y = \delta, T = T_s : u = 0.$$

Integrating (13.22), we get the following profile of velocities in the function on the vapor temperature in the film

$$u(T) = \frac{g_x \rho_f k_v}{q_m^2 \mu_v} \frac{T_w - T}{T_w - T_{sat}} \left[\int_{T_{sat}}^{T_w} \left(\int_T^{T_w} k_v dT \right) dT - \int_T^{T_w} \left(\int_T^{T_w} k_v dT \right) dT \right]. \tag{13.23}$$

Using (13.19), (13.20) and denoting $\theta = T/T_w$, we find that

$$\langle u\delta\rangle \equiv \int_0^{\delta} u dy = \frac{1}{q_m}\int_T^{T_w} k_v u dT, \tag{13.24}$$

$$k_v = k_{vw}\left(\frac{T}{T_w}\right)^n = k_{vw}\theta^n. \tag{13.25}$$

Substituting (13.23) and (13.25) into expression (13.24), we get after some algebra

$$\langle u\delta\rangle = \frac{g_x \rho_f T_w^3}{q_m^3} I, \tag{13.26}$$

where

$$I = \int_{\theta_{sat}}^{1}\left(\frac{1-\theta}{1-\theta_{sat}}\int_{\theta_{sat}}^{1} d\theta \int_{\theta}^{1} \theta^n d\theta - \int_{\theta}^{1} d\theta \int_{\theta}^{1} \theta^n d\theta\right) d\theta. \tag{13.27}$$

According to the above estimates, for $\theta_{sat} \ll 1$, we get

$$I \approx \frac{1}{(n+2)^2(2n+3)}. \tag{13.28}$$

Now from (13.26), we find that

$$\langle u\delta\rangle = \frac{1}{(n+2)^2(2n+3)} \frac{g_x \rho_f k_{vw}^3 T_w^3}{q_m^3 \mu_{vw}}. \tag{13.29}$$

Using (13.25) and taking into account $q \approx q_m$, Eq. (13.20) assumes the form

$$q_m = \frac{1}{n+1}\frac{k_{vw} T_w}{\delta}\left(1 - \theta_{sat}^{n+1}\right) \approx \frac{1}{n+1}\frac{k_{vw} T_w}{\delta}. \tag{13.30}$$

Hence, we have

$$\langle u\delta\rangle = \frac{(n+1)^3}{(n+2)^2(2n+3)} \frac{g_x \rho_f \delta^3}{\mu_v}. \tag{13.31}$$

From expression (13.30), we get

$$\frac{T_w}{T_{sat}} = \left[\frac{(n+1)q_m}{k_{vsat} T_{sat}}\right]^{1/(n+1)}. \tag{13.32}$$

Changing in (13.31) from μ_{vw} to μ_{vs} by the formula

$$\mu_{vw} = \mu_{vsat}\left(\frac{T_w}{T_{sat}}\right)^n, \tag{13.33}$$

and taking into account (13.32), we finally get

$$\langle u\delta\rangle = \frac{(n+1)^3}{(n+2)^2(2n+3)} \frac{g_x \rho_f \delta^{3-n/(n+1)}}{\mu_{vsat}[(n+1)q_m/(k_{vsat}T_{sat})]^{n/(n+1)}}. \tag{13.34}$$

Relation (13.34) for the average volumetric flow rate of vapor in the film in the notation of Fig. 13.4 is common both for a cylinder and for a plate. From this relation, one can find calculation relations for the heat transfer, which for these geometries are different. We give concrete results for bodies of various geometry.

13.5.2 Cylinder

We denote by φ the angle between the direction of the force of gravity and the cylinder radius (Fig. 13.4b). We can write $x = (D/2)\varphi$, $g_x = g\sin\varphi$. From (13.17) and (13.34), we get the expression for the thickness of the vapor film

$$\delta = N^{(n+1)/(2n+3)}\left(\frac{\varphi}{\sin\varphi}\right)^{(n+1)/(2n+3)}, \tag{13.35}$$

where

$$N = D\frac{(q_w - q^*)\mu_{vsat}}{p_w g \rho_f}\left[\frac{(n+1)q_m}{k_{vsat}T_{sat}}\right]^{n/(n+1)} \frac{(n+2)^2(2n+3)}{5(n+1)^3}. \tag{13.36}$$

The average (over the circumference) temperature of the cylinder is as follows:

$$\langle T_w\rangle = \frac{1}{\pi}\int_0^\pi T_w d\varphi. \tag{13.37}$$

Using (13.32) and (13.35), we get after some algebra

$$\langle T_w\rangle = f_1 T_{sat}\left[\left(\frac{D\mu_{vsat}(q_w - q^*)}{p_w g \rho_f}\right)^{1/3} \frac{q_w + q^*}{k_{vsat}T_{sat}}\right]^{3/(2n+3)}, \tag{13.38}$$

where $f_1 = f_0\left[\frac{(n+2)^2(2n+3)}{40}\right]^{1/(2n+3)}$, $f_0 = \frac{1}{\pi}\int_0^\pi\left(\frac{\varphi}{\sin\varphi}\right)^{1/(2n+3)} d\varphi$.

We have $T_w >> T_{sat}$, and hence the average temperature difference can be given as $\Delta T = \langle T_w \rangle - T_{sat} \approx \langle T_w \rangle$. The final expression for the average temperature difference with film boiling of He-II on horizontal cylinders reads as

$$\Delta T = f_1 T_{sat} \left[\left(\frac{D\mu_{vsat}(q_w - q^*)}{\rho_w g \rho_f} \right)^{1/3} \frac{q_w + q^*}{k_{vsat} T_{sat}} \right]^{3/(2n+3)}. \tag{13.39}$$

13.5.3 Plate

For a plate we have $g_x = g$, and hence, from (13.17) and (13.34), after integration, one easily finds the thickness of the vapor film

$$\delta = N_1^{(n+1)/(2n+3)} \eta^{(n+1)/(2n+3)}, \tag{13.40}$$

where η is the plate height, N_1 is defined from the expression

$$N_1 = \frac{2}{5} \frac{(n+2)^2(2n+3)}{(n+1)^3} \frac{(q_w - q^*)\mu_{vs}}{\rho_w g \rho_f} \left[\frac{(n+1)q_m}{k_{vsat} T_{sat}} \right]^{n/(n+1)}. \tag{13.41}$$

The average (over the height) temperature of the plate reads as

$$\langle T_w \rangle = \frac{1}{\eta} \int_0^\eta T_w dx. \tag{13.42}$$

Using (13.32), (13.35) and assuming that $\Delta T = \langle T_w \rangle$ (as in the case of a cylinder) one can find the sought-for expression for the average temperature difference for film boiling of He-II on the surface of a vertical flat plate

$$\Delta T = f_2 T_{sat} \left\{ \left(\frac{\eta\mu_{vsat}(q_w - q^*)}{\rho_w g \rho_f} \right)^{1/3} \frac{q_w + q^*}{k_{vsat} T_{sat}} \right\}^{3/(2n+3)}, \tag{13.43}$$

where

$$f_2 = \frac{1}{2} \left[\frac{(2n+3)^{2n+4}}{20(n+2)^{2n+1}} \right]^{1/(2n+3)}.$$

In (13.39) and (13.43), the vapor parameters μ_{vsat}, k_{vsat} are defined at the liquid saturation temperature. Using (13.25), (13.33) one can obtain expressions for ΔT,

in which these values are determined at the wall temperature: for the cylinder

$$\Delta T = f_3 \left(\frac{D\mu_{vw}(q_w - q^*)}{p_w g \rho_f} \right)^{1/3} \frac{q_w + q^*}{k_{vw}}, \tag{13.44}$$

and for the plate

$$\Delta T = f_4 \left(\frac{\eta\mu_{vw}(q_w - q^*)}{p_w g \rho_f} \right)^{1/3} \frac{q_w + q^*}{k_{vw}}. \tag{13.45}$$

Here

$$f_3 = k_1{}^{(2n+3)/3},\ f_4 = f_2{}^{(2n+3)/3}.$$

According to the above estimates, a variation of the exponent in the temperature dependences for μ_{vs}, k_{vs} in the range $0.6 \leq n \leq 1$ has practically no effect on the results of calculations. The resulting summary formulas (13.39), (13.43)–(13.45) are fairly simple and do not contain any empirical coefficients (obtained from experimental findings). From the practical point of view, (13.39) and (13.43) are more advantageous, because the quantities involved in them are determined at the saturation temperature T_s and the dependence $\Delta T(q)$ assumes an explicit form. These relations were obtained in the case when the heat transfer in the film is in the free-convection conditions.

13.6 Thermodynamic Principles of Superfluid Helium

13.6.1 Two-Fluid Model

Let us consider thermodynamic principles of superfluid helium in the frameworks of two-fluid flow model of He-II. Qualitatively, this model was first proposed by Tisza [12] and London [13]. Later to the quantitative level, it was extended by Landau [7]. Nowadays, when lengthy highly specialized studies are widespread, one cannot but be astonished by the fact that the novel ideas that pioneered the new intriguing direction in theoretical physics took only one page by Tisza, two pages by London, and three pages by Landau.

According to Tisza and London's conjecture, each particle of liquid contains a definite number of atoms of two sorts (normal "*n*" and superfluid "*s*" ones). Landau's model postulates the coexistence (at finite temperature) of two kinds of motion of liquid: superfluid flow and normal viscous flow. According to Landau's theory [7, 14], for small velocities, the flow of liquid is nondissipative (i.e., without manifestation of viscosity effects), while superfluid helium is looked upon as a simple thermodynamic system. In Landau's model, one also derives the critical flow velocity which triggers spontaneous nucleation of quasiparticles called "roton" (ring vortices of atomic size). Besides, the superfluid component loses momentum, which is equal to the momentum of emitted protons, and hence is decelerated. As the critical velocity is exceeded, the superfluidity of liquid becomes degenerated.

We note, however, that the experimentally obtained value of the velocity for which the superfluidity "breaks out" turned out to be smaller by an order than that of Landau. According to Feynman's conjecture [15], there exist states when the circulation of the superfluid component is not zero. In the simplest case of cylindrical geometry, such a state is a vortex. The motion velocity of the superfluid component (depending on the distance to the vortex axis) may exceed the Landau critical velocity. Hence, the core of the vortex is already filled not from the superfluid component, but rather from the normal component of liquid. According to Feynman, in real experiments the nucleation of vortices on inhomogeneities of the surface may occur at flow velocities smaller than the critical velocity predicted by Landau's theory, and hence, it may bound the domain of existence of superfluidity. In the equilibriums case, the internal relative motions of the normal "n" and superfluid "s" components of liquid are absent: $\vec{w} = \vec{\upsilon}_n - \vec{\upsilon}_s = 0$. Assume now that internal flows "$n - s$" are excited in liquid so that their relative velocity is not zero: $\vec{w} \neq 0$.

According to the two-fluid model, if the velocity of the internal "$n-s$"-counterflow is smaller than some definite values, then in liquid there is no special resulting dissipation. Under these conditions, the relaxation time of "$n - s$" flows should be quite large. This allows us to consider He-II as an "involved" thermodynamic system and to interpret states with "$n - s$" motions as quasi-equilibrium motions. As additional parameters of this involved system, one should take the characteristics of the internal "$n - s$"-counterflow [16]. Because of the exotic nature of superfluidity, one should expect that thermodynamics of the involved state of He-II should also be unconventional. According to [2, 6], the two-fluid model is a phenomenological (macroscopic) theory. At the same time, the closed formulation of the equations of the two-velocity hydrodynamics [16], as well as the analysis of the energy (entropy) transfer phenomena in He-II calls for the involvement of the microscopic theory.

Landau's two-fluid model [14] involves the following thermodynamic equation for the increment of the energy of a unit volume of liquid E_0

$$dE_0 = \zeta d\rho + T d(\rho s) + \vec{w} d\left(\rho_n \vec{w}\right). \tag{13.46}$$

Here, ζ is the chemical potential, ρ_n is the density of the normal component. It is worth specially noting that relation (13.46) is legitimate only in the special frame K_0 moving with the superfluid part of the liquid. In this frame by definition, we always have $\vec{\upsilon}_s = 0$, $\vec{w} = \vec{\upsilon}_n$. In [14], one can also find an expression relating the pressure P of liquid with the chemical potential

$$P = \rho\zeta - E_0 + T\rho s + \rho_n w^2. \tag{13.47}$$

Equations (13.46) and (13.47) underlie the construction of the specific thermodynamics of He-II given in [14]. From these relations, one derives the thermodynamic identity for the chemical potential

$$\rho d\zeta = -\rho s dT + dP - \rho_n \vec{w} d\vec{w}. \tag{13.48}$$

From (13.48), it follows that the vector velocity $\vec{w}$ and the vector momentum $\rho_n \vec{w}$ of normal motion are the conjugate parameters of the system, which reflect the internal "$n - s$" motions in this variant of He-II thermodynamics.

Donnelly [6] put forward a different variant of He-II thermodynamics. The mathematical description of [6] involves a system of differential equations of conservation of mass, total momentum of liquid, and entropy, as well as the equation of conservation of the "s"-momentum of the flow. The balance equation for the total kinetic "$n - s$" energy of motion in an arbitrary system of coordinates reads as

$$E_k = \frac{1}{2}\left(\rho_n v_n^2 + \rho_s v_s^2\right). \tag{13.49}$$

Donnelly [6] obtained the following expression for the total energy of a unit volume of liquid

$$E_0 = \frac{1}{2}\left(\rho_n v_n^2 + \rho_s v_s^2\right) + \rho e. \tag{13.50}$$

As a result of the detailed mathematical analysis, it was shown that E_0 satisfies the law of conservation of total energy only if the thermodynamic identities hold

$$de = Tds - Pd\left(\frac{1}{\rho}\right) + \frac{1}{2}w^2 dc_n \tag{13.51}$$

for the specific energy e,

$$d\zeta = -sdT + \frac{dP}{\rho} - \frac{1}{2}c_n d\left(w^2\right) \tag{13.52}$$

for the chemical potential ζ involved in the "s" motion equation. Here, $c_n = \rho_n/\rho$ is the mass fraction of the normal component. From postulates (13.51), (13.52) we have the following corollary

$$\zeta = e + \frac{P}{\rho} - sT - \frac{1}{2}c_n w^2. \tag{13.53}$$

According to (13.51)–(13.53), the scalar quantities c_n, $w^2/2$ are the conjugate parameters corresponding to the internal "n-s"-counterflow. To justify his postulates (13.51) and (13.52), Donnelly [6] had to adopt an additional postulate for the free Helmholtz energy based on the model of quasiparticles. Relations (13.49) and (13.50), which were introduced in [6], form a variant of He-II thermodynamics, which have a different appearance than the variant of [14], which is based on relations (13.46) and (13.47).

One important observation is worth mentioning here. Relations (13.49) and (13.50) were derived ab initio as explicit postulates, but the situation for relation (13.46) is less obvious. Equation (13.46) is clarified in [14] as follows: "...the last term shows that the derivative of the energy with respect to the momentum is the velocity". This statement is formally based on the phenomenological definition of the velocity of flow of liquid [14], and hence suggests that Eq. (13.46) is not a postulate. However, such a conclusion leads to the following questions.

- It is well-known that classical mechanics is based on Galileo's principle of relativity [17]: "In an inertial system physical processes proceed in the same manner, irrespective of whether or not the system is at rest or moves uniformly and rectilinearly". However, in this case, Eq. (13.46) has a "correct" appearance only in the system of coordinates K_0.
- The system under consideration is unconventional due to the fact that only a part of the liquid mass is involved in the "n" motion. For such a system, a formal application of a mechanical principle can be regarded per se as a postulate.
- The energy E_0 in the frame K_0 should include as an ingredient the kinetic energy of the "n"-motion. However, with this involved structure of energy, the justifiability of the thermodynamic identity is not evident a priori.

The above questions suggest that, in the actual fact, Eq. (13.46) is a postulate. Generally speaking, due to the unconventional character of such an involved system as He-II with internal "n – s" flows, it is difficult to believe in any progress in macroscopic justification of the corresponding thermodynamics.

13.6.2 Microscopic Analysis

Labuntsov [10] proposed a new method of construction of He-II thermodynamics from first principles. To this end, he used the concept of gas of quasiparticles, which consists of elementary excitations. This concept was proposed by Landau and Lifshitz [14] as a vivid way for the description of quantum mechanical laws of the collective motion of the atoms of helium in liquid at approximately $T = 0$. The author of [10] considers He-II in the Landau frame K_0, in which the superfluid component of the liquid is at rest, $\vec{v}_s = 0$. The microscopic analysis of [10] is based on the distribution function f_E over energies of the "gas of quasiparticles", which moves with the velocity of the normal component, $\vec{v}_n = \vec{w}$

$$f_E = \left\{ h^3 \left[\exp\left(\frac{(\varepsilon - \vec{p}\,\vec{w})}{(\varepsilon - \vec{p}\,\vec{w})} \right) - 1 \right] \right\}^{-1}. \qquad (13.54)$$

Here, $\varepsilon(\vec{p})$ is the energy of a quasiparticle as a function of its momentum $\vec{p}$, $\hbar$ is the Planck constant, k_B is the Boltzmann constant. The energy of a unit volume of the liquid considered in [14] is determined from the relation[2]

$$E_0 = \int \varepsilon f_E d^3\vec{p}. \tag{13.55}$$

Consider conditions when the velocities w of the internal counterflow are small in comparison with the velocity of the first sound in liquid. We expand f_0 in a Taylor series retaining only the terms quadratic in w

$$f_E = f - \frac{\partial f}{\partial \varepsilon}\vec{p}\,\vec{w} + \frac{1}{2}\frac{\partial^2 f}{\partial \varepsilon^2}\left(\vec{p}\,\vec{w}\right)^2. \tag{13.56}$$

Here, f is the equilibrium distribution function defined by (13.54) with $w = 0$.

Substituting (13.56) into (13.53) and averaging with respect to the angular coordinate, we note that the second (linear) term in (13.56) disappears. As a result, we get an expression for the energy, containing only the quadratic term of the expansion

$$E_0 = \int \varepsilon f d^3\vec{p} + \frac{1}{2}\int \varepsilon \frac{\partial^2 f}{\partial \varepsilon^2}\left(\vec{p}\,\vec{w}\right)^2 d^3\vec{p}. \tag{13.57}$$

Before proceeding with the further analysis, we prove that

$$\varepsilon \frac{\partial^2 f}{\partial \varepsilon^2} = -\left[T\frac{\partial}{\partial T}\left(\frac{\partial f}{\partial \varepsilon}\right) + \frac{\partial f}{\partial \varepsilon}\right]. \tag{13.58}$$

To this end, we take into account that f depends only on one dimensionless variable ε/kT. Indeed, we have

$$\frac{\partial^2 f}{\partial \varepsilon^2} = \frac{\ddot{f}}{(kT)^2}, \tag{13.59}$$

where the dot denotes the total derivative. Next, it is easily checked that

$$\frac{\partial}{\partial (kT)}\left(\frac{\dot{f}}{kT}\right) = -\varepsilon\frac{\ddot{f}}{(kT)^3} - \frac{\dot{f}}{(kT)^2}, \tag{13.60}$$

$$\frac{\partial f}{\partial \varepsilon} = \frac{\dot{f}}{kT}. \tag{13.61}$$

---

[2]The energy E_0, as established by (13.55), is given up to the energy of ground state at $T = 0$. However, the latter energy is immaterial for the analysis that follows.

Substituting (13.59) into (13.60) and taking into account (13.61), we get the required equality (13.58).

We further need the following considerations

- The integral

$$-\frac{1}{2}\int (\vec{p}\,\vec{w})^2 \frac{\partial f}{\partial \varepsilon} d^3\vec{p}$$

after another integration with respect to the angular coordinate of the momentum space, assumes the form

$$\frac{w^2}{2}\left[\frac{4\pi}{3}\int_0^P \left(-\frac{\partial f}{\partial \varepsilon}\right) p^4 dp\right].$$

- By definition, the expression in square brackets is the density of normal motion ρ_n [14].

Consequently, the energy E_0 can be completely determined. We finally have

$$E_0 = E_{eq} + \frac{1}{2}w^2\left(\rho_n + T\frac{\partial \rho_n}{\partial T}\right). \tag{13.62}$$

Let us consider the physical meaning of the terms on the right of (13.62).

- The first term is the internal energy of liquid in a state of complete equilibrium (i.e., without internal motions) per unit volume.
- The second term is the kinetic energy of translational motion of the "n"-component.
- The third term is the specific part of the He-II energy related to "$n - s$" motions. This part controls the structure of the principal thermodynamic identity. The derivative $\partial \rho_n / \partial T$ is taken with fixed w^2, ρ. It should be remembered that the density ρ is a parameter governing the energy spectrum of the gas of quasiparticles.

In Donnelly's thermodynamics [6], the quantity e is the internal energy, which is defined as the total energy minus the kinetic energies of "$n - s$" motions. The quantities e and E_0 are related as follows

$$\rho e = E_0 - \frac{1}{2}\rho_n w^2. \tag{13.63}$$

From (13.63), we get

$$e = e_{\text{eq}} + \frac{1}{2}w^2 T\left(\frac{\partial c_n}{\partial T}\right)_{\rho, w^2}. \tag{13.64}$$

Here, e_{eq} is the internal energy of a unit mass of liquid at equilibrium in the absence of internal motions ($w = 0$).

From expression (13.64), we readily get that the so-defined internal energy of He-II with "n-s"-motions satisfy Galileo's principle of relativity [17]. Let us now find to which structure of the principal thermodynamic identity there corresponds expression (13.64). To this end, we write

$$de^* = Tds - Pd\left(\frac{1}{\rho}\right) + Adb. \tag{13.65}$$

Here, e^* is the sought-for internal energy, A, b are the unknown conjugate parameters, which characterize the internal "$n - s$"-counterflow. Dividing both sides of (13.65) by T and rearranging, we get

$$d\left(\frac{e^* - Ts - Ab}{T}\right) = \left(e^* - Ab\right)d\left(\frac{1}{T}\right) - \frac{P}{T}d\left(\frac{1}{\rho}\right) - \frac{b}{T}dA. \tag{13.66}$$

Let us now dwell upon the Gibbs's thermodynamic potential method [18]. At present, this is the commonly accepted language of the macroscopic theory, which has abilities for finding meeting points between thermodynamics and statistical mechanics. According to Gibbs's method, all properties of equilibrium thermodynamic systems can be obtained if one knows at least one thermodynamic potential. The well-known Maxwell relations are included in the Gibbs's theory of potentials. In the context of the above conditions, we can write the following Maxwell relation

$$\left.\frac{\partial(e^* - Ab)}{\partial A}\right|_{T,p} = -\left.\frac{\partial(b/T)}{\partial(1/T)}\right|_{\rho,A}, \tag{13.67}$$

or

$$\left(\frac{\partial e^*}{\partial A}\right)_{T,p} - A\left(\frac{\partial b}{\partial A}\right)_{T,p} = T\left(\frac{\partial b}{\partial T}\right)_{\rho,A}. \tag{13.68}$$

Consider in the linear approximation the increment e^* as the parameter A is varied. From (13.68), it follows that

$$\Delta e^*\big|_{T,p} = T\left(\frac{\partial b}{\partial T}\right)_{\rho,A} A. \tag{13.69}$$

A comparison of (13.69) and (13.64) shows that if the principal thermodynamic identity for He-II has the structure of (13.65), then in it one necessarily has

$$e^* = e,\ A = \frac{w^2}{2},\ b = c_n \equiv \frac{\rho_n}{\rho}. \tag{13.70}$$

It follows directly that conditions (13.70) establish the correctness of identity (13.51) and justify it on the quasiparticle level of analysis. This establishes the agreement between the macroscopic (13.64) and microscopic (13.70) levels of description of the internal energy e. It is worth pointing out from (13.63), it follows that the expressions (13.47) and (13.53) are identical for the chemical potential μ, and hence (13.48) and (13.52) are in a qualitative agreement. It can be verified that each of the expressions (13.46) and (13.51) can be obtained from the other expression. Thus, the following fundamental result is obtained: although the external appearance is different, the variants [6] and [14] of He-II specific thermodynamics are consistent.

References

1. Kapitza P (1938) Viscosity of liquid helium below the λ-point. Nature 141:74
2. Schmitt A (2014) Introduction to superfluidity: field-theoretical approach and applications. Springer
3. Audi G, Bersillon O, Blachot J, Wapstra AH (2003) The NUBASE evaluation of nuclear and decay properties. Nucl Phys A 729:3–128
4. Chanson H (2009) Applied hydrodynamics: an introduction to ideal and real fluid flows. CRC Press
5. Negele JW, Orland H (1998) Quantum many-particle systems. Westview Press
6. Donnelly RJ (1991) Quantized vortices in helium II. Cambridge University Press
7. Landau L (1941) Theory of the superfluidity of helium II. Phys Rev 60:356–358
8. Khalatnikov I (1989) An introduction to the theory of superfluidity. Addison-Wesley
9. Rivers WJ, McFadden PW (1966) Free convection in helium II. ASME J Heat Transfer 88:343–349
10. Labuntsov DA (2000) Physical foundations of power engineering. Selected works, Moscow Power Energetic Univ (Publ) (In Russian)
11. Muratova TM, Labuntsov DA (1969) Kinetic analysis of evaporation and condensation processes. High Temp 7(5):959–967
12. Tisza L (1938) Transport phenomena in helium II. Nature 141:913
13. London F (1938) The λ-phenomenon of liquid helium and the Bose-Einstein degeneracy. Nature 141:643–644
14. Landau LD, Lifshitz EM (2013) Statistical physics (Course of theoretical physics, vol 5. 3rd edn). Elsevier
15. Feynman RP, Cohen M (1956) Energy spectrum of the excitations in liquid helium. Phys Rev 102:1189–1204
16. Barenghi CF, Donnelly RJ, Vinen W (2001) Quantized vortex dynamics and superfluid turbulence. Springer
17. de Saxcé G, Valleé C (2016) Galilean mechanics and thermodynamics of continua. Wiley-ISTE
18. Bejan A (2006) Advanced engineering thermodynamics, 3rd edn. Wiley

Chapter 14
Pseudoboiling

Abbreviations

SCP	Supercritical pressures
SCF	Supercritical fluid

Symbols

c_f	Friction coefficient
c_p	Isobaric heat capacity
H	Enthalpy
k	Thermal conductivity
M	Mach number
p	Pressure
Pr	Prandtl numberμ
Re	Reynolds number
T	Temperature
t	Time

Greek Letter Symbols

μ	Dynamic viscosity
ν	Kinematic viscosity
η	Similarity variable
ρ	Density
Ω	Heat function
τ	Shear stress
ψ_τ	Relative law of friction
ψ_q	Relative law of heat transfer

Subscripts

$*$	Reference case
c	State at critical point
mod	Model

Y. B. Zudin, *Non-equilibrium Evaporation and Condensation Processes*,
Mathematical Engineering, https://doi.org/10.1007/978-3-030-67553-0_14

m	State at pseudophase transition
∞	State at infinity
w	State at wall

14.1 Area of Supercritical Pressures

14.1.1 *The Relevance of the Problem*

In the area of pressures above the thermodynamic critical ($p > p_c$), pure substances are known to behave like single-phase liquids with locally equilibrium properties. This range of parameters is called the area of Supercritical Pressures (SCP), and the medium in it, a Supercritical Fluid (SCF) [1]. The area of Supercritical Pressures (SCP) had become the subject of interest in thermal engineering in connection with the attempts to solve the principal problem of enhancing the initial vapor parameters in the 1960s. SCFs are considered as promising coolants due to their specific properties, and at present SCP power units play a key role in heat power engineering of advanced countries.

The problem of transition to the SCP area has been discussed in atomic power engineering since the early 2000s [2] in connection with the task of development of fourth-generation nuclear reactors cooled by "light water" with supercritical parameters. According to forecasts, direct-cycle nuclear power plants would be capable of considerably increasing the performance index of electric power generation and to drastically reduce the discharge of water through the reactor. It also seems promising to cool the core of fast neutron reactors with supercritical water. To cope with the SCP in atomic power engineering, it is required to able to calculate the heat output in the reactor core, and in the first place, to find the zone of safe thermal loads. With available theoretical recommendations, one could avoid transiting to aggravated heat transfer regimes, which may result in reactor accidents.

However, due to the great variability of SCFs' thermal–physical properties and consequent uncertainty in their thermo-hydrodynamics, such regimes have not been thoroughly considered at present. However, in the literature, there is only a limited bundle of experimental data and a number of empirical design formulas for SCP, and what is more, these formulas are known to hold only for very limited ranges of operating parameters. In addition, most of such information had been obtained before the mid-1980s and essentially only for the heat transfer setting. Such a judgment can be made based on the bibliographic survey [1] containing more than 400 titles. The bank of available data should be thoroughly reconsidered, because in the early 2000s the properties of water in the SCP area have been considerably refined.

Notwithstanding the fact that an SCF is a single-phase medium, its thermophysical behavior partially resembles that of a "liquid–vapor" two-phase medium. A classical phase transition is a transition of a thermodynamic system from one aggregate state to another. This transition is accompanied by a discontinuous variation of thermal–physical properties (density, dynamic viscosity, thermal conductivity), and moreover, the energy (called the phase transition heat) is extracted (for evaporation) or produced (for condensation).

Thermodynamically, this quantity is a jump of enthalpies realized at the phase translation temperature. Similarly, in the SCP area for each isobar one can define the temperature of "pseudophase transition" T_m. Near T_m an abrupt change of density and dynamic viscosity of the SCF is manifested, and its thermal conductivity and isobaric heat capacity attain their maximum values. To illustrate this, in Fig. 14.1 we show the dependences of thermophysical properties of helium (density ρ, thermal conductivity k, dynamic viscosity μ, specific heat c_p) on the enthalpies in the SCP area at pressure 0.25 MPa. With each value of T_m, we can associate in a unique way the related enthalpy of pseudophase transition H_m (for $p/p_c < 1.5$ the values of H_m are hardly affected by the pressure).

For the analysis of thermohydraulic flow laws in the SCP area, it is expedient to split the enthalpy range into three regions.

- The pseudoliquid state zone ($H < H_m$).
- The zone of pseudophase transition ($H \approx H_m$), which encompasses the enthalpy of maximum heat capacity.

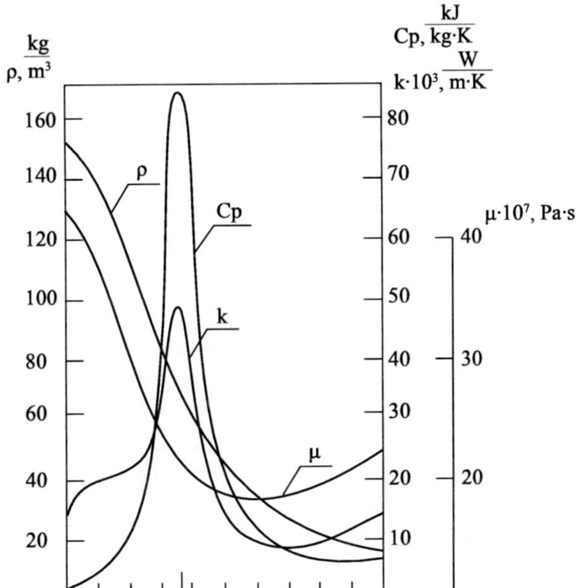

Fig. 14.1 Thermophysical properties of helium versus enthalpy in the supercritical pressures area at pressure 0.25 MPa

- The pseudogaseous state zone ($H > H_m$). The boundaries of zones can be determined from the dependence of the thermal expansion work on the enthalpy $E_q = (pdV/dh)_p$ [3].

In a direct flow power unit, the enthalpy of water can vary in the range 200–3500 kJ/kg, which covers the range of pseudophase transition and which compares favorably water against other coolants. So, for example, for the carbon dioxide, already at room temperature we have $H > H_m$, which corresponds to the pseudogaseous zone. On the opposite side, halocarbons are useful for modeling the heat transfer in the pseudoliquid zone $H < H_m$. In the region of pseudoliquid state, in analogy with capillary liquids we have $E_q \approx 10^{-2}$. When changing through H_m, the quantity E_q increases approximately by one order, starting from 0.02–0.03 to the level $E_q \approx R_g/c_p$, which is characteristic of a perfect gas. For $H < H_m$, the properties of an SCF are similar to those of a capillary liquid with $p < p_c$. Here the key role is played mostly by the temperature dependence of the viscosity.

The thermohydraulic characteristics of flows in the pseudoliquid zone are practically the same as for a capillary liquid with variable viscosity. For an upflow motion of an SCF in heated vertical large-diameter tubes, for low Reynolds numbers, peaks of temperature of the wall can be observed at the entrance region of the channel. Critical deterioration of heat transfer occurs as the wall temperature attains the level $T \geq T_m$. In short small-diameter tubes, a significant improvement of the heat transfer is observed when passing through T_m from the pseudoliquid zone, this improvement is accompanied by high-frequency thermoacoustic oscillations. For the zone of pseudophase transition, two main regimes of heat transfer are typical: the normal and the aggravated ones [1].

The principal characteristics of the normal regime are as follows

- monotone variation of the wall temperature along the wall length,
- stabilization of SCF thermohydraulic parameters,
- Independence of the heat transfer of the tube orientation and the inlet enthalpy beyond the original section.

A transition to the aggravated heat transfer occurs, as a rule, when some critical thermal load is exceeded. This is accompanied by an abrupt decrease of heat transfer with the formation of wall temperature peaks (or elongated areas of its superheating). In the pseudogaseous zone, the SCF acquires properties of a perfect gas as the temperature increases: the viscosity, thermal conductivity, and heat capacity are increasing, while the Prandtl number is of order 1. For water, due to constraints on the admissible wall temperature values in this zone, normal heat transfer regimes are implemented as a rule.

14.1.2 Theoretical Studies of Heat Transfer

Aggravation phenomena of heat transfer in the SCP have become the subject of intensive studies after reports on steam generator failures on supercritical pressure heat-and-power stations in the 1960s. Theoretical description of such regimes was carried out on the basis of the classical theory of turbulent heat transfer with the involvement of various modifications: introduction of generalized variables, analogies with blowing flows through a permeable surface, generalizations of the Prandtl's mixing-length theory. The following conclusions have been made after the critical interpretation of the information collection in the course of early studies.

- Heat transfer aggravation phenomena are of hydrodynamic nature, they are not related to local distributions of the SCF parameters in the dangerous section of the channel.
- The principal root cause of aggravation of the heat transfer in the SCP area is the significant change (under non-isothermal conditions) of the hydrodynamic parameters: composition and magnitude of operating forces, average flow parameters, oscillatory of the turbulent flow.
- A decrease in the intensity of the turbulent exchange is preceded by a substantial deformation of the averaged flow.

These observations support the relevance of the detailed investigation of regimes of aggravation of the heat transfer in the SCP area and the development of reliable methods of their calculations. The physical analogy between the phase and pseudophase transitions provided the basis for the "pseudoboiling" model of [4]. This model was based on the experimentally observed dependences of the heat transfer coefficient on the heat flow density in the SCP area. To evaluate the heat transfer in such regimes, papers [4, 5] used the empirical relations, which were obtained, respectively, for the bubble and film boiling of liquid. In [4, 5], an SCF was looked upon as a mixture of two components: the heavy and light ones. Taking into account that the medium density depends on the temperature T (the enthalpy H), it was found that during heating the light component will be concentrated near the wall, and the heavy one, in the flow core. The picture is opposite in the process of culling.

Further theoretical investigations of the heat transfer in the SCP area were carried out mostly by numerical methods, which gave important information about the mechanisms of hydrodynamics and heat transfer under these conditions. However, a numerical study describes only some or other partial variant of the problem for concrete ranges of variation of parameters. As a result, the statement that numerical solutions are the most advantageous, which sometimes occurs in the literature, cannot be considered legitimate. In the actual fact, analytical solutions to hydrodynamic and heat transfer problems play a considerable role even in our computerized age.

They have the following important advantages over numerical methods [6].

- The merit of the analytical approach lies in the possibility of a closed qualitative description of a process, detection of dimensionless parameters, and their classifications in the order of importance.

- Analytical solutions have the required generality, and parametric studies can be carried out by varying the initial and boundary conditions in analytical solutions.
- For the purpose of testing of numerical solutions of complete equations, one has to have basis analytical solutions of simplified equations (which are obtained after assessment and rejection of some terms in the original equations).
- Analytical solutions can be used for the validity check of statements of numerical investigations of each specific problem.

The present Chapter puts forward an analytical model of turbulent exchange in the SCP area. This model is capable of assessing the effect of variable thermophysical properties on the friction heat transfer.

14.2 Surface Renewal Model

14.2.1 Periodic Structure of Near-Wall Turbulent Flow

According to [7–9], the turbulent flow in the boundary layer is characterized by periodic "penetrations" of accelerated portions of liquid from the flow core into the near-wall region and "surges" of decelerated liquid masses from the wall. Such structure of near-wall turbulent flow was experimentally revealed (and then supported many times) and is now known as "bursting" [8, 9]. The discovery of the "bursting" phenomenon helped to detect the existence of stable low-frequency pulsations against the background of the stochastic noise. These pulsations look like horseshoe-shaped eddies elongated along the flow. It turned out that the pulsations of the velocity in the viscous sublayer of the turbulent flow are comparable in magnitude with the average flow velocity. It was also found that the principal contributors to the generation and dissipation of turbulent energy are also concentrated in the narrow near-wall zone.

The available experimental information about the "bursting" phenomenon indicated the possibility of approximation of the involved pulsation pattern in the turbulent boundary layer by simple monoharmonic oscillations. First monoharmonic models of the velocity field in the viscous sublayer, which were based on the linearized Navier–Stokes equations, were put forward in [10–12]. In the model of [10], it was assumed that the velocity perturbations come from outside in the liquid sublayer as discrete portions and then decay there. In [11], pulsations continuously generated on the outer boundary of the liquid sublayer were considered. In [12], the linearized convective terms of the Navier–Stokes equations for velocity pulsations were additionally taken into account.

The "surface renewal model" was employed in [13] to describe the turbulent mass flow through the interfacial boundary. In essence, in this model, the mass-transfer coefficient is calculated on the basis of the solution of the heat shock problem of the transient heat conduction theory [14]. In [15, 16], the surface renewal model was used as a "spacer" in a specific semi-empirical turbulence model. It was used to estimate the effect of several factors on the turbulent friction: the unsteadiness of the boundary layer, the viscous dissipation, and the turbulent Prandtl number.

The method of [15, 16] involved the following stages.

- from the solution of one-dimensional time-dependent equations of conservation of momentum and energy, one finds the distributions of the velocities and temperatures in the near-wall region.
- The time-dependent profiles of these quantities are averaged over the pulsation period.
- The averaged near-wall distributions of the velocities and temperatures are glued with the turbulent flow core.

In the present Chapter, we develop the modified surface renewal model constructed on the basis of the "relative correspondence method". This model is capable of describing the effects of the variations of the medium thermophysical properties on the friction and heat transfer in the turbulent boundary layer.

14.2.2 Relative Correspondence Method

The relative correspondence method was first proposed in [16] for approximate solution of thermohydraulic problems. The essence of the method is briefly outlined below. Assume that we need to calculate the friction (c_f) and heat transfer (h) coefficients for a turbulent medium flow (liquid or gas), whose thermophysical properties are functions of the temperature (the enthalpy). If the medium properties are imaginatively frozen at the flow core temperature T_∞, then we get the "reference" case of constant properties, for which the expressions for c_{f*}, h_* are available in the literature. As reference values, one can also use such well-proven relations as Blasius's friction law [17], and Petukhov and Kirillov's formula for heat transfer [18], etc. Following [19], we write the relative law of friction

$$\psi_\tau = \left(\frac{c_f}{c_{f*}}\right),$$

and the relative law of heat transfer

$$\psi_q = \left(\frac{h}{h_*}\right),$$

which are calculated with the fixed Reynolds number (Re = idem). Here, h is the heat transfer coefficient.

Assume that we now consider a simplified model of a real process that takes into account its principal physical features. It is supposed that in the frames of the simplified model one can calculate both the reference values c_{f*}, h_* and their values in the case when c_f, h can vary. Hence, the relative model laws

$$\psi_{\tau,m} = \left(\frac{c_f}{c_{f*}}\right)_m, \psi_{q,m} = \left(\frac{h}{h_*}\right)_m$$

are also known. Briefly, the crux of the relative correspondence method is reduced to the following conjecture: the quantities ψ, as written for a real turbulent friction and heat transfer, can be approximately replaced by their corresponding model values

$$\psi = \psi_m. \tag{14.1}$$

Then for the turbulent under study, c_f and h can be evaluated as

$$c_f = c_{f*}\psi_{\tau,m}, h = h_*\psi_{q,m}. \tag{14.2}$$

So, the relative correspondence method can be looked as a "shell" which should be "filled" by the calculation model for ψ_m. As such "spacer", we shall use the model based on the periodic structure of the near-wall turbulent flow.

14.3 Mathematical Description

14.3.1 Conservation Equations

Let us consider the heat transfer problem when a semi-infinite volume of hot (cold) liquid $0 \leq y < \infty$ with homogeneous initial temperature $T = T_\infty$ is in contact with the heat (cold) wall. On the surface $y = 0$, at the initial time $t = 0$, a stepwise increase (decrease) of temperature to the value $T = T_w$ is specified. It is assumed that when the time-dependent heating (cooling) process is finished, the fluid volume located near the wall is replaced by a new portion of the fluid that arrived from the flow core. If the medium density depends on temperature, then, in all sections of the volume $y \geq 0$, a convection mass-transfer mechanism takes place, the boundary $y = 0$ itself remaining impermeable. The thermophysical characteristics ρ, k, c_p of the medium are assumed to be arbitrary functions depending only on the temperature (or on the enthalpy, which is uniquely related to the temperature). So, we consider a one-dimensional time-dependent heat conduction problem, which describes the thermal shock for a medium with variable thermophysical properties [20].

The mathematical description of the problem involves the continuity equation

$$\frac{\partial \rho}{\partial t} + \frac{\partial J}{\partial y} = 0, \tag{14.3}$$

and the energy equation

$$\rho\frac{\partial H}{\partial t}+J\frac{\partial H}{\partial y}=\frac{\partial}{\partial y}\left(\frac{k}{c_p}\frac{\partial H}{\partial y}\right). \tag{14.4}$$

Here, J is the mass rate of thermal expansion (contraction) of liquid, H is the specific medium enthalpy, y is the transverse coordinate measured from the wall toward the volume. The uniformity conditions for the system of Eqs. (14.3), (14.4) read as

$$\left.\begin{aligned} t=0(y=0):H=H_{\infty},\\ y=0(t>0):H=H_{w}.\end{aligned}\right\} \tag{14.5}$$

The index "w" means the conditions of the wall, and "∞", the conditions at infinity. The system of Eqs. (14.3)–(14.5) describes the heat transfer as the semi-infinite volume of medium with variable thermophysical properties interacts with the wall.

Introducing the potential of the mass velocity ($J\equiv\partial\varphi/\partial y$), we rewrite the original system of Eqs. (14.3), (14.4) as

$$\frac{\partial\varphi}{\partial t}+\frac{\partial^2\varphi}{\partial y^2}=0, \tag{14.6}$$

$$\rho\frac{\partial H}{\partial t}+\frac{\partial\varphi}{\partial y}\frac{\partial H}{\partial y}=\frac{\partial}{\partial y}\left(\frac{k}{c_p}\frac{\partial H}{\partial y}\right). \tag{14.7}$$

Using the similarity variable $\eta=y/\sqrt{2t}$, we can transform the partial differential equations (14.6), (14.7) to the system of ordinary differential equations

$$-\eta\frac{d\rho}{d\eta}+\frac{d^2\varphi}{d\eta^2}=0, \tag{14.8}$$

$$-\eta\rho\frac{dH}{d\eta}+\frac{d\varphi}{d\eta}\frac{dH}{d\eta}=\frac{d}{d\eta}\left(\frac{k}{c_p}\frac{dH}{d\eta}\right). \tag{14.9}$$

Differentiating both sides of (14.9) with respect to η, this using (14.8) gives

$$\rho+\frac{d}{d\eta}\left[\frac{d}{d\eta}\left(\frac{k}{c_p}\frac{dH}{d\eta}\right)\right]=0.$$

Let the enthalpy H be the independent variable. As the dependent variable, we shall seek the heat function $\Omega=-\frac{k}{c_p}\frac{dH}{d\eta}$. The quantity Ω is related to the heat flow density

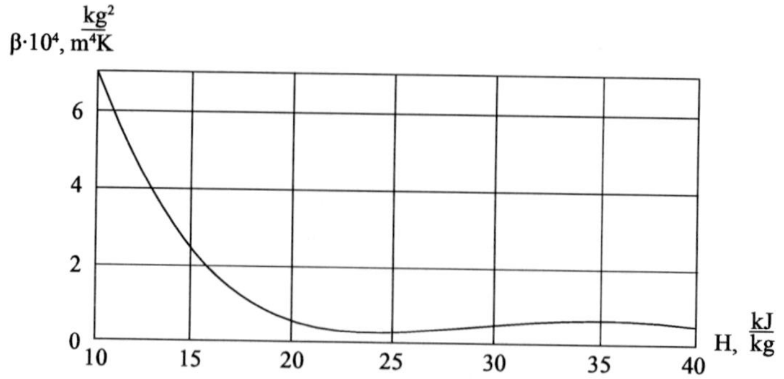

Fig. 14.2 Dependence of the generalized thermophysical properties parameter for supercritical helium at pressure 0.25 MPa

$$q \equiv -k\frac{\partial T}{\partial y} = -\frac{k}{c_p}\frac{\partial H}{\partial y}$$

by the clear relation: $\Omega = -\sqrt{2t}q$. Changing from the variables η, H to the new variables H, Ω, we get from (14.5), after some algebra, the following main equation

$$\Omega\frac{d^2\Omega}{dH^2} + \beta = 0, \tag{14.10}$$

where $\beta = \frac{\rho k}{c_p}$ is the generalized parameter of the medium thermophysical properties. From (14.10) it follows that, in the framework of the surface renewal model, the heat capacity, the thermal conductivity, and the density of the medium have an effect on the heat transfer not individually (as one could imagine from the form of Eqs. (14.3), (14.4)), but rather as a unified complex of properties. Figure 14.2 shows the dependence $\beta(H)$ for supercritical helium at pressure 0.25 MPa.

14.3.2 Boundary Conditions

The main Eq. (14.10) should be augmented with two boundary conditions. The first equation is physically clear and should describe the condition of the heat flux attenuation at infinity of the volume

$$\eta \to \infty : \Omega = 0. \tag{14.11}$$

To write down the second boundary condition, we argue as follows. We start with the condition of the wall impenetrability

$$y = 0(t > 0) : J = 0.$$

It follows that

$$\eta = 0 : \frac{d\varphi}{d\eta} = \sqrt{2t}\frac{d\varphi}{dy} = 0.$$

Therefore,

$$\eta = 0 : \frac{d}{d\eta}\left(\frac{k}{c_p}\frac{dH}{d\eta}\right) = \frac{d\Omega}{d\eta} = 0.$$

Hence $\eta = 0 : \Omega\frac{d\Omega}{dH} = 0$. The following inequalities are clear: $\eta = 0 : \Omega \neq 0$. As a result

$$\eta = 0 : \frac{d\Omega}{dH} = 0. \tag{14.12}$$

The boundary conditions (14.5), (14.12) are written down with respect to the similarity variable η. To change to the enthalpy H, we have from the uniformity conditions (14.7)

$$\eta = 0 : H = H_w,$$

$$\eta \to \infty : H = H_\infty.$$

This gives us the following boundary conditions for the main Eq. (14.10)

$$H = H_\infty : \Omega = 0, \tag{14.13}$$

$$H = H_w : \frac{d\Omega}{dH} = 0. \tag{14.14}$$

14.3.3 Dimensionless Variables

The main Eq. (14.10) with boundary conditions (14.13), (14.14) describes the propagation of an expansion (contraction) heatwave in a medium, whose properties are arbitrary functions of the temperature (enthalpy). Let us introduce the following dimensionless variables

The enthalpy.

$$\tilde{H} \equiv \frac{H - H_\infty}{H_w - H_\infty},$$

the thermophysical property parameter

$$\tilde{\beta}\left(\tilde{H}\right) \equiv \frac{\beta}{\beta_\infty} = \left(\frac{\rho k}{c_p}\right)\left(\frac{\rho_\infty k_\infty}{c_{p\infty}}\right)^{-1},$$

and the heat function

$$\omega \equiv \frac{\Omega}{(H_w - H_\infty)\sqrt{\beta_\infty}}.$$

It should be noted that the choice of the scaling parameter for the thermophysical properties of ρ, k, c_p the liquid is not predetermined by the problem. For definiteness, let us suppose that all properties should be assumed to be the properties of the medium at infinity of the volume with temperature T_∞ (i.e., with the enthalpy H_∞). Taking into account the above dimensionless parameters, we can rewrite Eq. (14.10) as

$$\frac{d^2\omega}{d\tilde{H}^2} = -\tilde{\beta}. \tag{14.15}$$

The boundary conditions for Eq. (14.15) follow from (14.13), (14.14)

$$\tilde{H} = 0 : \omega = 0, \tag{14.16}$$

$$\tilde{H} = 1 : \frac{d\omega}{d\tilde{H}} = 0. \tag{14.17}$$

Figure 14.3 shows the dependences $\tilde{\beta}\left(\tilde{H}\right)$ for helium at pressure 0.25 MPa separately for the heating and cooling problems.

14.4 Solution of the Main Equation

14.4.1 Exact Solution

Despite the relatively simple form of Eq. (14.15) with homogeneous boundary conditions (14.16), (14.17), it fails in general to have an exact solution. As far as we know,

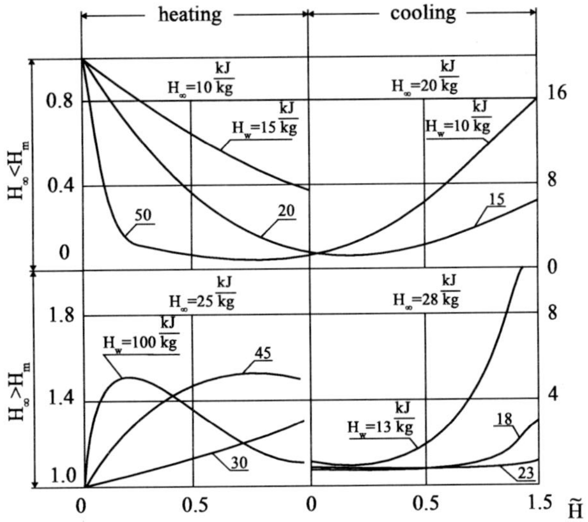

Fig. 14.3 Dependences $\tilde{\beta}\left(\tilde{H}\right)$ for helium at pressure 0.25 MPa separately for the heating and cooling problems

it has only one partial solution with $\tilde{\beta}=1$, which is briefly outlined below. Multiplying both sides of Eq. (14.15) by $\tilde{f}\,\frac{d\tilde{f}}{d\tilde{H}}$ and integrating with respect to $\tilde{H}$, we find that

$$\frac{d\omega}{d\,\tilde{H}}=\sqrt{2\ln\frac{\omega_w}{\omega}},$$

where ω_w is the required heat flow value on the wall $\left(\eta=0,\,\tilde{H}=1\right)$. This equation possesses the following integral

$$\tilde{H}=\sqrt{\frac{\pi}{2}}\omega_w\mathrm{erfc}\left(\sqrt{\ln\frac{\omega_w}{\omega}}\right), \tag{14.18}$$

where erfc(x) is the complementary error function. For $\omega=0$, expression (14.18) describes the boundary condition (14.16). Making $\tilde{H}=1,\,\omega=\omega_w$ in (14.18), this gives $\tilde{f}_w=\sqrt{\frac{2}{\pi}}$. Using the above relations, we write the expression for the heat flow density on the wall

$$\eta=0,\,\tilde{H}=1,\,q_w=\sqrt{\frac{k_\infty c_{p\infty}\rho_\infty}{\pi t}}(T_w-T_\infty).$$

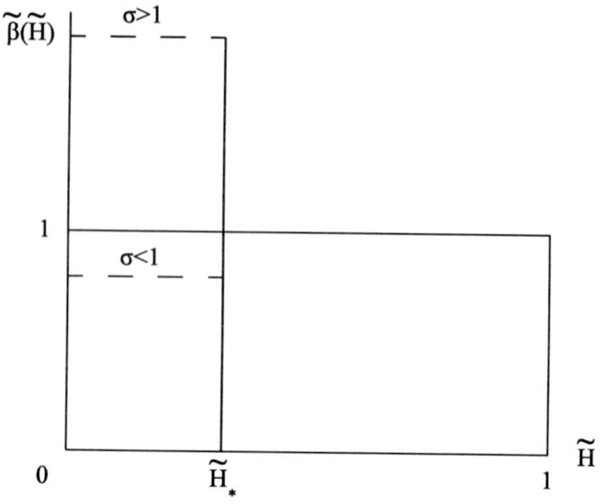

Fig. 14.4 Stepwise law of variation of the generalized thermophysical property parameter

This expression coincides with the solution of the "heat shock" problem in the thermal conductivity theory, as was obtained in the case of constant properties $\left(c_p = \text{const}, k = \text{const}, \rho = \text{const}\right)$ in [14]. Hence, the above solution describes the case of variable properties, provided the latter are related as $\beta \equiv \rho k / c_p = \text{const}$. In the [21, 22], another exact solution of the main Eq. (14.15) was found for the discontinuous (stepwise) law of variation of the thermophysical property parameter (Fig. 14.4)

$$\tilde{\beta} = \begin{cases} 1, 0 \leq \tilde{H} \leq \tilde{H}_*, \\ \sigma, \tilde{H}_* \leq \tilde{H} \leq 1. \end{cases}$$

From this solution (which is quite cumbersome), we get the following expression for the heat flow on the wall

$$\omega_w = \left(b^{1-\sigma} \operatorname{erfc} \sqrt{-a \ln b} + \frac{1}{\sqrt{\sigma}} \operatorname{erf} \sqrt{-a \ln b} \right)^{-1},$$

$$\tilde{H}_* = \omega_w b^{1-\sigma} \operatorname{erfc} \sqrt{-a \ln b},$$

$$b = \frac{\omega_*}{\omega_w}.$$

Here, $\omega_* \equiv \omega\left(\tilde{H}_*\right)$ is the heat flow at the point of discontinuity $\tilde{H} = \tilde{H}_*$ of the stepwise function $\tilde{\beta}\left(\tilde{H}\right)$.

14.4.2 Approximate Solution

Let us now consider the general case of an arbitrary dependence $\tilde{\beta}\left(\tilde{H}\right)$ and apply Galerkin's variational method [23] to solve the main Eq. (14.6a). According to this method, the exact solution is approximated by basis functions that precisely satisfy the boundary conditions and approximately satisfy (in the mean) the equation itself. The simplest variant of the method involves a unique basis function of the form

$$\tilde{\beta}\left(\tilde{H}\right) = \omega_w \sin\left(\frac{\pi}{2}\tilde{H}\right).$$

Multiplying the left-and right-hand sides of (14.15) by ω and integrating from $\tilde{H} = 0$ to $\tilde{H} = 1$, we get

$$\omega_w^2 = \frac{3}{\pi}\int_0^1 \tilde{\beta} \sin\left(\frac{\pi}{2}\tilde{H}\right) \mathrm{d}\tilde{H}.$$

Putting $\tilde{\beta} = 1$ in the integrand, we get a solution for the "rigid medium" (i.e., with constant thermophysical properties): $\omega_w^2 = 6/\pi^2$. In order that the approximate solution be equal to the exact one, it should be multiplied by $\pi/3$. This gives us the following relation

$$\omega_w^2 = \int_0^1 \tilde{\beta} \sin\left(\frac{\pi}{2}\tilde{H}\right) \mathrm{d}\tilde{H}. \qquad (14.19)$$

In Fig. 14.5, we compare the results of calculation based on the approximate and exact solutions, as given in the form of dependences $\psi\left(\tilde{H}_*\right)$. Calculations show that for $\sigma > 0$ the relative error of calculation by (14.19) is at most 4% and only for the extremal case $\sigma = 0$ it attains the value 10%. It should be noted that physically the stepwise law of variation of the generalized thermophysical property parameter represents some limiting case. In practice, one should expect more smooth behavior of the functions $\tilde{\beta}\left(\tilde{H}\right)$, so that the expected error of the approximate solution (14.19) should also be much smaller.

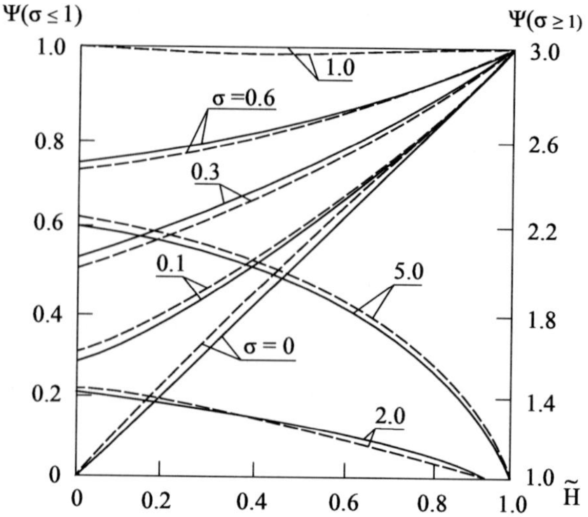

Fig. 14.5 The stepwise function $\tilde{\beta}\left(\tilde{H}\right)$. Comparison of calculation results based on the approximate (dashed line) and exact (solid line) solutions

14.5 Heat Transfer and Friction in the Turbulent Boundary Layer

14.5.1 Mathematical Description

Figure 14.1 shows that the behavior of the medium thermophysical properties in the SCP area is very involved and even exotic to a certain degree. This suggests the problem of validation of the surface renewal method by using problems with available solutions. Let us consider in this connection the case of flow of a viscous compressible gas in a turbulent boundary layer. Under isobaric conditions, the dependence of the gas thermophysical properties on the temperature can be approximated by the relations

$$c_p = \text{const}, \ \rho T = \text{const}, \ \frac{\mu}{T} = \text{const}, \ \text{Pr} \equiv \frac{\mu c_p}{k} = 1.$$

The heat and dynamical interactions of the gas volume with the wall with homogeneous initial distributions of the longitudinal velocity and temperature are described by the continuity equation

$$\frac{\partial \rho}{\partial t} + \frac{\partial J}{\partial y} = 0, \tag{14.20}$$

the equation of conservation of the momentum longitudinal component

$$\rho \frac{\partial U}{\partial t} + J \frac{\partial U}{\partial y} = \frac{\partial}{\partial y}\left(\mu \frac{\partial U}{\partial y}\right), \tag{14.21}$$

the energy conservation equation

$$\rho\frac{\partial H}{\partial t}+J\frac{\partial H}{\partial y}=\frac{\partial}{\partial y}\left(\mu\frac{\partial H}{\partial y}\right)+A\mu\left(\frac{\partial U}{\partial y}\right)^2, \tag{14.22}$$

and the uniformity conditions

$$t=0:(y>0), U=U_\infty, H=H_\infty, \tag{14.23}$$

$$y=0:(t>0), U=0, H=H_w, J=0. \tag{14.24}$$

Here, $A=\left(\frac{c_{p\infty}}{c_{v\infty}}-1\right)\mathrm{M}^2$ is the parameter of compressibility, $\mathrm{M}=\frac{U_\infty}{\sqrt{c_{p\infty}/c_{v\infty}R_gT_\infty}}$ is the Mach number.

Equation (14.20) coincides with the continuity Eq. (14.3). As distinct from Eq. (14.4), in the derivation of the energy Eq. (14.22) we took into account the energy dissipation (the last term on the right). Let us change to the dimensional values of the velocity$u=\frac{U}{U_\infty}$, the mass flow $j=\frac{J}{\rho_\infty}\sqrt{\frac{2t}{\nu_\infty}}$, the enthalpy$h=\frac{H}{H_\infty}$, and the thermophysical properties$\tilde{\rho}=\frac{\rho}{\rho_\infty}$, $\tilde{\mu}=\frac{\mu}{\mu_\infty}$, $\tilde{\nu}=\frac{\nu}{\nu_\infty}$. Introducing the similarity variable $\eta=\frac{y}{\sqrt{2\nu_\infty t}}$, we can rewrite the system of Eqs. (14.20)–(14.22) as

$$-\eta\frac{d\tilde{\rho}}{d\eta}+\frac{dj}{d\eta}=0, \tag{14.25}$$

$$-\tilde{\rho}\,\eta\frac{du}{d\eta}+j\frac{du}{d\eta}=\frac{d}{d\eta}\left(\tilde{\mu}\frac{du}{d\eta}\right), \tag{14.26}$$

$$-\tilde{\rho}\,\eta\frac{dh}{d\eta}+j\frac{dh}{d\eta}=\frac{d}{d\eta}\left(\tilde{\mu}\frac{dh}{d\eta}\right)+A\,\tilde{\mu}\left(\frac{du}{d\eta}\right)^2. \tag{14.27}$$

The uniformity conditions (14.23), (14.24) assume the form

$$\eta=0:u=0, h=F, \tag{14.28}$$

$$\eta\to\infty, u=1, h=1, j=0, \tag{14.29}$$

where

$$F=\frac{H_w}{H_\infty}=\frac{T_w}{T_\infty}$$

is the temperature factor. It can be shown that Eq. (14.27) has an integral relating the fields of the enthalpy and longitudinal velocity

$$h = F + \left(1 + \frac{A}{2} - F\right)u - \frac{A}{2}u^2. \tag{14.30}$$

Relation (14.30) extends the well-known Crocco integral [17] in the laminar boundary layer theory to the time-dependent problem under consideration. It can be written in the following more transparent form

$$\left(\frac{dT}{dU}\right)_w = \frac{T_0 - \Delta T}{U_\infty}.$$

Here, $\Delta T = T_w - T_\infty$ is the "wall–flow core" temperature difference, $T_0 = \frac{U_\infty^2}{2c_{p\infty}}$ is the flow stagnation temperature.

14.5.2 *Effect of Variable Thermophysical Properties*

Let us consider for definiteness the case of gas heating, $\Delta T > 0$. From the general problem of friction and heat transfer associated with the flow of a viscous compressible gas, we single out the following particular problems [19, 24]:

- The unisothermality problem. $\Delta T \gg T_0 : \left(\frac{dT}{dU}\right)_w = -\frac{\Delta T}{U_\infty} < 0$. The distribution of temperatures is similar to that of velocities.
- The compressibility problem. $\Delta T = T_0 : \left(\frac{dT}{dU}\right)_w = 0$.The intensity of the heat source formed due to dissipation of the kinetic energy of the decelerating gas volume precisely offsets the heat flow incoming to the gas from the hot wall. As a result, the high-speed gas passes the thermally insulated wall.

Hence, we get the expression for the temperature factor, which corresponds to the gas flow with sound speed

$$F_* \equiv \frac{T_{w*}}{T_\infty} = \frac{1}{2}\left(\frac{c_{p\infty}}{c_{v\infty}} + 1\right).$$

For $F < F_*$ we have a subsonic flow, and for $F_* > 1$, a supersonic flow. The inversion of the heat flow is as follows $\Delta T < T_0 : \left(\frac{dT}{dU}\right)_w > 0$. The heat produced in the gas volume due to dissipation forms the "back-flow" of heat toward the wall, which exceeds the heat flow from the overheated wall. Moreover, the total heat flow changes its sign, and as a result, the heat problem is replaced by the cooling problem. Below we shall be concerned only with the first two problems.

Using relation (14.30), one can exclude Eq. (14.27) and reduce the mathematical description of the problem to the system of Eqs. (14.25), (14.26). Applying to these equations the procedure similar to that of Sect. 14.3, we arrive at the following analogue of Eq. (14.5)

$$\tilde{\rho} + \frac{d}{d\eta}\left[\frac{d}{d\eta}\left(\tilde{\mu}\,\frac{du}{d\eta}\right)\right] = 0.$$

Let us introduce the dimensionless momentum flow

$$g = \tilde{\mu}\,\frac{du}{d\eta} \tag{14.31}$$

and change from the variables η, u to the new variables u, g. Now, proceeding as in Eq. (14.10), we get the following main equation

$$g\frac{d^2 g}{du^2} + \gamma = 0. \tag{14.32}$$

Here, $\gamma = \tilde{\rho}\tilde{\mu}$ is the generalized complex of thermophysical properties depending on the medium temperature (similarly to the parameter $\tilde{\beta}$). The boundary conditions for Eq. (14.32) are similar to the boundary conditions (14.13), (14.14)

$$u = 0 : \frac{dg}{du} = 0, \tag{14.33}$$

$$u = 1 : g = 0. \tag{14.34}$$

From the dependences of the gas thermophysical properties on the temperature it follows that $\gamma \equiv \tilde{\rho}\tilde{\mu} = 1$. This means that Eq. (14.32) has the same solution as in the reference case of constant properties

$$u = \text{erf}(\text{z}). \tag{14.35}$$

Here, $z = \sqrt{\ln \frac{g_w}{g}}$ and $\text{erf}(z)$ is the error function.

For the dimensionless momentum flow on the wall, we have $u = 0$, $g_w = \sqrt{2/\pi}$. In view of (14.31), (14.35), the expression for the momentum flow on the wall ($\xi = 0, u = 0$) assumes the form

$$\tau_w = \sqrt{\frac{\mu_\infty \rho_\infty}{\pi t}} U_\infty.$$

By definition, the relative law of friction reads as

$$\psi_m = \psi_{\text{var}} \psi_{\text{t}}. \tag{14.36}$$

The first component describes the effect of variability of properties

$$\psi_{\text{var}} = \frac{(g_{\text{w}})_{\text{var}}}{(g_{\text{w}})_{\text{const}}}. \tag{14.37}$$

The second factor reflects the effect of thermal expansion (contraction)

$$\psi_{\text{t}} = \sqrt{\frac{t_{\text{const}}}{t_{\text{var}}}}. \tag{14.38}$$

Here, $t_{\text{const}}, t_{\text{var}}$ is the mole travel time to the wall, respectively, for the cases of constant and variable properties. Since $\gamma = 1$, in the case under consideration we have $(g_w)_{\text{var}} = (g_w)_{\text{const}}, \psi_{\text{var}} = 1$. So, in the framework of the surface renewal model, the variability of the gas thermophysical properties does not introduce any changes in the momentum flow of gas on the wall.

14.5.3 *Effect of Thermal Expansion/Contraction*

Thermal expansion (contraction) results in the appearance in the gas volume of the transverse velocity, which can be evaluated by solving Eqs. (14.25), (14.27)

$$\text{V}\sqrt{\frac{t}{\nu_\infty}} = \frac{1/\sqrt{\pi}\left(1-\exp\left(-z^2\right)\right)(F-1)+}{\left[1/\sqrt{2\pi}\operatorname{erf}\left(\sqrt{2}z\right) - 1/(2\sqrt{\pi})\left(1-\exp\left(-z^2\right)\right) - 1/\sqrt{\pi}\left(1-\exp\left(-z^2\right)\right)\right]A}$$

The absolute value of V increases from zero on the impermeable wall ($\eta = u = 0$) to the maximal value at the infinity of the volume ($\eta \to \infty, u \to 1$)

$$\text{V}_\infty = \sqrt{\frac{\nu_\infty}{\pi t}}\left(F - 1 + \frac{\sqrt{2}-1}{2}A\right). \tag{14.39}$$

The expression for the heat flux through the wall reads as

$$q_w \equiv -\left(k\frac{\partial T}{\partial y}\right)_w = c_{p\infty}T_w\sqrt{\frac{\mu_\infty \rho_\infty}{\pi t}}\left(F - 1 - \frac{1}{2}A\right). \tag{14.40}$$

Formulas (14.39), (14.40) show that the signs of V_∞ and q_w are determined by combinations of the temperature factor $F = T_w/T_\infty$ and the compressibility factor$A = 0.4\text{M}_\infty^2$.[1]

Let us examine the effect of these factors.

---

[1]The expression for the compressibility factor is written for the case of a diatomic gas.

- For small values of the Mach number ($M \to 0$), the signs of the transverse velocity and the heat flow are equal. We have: (a) in the range $0 < F < 1$, the thermal contraction ($V_\infty < 0$) and heating of the wall ($q_w < 0$), (b) in the range $1 < F < \infty$, the thermal expansion ($V_\infty > 0$) and cooling of the wall ($q_w > 0$). This means that for $F = 1$ the inversion of both dependences takes place simultaneously (14.39), (14.40).
- As the Mach number increases, the thermal contraction takes place with more and more smaller value of the temperature factor. For $M \geq 3.47$, the thermal expansion of the gas volume takes place even in the hypothetical limit case $F = 0$ ($T_w = 0$). At the same time, the range of F corresponding to the wall cooling expands more and more and for $M = 3.47$ it becomes $0 < F < 3.41$. Physically, such a trend is due to an increasing dissipation of energy (the last term on the right of Eq. (14.22). As a corollary, a heat source appears in the near-wall region, whose intensity increases with increasing M. The heat flow from the source propagates both toward the wall (causing the heating of the wall) and from the wall (causing the thermal expansion of gas).
- To each Mach number, there corresponds the boundary value of the temperature factor $F_* = 1 + 0.2M_\infty^2$ corresponding to the thermally insulated wall ($q_w = 0$). Besides, the volume will expand with the rate $V_\infty = 0.16M\sqrt{\frac{\nu_\infty}{t}}$. For $F < F_*$ the wall is heated, $q_w < 0$, while for $F > F_*$ the wall is cooled down, $q_w > 0$.

The surface friction accompanying the gas flow in the turbulent boundary layer is given by the relation

$$\tau_w = c_f/2 \rho_\infty U_\infty^2. \tag{14.41}$$

Here,

$$c_f/2 = (2.5\ln\mathrm{Re} + 3.8)^{-2}$$

is the friction coefficient which can be found from Karman's formula [19], Re is the Reynolds number, which is calculated from the loss of momentum thickness.

According to the two-layer model of near-wall turbulence [17], the velocity at the boundary of the viscous sublayer of thickness δ is given by $U_\delta = 12.7\upsilon_*$. Here, $\upsilon_* = \sqrt{\tau_w/\rho_\infty}$ is the "friction velocity" defined as $\upsilon_* = \sqrt{c_f/2}U_\infty$ [18]. The momentum flow transported under a turbulent flow between the core and the viscous sublayer reads as $\tau_w = \rho_\infty W(U_\infty - U_\delta)$. From this expression, one can find the turbulent transport velocity (the "Reynolds velocity") [24]

$$W = \frac{c_f/2}{1 - 12.7\sqrt{c_f/2}} U_\infty.$$

In the frameworks of the surface renewal model, the exchange of momentum and heat between the medium and the wall is affected by periodic contacts with the wall of volumes (turbulent moles) incoming from the flow turbulent core. As in the classical Prandtl's mixing-length model, we introduce the characteristic length traveled by a turbulent mole as it moves to the wall

$$l \equiv W t_{\text{const}} = W t_{\text{var}} - \int_0^{t_{\text{var}}} V_\infty dt. \tag{14.42}$$

We shall assume that the length of the mixing path is independent of the flow parameters, but is rather geometric characteristics of the flow: l = idem. For the reference case of constant properties, we have

$$V_\infty = 0, t_{\text{var}} = t_{\text{const}}.$$

The thermal expansion of the near-wall volume ($V_\infty > 0$) will result in the deceleration of the mole traveling from the turbulent core of the flow. As a corollary, the motion time of a turbulent mole to the wall is increasing, $t_{\text{var}} > t_{\text{const}}$. In turn, the thermal contraction ($V_\infty < 0$) has an accelerating effect, and in this case, we have $t_{\text{var}} < t_{\text{const}}$. From the continuity considerations of the continuous medium, it follows that the residence time of a model decelerating near the wall will be equal to the travel time of an accelerated turbulent mole from the core to the wall. Substituting the above expressions for V_∞, W, $c_f/2$ into (14.42), we get

$$\Psi_{\text{mod}} = \sqrt{B^2 + 1} - \text{B}, \tag{14.43}$$

where

$$B = 0.5\gamma\left(1 - 12.7\sqrt{c_f/2}\right)\left(F - 1 + 0.16\text{M}_\infty^2\right),$$

γ is a constant number.

In view of formula (14.41) for the reference friction coefficient, relation (14.43) can be looked upon as a final result in the case of the flow of a viscous compressible gas. With the help of this relation, the relative laws of friction were calculated in the case of a flow of a viscous compressible gas in the turbulent boundary layer. The calculation results have been compared with the theoretical results of [19], which were obtained on the basis of the "conservative boundary layer" theory.

Figure 14.6 compares the calculation results of the relative law of friction for the unisothermality problem with the solution of [19]. The figure shows that with the value $\gamma = 0.855$ the curves $\Psi(F)$ compare satisfactorily. It is worth noting that under this approach the effect of the Reynolds number is reproduced qualitatively in a broad range of its variation: Re $= 10^3 \ldots 10^6$. The quantity ψ is shown to decrease as each of the parameters F, Re is increasing.

Note that for the compressibility problem (Fig. 14.7), only qualitative agreement of the dependences ψ(M, Re), as calculated by (14.43), was achieved with the results

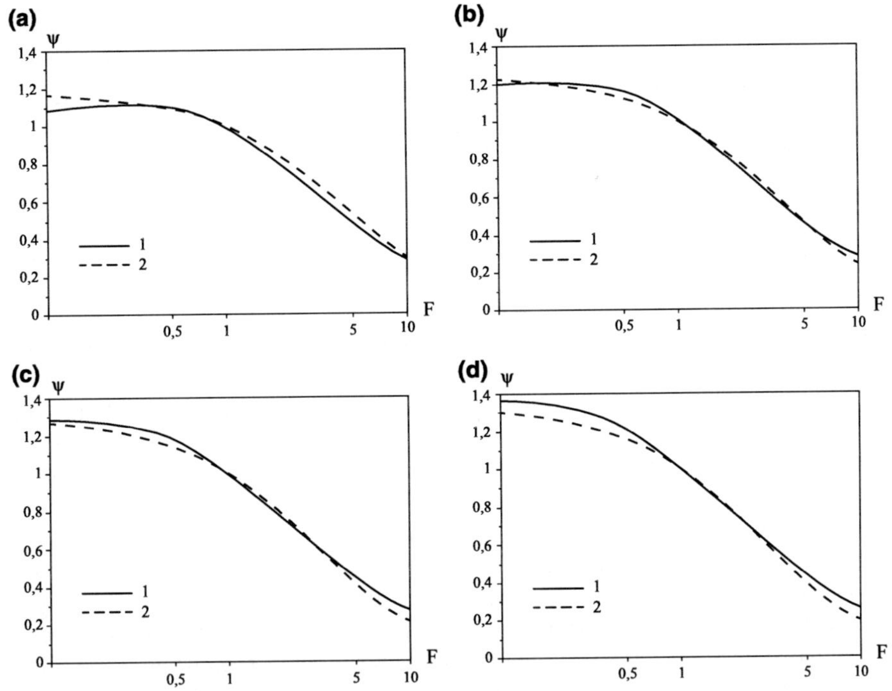

Fig. 14.6 Dependence of the relative law of friction on the temperature factor for the unisothermality problem. 1 Results of [19], 2 calculation by Eq. (14.43) **a** $\mathrm{Re} = 10^3$, **b** $\mathrm{Re} = 10^4$, **c** $\mathrm{Re} = 10^5$, **d** $\mathrm{Re} = 10^6$

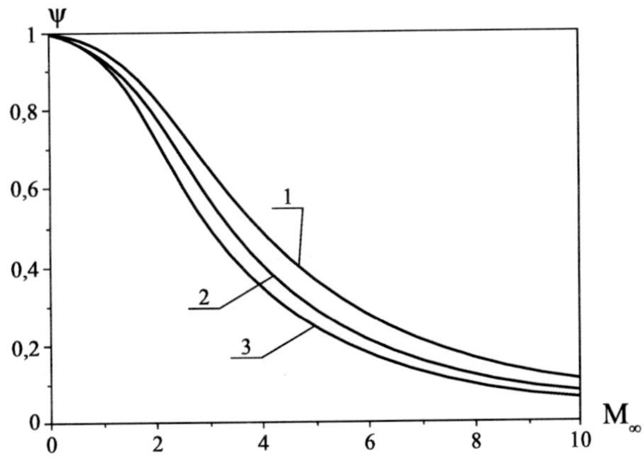

Fig. 14.7 Dependence of the relative law of friction on the Mach number for the compressibility problem. 1 $\mathrm{Re} = 10^3$, 2 $\mathrm{Re} = 10^6$, 3 $\mathrm{Re} = \infty$

of the theory of [19]. Here it should be noted that the scattering of experimental data in the zone of supersonic gas flow is very large and exceeds that that of the corresponding theoretical curves. So, the modified surface renewal model is capable of satisfactorily describing the case of flow of a viscous compressible gas in the turbulent boundary layer.

14.6 Wall Blowing/Suction

Blowing and suction of medium through a permeable surface is used in a number of engineering applications. Blowing is an efficient method of protection of the structure of a flight vehicle from high thermal loads. A distributed blowing into the boundary layer of a gaseous coolant passing through porous material is also useful for thermal protection of jet blades and walls of combustion chambers of liquid-propelled launch vehicle engines. Suction is used to enhance the aerodynamic characteristics of light vehicles by efficiently influencing the gas-dynamic structure of the flow.

Hydrodynamically, the blowing and suction are analogues to the evaporation and condensation processes. In [25], a model of liquid flow in the turbulent boundary layer on a permeable surface was proposed. This model is based on the relation between the components of the tensor of Reynolds stresses and the averaged velocity profile. In the book [19], the following relative law of friction

$$\psi = (1 - \alpha)^2 \tag{14.44}$$

was obtained in the case of a flow past a permeable surface, where $\alpha = \frac{1}{2c_f}\frac{V_w}{U_\infty}$ is the parameter of blowing.

Let us apply the model developed above for the calculation of the wall permeability effect on the model relative law of friction. Assume that a liquid volume moves with homogeneous velocity U_∞ along the wall and that it is in contact with it ab initio. Then in the liquid there develops the time-dependent process of liquid viscous drag. Let us consider the case when through the permeable wall the same liquid is blown (sucked off) with velocity V_w. Let us write the equation of momentum conservation for the liquid volume as

$$\frac{\partial \tilde{U}}{\partial t} + \tilde{V}_w \frac{\partial \tilde{U}}{\partial y} = \frac{\partial^2 \tilde{U}}{\partial y^2}. \tag{14.45}$$

Here, $\tilde{U} = U/U_\infty$, $\tilde{V}_w = V_w/U_\infty$ are the dimensionless velocities of the volume of blown (sucked) liquid. For $\tilde{V}_w > 0$, we have the blowing, and for $\tilde{V}_w < 0$, the suction. In analogy with formula (14.39), we define the velocity of blowing in the following similarity form: $V_w = \epsilon\sqrt{\frac{\nu}{t}}$. Introducing the similarity variable $\eta = \frac{y}{\sqrt{2\nu t}}$, we rewrite Eq. (14.45) as

$$(\epsilon - \eta)\frac{d\tilde{U}}{d\eta} = \frac{d^2\tilde{U}}{d\eta^2}. \tag{14.46}$$

The boundary conditions for Eq. (14.46) are as follows

$$\eta = 0 : \tilde{U} = 0,$$

$$\eta \to \infty : \tilde{U} = 1.$$

The integral for Eq. (14.31a) with given boundary conditions assumes the form

$$\tilde{U} = \frac{\operatorname{erf}\left(\epsilon/\sqrt{2}\right) + \operatorname{erf}\left(\left(\begin{smallmatrix}\eta \\ -\epsilon\end{smallmatrix}\right)/\sqrt{2}\right)}{1 + \operatorname{erf}\left(\epsilon/\sqrt{2}\right)}.$$

Let us find the momentum flow on the wall (with viscous frequency)

$$\tau_w = \left(\mu\frac{\partial U}{\partial y}\right)_w = \sqrt{\frac{\mu\rho}{\pi t}}U_\infty\frac{\exp\left(-\epsilon^2/2\right)}{1 + \operatorname{erf}\left(\epsilon/\sqrt{2}\right)}.$$

For $\epsilon = 0$ we get the reference value of the friction coefficient

$$\left(c_f/2\right)_* = \frac{1}{U_\infty}\sqrt{\frac{\nu_\infty}{\pi t}},$$

and the formula for the model parameter of blowing $\epsilon = 4k/\sqrt{\pi}\alpha$, where k is a constant number. As a result, we get the following model relative law of friction in the case of a permeable wall

$$\psi_{\mathrm{mod}} = \frac{\exp\left(-\epsilon^2/2\right)}{1 + \operatorname{erf}\left(\epsilon/\sqrt{2}\right)}. \tag{14.47}$$

Figure 14.8 compares the results of calculation by formula (14.47) with $k = 1.16$, and for the range $-0.5 < \alpha < 1$. The above range contains almost completely the entire bank of experimental data. The figure shows that the curves $\psi(\alpha)$ and $\psi_{\mathrm{mod}}(\alpha)$ are practically equal (the maximal departure is at most $\pm$ 2%). It should be noted that the vanishing of ψ with $\alpha = 1$, which is predicted by formula (14.44), appears to be unphysical. It can be assumed that this is a consequence of unnecessarily stringent original assumptions adopted in the model of [19]: the change of the logarithmic distribution of velocities by a power-law dependence, the assumption that the viscous sublayer has zero thickness as Re $\to \infty$. At the same time, from (14.47) it follows

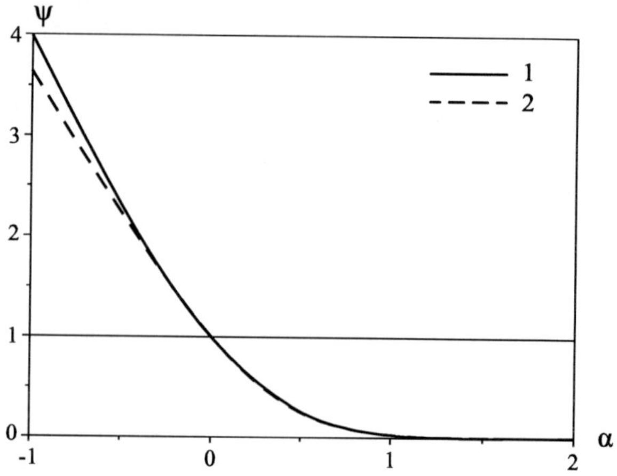

Fig. 14.8 Dependence of the relative law of friction on the parameter of blowing. 1 Results of [19], 2 calculation by Eq. (14.47)

that the friction is decaying monotonically with infinitely intensive blowing, $\alpha \to \infty$

$$\psi_{\mathrm{m}} = \frac{1}{2}\exp\left(-\frac{\epsilon^2}{2}\right) = 0.5\exp\left(-0.264\alpha^2\right).$$

This smooth behavior is in qualitative agreement with the results of [25].

For the suction conditions ($\beta < 0$), the solution (14.47) can be conveniently written in the from

$$\psi_{\mathrm{mod}} = \frac{\exp\left(-|\epsilon|^2/2\right)}{\mathrm{erfc}\left(|\epsilon|/\sqrt{2}\right)}, \tag{14.48}$$

where $|\epsilon| = -\epsilon$.

From (14.48), we get the following asymptotics for the suction of infinite intensity

$$|\alpha| \to \infty : \psi_{\mathrm{mod}} = \sqrt{\frac{\pi}{2}}|\epsilon| = 1.45|\alpha|.$$

We note that this behavior is drastically different from the asymptotics of solution (14.44) $|\alpha| \to \infty : \psi = |\alpha|^2$. So, the modified surface renewal model is capable of satisfactorily describing the effect of permeability of a surface past on the laws of friction in the turbulent boundary layer.

14.7 Heat Transfer Evaluation

14.7.1 Relative Law of Heat Transfer

Following the above validation of the surface renewal model, let us proceed with the evaluation of the relative law of turbulent flow heat transfer in the SCP area. The Fourier low for the wall conditions reads as $q_w \equiv \left(-k\frac{\partial T}{\partial y}\right)_w = \left(-\frac{k}{c_p}\frac{\partial H}{\partial y}\right)_w$. In the above notation, $\omega = \frac{\tilde{k}}{\tilde{c}_p}\frac{d\tilde{H}}{d\eta}, \tilde{H} = \frac{H}{H_w - H_\infty}, \eta = \frac{y}{\sqrt{2a_\infty t}}$, and so

$$q_w = \sqrt{\frac{k_\infty c_{p\infty}\rho_\infty}{2t}}\frac{H_w - H_\infty}{c_{p\infty}}\omega_w.$$

Let us introduce the coefficients of heat transfer for the cases of constant and variable properties, respectively

$$h_{\text{const}} = \sqrt{\frac{k_\infty c_{p\infty}\rho_\infty}{2t_{\text{const}}}}(\omega_w)_{\text{const}},\ h_{\text{var}} = \sqrt{\frac{k_\infty c_{p\infty}\rho_\infty}{2t_{\text{var}}}}\frac{\langle c_p\rangle}{c_{p\infty}}(\omega_w)_{\text{var}}.$$

Here, $\langle c_p\rangle \equiv \frac{H_w - H_\infty}{T_w - T_\infty}$ is the isobaric heat capacity averaged over the temperature range $T_w - T_\infty$. Now the relative law of heat transfer can be written as

$$\psi = \psi_{\text{var}}\psi_{\text{t}}\psi_{\text{c}_\text{p}}. \tag{14.49}$$

From (14.49), it follows that the quantity ψ is a product of three factors.

- The quantity $\psi_{\text{var}} = \left\{\frac{2}{\pi}\int_0^1 \tilde{\beta}\sin\left(\frac{\pi}{2}\tilde{H}\right)d\tilde{H}\right\}^{1/2}$ defined by (14.19) describes the effect of variability of thermophysical properties.
- The function $\psi_t = \sqrt{\frac{t_{\text{const}}}{t_{\text{var}}}}$ is responsible for the mechanism of thermal expansion (contraction) of the near-wall volume of the medium in the course of its thermal interaction with the superheated (supercooled) wall.
- The quantity $\psi_{c_p} = \frac{\langle c_p\rangle}{c_{p\infty}}$ is the ratio of the averaged and the reference heat capacities. It appears from the condition $T_w - T_\infty =$ idem in the construction of the relative law of heat transfer.

14.7.2 Deteriorated and Improved Regimes

Formula (14.49) can be used to detect regimes of heat transfer under a turbulent flow of a coolant in the SCP are in the channel. Let us calculate the heat transfer for helium when enthalpy of the coolant ranges in the interval 10 kJ/kg $< H <$ 40 kJ/kg. Let the enthalpy in the liquid flow be fixed, H_∞= 10 kJ/kg. For the model case, this corresponds to the conditions as $y \to \infty$. We shall monotonically increase the wall temperature T_w (and therefore, the enthalpy H_w, which is in a one-to-one correspondence with T_w). According to the above classification of the zones of parameters, this means that the coolant is in the pseudoliquid zone, the wall temperature is monotone increasing passing from the pseudoliquid zone to the pseudogaseous zone.

Figure 14.9 shows that, under these conditions, calculation by (14.49) gives better heat transfer ($\psi > 1$) in a bigger part of the range under consideration. The maximum is attained near the enthalpy of the pseudophase transition $H_w \approx 23.3$ kJ/kg, which practically agrees with the peak of the heat capacity (Fig. 14.1). The level of the heat transfer decreases as H_w is further increasing, and for $H_w = 40$ kJ/kg it attains its minimum.

Figure 14.9 shows that with increasing Reynolds number the heat transfer augmentation ratio is decreasing, so that in the range Re = $10^5 - 10^6$ and as the enthalpies of the wall $H_w \to 40$ kJ/kg, a marked deterioration of the heat transfer can be observed ($\psi < 1$). Let us now fix the enthalpy corresponding to the wall temperature T_w on the upper boundary of the pseudogaseous zone (H_w = 40 kJ/kg) and start to monotonically increase the enthalpy of the liquid flow H_∞.

According to Fig. 14.10, on the lower boundary of the pseudoliquid zone (H_∞= 10 kJ/kg) for Re = 10^4, a significant improvement of the heat transfer ($\psi > 1$) is observed. However, already for $H_\infty \approx 13$ kJ/kg this effect is offset, and with a further increase of the enthalpy of the coolant we can see an extended stage of aggravated heat transfer ($\psi < 1$). In the range Re = $10^5 \ldots 10^6$, only aggravated regimes are implemented in the entire range of variation of the fluid enthalpy (Fig. 14.10). This being so, using the modified surface renewal model it can is possible to analytically calculate the averaged thermohydraulic characteristics for near-wall turbulent flow (the boundary layer, the channel) in the general case of variable thermophysical properties.

The model is based on physically clear assumptions and was validated for the two standard problems available in the literature. It is worth specially mentioning that

Fig. 14.9 Helium in the supercritical pressures area at pressure 0.25 MPa. The relative law of heat transfer, the enthalpy at infinity is fixed, the enthalpy on the wall varies. 1 Re = 10^4, 2 Re = 10^5, 3 Re = 10^6

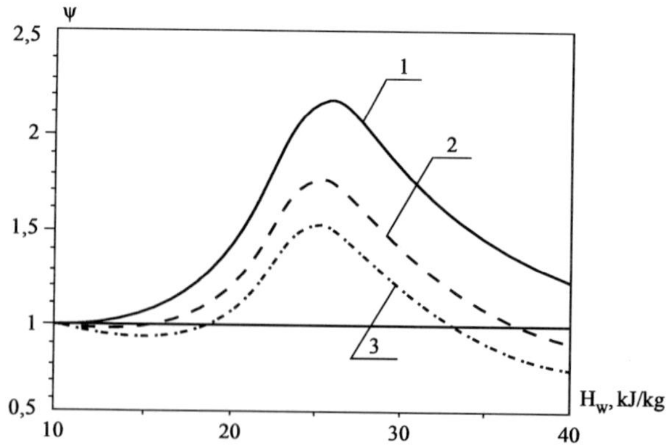

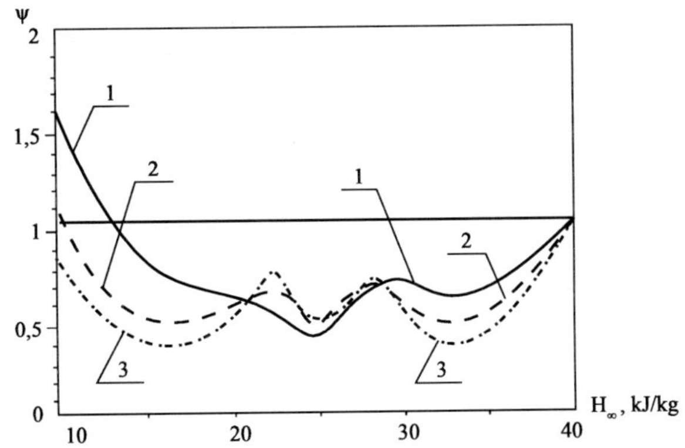

Fig. 14.10 Helium in the supercritical pressure area at pressure 0.25 MPa. The relative law of heat transfer, the enthalpy at the wall is fixed, the enthalpy at infinity varies. 1 $\mathrm{Re} = 10^4$, 2 $\mathrm{Re} = 10^5$, 3 $\mathrm{Re} = 10^6$

numerical studies of turbulent flow in the SCP area require very laborious calculations of the effects of various factors: the character and scale of the deformation of the profile of tangential stress, the thermal acceleration of the flow, the effect of buoyant forces, peculiarities of the heat transfer under the M-shaped velocity profile, generation and diffusion of turbulent energy, rearrangement of the flow structure under upflow and downflow motions. Based on the above, a conclusion can be made that it is desirable to have an optimal combination of numerical and analytical studies in this field, which would be instrumental in solving a number of practically important problems. In particular, it is a relevant task to develop a reliable complex of formulas for forecasting various regimes of heat transfer in the SCP area (the normal, improved, and aggravated ones).

14.7.3 Comparison with Experimental Data

As was already pointed out, the model developed above is based on the relative correspondence method. This method depends largely on the choice of a reference value for the friction and heat transfer coefficient. At present, there are several empirical formulas capable of satisfactorily describing experimental data on normal teat transfer to coolants in the area of SCP. However, one should be careful in dealing with the previously tried relations because of the appearance of refined thermophysical properties for water and water vapor. According to estimates, the use of these formulas can result in an overestimation of the heat transfer coefficient (up to 80%) for elongated ranges of enthalpy variation. This leads to the problem of reasonable "refurbishment" of old-generation empirical formulas. In the light of what we have said, it seems expedient to use as a reference relation the classical Petukhov and Kirillov's relation [18] for the heat transfer under the turbulent flow of a capillary liquid in a tube

$$\mathrm{Nu} = \frac{\mathrm{RePr}\xi/8}{1 + 900/\mathrm{Re} + 12.7\sqrt{\xi/8}\left(\mathrm{Pr}^{2/3} - 1\right)}. \tag{14.50}$$

Here, Nu is the Nusselt number.

Figures 14.11, 14.12 compare the results of calculation by (14.49), (14.50) with experimental data on heat transfer for water [26] and helium [27] flows in the SCP area. It can be concluded that the method developed above is capable of describing the above results at least not worse than the available numerical recommendations.

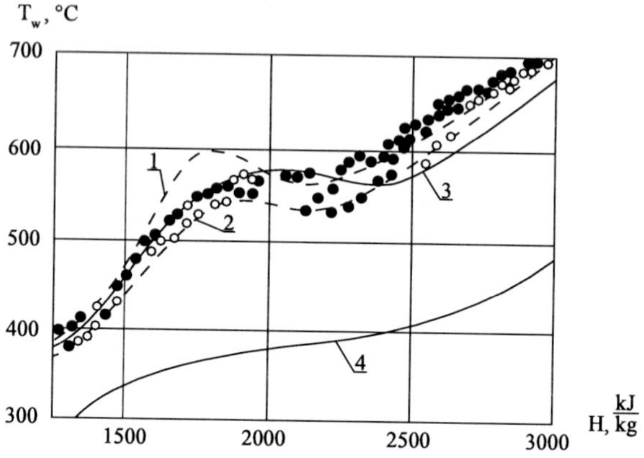

Fig. 14.11 Comparison of the results of calculation by (14.34), (14.35) with experimental data [26] on heat transfer for water flow in the supercritical pressure area. 1 Ascending flow, 2 descending flow, 3 calculation by Eq. (14.49), 4 mass average temperature of the coolant

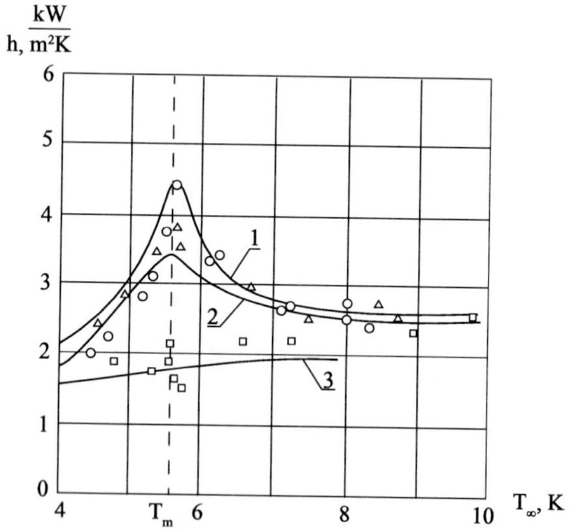

Fig. 14.12 Comparison of the results of calculation by (14.34), (14.35) with experimental data [27] on heat transfer for helium flow in the supercritical pressures area. 1 q = 1700 Wt/m^2, 2 q = 3840 Wt/m^2, 3 q = 7850 Wt/m^2

14.8 Conclusions

The heat transfer problem under a turbulent flow in a coolant channel in the zone of supercritical pressures was considered. The modified surface renewal model (the "spacer") was developed capable of calculating the effect of variable thermophysical properties on the friction and heat transfer. The model is based on the time-dependent periodic structure of near-wall turbulent flow and is applied in the frameworks of the relative correspondence method (the "shell"). The main equation is derived involving the entire complex of thermophysical properties: the heat capacity, thermal conductivity, and density of a supercritical fluid. An exact solution of the main equation is obtained for a discontinuous (stepwise) law of variation of the generalized parameter. The approximation solution is shown as being legitimate in describing the general case of variation of thermophysical properties. The model was validated on problems with available solutions: flow in a turbulent boundary layer of viscous compressible gas, a permeable wall past incompressible fluid. The law of turbulent flow heat transfer in the channel in the zone of supercritical pressures was calculated. The results obtained are compared with the available data on heat transfer. A conclusion was made that is desirable to have an optimal combination of numerical and analytical studies in this field, which would be instrumental in solving a number of practically important problems.

References

1. Pioro IL, Duffey RB (2007) Heat transfer and hydraulic resistance at supercritical pressures in power engineering applications. ASME Press, N-Y
2. Oka Y (2003) Research and development of the supercritical pressure light water cooled reactors. In: Proceedings of the 10-th international topical meeting on nuclear thermal hydraulics (NURETH-10). Seoul, Korea, 5–9 Oct 2003
3. Kurganov VA (1998) Heat transfer and pressure drop in tubes under supercritical pressure of the coolant. Part 1: specifics of the thermophysical properties, hydrodynamics, and heat transfer of the liquid. Regimes of normal heat transfer. Therm Eng 45(3):177–185
4. Goldman K (1954) Heat transfer to supercritical water and other fluids with temperature dependent properties. Chem Eng Prog Symp Ser 50(11):106–110
5. Hendricks RC, Graham RW, Hsu Y, Medeiros AA (1962) Correlation of hydrogen heat transfer in boiling and supercritical pressure states. ARS J 32(2):244–252
6. Weigand B (2015) Analytical methods for heat transfer and fluid flow problems. Springer
7. Kline SJ, Reynolds WC, Schraub FA, Runstadler PW (1967) The structure of turbulent boundary layers. J Fluid Mech 30:741–773
8. Corino ER, Brodkey RS (1969) A visual investigation of the wall region in turbulent flow. J Fluid Mech 37:1–30
9. Blackwelder RF, Kaplan RF (1976) On the wall structure of the turbulent boundary layer. J Fluid Mech 76(Pt. 1):89–112
10. Sternberg JA (1962) Theory for the viscous sub-layer along a smooth boundary. J Fluid Mech 13:241–271
11. Schubert G, Corcos GM (1967) The dynamics of turbulence near a wall according to a linear model. J Fluid Mech 29:113–135

12. Danckwetrs PV (1951) Significance of liquid-film coefficients in gas absorptions. Ind Eng Chem 43:1460–1467
13. Tomas LC, Cingo PI, Chung B (1975) The surface rejuvenation model for turbulent convective transport. Chem Eng Sci 30:1239–1242
14. Carslaw HS, Jaeger JC (1992) Conduction of heat in solids. Clarendon Press, London, Oxford
15. Tomas LC (1976) The surface renewal approach to turbulence. Chem Eng Sci 31:787–794
16. Gudkov VI, Motulevich VP (1984) Relative correspondence method and its application in measurement practice. J Eng Phys 47(2):922–928
17. Schlichting H, Gersten K (1997) Grenzschicht-Theorie. Springer, Berlin, Heidelberg, New York
18. Taler D (2017) Simple power-type heat transfer correlations for turbulent pipe flow in tubes. J Therm Sci 26(4):339–348
19. Kutateladze SS, Leontiev AI (1964) Turbulent boundary layers in compressible gases. Academic Press and Arnold (translated and exquisitely commented by D. B. Spalding)
20. Yener Y, Kakac S (2008) Heat conduction, 4th edn. CRC Press
21. Zudin YB (2012) Theory of periodic conjugate heat transfer, 2 edn. Springer
22. Zudin YB (2016) Theory of periodic conjugate heat transfer, 3 edn. Springer
23. Kantorovich LV, Krylov VI (1958) Approximate methods of higher analysis. P. Noordho, Groningen, The Netherlands
24. Spalding DB (1963) Convective mass transfer—an introduction. McGraw Hill
25. Vigdorovich II (2015) New formulations of the temperature defect law for turbulent boundary layers on a plate. Int J Heat Mass Transf 84(5):653–659
26. Barulin YD, Vikhrev YV, Dyadyakin BV, Koblyakov AN, Sinitsyn IT (1971) Heat transfer during turbulent flow in vertical and horizontal tubes containing water with supercritical state parameters. J Eng Phys 20(5):665–666
27. Giarratano PJ, Jones MC (1975) Deterioration of heat transfer to supercritical helium. Int J Heat Mass Transf 18(5):649–653

Chapter 15
Bubble Rising in a Liquid

Symbols

F	Force
Fr	Froude number
r	Radial coordinate
z	Axial coordinate
Re	Reynolds number

Greek Letter Symbols

φ	Velocity potential
μ	Dynamic viscosity
ν	Kinematic viscosity
ρ	Density
σ	Surface tension

Subscripts

f	Fluid
p	Particle

15.1 Bubble Rising in Liquid Column

15.1.1 Spherical Bubbles

Gas bubbles rising in a liquid at rest under the gravity force may have various forms: sphere, oblate spheroid, and spherical cap. Depending on the form of the bubble, its trajectory can be either rectilinear, zigzag, or spiral. The unusual character of the form and the buoyancy trajectory of rising gas bubbles was first analyzed already by

Y. B. Zudin, *Non-equilibrium Evaporation and Condensation Processes*,
Mathematical Engineering, https://doi.org/10.1007/978-3-030-67553-0_15

Leonardo da Vinci [1, 2], who tried to explain it (at the level of development of the hydrodynamics available at that time).

When considering the buoyancy of small spherical bubbles it is natural to use the analogy with the motion of a rigid sphere in a liquid medium. The classical solution of this problem was obtained in 1851 by Stokes [3] by approximating the "creeping flow": $\mathrm{Re}_p \ll 1$. Here $\mathrm{Re}_p = \frac{2au_\infty|\Delta\rho|}{\mu_f}$ is the Reynolds number,[1] a is the radius of the sphere, u_∞ is the main-stream velocity of the liquid at the sphere at infinity (in the reference in which the sphere is at rest), $\Delta\rho = \rho_p - \rho_f$, ρ_p, ρ_f is the density of the rigid sphere and liquid, respectively, r, μ_f is the dynamic viscosity of liquid. The basic assumption about the creeping flow is as follows: the inertia terms are negligible in the momentum equation. The drag force occurring when the moving sphere moves in the liquid is determined by integration over its surface of the normal stress $\left(F_n = 2\pi a\mu_f u_\infty\right)$ and the shear stress $\left(F_\tau = 4\pi a\mu_f u_\infty\right)$. Summing up, we get the Stokes' drag (see [4])

$$F_d = F_n + F_\tau = 6\pi a\mu_f u_\infty. \tag{15.1}$$

Using Stokes' formula, one can find the velocity of free stationary fall of solid balls in liquid or gaseous medium. The equation of force balance

$$\frac{4\pi}{3}\rho_p g a^3 = F_d + F_A \tag{15.2}$$

implies the well-known Stokes' law

$$u_\infty = \frac{2}{9}\frac{a^2 g|\Delta\rho|}{\mu_f}. \tag{15.3}$$

Here $F_A = \frac{4\pi}{3}\rho_f g a^3$ is the buoyant force, g is the gravitational acceleration.

In contrast to the motion in a medium of solid particles, the rise of a bubble in a liquid is much more involved process.

- During the rise, the bubbles may change their form.
- The flow past a moving phase interface is responsible for gas circulation inside the bubble, which has effect on the character of its motion.

[1] Here and below, for definiteness, we shall always consider the positive difference of the densities ($|\Delta\rho| > 0$).

- Surfactant molecules of complicated structure may accumulate on the surface of a gas bubble as it moves in the liquid. As a result, the phase interface assumes certain rigidity properties without losing its flexibility.

An exact analytical solution of the problem of spherical particle motion in liquid medium with $\mathrm{Re}_p \ll 1$ was obtained in 1911 independently by Hadamard [5] and Rybczynski [6]. In [5, 6], the Navier–Stokes equations were considered for a particle and an incident medium with due account of the matching conditions on the interphase boundary: the equality of the tangential velocities and the shear stresses. The solution obtained by Hadamard and Rybczynski reads as

$$u_\infty = \frac{2}{9}\frac{\left(\mu_p + \mu_f\right)}{\mu_p + 2/3\mu_f}\frac{a^2 g|\Delta\rho|}{\mu_f}. \tag{15.4}$$

Here the subscript "p" refers to the moving particle, and "f" the surrounding liquid.

Relation (15.4) is universal, it describes all possible combinations of the phases in contact: a drop in the liquid, a bubble in the liquid, a drop in the gas. If the viscosity of the inner medium is much larger than that of the external medium $\left(\mu_p \gg \mu_f\right)$, then (15.4) becomes Stokes' law (15.3). In the other limit case $\left(\mu_p \ll \mu_f\right)$, we get

$$u_\infty = \frac{2}{9}\frac{a^2 g|\Delta\rho|}{\mu_f}. \tag{15.5}$$

For a spherical gas bubble rising in a liquid, $\mu_p \ll \mu_f$, $\rho_p \ll \rho_f$, $|\Delta\rho| = \rho_f - \rho_p \approx \rho_f$. Here the weight of the bubble is negligible relative to the buoyant force, and Eq. (15.2) simplifies to read (the positive velocity is directed from top to bottom)

$$F_d = F_A. \tag{15.6}$$

Hence for the rise velocity of a spherical gas bubble for $\mathrm{Re}_p \ll 1$, we have

$$u_\infty = \frac{2}{9}\frac{a^2 g}{\nu_f}, \tag{15.7}$$

where $\nu_f = \mu_f/\rho_f$ is the kinematic viscosity of liquid.

Experimental studies of a bubble rise have been performed in the past 120 years [7–23]. A detailed analysis of the existing bank of experimental data, as well as semi-empirical and empirical correlations, can be found in the papers [24, 25] and in the books [26, 27]. The forms of air bubbles rising in water were first classified in [10, 11]. Since a rising bubble may assume various forms, the analysis of peculiarities of its motion should reasonably include as the linear size the "equivalent radius" r_e, which is the radius of the sphere of the same volume as the bubble. In [10, 11], an experimental dependence $u_\infty(r_e)$ for rising air bubbles in water was given (Fig. 15.1).

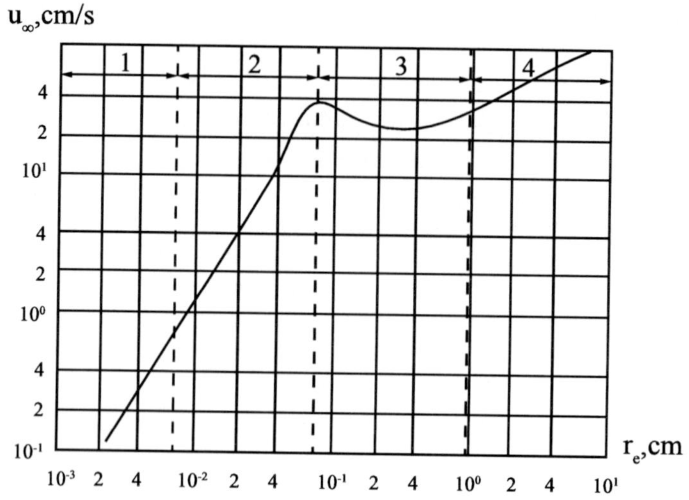

Fig. 15.1 Experimental dependencies of the rise velocity of air bubbles in water

A change in r_e has a considerable effect on the main characteristics of the process: the rise velocity, the shape, and the trajectory of a bubble (see Table 15.1 from [26] for air bubbles rise in water).

By analyzing the experimental curve $u_\infty(r_e)$, one can single out five conventional regions that differ in the bubble shape and the character of the flow past the liquid. These regions are indicated in Fig. 15.1 by numbers 1–5. Figure 15.2 shows the approximate forms of bubbles and the nature of the flow in each of these regions.

Region 1 corresponds to the lift motion of very small spherical bubbles for $\mathrm{Re}_p \ll 1$. For the motion of air bubbles in water, this corresponds to the condition: $r_e = a < 0.1$ mm. Since, for bubbles in this region, the Reynolds number is smaller than 1 and the form of the bubbles is spherical, the inertial and capillary forces should have no effect on their motion. Hence the analysis of dimensions gives a unique combination of the gravitation and viscous effects, which is expressed by the dependence of the form (15.5).

Table 15.1 Forms and trajectories of gas bubbles rising through stagnant liquid (see [26])

Region	Equivalent radius	Reynolds number	Buoyancy characteristics
1	$a < 0.4$ mm	$\mathrm{Re}_p < 70$	Spherical bubbles with rectilinear trajectory
2	0.4 mm $< a <$ 0.9 mm	$70 < \mathrm{Re}_p < 650$	Spherical or ellipsoidal bubbles with rectilinear trajectory
3	0.9 mm $< a <$ 9 mm	$650 < \mathrm{Re}_p < 5000$	Ellipsoidal bubbles with zigzag and spiral trajectories
4	$a > 9$ mm	$\mathrm{Re}_p > 5000$	Spherical-cap bubbles with rectilinear trajectory

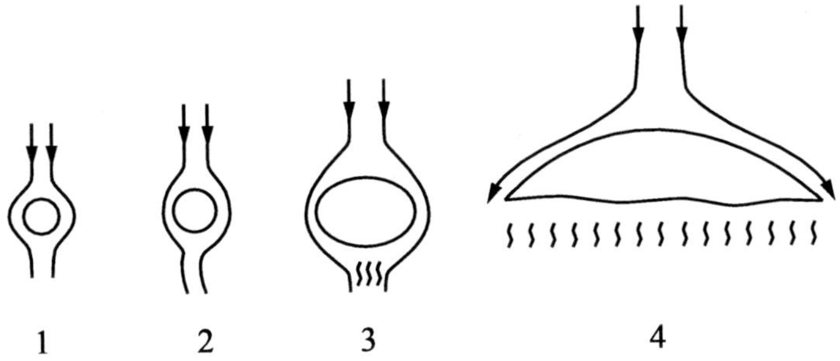

Fig. 15.2 Approximate forms of gas bubbles in liquid and the nature of the flow

Region 2 corresponds to the motion of spherical bubbles in region $1 < \text{Re}_p \le 300$, which corresponds to the variation of the equivalent radius in the range $0.1\,\text{mm} < r_e < 0.6\,\text{mm}$. The nature of the flow past the bubble in this case is similar to region 1, the dependence $u_\infty(r_e)$ is described by formula (15.5) with a larger number coefficient.

Region 3 is characterized by rectilinear motion of bubbles in the form of oblate spheroids oblate in the rise direction. In this narrow region $\left(300 < \text{Re}_p \le 500, 0.6\,\text{mm} < r_e < 0.8\,\text{mm}\right)$, the dependence $u_\infty(r_e)$ attains its maximum. Since the density and viscosity of gas in the bubble is much smaller than the density and viscosity of the liquid, the nature of the flow past ellipsoidal bubbles remains separation-free even outside the midsection. Here the narrow separation zone exists only in the neighborhood of the aft critical point. The most intriguing region 4, in which the condition $\text{Re}_p \gg 1$ holds, has been the least investigated. In this region, the form of a bubble evolves monotonically from an oblate spheroid ($r_e \approx 0.8$ mm) to a spherical cap ($r_e \approx 10$ mm).

The characteristic features of the motion of bubbles in this region are as follows.

- Pulsations of the form under the action of the surface tension force due to oscillations of the curvature of the phase interface.
- Existence of the considerable flow detachment zone in the wake of the bubble.
- Unstable (zigzag or spiral) buoyancy trajectory.

The rise velocity in region 4 attains its minimum and varies in the narrow range of $0.24\,\text{m/s} < u_\infty \le 0.32\,\text{m/s}$. For the initial zone of region 4, in [10] the formula is recommended

$$u_\infty \approx 1.9\sqrt{\frac{\sigma}{\rho_f r_e}}. \tag{15.8}$$

Region 5 corresponds to the gas bubbles of diameter $r_e > 10$mm. Photographies obtained in liquids with quite different physical properties (water, glycerin, oils, alcohols, liquid metals) show (see [24]) that the front part of such bubbles is a smooth spherical surface of radius R, while the aft part is practically flat. So, the bubbles in this region have the form of a practically regular spherical cap. For larger Reynolds numbers, in the aft zone of the bubble one can clearly see the turbulent wake characteristic of a separated flow of liquid around discs and spheres [28]. Motion of large air bubbles in very viscous liquids (glycerine, mineral oil) shows the following quite unusual picture [24]: rise of a practically regular "two-phase sphere" of radius R. Its upper (gas) part is occupied by the bubble itself, while in the lower (liquid) part one can vividly see close flow lines resembling a spherical Hill vortex [29]. In some cases, the aft part is elongated downwards forming a peculiar asymmetric film (a "skirt") shed from the bubble [12]. In region 5, the rise velocity is described by the empirical dependence [26]

$$u_\infty \approx c\sqrt{gr_e}. \tag{15.9}$$

It is interesting to note that in spite of quite different conditions (the liquid viscosity and surface tension on the interphase boundary vary in a wide range) the numerical coefficient in formula (15.9) varies in a very small interval $0.95 < c < 1.05$. With the development of computational methods, it became possible to use direct numerical modeling to solve the problem of a single bubble rise. The corresponding machinery is based on the use of the system of three-dimensional Navier–Stokes equations for a liquid flowing past a closed gas volume (see, for example, [30–33]). By numerical modeling, it proved possible to obtain very important information about the local characteristics of the process for specific ranges of regime parameters.

One should note, however, that numerical modeling of the motion of an ascending bubble is a very difficult task. These difficulties increase markedly for the problem of unstable bubble ascent. So, the paper [33] presents a solution based on the combined Euler–Lagrange approach, in which the bubbles exhibit an oscillating shape and unstable trajectories. The developed model makes it possible to calculate the direction, shape, as well as zigzag or spiral trajectories of the bubbles floating up. However, parametric study was carried only for a bubble with an equivalent diameter of 3 mm. It is clear that the appropriate model constants chosen for some or other specific case cannot be considered appropriate for describing the general pattern of this complex process.

15.1.2 Force Balance

The use of the force balance provides a simplest way for the description of spherical bubbles. However, despite the apparent triviality of the object of research, this question is not so obvious as, for example, the case of the motion of a rigid sphere in liquid. In fact, gas bubbles can be considered as "cavities" in liquid—this is supported by the traditional use of the term "void fraction" to denote the volumetric gas content of bubbles in two-phase flows. Next, a boundary bubble is a "free surface". As is known [29], the free surface is the surface of a flowing fluid that is subjected to both zero perpendicular normal stress and parallel shear stress. Therefore, calculation of free surfaces is difficult due to the uncertainty of the position of the interface.

Note also that, despite the external similarity, a gas bubble is not similar to a balloon floating in a liquid. The essential difference is that the only mechanical connection between the bubble and the environment is the surface tension on the interface. A change in the external pressure on any part of the surface leads to its deformation, but is not transmitted to the center of mass of the bubble. As a result, the basic concepts of mechanics of nondeformable solids like “a force applied to the center of mass”, “the volumetric action of surface forces” cannot be applied in the case of a bubble in a liquid. Therefore, the use of relations based on the balance of forces acting on the bubble in the analysis is actually reduced to qualitative estimates similar to the analysis of dimensions. In rigorous analysis, the shape of the bubble surface is unknown in advance and should be determined in the process of the solution of the problem based on the analysis of the velocity and pressure fields in the surrounding liquid.

15.1.3 Energy Balance

The energy balance provides a more rigorous method of the description of rising bubbles. Consider the motion of a rigid sphere in liquid with the following conditions: $\mathrm{Re}_p \ll 1$, $\rho_p \ll \rho_f$, $|\Delta\rho| = \rho_f - \rho_p \approx \rho_f$. Let the origin of the polar coordinate (r, θ) be located at the sphere center. The radial velocity u_r and the tangent velocity u_θ in the surrounding liquid are described by the relations [34]

$$\left.\begin{aligned} u_r &= u_\infty\left(1 - 2\tfrac{\gamma_1}{r^3} - 2\tfrac{\gamma_2}{r}\right)\cos\theta, \\ u_\theta &= u_\infty\left(-1 - \tfrac{\gamma_1}{r^3} + \tfrac{\gamma_2}{r}\right)\sin\theta, \end{aligned}\right\} \tag{15.10}$$

which show that the field of velocities satisfies the boundary condition

$$r \to \infty : u_r = u_\infty \cos\theta,\, u_\theta = -u_\infty \sin\theta. \tag{15.11}$$

Expressions (15.10) are universal and describe, in particular, the inviscid flow around the sphere

$$\left.\begin{aligned} u_r &= u_\infty\left(1 - \tfrac{a^3}{r^3}\right)\cos\theta, \\ u_\theta &= u_\infty\left(-1 - \tfrac{1}{2}\tfrac{a^3}{r^3}\right)\sin\theta. \end{aligned}\right\} \tag{15.12}$$

Formula (15.12) is a particular case of (15.10) with $\gamma_1 = 1/2$, $\gamma_2 = 0$.

The motion of a body in a viscous liquid is accompanied by energy dissipation—the mechanical energy is transformed into thermal energy. According to [34], the dissipation rate of a volume unit (the dissipation function) in polar coordinates is as follows

$$\Phi = 2\left[\left(\frac{\partial u_r}{\partial r}\right)^2 + \left(\frac{1}{r}\frac{\partial u_\theta}{\partial \theta} + \frac{u_r}{r}\right)^2 + \left(\frac{u_r + u_\theta \cot\theta}{r}\right)^2\right]$$
$$+ \left[r\frac{\partial}{\partial r}\left(\frac{u_\theta}{r}\right) + \frac{1}{r}\frac{\partial u_r}{\partial \theta}\right]^2 - \frac{2}{3}\left[\frac{1}{r^2}\frac{\partial}{\partial r}(r^2 u_r) + \frac{1}{r\sin\theta}\frac{\partial}{\partial\theta}(u_\theta \sin\theta)\right]^2. \quad (15.13)$$

The full rate of energy dissipation in the entire volume of the surrounding liquid is given by the integral

$$\mathcal{D} = \mu \int_a^\infty 2\pi r dr \int_0^\pi \Phi \sin\theta d\theta. \quad (15.14)$$

Substituting Φ from (15.13) into (15.14) and taking into account (15.12), we get

$$\mathcal{D} = 16\pi \mu u_\infty^2 \left(3\frac{\gamma_1^2}{a^5} + 2\frac{\gamma_1\gamma_2}{a^3} + \frac{\gamma_2^2}{a}\right). \quad (15.15)$$

Using (15.6), the drag force preventing the motion of the gas bubble can be found from the energy balance: the work of the buoyant force per unit time is equal to the dissipation rate of the kinetic energy

$$\mathcal{D} = F_d u_\infty. \quad (15.16)$$

From (15.16), we get the following relation for the drag force

$$F_d = 16\pi \mu u_\infty \left(3\frac{\gamma_1^2}{a^5} + 2\frac{\gamma_1\gamma_2}{a^3} + \frac{\gamma_2^2}{a}\right). \quad (15.17)$$

The values of the coefficients γ_1, γ_2 in (15.17) will be determined by the boundary conditions on the surface of the sphere. In the approximation of "creeping flow" under consideration, the two following limit cases are possible.

- To the motion of the hard sphere, there corresponds the no-slip boundary condition

$$r = a : u_r = u_\theta = 0, \quad (15.18)$$

  which gives the classical Stokes' drag (see (15.1)).
- In the cases of the bubble motion, the tangent stresses on the phase interface should be absent Hence, we get the slip boundary condition

$$r = a : u_r = 0, \frac{\partial u_\theta}{\partial r} = 0 \quad (15.19)$$

for which the drag force is smaller

$$F_d = \frac{9\pi}{2} a\mu_f u_\infty. \tag{15.20}$$

- From (15.20), it follows that the resulting drag force is 3/4 of the Stokes force. Indeed, the slip condition on the surface of the sphere in a stream reduces the velocity gradient along the normal vector to the surface. Hence the surface friction will be smaller in this case than for the no-slip boundary condition.

15.1.4 Semiempirical Models

In [35], Levich proposed a novel model based on the energy balance for evaluation of the bubble rise velocity for $\mathrm{Re}_p \gg 1$. The basic assumption in Levich's model is that on the surface there exists a bubble in the boundary layer of thickness $a\mathrm{Re}_p^{-1/2}$. The specific property of this boundary layer is that, in contrast to the Prandtl classical boundary layer with the no-slip boundary condition (15.18), the slip boundary condition (15.19) is used. The flow outside the boundary layer is considered irrotational and is described by the velocity field (15.11).

However, further arguments of Levich [35] are controversial and contradictory.

- Because of the "softness" of the slip boundary condition (15.19) (in comparison with the "hard" no-slip boundary condition (15.18)), the dissipative function inside the boundary layer is assumed to be negligible.
- It is assumed that for $\mathrm{Re}_p \gg 1$, the velocity field in an inviscid flow is practically the same as the exact distribution of the velocities satisfying the Navier–Stokes equations for a viscous liquid.
- The expressions for the velocity components of the inviscid flow (15.12) are substituted into (15.15). This gives us the full rate of dissipation of the energy in the entire liquid volume surrounding the bubble.
- The drag force is found by formula (15.15)

$$F_d = 12\pi a\mu_f u_\infty. \tag{15.21}$$

It is easily seen that the value is twice larger than the drag force obtained from the Stokes' drag (see (15.1)). Table 15.2 gives the values of γ_1, γ_2, F_d calculated for the above three cases.

- From (15.6) the required bubble rise velocity is determined

$$u_\infty = \frac{1}{9}\frac{gr_e^2}{\nu_f}. \tag{15.22}$$

Table 15.2 Evaluation of the drag force for a rising sphere in liquid ($\rho_p \ll \rho_f$)

	Viscous flow No-slip boundary condition	Viscous flow Slip boundary condition	Inviscid flow
γ_1	$-a^3/3$	$a^3/8$	$1/2$
γ_2	$3a/4$	$3a/8$	0
F_d	$6\pi a\mu u_\infty$	$9\pi/2a\mu_f u_\infty$	$12\pi a\mu u_\infty$

Levich's method was extended by Moore [36], who in addition considered the energy dissipation in the boundary layer and in the wake behind the bubble in the flow. The expression of [36] for the drag force reads as

$$F_d = 12\pi a\mu_f u_\infty\left(1 - \frac{2.21}{\sqrt{\mathrm{Re}_p}}\right). \tag{15.23}$$

However, in [37] it is claimed that the method of the papers [35, 36] is incomplete due to the neglection of the viscous effects. As is known, potential flows are solutions of the Navier–Stokes equations for a viscous incompressible fluid, for which the vorticity is identically zero. In this case, the viscous term in the equation disappears. However, the viscous contribution to stress in an incompressible fluid is retained. In [37], it was shown that the viscosity in the potential flow from the boundary layers enters into the Prandtl boundary layer equations. In [37], it was shown that the viscosity should be considered under dynamic conditions expressing the continuity of normal stress at the gas–liquid interface.

Summarizing the results of [35–37], we should point out the internal inconsistency of the Levich method [35]: at first, the liquid flowing past the bubble is considered as inviscid (see formulas (15.12)), but later the resulting flow is formally "attributed" the viscosity property. Papers [36, 37], which propose small corrections to the Levich solution, also violate the logic of classical hydrodynamics. Moreover, the resulting formula (15.22) obtained in this approach is a direct consequence of the dimensionality considerations.

In this regard, let us briefly discuss the value of the classical Prandtl's theory [38]. The boundary layer theory was founded in 1904 at the International Mathematical Congress in Heidelberg, when Ludwig Prandtl give a lecture entitled "Über Flüssigkeitsbewegung bei sehr kleiner Reibung" ("On fluid flow with very little friction"). He explained that the viscosity of a fluid plays a role in a very thin layer adjacent to the surface, which he called "Übergangsschicht" or "Grenzschicht". Translated into English, the latter led to the term "boundary layer". This lecture, which radically expanded the understanding of fluid flow, included the significant phrase: "Wenn man von der Bewegung mit Reibung zur Grenze der Reibungslosigkeit übergeht, so erhält man etwas ganz anderes als die Dirichlet-Bewegung".

For example, D'Alembert's paradox, which states that a body placed in a potential stream suffers no resistance from the environment, has been resolved. Based on the properties of the boundary layer, it was subsequently possible to explain why airplanes can fly. The concept of the boundary layer served as an impetus for the creation of a new branch of mathematics: the theory of singular perturbations. Outside the boundary layer, the flow can be considered inviscid. The general flow field is found by connecting the boundary layer and the inviscid outer region. The boundary layer theory is a unique method of description of a very large class of problems. This generic method is more accurate the more large is the Reynolds number. The boundary layer theory has made it possible to formulate problems in a sufficiently simple form to obtain analytical solutions. At the same time, despite the approximate nature of the theory, it gives results with high accuracy.

Let us give some important corollaries of the boundary layer theory in application to the problem when a viscous liquid with velocity U pasts a body with linear scale L [39].

- The boundary layer has a thickness of order $\sim \sqrt{\nu L/U}$.
- The tangent velocity of liquid increases from zero (on the hard wall) to the external stream velocity (on the boundary of the boundary layer).
- With increasing U (i.e., with increasing Reynolds number $\mathrm{Re} = UL/\nu$), the thickness of the boundary layer decreases, but always remains finite.
- The process of kinetic energy dissipation into thermal energy (i.e., the process of dissipation) occurs almost completely within the boundary layer.
- The gradient of the tangent velocity along the normal vector to the surface increases with decreasing viscosity of the liquid. As a result of this compensation, the dissipative function within the layer always remains finite.

It is easy to see that none of the above conditions are met in Levich's method [35]. It seems that to justify the correctness of this method, it is necessary to conduct a direct numerical simulation of the flow of a viscous liquid around the bubble in a wide range of Reynolds numbers. So, the contradictory assumptions of Levich's method can be tested only on the basis of the results of numerical solutions of the Navier–Stokes equations.

15.1.5 Empirical Models

For the case of an ellipsoidal bubble (region 4 in the classification of [10, 11]), a popular instrument is the Mendelson empirical formula [40], based on the "wave analogy"

$$u_\infty = \sqrt{\frac{\sigma}{\rho_f r_e} + g r_e}. \tag{15.24}$$

Consider the physical mechanism of generation of waves on an initially planar interface of two media [34]. Assume that a liquid of density ρ_f is the lower medium and a gas of density ρ_g is the upper medium, $\rho_g \ll \rho_f$. An external perturbation results in a distortion of the interface surface, which is responsible for the appearance of forces trying to recover the initial state: the capillary force prevents the increase in the phase interface area, the gravity force tries to bring the surface to the initial planar form.

Liquid particles, which tend to an equilibrium under the action of the stabilizing forces, will pass by inertia through this state and again deviate from the equilibrium to fall under the influence of the restoring forces, and so on. Hence, if the system is viscosity-free, in it there appear self-sustained oscillations running along the free surface of the liquid (progressive surface waves). If the appearing perturbation will not increase in time, then the phase interface will preserve its stability. In the case of unbounded increase of the amplitude in time, the system will be unstable. Consider the linear approximation when the amplitude of oscillations of the surface is negligible relative to the wavelength of perturbation λ. Form the instability analysis, we get the following relation for the phase velocity of propagation of the surface wave

$$u_w = \sqrt{\frac{\sigma}{\rho_f l} + gl}, \tag{15.25}$$

where $l = \lambda/2\pi$.

From (15.25), it is seen that the phase velocity has a minimum with $l = b \equiv \sqrt{\sigma/\rho_f g}$.

The quantity b is known as the "capillary constant". This value is a fundamental length scaling factor that relates gravity and surface tension. According to (15.25), the smallest value of the phase velocity is $u_{w\min} = \sqrt[4]{2\sigma g/\rho_f}$. In the presence of perturbations, each particle in the surface wave moves along a circle with respect to its equilibrium position. Hence the phase velocity of the wave is just the "motion" velocity of the crest (or trough) chosen for observation, but not the velocity of the matter.

Consider the asymptotics of expression (15.25).

- In the case of long (gravitation) waves, the phase velocity is determined by the gravity acceleration and increases with increasing wavelength: $l \to \infty$, $u_w \to \sqrt{gl}$.
- For short (capillary) waves, $l \to 0$, $u_w \to \sqrt{\sigma/\rho_f l}$. The phase velocity of capillary waves is controlled by the surface tension and increases with decreasing the wavelength.

A comparison of (15.24) and (15.25) shows that in essence the wave analogy [40] is reduced to postulating the equalities $r_e = l$, $u_w = u_\infty$. In other words, it is assumed without proper physical justification that the dependence of the bubble rise velocity on its equivalent radius agrees with the dependence of the phase velocity of the perturbation wave propagation along the liquid surface on its wavelength.

Paper [41] gives an empirical method of calculation of the bubble rise velocity for the entire range of variation of its equivalent radius. The expression for the rise velocity is sought in the form

$$u_\infty = \left(u_1^2 + u_2^2\right)^{-1/2}. \tag{15.26}$$

The component u_1 of the velocity responsible for the left branch of the experimental curve $u_\infty(r_e)$ (Fig. 15.1) is found from the force balance (15.6)

$$12\pi\mu u_1\left(1 - \frac{2.21}{\mathrm{Re}^{1/2}}\right) = \frac{4\pi}{3}\rho_f g r_e^3. \tag{15.27}$$

The left-hand side of (15.27) corresponds to the Moore solution (15.23) with the Reynolds number $\mathrm{Re} = r_e u_1/\nu_f$. Solving Eq. (15.27) with respect to u_1, it was shown in [41] that

$$u_1 \approx u_*\left[1 + 1.04\frac{(gr_e)^{1/2}}{u_*}\right]^{1/2}. \tag{15.28}$$

The component u_2 describing the region to the right of the maximum is given by (15.24) with corrected number coefficient

$$u_2 = \sqrt{\frac{3}{2}\frac{\sigma}{\rho_f r_e} + g r_e}. \tag{15.29}$$

Substituting u_1, u_2 from (15.28), (15.29) into (15.26) gives the required dependence $u_\infty(r_e)$ (which is too complicated to present here).

15.1.6 Instability of Buoyancy Trajectory for a Bubble

An unstable (zigzag or spiral) buoyancy trajectory of a bubble in region 4 has been the subject of experimental investigations starting from Leonardo da Vinci [1, 2]. In this connection, we mention, in particular [15–23]. Experiments show that depending on the initial buoyancy conditions a bubble may move in a clockwise or a counterclockwise spiral. The following experimentally observed effect is worth mentioning: the motion of a bubble may change from a zigzag to a spiral motion, however, the inverse transition never takes place. In experiments, it was also found that the transition from a zigzag to a spiral trajectory may occur as the size of the bubble increases. From detailed observations, one can distinguish two different modes corresponding to horizontal oscillations of the velocity of the bubble in mutually orthogonal directions. These modes have the same frequency and the relative phase shift by $\pi/2$ in the case of a spiral motion [19]. Moreover, the amplitude of horizontal pulsations of a bubble can be quite large. For example, the value 4.3 mm of the amplitude was obtained in experiments [21] for a bubble of $d = 2.5$ mm.

Analysis of experimental data [15–23] suggests that the zigzag motion is related to regular formation and release of oppositely oriented vortex structures. The second

Table 15.3 Characteristics of an unstable trajectory of gas bubbles emerging in an unbounded fluid

	Straight path	Spiral path	Zigzag path
Haberman and Morton [11]	$a < 0.5$ mm $\mathrm{Re}_p < 300$	$0.5\ \mathrm{mm} < a < 5$ mm $300 < \mathrm{Re}_p < 3000$	$a > 5$ mm $\mathrm{Re}_p > 3000$
Hartunian and Sears [16]	$a < 0.3$ mm $\mathrm{Re}_p < 200$	$a > 0.3$ mm $\mathrm{Re}_p > 200$	$a > 0.3$ mm $\mathrm{Re}_p > 200$
Saffman [15]	$a < 0.7$ mm $\mathrm{Re}_p < 400$	$a > 1$ mm $\mathrm{Re}_p > 600$	$a > 0.7$ mm $\mathrm{Re}_p > 400$
Aybers and Tapucu [17]	$a < 0.65$ mm $\mathrm{Re}_p < 550$	$0.65\ \mathrm{mm} < a < 1$ mm $550 < \mathrm{Re}_p < 880$	$a \approx 1$ mm $\mathrm{Re}_p \approx 880$
Duineveld [18]	$a < 0.9$ mm $\mathrm{Re}_p < 660$		$a > 0.9$ mm $\mathrm{Re}_p > 660$
De Vries et al. [22]		$a \approx 1$ mm $\mathrm{Re}_p \approx 660$	$a \approx 1$ mm $\mathrm{Re}_p \approx 660$

mode of oscillations (in which the bubble form changes) is developed due to the instability of the zigzag plane. As a result, a special spiral buoyancy trajectory of a bubble is formed. It can be assumed that the development of the instability by the emergence of an ellipsoidal bubble is produced by the following two factors: the character of the turbulent wake down the bubble and the variation of its form. Table 15.3 gives the experimentally obtained instability characteristics of an unstable trajectory of gas bubbles.

The number of numerical investigations of unstable trajectories of bubble rise is relatively small [30–33]. We mention some distinctive features of this involved process, which were identified in these papers.

- Nucleation of vortices in the aft part of an ellipsoidal bubble and their successive asymmetric separation from the surface.
- Formation of a vortex path in the wake behind the bubble.
- Intensification of the instability in the wake behind the bubble as its size increase.
- Deformation of the bubble form and oscillations of its form.
- Development of stable oscillations for bubbles of moderate sizes.
- Softening of the interaction of the separating vortexes for large bubbles.
- Transition from a rectilinear to a zigzag trajectory as the bubble size increases.
- Development of the instability of a planar zigzag trajectory of bubble rise.
- Transition to a spatial spiral trajectory of bubble rise.
- Development of double pairs of vortex lines along the flow, which are wound on the spiral trajectory and join the bubble base.

However, it should be noted that numerical methods are still incapable of providing a complete physical picture of bubble rise.

15.1.7 *Base Underpressure Model*

15.1.7.1 Principal Peculiarities

According to the experimental studies [15–23], regions 4, 5 are characterized by the two principal peculiarities.

- Bubble rise velocity is practically independent of the liquid viscosity.
- The flow past a bubble involves an extensive zone of separation.

It can be asserted that at present there is no correct solution of the problem of bubble rise in regions 4, 5 (see Fig. 15.1). One can mention only empirical correlations like (15.8), (15.9), (15.24), (15.26) capable of describing the experimental data for certain ranges of parameters.

Below we shall present the semi-empirical model of a bubble flow past by an inviscid liquid with midsection flow separation. This model, which was first proposed in [42–44], is based on the concept of "base underpressure" put forward in the book [28]. It is assumed that the volume of the bubble is theoretically separated into two components with a substantially different type of their interaction with the liquid flow (Fig. 15.3). The front ("hydrodynamic") part has the form of a spherical cap of radius R and solid angle θ_0. The aft ("hydrostatic") part is bounded from below by an equilibrium surface formed under the joint action of the gravitational and capillary forces.

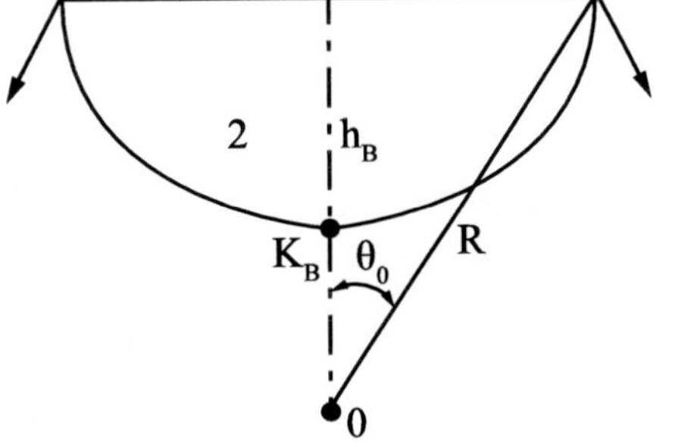

Fig. 15.3 The scheme of a composite bubble

15.1.7.2 Frontal Surface

We consider an irrotational flow of liquid in the frame with the origin at the curvature center of the spherical cap (Fig. 15.3). In this frame, the liquid velocity away of the bubble is u_∞ and coincides with the bubble rise velocity. If the flow around the sphere is inviscid, the tangent velocity on the surface can be determined from the relation (see [29])

$$u_\theta = \frac{3}{2} u_\infty \sin\theta. \tag{15.30}$$

Here θ is the angular coordinate counted from the frontal critical point K. For $\rho_g \ll \rho_f$, the pressure in the entire volume of the bubble can be assumed to be p_g. The continuity condition of the normal momentum component on the bubble surface implies that the same pressure will also be at the point K. Hence

$$p_K = p_g. \tag{15.31}$$

For the flow line on the frontal surface of the bubble, the Bernoulli law holds

$$p_K = p(\theta) + \frac{9}{8}\rho_f u_\infty^2 \sin^2\theta - \rho_f g R(1 - \cos\theta).$$

The last term on the right takes into account the difference of the levels between the point A and an arbitrary point on the surface. Expanding the function $(1 - \cos\theta)$ in a Maclaurin series for $\theta \ll 1$, we get

$$1 - \cos\theta = \left(1 - \sin^2\theta\right)^{1/2} = 1 - 1 + \frac{1}{2}\sin^2\theta + \frac{1}{8}\sin^4\theta + \cdots \approx \frac{1}{2}\sin^2\theta.$$

Using this result, as well as (15.31) in the Bernoulli law, we get the relation connecting the radius of the spherical cap and the bubble rise velocity

$$R = \frac{9}{4}\frac{u_\infty^2}{g}. \tag{15.32}$$

So, by assuming ab initio that the frontal surface of the bubble is a part of a sphere of radius R, we get a relation for the value of this radius. In turn, formula (15.32) justifies the sphericity of this part of the surface of the bubble: to an irrotational flow past of the frontal part of the bubble, there corresponds a spherical surface of radius R. The result, as expressed by formula (15.32), was first established in 1950 by Davies and Taylor [9]. However, if the value of the angle θ_0 is unknown, formula (15.32) does not link the rise velocity u_∞ with the bubble volume (or with the equivalent radius r_e).

15.1.7.3 Aft Surface

We consider a hydrostatic surface formed under the joint action of gravitational and capillary forces. Use is made of the theoretical results of [45] on the study of equilibrium forms of the free surface of the liquid, which has a common boundary with a fixed gaseous cavity on the horizontal wall. The hydrostatic analysis is based on local energy balance in the absence of motion. The components of the energy balance are as follows.

- the surface energy on the interface boundary of three media ("liquid–solid body", "gas–solid body", "liquid–gas"),
- the potential energy of phases in the gravity field.

Equilibrium forms are of considerable interest in applications related to the low (near zero) intensity of the field of mass forces, when the governing role is played by the surface tension effects. Consider a region of a two-phase system with the phase interface defined by the relation $z = f(x, y)$ (Fig. 15.4).

By Laplace equation

$$p_g(z) - p_f(z) = 2\sigma H(z). \tag{15.33}$$

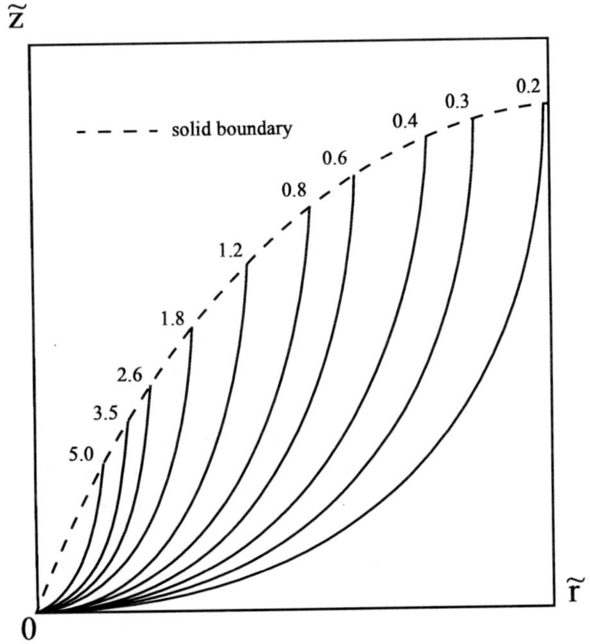

Fig. 15.4 Hydrostatic equilibrium forms

Here $H(z)$ is the curvature of the phase interface at a given point of the surface at the level z, the subscripts g, f indicate gas and liquid, respectively.

For a hydrostatic equilibrium, the pressure in the phases at rest changes linearly with height by the influence of the field of gravity forces. The equations of momentum balance for the contacting phases read as

$$\frac{dp_g}{dz} = -g\rho_g, \frac{dp_f}{dz} = -g\rho_f$$

with zero velocity. Integrating with respect to the z-coordinate, this gives

$$\left.\begin{array}{l} p_g(z) = p_g(0) - g\rho_g z, \\ p_f(z) = p_f(0) - g\rho_f z. \end{array}\right\} \tag{15.34}$$

Substituting (15.34) into the original Eq. (15.33), we get the general equation of hydrostatic equilibrium

$$2\sigma H(z) = p_g(z) - p_f(z) = (\rho_f - \rho_g)gz + C. \tag{15.35}$$

Here the constant $C = p_g(0) - p_f(0)$ is determined from the boundary conditions of a specific problem. The curvature of a surface is the second-order nonlinear differential operator

$$H(z) = \operatorname{div}\left(\frac{\nabla f}{\sqrt{1 + (\nabla f)^2}}\right). \tag{15.36}$$

In view of (15.36), Eq. (15.35) is a nonlinear second-order differential equation. Integrating this equation, one can find the form of the equilibrium surface. Since this boundary is ended on the hard surface, as two necessary boundary conditions one may specify: the conditions of tangency of a rigid body with a given contact angle θ and the total volume of the bubble.

The equation of hydrostatic equilibrium has attracted the attention of researchers for more than a century. The first attempt to solve it was made by Bashforth and Adams [46] in 1883, who published tables obtained by numerical integration of Eq. (15.35). These tables relate the volume of a body of revolution, its height, radius of the base, and contact angle. However, the gigantic manual calculation work done performed in [46] does not provide complete information about the shape of the equilibrium surfaces of phase interfaces with various values of θ_0.

It is worth pointing out that the integration of Eq. (15.35) does not mean a complete solution of this problem. The phase interface should not only satisfy the conditions of hydrostatic equilibrium, but should also be stable with respect to perturbations of the equilibrium form. For any arbitrary small departure of the form from equilibrium, a stable system should return to its initial state. On the other hand, for unboundedly growing perturbations, the system will be absolutely unstable. The stability analysis of equilibrium forms of the interphase surface is a separate nontrivial problem even in the simplest (linear) approximation. This investigation (which is much more difficult than the numerical integration itself) was performed in [45]. As a result, maximal stability parts of the integral curves were identified.

In Fig. 15.4, we show the family of equilibrium forms (as calculated in the book [45]), whose contact angle with the hard wall is $\pi/2$. Estimates show that these surfaces can be well approximated by the half of an oblate spheroid. The characteristic dimension here is given by the semi-major axis A of the spheroid. For $A \ll b$, the capillary forces are prevailing and the free surface is practically equal to the regular hemisphere. However, the gravity force becomes prevailing as $A \to \infty$, the semi-minor axis B of the spheroid tends to its limit maximum value: $B \to \sqrt{2}b$. This asymptotic describes the gas film of an infinite radius "spread" along the hard wall. The role of the surface tension here is only to "retain" the peripheral part of the film near the hard surface. Hence the semi-major axis is related to the semi-minor axis as follows

$$B = \frac{\sqrt{2}Ab}{\sqrt{2}b + A}. \tag{15.37}$$

The curvature radius of the surface at the point of symmetry K_B is

$$R_B = \frac{A^2}{B}. \tag{15.38}$$

15.1.7.4 Base Underpressure

In accordance with the classical scheme of the Helmholtz–Kirchhoff flow [28], we shall assume that the flow around the frontal part of the bubble involves the separation along the midsection and a formation of a stagnant zone behind its aft part (Fig. 15.3). Experiments on flows of a rigid body with a large Reynolds number show that the pressure in the turbulent wake behind the body is much smaller than the pressure p_∞ in the incident flow away from the body. Moreover, the pressure in the vicinity of the aft surface (near the wake) is characterized by a practically constant pressure, $p_B \approx$ const. These two quantities can be related via the base underpressure coefficient k [28]

$$p_\infty - p_B = \frac{1}{2}k\rho_f u_\infty^2, \tag{15.39}$$

for a flow of discs and spheres $k \approx 0.4$. Note that this estimate is supported by the recent experimental study of the flow in the wake of a rigid sphere [47]. In the framework of the Helmholtz–Kirchhoff flow, equality (15.39) means that in the transition through the free streamline the pressure in the near wake is discontinuously reduced by a fraction k of the dynamic pressure of liquid, $1/2\rho_f u_\infty^2$.

Consider the flow of an unperturbed liquid with parameters u_∞, p_∞ incident on the bubble at rest along the central flow line. At the frontal stagnation point K the velocity is zero, the pressure p_K is determined by the Bernoulli equation

$$p_K = p_\infty + \frac{1}{2}\rho_f u_\infty^2. \tag{15.40}$$

Continuing mentally the central flow line inside the bubble, we have $p_g = p_K$ by (15.31). When transiting through the bubble surface into the liquid, the "base pressure" at the aft stagnation point B in the liquid will discontinuously decrease by the value of the Laplace pressure jump

$$p_B = p_\infty + \frac{1}{2}\rho_f u_\infty^2 - 2\frac{\sigma}{R_B}. \tag{15.41}$$

Here R_B is the curvature radius on the symmetry axis of the bubble aft surface.

Let us now look at the variation of the pressure in liquid. On the one hand, the pressure increment due to the gravity force in the liquid column between the levels K and K_B is $\rho_f g(h + h_B)$. Here h, h_B is the height of the front and aft part of the bubble, respectively. On the other hand, pressure loss due to the "base underpressure" effect is $1/2k\rho_f u_\infty^2$. This gives us the second expression for the base pressure

$$p_B = p_\infty + \rho_f g(h + h_B) - \frac{1}{2}k\rho_f u_\infty^2. \tag{15.42}$$

Excluding from (15.41), (15.42) the pressure difference $p_B - p_\infty$, we get

$$\frac{1}{2}(k+1)\rho_f u_\infty^2 = \rho_f g(h + h_B) + 2\frac{\sigma}{R_B}. \tag{15.43}$$

Expressing the rise velocity from (15.32) and substituting into (15.43), we get the main equation of the "base underpressure" model

$$\frac{2}{9}(k+1)Rg = (h + h_B)g + 2\frac{\sigma}{\rho_f R_B}. \tag{15.44}$$

15.1.8 *Gravitational Asymptotics*

Consider the rise of a sufficiently large bubble ($r_e \to \infty$). From physical considerations, it is clear that the aft surface of the bubble will be practically flat. We have $R \to \infty$, $R_B \to \infty$, $h_B \to 0$. In this limit case, the surface tension effects can be neglected and the bubble rise will be completely determined by gravity. Now Eq. (15.44) simplifies to read

$$\frac{2}{9}(k+1)R = h. \tag{15.45}$$

Let us write down the geometric relation for the spherical cap, which relates its curvature radius R, the height h, and the base radius r_0 (Fig. 15.3)

$$r_0^2 + h^2 = 2Rh. \tag{15.46}$$

The volume of its spherical cap is

$$V_1 = \frac{\pi}{3}h^2(3R - h).$$

Hence we get the expression for the bubble equivalent radius

$$r_e = \left[\frac{1}{4}h^2(3R - h)\right]^{1/3}. \tag{15.47}$$

The remarkable feature of the system of Eqs. (15.43)–(15.47) is that it involves only linear dimensions, its solution gives the following relation between the geometric parameters of the bubble

$$R = 2.486r_e, h = 0.7735r_e, r_0 = 1.802r_e. \tag{15.48}$$

The scales are related as follows:

$$\frac{r_0}{R} = 0.7249, \frac{h}{R} = 0.3111, \frac{h}{r_0} = 0.4292.$$

As a result, the large bubbles are geometrically similar. The space angle of the spherical cap is $\theta_0 = \arcsin\left(\frac{r_0}{R}\right) = 0.8108$ radian $= 46.46°$. In the "gravitational asymptotics" under consideration, the bubble rise velocity id described by formula (15.32) for region 5 (Fig. 15.1) with $k = 1.051$

$$u_\infty = 1.051\sqrt{gr_e}. \tag{15.49}$$

15.1.9 Capillary Asymptotics

Consider now the hypothetical case of the rise of a bubble of a vanishingly small radius ($r_e \to 0$). Note that in reality, we have the region of dominating influence of the viscous effects, for which the exact solution was obtained by Hadamard and Rybczynski (15.4) (or, more precisely, its particular case (15.7)). This hypothetical asymptotics, however, is important for the subsequent construction of a general solution for region 4.

The form of the aft surface of the bubble will be sought in the family of surfaces in Fig. 15.4. Mentally "removing" the solid wall from the drawing, we get a "composite bubble". Here the contact angle between the periphery of the aft part with the horizontal surface remains unknown. To choose a concrete surface, we take it equal to the corresponding contact angle from the side of the frontal surface: $\theta_0 = 46.46°$. As $r_e \to 0$ there will be no effect on the formation of the free surface, under the action of capillary forces it should be a part of the sphere.

Now the main equation of model (15.44) simplifies to read

$$\frac{2}{9}(k+1)Rg = +\frac{2\sigma}{\rho_f R_B}. \tag{15.50}$$

The geometric relation for the spherical cap (15.46) and the expression for the bubble equivalent radius (15.47) now read, respectively, as

$$r_0^2 + h_B^2 = 2R_B h_B, \tag{15.51}$$

$$r_e = \left[\frac{1}{4}h_B^2(3R_B - h_B)\right]^{1/3}. \tag{15.52}$$

It is interesting to note that under the above assumptions we get here the same spherical cap as in the above gravitational asymptotics, but whose convex surface is directed downwards. From Eqs. (15.51), (15.52), we get relations between the geometric parameters of the bubble

$$R_B = 2.486 r_e,\ h_B = 0.7735 r_e,\ r_0 = 1.802 r_e. \tag{15.53}$$

Formulas (15.53) are similar to formulas (15.48) with the following change: $R \to R_B, h \to h_B$.

As before, the flow past the frontal surface is described by relation (15.32). Using it together with (15.50), we get the relation for the "capillary asymptotics"

$$u_\infty = 1.07\sqrt{\frac{\sigma}{\rho_f r_e}}. \tag{15.54}$$

Formula (15.54) has the same structure as the empirical formula (15.8) with the corresponding modification of the numerical constant.

15.1.9.1 General Solution

Consider now the general case of joint influence of the gravitational and capillary forces. Here the aft surface constitutes a part of the equilibrium form of the oblate spheroid of volume

$$V_2 = \pi A^2 B\left[\left(\frac{h_B}{B}\right)^2 - \frac{1}{3}\left(\frac{h_B}{B}\right)^3\right].$$

Summing the volumes of the front and aft parts, we get the expression for the bubble equivalent radius

$$r_e = \left[\frac{1}{4}h^2(3R-h) + \frac{3}{4}A^2 B\left[\left(\frac{h_B}{B}\right)^2 - \frac{1}{3}\left(\frac{h_B}{B}\right)^3\right]\right]^{1/3}. \tag{15.55}$$

Formula (15.55) involves the following relations for the linear dimensions: the base radius of the circle separating the front and aft parts is

$$r_0 = \frac{AF\sin\theta_0}{\sqrt{\cos^2\theta_0 + F^2\sin^2\theta_0}}, \tag{15.56}$$

the height of the aft part is

$$h_B = \left(1 - \frac{\cos\theta_0}{\sqrt{\cos^2\theta_0 + F^2\sin^2\theta_0}}\right). \tag{15.57}$$

Here $F = 1 + \frac{1}{\sqrt{2}}\frac{B}{b}$.

Let us introduce the dimensionless quantities of the equivalent radius $\tilde{r}_e$ and the rise velocity$\tilde{u}_\infty$. As the length scale, we take the capillary constant $b \equiv \sqrt{\sigma/\rho_f g}$, and as the velocity scale we take the quantity $u_* = \sqrt{gb} = \sqrt[4]{\sigma g/\rho_f}$. In Fig. 15.5, we give the results of calculations of the dependence $\tilde{u}_\infty\left(\tilde{r}_e\right)$ by the empirical formula (15.24) and by the base underpressure model. The calculation curves are seen to be in a very satisfactory agreement. The coordinate of the minimum of the velocity is shifted relative to dependence (15.24) by 16%, and its value, by 4%. Here one should also mention the considerable scatter of experimental data in regions 4, 5 (in this regard, see also [20]).

Note that this model is semi-empirical—it involves two empirical constants: the base underpressure coefficient" Q and the contact angle between the front and aft parts of the composite bubble $\theta_0 = 46.46°$ base underpressure. Nevertheless, the base underpressure model gives a clear physical picture of laws of bubble rise in regions 3, 4.

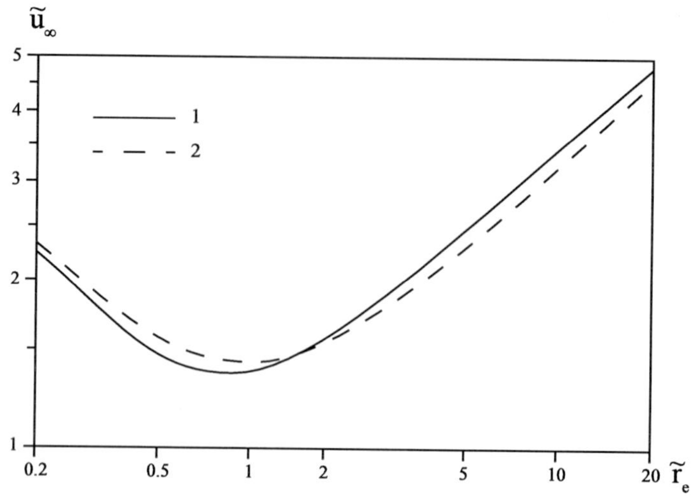

Fig. 15.5 Theoretical dependence of the rise velocity on the equivalent radius

15.2 Taylor Bubble Rising in Vertical Tube

15.2.1 Theoretical Solutions

One of the classical two-phase flows is the "slug flow". This flow is characterized by the periodic structures in the form of the large gas bubbles (Taylor bubbles) rising in a vertical round tube under the influence of the gravitational force [48]. The Taylor bubbles have an approximately spherical front part and their length can reach up to 10 to 15 diameters of the tube. It was experimentally established that the rise velocity of the Taylor bubble can be described by the following equation

$$\mathbb{U} \approx 0.5\sqrt{gR_0}, \tag{15.58}$$

with the radius of its front part being

$$R_k \approx 0.7R_0, \tag{15.59}$$

where g is the gravity acceleration, R_0 is the tube radius.

The first theoretical investigation of this problem was performed in 1943 by Dumitrescu [49], a student of Prandtl. Then the work of Davies and Taylor [9] appeared in 1950. He was actually the author, who gave the name to the bubble. Unfortunately, the mathematical description of this problem in both pioneer works was limited to the use of divergent infinite series. Such an obvious incorrectness of the mathematical description of this old problem, being though rather strange, remained nevertheless "not noticed", and all literature sources known to the author cite namely the works [9, 49]. Therefore, the author of the present book believed necessary to perform a correct analysis of the problem looking into modeling of the rise of the Taylor bubble in a round tube.

Let us consider in brief the problem statement as given in the classical works [9, 49]. Owing to the absence of the velocity shear in the liquid on the bubble surface, it is possible to neglect the viscous terms in the Navier–Stokes equation for the liquid, as well as capillary pressure for the sufficiently large bubbles. Then the problem is reduced to finding out the shape and velocity of the bubble rising in a vertical round tube filled with an ideal liquid. The velocity potential ϕ of such a flow should satisfy the Laplace equation

$$\frac{\partial^2 \varphi}{\partial z^2} + \frac{1}{r}\frac{\partial}{\partial r}\left(r\frac{\partial \varphi}{\partial r}\right) = 0. \tag{15.60}$$

Here, z, r are dimensionless axial and radial coordinates, respectively. As the linear length scale, the tube radius R_0 is accepted. An application of the method of separation of variables to Eq. (15.60) yields two alternative types of solutions

$$\varphi = \sum_{k=1}^{\infty} (-1)^{k+1} \alpha_k \mathrm{J}_0(\beta_k r) \exp(\beta_k z), \tag{15.61}$$

$$\varphi = \sum_{k=1}^{\infty} (-1)^{k+1} \alpha_k \mathrm{I}_0(\beta_k r) \cos \beta_k z. \tag{15.62}$$

Here, $\mathrm{J}_0(x), \mathrm{I}_0(x)$ are the Bessel function and the modified Bessel function of the zero order [50], respectively, α_k, β_k are unknown coefficients (eigenvalues of the boundary value problem).

The flow is considered in the coordinate system in which the bubble is in rest, while the liquid flows onto it, with a front stagnation point being formed on the bubble. On the streamline restricting the bubble, the condition of constant pressure (a free surface condition [28]) should be satisfied. Then the following relation can be derived from the Bernoulli equation for the point on the bubble surface located on the distance $|z|$ from the front stagnation point

$$U_s^2 + V_s^2 = 2g|z|. \tag{15.63}$$

Here, U_s, V_s are the axial and radial velocity components on the bubble surface, respectively. On the tube wall, the condition of non-permeability should be satisfied

$$V|_{r=1} = 0. \tag{15.64}$$

As a result, we formulated the Neumann boundary value problem for the Laplace equation (15.60). For the flow description in the entire range of variation of the axial coordinate $0 < z < \infty$, the first type of the solution, Eq. (15.61), is used in [49]. Then the radial velocity will be equal to

$$V \equiv \frac{\partial \varphi}{\partial r} = \sum_{k=1}^{\infty} (-1)^{k+1} \alpha_k \beta_k \mathrm{J}_1(\beta_k r) \exp(\beta_k z), \tag{15.65}$$

where $\mathrm{J}_1(x)$ is the Bessel function of the first order [50]. It follows from Eqs. (15.64) and (15.65) that the eigenvalues β_k are the solutions of the Bessel equation (or zero values of the Bessel function of the first order)

$$\mathrm{J}_1(\beta_k) = 0. \tag{15.66}$$

It is important to accentuate that, in accordance with Eq. (15.66), at any choice of the direction of the z-axis the relation (15.61) will always include the infinite alternating series, with each term of them, in turn, being exponentially diverging. This obvious incorrectness of the solution was circumvented in the works [9, 49], because only several first terms of the infinite series (15.61) were used in calculations. Thus, it is possible to assert that mathematical expressions having double divergence were used in the works [9, 49]: each k-term of the solution (15.61) diverges. The infinite series is compounded of the diverging terms and, naturally, it also diverges.

15.2.2 Correct Statement of the Problem

The correct approximate statement of the problem was offered in the work [51]. The rise of the Taylor bubble in a tube is replaced with a problem of modeling a flow over a solid body. Free surface conditions hold not on the entire surface (as it would be necessary to do in a hypothetical exact problem treatment), but only at specially chosen points. The overall flow is assumed to result from a superposition of three elementary flows. Let us place in the origin of the coordinate system a source of mass with the intensity Q (Fig. 15.6a). The velocity potential for such a source will be

$$\varphi_1 = -\frac{Q}{4\pi} \frac{1}{\left(z^2+r^2\right)^{1/2}}. \tag{15.67}$$

The axial and radial velocity components will look like

$$U_1 \equiv \frac{\partial \varphi_1}{\partial z} = -\frac{Q}{4\pi} \frac{z}{\left(z^2+r^2\right)^{3/2}}, \tag{15.68}$$

$$V_1 \equiv \frac{\partial \varphi_1}{\partial r} = \frac{Q}{4\pi} \frac{r}{\left(z^2+r^2\right)^{3/2}}. \tag{15.69}$$

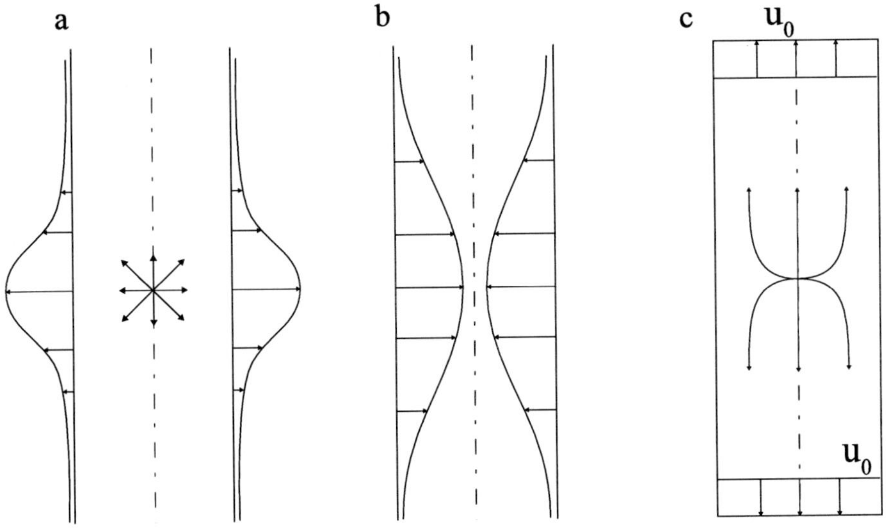

Fig. 15.6 Schematic of the flow superposition. **a** Source of mass in the infinite space, **b** injection flow, **c** source of mass in an impermeable tube

If such a flow from the source is placed in an imaginary cylinder imitating a tube, its wall will thus be permeable. To make this wall impermeable, or in other words to satisfy the physically required condition of non-permeability on the tube wall, let us construct an injection flow possessing φ_2 the following property

$$V_1|_{r=1} + V_2|_{r=1} = 0. \tag{15.70}$$

The potential of the injection flow φ_2 will be searched in the following form

$$\varphi_2 = \int_0^\infty A(y)\mathrm{I}_0(yr)\cos yz\,dy. \tag{15.71}$$

Equation (15.71) can be considered as an integral analogue to the discrete expression, Eq. (15.62). The condition (15.70) will be then rewritten as

$$\frac{Q}{4\pi}\frac{1}{\left(z^2+1\right)^{3/2}} = \int_0^\infty A(y)\mathrm{I}_1(y)\mathrm{y}\cos yz\,dy. \tag{15.72}$$

Introducing notations

$$\frac{Q}{4\pi}\frac{1}{\left(z^2+1\right)^{3/2}} = f(z),\ A(y)\mathrm{I}_1(y)y = A_1(y), \tag{15.73}$$

let us rewrite Eq. (15.72) to the following form

$$f(z) = \int_0^\infty A_1(y) \cos yz \, dy. \tag{15.74}$$

Expression (15.74) is a representation of the function $f(z)$ in the form of the Fourier cosine integral [52]. Then the function $A_1(y)$ will be defined from the inverse Fourier transformation

$$A_1(y) = \frac{2}{\pi} \int_0^\infty f(\chi) \cos \chi y d\chi = \frac{Q}{2\pi^2} \int_0^\infty \frac{\cos \chi y}{\left(1 + \chi^2\right)^{3/2}} d\chi. \tag{15.75}$$

Having found the tabulated integral in Eq. (15.75), one can obtain

$$\int_0^\infty \frac{\cos \chi y}{\left(1 + \chi^2\right)^{3/2}} d\chi = y\mathrm{K}_1(y), \tag{15.76}$$

where $\mathrm{K}_1(x)$ is the first-order modified Bessel function of the second kind [50]. From Eqs. (15.75), (15.76), one can derive

$$A_1(y) = \frac{Q}{2\pi^2} y\mathrm{K}_1(y),\ A(y) = \frac{Q}{2\pi^2} \frac{\mathrm{K}_1(y)}{\mathrm{I}_1(y)}. \tag{15.77}$$

Let us transform Eq. (15.71) with allowance for Eq. (15.77) for the injection flow as

$$\varphi_2 = \frac{Q}{2\pi^2} \int_0^\infty \frac{\mathrm{K}_1(y)}{\mathrm{I}_1(y)} \mathrm{I}_0(yr) \cos yz \, dy. \tag{15.78}$$

An important feature of Eq. (15.60) is its linearity, which allows a superposition of its solutions. Summarizing the flows from the source φ_1 and injection φ_2, one can obtain the required flow from a source in an impermeable tube $\varphi_1 + \varphi_2$ (Fig. 15.6b). At $z \to \mp\infty$, it represents a homogeneous flow, whose velocity U_0 is connected with the source intensity $U_0 = Q/2\pi$ by the mass conservation law. To obtain a picture of the flow over some axisymmetric body in an impermeable tube, it is necessary to "impose" a homogeneous flow on the flow from a source constructed above (Fig. 15.6c)

$$\varphi_3 = -U_\infty z. \tag{15.79}$$

The velocity potential from the combined flow obtained by a superposition of the three elementary flows ($\varphi = \varphi_1 + \varphi_2 + \varphi_3$) can be written in the following form

$$\varphi = -\frac{U_0}{2}\frac{1}{\left(z^2+r^2\right)^{1/2}} - \frac{U_0}{\pi}\int_0^\infty \frac{\mathrm{K}_1(y)}{\mathrm{I}_1(y)}\mathrm{I}_0(yr)\cos yz\,dy - U_\infty z. \tag{15.80}$$

Knowing the velocity potential of this flow, it is possible to find out the axial $U = \partial\varphi/\partial z$ and the radial $V = \partial\varphi/\partial r$ components of the velocity

$$U = \frac{\partial\varphi}{\partial z} = \frac{U_0}{2}\frac{z}{\left(z^2+r^2\right)^{3/2}} + \frac{U_0}{\pi}\int_0^\infty \frac{\mathrm{K}_1(y)}{\mathrm{I}_1(y)}\mathrm{I}_0(yr)\sin yz\,dy - U_\infty, \tag{15.81}$$

$$V = \frac{\partial\varphi}{\partial r} = \frac{U_0}{2}\frac{r}{\left(z^2+r^2\right)^{3/2}} - \frac{U_0}{\pi}\int_0^\infty \frac{\mathrm{K}_1(y)}{\mathrm{I}_1(y)}\mathrm{I}_1(yr)\cos yz\,dy. \tag{15.82}$$

The stream function of the obtained combined flow will be equal to

$$\psi = \frac{U_0}{2}\left[1 - \frac{z}{\left(z^2+r^2\right)^{1/2}}\right] + \frac{U_0}{\pi}\int_0^\infty \frac{\mathrm{K}_1(y)}{\mathrm{I}_1(y)}r\mathrm{I}_1(yr)\sin yz\,dy - \frac{U_\infty}{2}r^2. \tag{15.83}$$

Assuming $\psi = 0$, one can obtain an equation defining a contour of a body placed in a tube with an ideal liquid flow in it (Fig. 15.7)

$$r_s^2 = \frac{1}{\varepsilon}\left\{1 - \frac{z_s}{\left(z_s^2+r_s^2\right)^{1/2}} + \frac{2}{\pi}\int_0^\infty \frac{\mathrm{K}_1(y)}{\mathrm{I}_1(y)}r_s\mathrm{I}_1(yr_s)\sin yz_s\,dy\right\}, \tag{15.84}$$

where

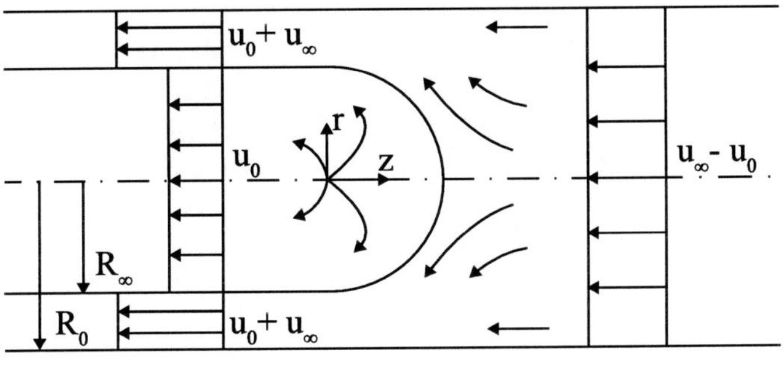

Fig. 15.7 Flow of an ideal fluid over an axisymmetric body in a tube

$$\varepsilon = \frac{U_\infty}{U_0} \tag{15.85}$$

is the parameter characterizing relative intensity of the source.

The flow velocity over a body at $z \to \infty$ is equal to $U_\infty - U_0 = U_0(\varepsilon - 1)$. The flow velocity between the body and the tube at $z \to -\infty$ is equal to $U_\infty + U_0 = U_0(\varepsilon + 1)$. Then from the mass conservation law

$$U_\infty - U_0 = (U_\infty + U_0)r_\infty^2, \tag{15.86}$$

one can obtain the asymptotic body radius at $z \to \infty$

$$r_\infty = \left(\frac{2}{\varepsilon + 1}\right)^{1/2}. \tag{15.87}$$

It is interesting to point out that inside the body the flow with a stagnation point and an asymptotic velocity U_0 at $z \to -\infty$ also takes place. However, within the frames of the problem under investigation, we are interested only in the outer flow over the body. From the condition of zero velocity in the stagnation point, one can obtain

$$\varepsilon(x) = \frac{1}{2x^2} + \frac{1}{\pi}\int_0^\infty \frac{\mathrm{K}_1(y)y}{\mathrm{I}_1(y)} \sin yx\, dy, \tag{15.88}$$

where x is the distance from the stagnation point to the source.

15.2.3 Stagnation Point

From the very beginning, we act in frames of the approximate solution. Therefore, it is basically impossible to fulfill the condition of the free surface for the class of the bodies found above. This condition is possible to be fulfilled asymptotically in the vicinity of the stagnation point. Let us consider the flow in the vicinity of the stagnation point on the body. Let us present velocity components with the first terms of the expansion in the Taylor series

$$U = -U_z'\hat{z} + U_\eta'\eta, \tag{15.89}$$

$$V = V_r'r. \tag{15.90}$$

Here, $\eta \equiv r^2$, $\hat{z} \equiv x{-}z$ is the axial coordinate (counted from the stagnation point toward the source), the primes denote derivatives with respect to the corresponding coordinate denoted by the subscript.

With account of the continuity equation $V_r' = U_z'/2$, one can rewrite this equation in the following form: $V = \left(U_z'/2\right)r$. In the vicinity of the front stagnation point, the equation of the contour of the body placed in the flow can be presented as the first term of the Taylor series with respect to the coordinate $\hat{z}$ (i.e., otherwise, in the form of a quadratic parabola)

$$\eta_s = 2r_k\hat{z}. \tag{15.91}$$

It is convenient to write the condition of equality to zero of the stream function on the body surface in the following integral form

$$\psi = \int_0^\infty Urdr = \frac{1}{2}\int_0^\infty Ud\eta. \tag{15.92}$$

A substitution of Eq. (15.91) into Eq. (15.92) yields an expression for the dimensionless curvature radius of the body

$$r_k = \frac{U_z'}{U_\eta'}. \tag{15.93}$$

Taking into account the equality $U_\eta' = -U_{zz}''/4$, whose validity can be verified via direct calculations, one can rewrite Eq. (15.93) in the following form

$$r_k = -4\frac{U_z'}{U_{zz}''}. \tag{15.94}$$

Let us present the square of the full velocity of the liquid on the body surface with the first term of the Taylor series

$$U_s^2 + V_s^2 = -2\frac{\left(U_z'\right)^3}{U_{zz}''}\hat{z}. \tag{15.95}$$

Equation (15.95) (as well as all other previous relations) describes a flow of an ideal liquid onto an axisymmetric solid body located in an impermeable tube. Let us pass now to the free surface. To pass to the case of the flow of a liquid over a bubble, it is necessary to substitute Eq. (15.95) into Eq. (15.63) (the free surface condition)

$$-\frac{\left(U_z'\right)^3}{U_{zz}''} = gR_0. \tag{15.96}$$

Let us introduce the function $\epsilon(x) \equiv \varepsilon(x) - 1$, or, with allowance for Eq. (15.88).

$$\epsilon(x) \equiv \varepsilon(x) - 1 = \frac{1}{2x^2} - 1 + \frac{1}{\pi} \int_0^\infty \frac{\mathrm{K}_1(y)y}{\mathrm{I}_1(y)} \sin yx \, dy. \tag{15.97}$$

It follows then from Eqs. (15.81), (15.82), (15.85), (15.97) that

$$U_z^{'} = U_0 \epsilon^{'}, U_{zz}^{''} = U_0 \epsilon^{''}. \tag{15.98}$$

In the coordinate system, where the bubble rises in the tube filled with the liquid under the influence of the buoyant force, its velocity $\mathbb{U}$ will be equal to the velocity of the flow onto the body at $z \to \infty$

$$\mathbb{U} = U_\infty - U_0 = U_0 \epsilon. \tag{15.99}$$

Let us define the Froude number with the following relation

$$\mathrm{Fr} = \frac{\mathbb{U}}{\sqrt{g R_0}}. \tag{15.100}$$

Taking into account equalities (15.97), one can rewrite Eqs. (15.87), (15.94), (15.87) in the following form

$$r_\infty = \sqrt{\frac{2}{2 + \epsilon}}, \tag{15.101}$$

$$r_k = -4 \frac{\epsilon^{'}}{\epsilon^{''}}, \tag{15.102}$$

$$\mathrm{Fr} = \sqrt{- \frac{\epsilon^2 \epsilon^{''}}{\left(\epsilon^{'}\right)^3}}. \tag{15.103}$$

Thus, we have modeled the flow pattern in the liquid over the free surface in the gravity field, where the free surface condition is asymptotically satisfied in the vicinity of the stagnation point. However, this body cannot be considered as the bubble model yet.

15.2.4 Method of Collocations

As it was already mentioned, the free surface condition (15.63) holds only asymptotically at $\hat{z} \to 0$. In the mathematical sense, it means non-uniqueness of the solution:

it represents a description of an infinite set of bubbles with the free parameter x. Therefore, in order to close the problem description, it is necessary to find out an additional condition. One of the possible ways to close the problem is offered in the work [51]. To make it possible to exactly fulfill Eq. (15.63) over the entire body surface, one can place a continuously distributed system of mass sources (with the respective flow injection) along the $\hat{z}$ axis. Since, however, the form of the free surface is unknown in advance and should be found out via the solution, such a way is believed to be unreal in the mathematical sense.

To follow the logic of the analysis developed above, one should remain within the limits of the one-parametrical set of the bodies. Then one needs, in addition to the stagnation point, to find out one more point on the body surface and to require performance of the free surface condition (15.63) in that point. One should point out that this condition corresponds to the known direct method of the variation calculus (the method of collocations) [53]. Choosing the "source coordinate" $\left(z = 0, \hat{z} = x\right)$ as the second point, one can obtain the missing condition

$$\left(U_s^2 + V_s^2\right)\big|_{z=0} = g R_0 x. \tag{15.104}$$

As a result, one can derive the following relations for the key parameter: $x = 0.58$, $\epsilon = 0.92$. It follows from here that

$$r_\infty = 0.827, \tag{15.105}$$

$$r_k = 0.69, \tag{15.106}$$

$$\mathrm{Fr} = 0.488. \tag{15.107}$$

Comparing Eqs. (15.106), (15.107) with the experimental values (Eqs. (15.45), (15.57)), one can assure in their practical coincidence. Let us find out the asymptotical forms of the above relations at $x \to 0$. It follows from Eq. (15.97) that

$$\epsilon = \frac{1}{2x^2}. \tag{15.108}$$

Further, it can be obtained from (15.88)–(15.103) that

$$r_\infty = 2x, \tag{15.109}$$

$$r_k = \frac{4}{3}x, \tag{15.110}$$

$$\mathrm{Fr} = \sqrt{\frac{3}{4}x}. \tag{15.111}$$

Thus, the limiting case of $x \to 0$ corresponds to the flow over an infinitely thin body placed in the tube. In this case, the source intensity is negligibly small ($U_0 \ll U_\infty$), and the Froude number also tends to zero.

15.2.5 Asymptotical Solution

In the work [44], we offered a method for the approximate analytical solution of the problem based on the investigation into the asymptotical cases of the non-unique family of the solutions at $x \to \infty$. Let us write down the Taylor series expansion of the modified Bessel function of the first order

$$\mathrm{I}_1(y) = \sum_{k=0}^{\infty} \frac{y^{k+1}}{2^{2k+1} k!(k+1)!}. \tag{15.112}$$

Let us find out the function

$$E(x) \equiv \int_0^\infty \frac{\mathrm{K}_1(y)y}{\mathrm{I}_1(y)} \sin yx \, dy, \tag{15.113}$$

connected with the function

$$\epsilon(x) \equiv \varepsilon(x) - 1 = \frac{1}{2x^2} - 1 + \frac{1}{\pi} \int_0^\infty \frac{\mathrm{K}_1(y)y}{\mathrm{I}_1(y)} \sin yx \, dy$$

from Eq. (15.88) via the relation

$$\varepsilon(x) = \frac{1}{2x^2} + \frac{2}{\pi} E(x). \tag{15.114}$$

Calculating even derivatives $E^{(2k)} = \partial^{2k} y / \partial x^{2k}$ from Eq. (15.113), one can obtain from Eq. (15.113), one can obtain

$$\left.\begin{aligned}
E^{(2)} &= -\tfrac{1}{2}\int\limits_0^\infty \tfrac{K_1(y)}{I_1(y)} y^3 \sin yx\,dy,\\
E^{(4)} &= \tfrac{1}{2}\int\limits_0^\infty \tfrac{K_1(y)}{I_1(y)} y^5 \sin yx\,dy,\\
E^{(6)} &= \tfrac{1}{2}\int\limits_0^\infty \tfrac{K_1(y)}{I_1(y)} y^7 \sin yx\,dy,\\
&\cdots\cdots\cdots\cdots\cdots\cdots\cdots\cdots,\\
E^{(2k)} &= \tfrac{(-1)^k}{2}\int\limits_0^\infty \tfrac{K_1(y)}{I_1(y)} y^{2k+1} \sin yx\,dy,\\
k &= 0, 1, 2, 3, \ldots
\end{aligned}\right\} \tag{15.115}$$

Multiplying the left-hand and right-hand parts of the recurrence relation (15.115) by $\frac{(-1)^k}{2^{2k+1}k!(k+1)!}$ and summing them term by term, the terms $\frac{(-1)^k E^{(2k)}}{2^{2k+1}k!(k+1)!}$ one can obtain

$$\sum_{k=0}^{\infty} \frac{(-1)^k}{2^{2k+1}k!(k+1)!} E^{(2k)} = \int\limits_0^\infty \frac{K_1(y)}{I_1(y)} \left\{ \sum_{k=0}^{\infty} \frac{y^{2k+1}}{2^{2k+1}k!(k+1)!} \right\} \sin yx\,dy. \quad (15.116)$$

As follows from Eq. (15.112), the expression in brackets under the integral in Eq. (15.116) is, namely, the Taylor series expansion of the modified Bessel function of the first order. Therefore Eq. (15.115) can be rewritten in the following form:

$$\sum_{k=0}^{\infty} \frac{(-1)^k}{2^{2k+1}k!(k+1)!} E^{(2k)} = \int\limits_0^\infty K_1(y) \sin yx\,dy. \quad (15.117)$$

The integral in the right-hand part of Eq. (15.117) is tabular [50] and looks as

$$\int\limits_0^\infty K_1(y) \sin yx\,dy = \frac{\pi}{2} \frac{x}{\sqrt{1+x^2}}. \quad (15.118)$$

As a result of the transformations performed above, we have obtained the linear infinite-order inhomogeneous differential equation with respect to the function $E(x)$

$$\sum_{k=0}^{\infty} \frac{(-1)^k}{2^{2k+1}k!(k+1)!} E^{(2k)} = \frac{\pi}{2} \frac{x}{\sqrt{1+x^2}}. \quad (15.119)$$

As the zero derivative in Eq. (15.119), the function itself is meant: $E^{(0)} \equiv E$. Coming back now to the function $\varepsilon(x)$, which is of interest to us, we can obtain from Eqs. (15.113), (15.118) that

$$\sum_{k=0}^{\infty}\frac{(-1)^k}{2^{2k+1}k!(k+1)!}\varepsilon^{(2k)}=\sum_{k=0}^{\infty}\frac{(-1)^k(2k+1)!}{2^{2k+1}k!(k+1)!}\frac{1}{x^{2(k+1)}}+\frac{x}{\sqrt{1+x^2}}. \tag{15.120}$$

Let us search for the solution of Eq. (15.120) at $x \to \infty$. Let us write down a power series expansion of the last term in its right-hand part at $x \gg 1$

$$\frac{x}{\sqrt{1+x^2}} \approx 1-\frac{1}{2}\frac{1}{x^2}+\frac{3}{8}\frac{1}{x^4}-\frac{5}{16}\frac{1}{x^6}+\cdots\frac{(-1)^{k+1}(2k+1)!}{2^{2k+1}k!(k+1)!}\frac{1}{x^{2(k+1)}}. \tag{15.121}$$

Substituting Eq. (15.121) into Eq. (15.120), one can ascertain that the infinite power series in the right-hand part of the latter equation is "mutually compensated". This circumstance represents a remarkable property of the asymptotical solution. As a result, one can obtain an infinite-order homogeneous differential equation with respect to the function $\epsilon(x) \equiv \varepsilon(x) - 1$

$$\sum_{k=0}^{\infty}\frac{(-1)^k}{2^{2k+1}k!(k+1)!}\epsilon^{(2k)}=0. \tag{15.122}$$

The solution of Eq. (15.122) can be formally written down in the form of the infinite exponential series

$$\epsilon=\sum_{k=1}^{\infty}C_k\exp(\mp\beta_k x). \tag{15.123}$$

Here, β_k are the points where the Bessel function of the first order takes zero values (see Eq. (15.66)), C_k are free numerical coefficients. As soon as we are interested in the solution converging at $x \to \infty$, we choose in the relation (15.123) the exponential functions with the negative exponents

$$\epsilon=\sum_{k=1}^{\infty}C_k\exp(-\beta_k x). \tag{15.124}$$

It should be pointed out, however, that the coefficients C_k can not be found in principle via such an approach. Indeed, they should be determined from the solution of the corresponding Cauchy problem from an infinite set of the boundary conditions at $x = 0$. But a substitution of the solution (15.123) into the condition at $x = 0$ is not allowed, as Eq. (15.122) itself holds only at $x \gg 1$. Fortunately, here the asymptotical analysis serves as the aid. Indeed, at $x \to \infty$, it is possible to limit ourselves to only the first term in Eq. (15.124)

$$\epsilon=C_1\exp(-\beta_1 x), \tag{15.125}$$

where $\beta_1 = 3.83170597$ is the first zero of the Bessel function of the first order.

It follows from Eqs. (15.109)–(15.111) that

$$r_\infty \approx 1 - \frac{C_1}{4}\exp(-\beta_1 x) \approx 1, \tag{15.126}$$

$$r_k = \frac{4}{\beta_1} \approx 1.04, \tag{15.127}$$

$$\mathrm{Fr} = \sqrt{\frac{1}{\beta_1}} \approx 0.511. \tag{15.128}$$

So, after rather refined transformations, whose mathematical strictness the author himself does not undertake to estimate if the full entirety, the asymptotical problem solution was obtained. Equation (15.128) for the Froude number obtained above by less than 5% differs from the numerical solution (15.107). Thus, at the calculation of the parameters of the Taylor bubble for $x \to \infty$, the coefficient C_1 is reduced. Formally speaking, the asymptotical problem solution appears homogeneous with respect to C_1. Playing upon words, it is possible to say that the physical features of the problem have helped to "bypass" the mathematical difficulties. As seen from Eq. (15.126), the limiting case $x \to \infty$ corresponds to the flow in the tube over the bubble, whose cylindrical part's radius is equal to the tube radius: $r_\infty \to 1$. In this case, the source intensity is equal to the homogeneous flow intensity ($U_0 \to U_\infty$), and the Froude number also tends to the largest possible value: $\mathrm{Fr} \to 0.511$.

15.2.6 Plane Taylor Bubble

As appears from Eq. (15.103), the Froude number is a function of the parameter x, which is the distance from the stagnation point to the source point. The asymptotical cases at $x \to 0$, Eq. (15.111), and at $x \to \infty$, Eq. (15.128), investigated above allow assuming a monotonous character of the dependence $\mathrm{Fr}(x)$. However, an attempt to calculate the quadrature (15.113) encounters with considerable complications. Therefore, for the qualitative analysis of the axisymmetric problem, it is expedient to consider a corresponding two-dimensional (i.e., "flat") case.

An investigation into the problem of the rise of the plane Taylor bubble has begun in 1957 [28, 54] and lasts till now. Certainly, the flat case describing the rise of a bubble in a space between two infinite plates (a flat gap) with a cross-sectional width of $2R$ is a mathematical abstraction. Contrary to the axisymmetric case, it has no physical analogue. However, since 1950, the problem of the rise of the plane Taylor bubble drew the attention of the mathematicians, as it can be investigated by the methods of the theory of functions of a complex variable [54–56]. It is of interest to investigate also the flat problem by the method developed above at the analysis of the axisymmetric problem. It is possible to use it then as a benchmark problem, whose solution is based on the powerful methodology of the theory of functions of a complex variable [56].

So, let us return to the approximate method. Instead of Eq. (15.60), we will have a two-dimensional Laplace equation

$$\frac{\partial^2\varphi}{\partial z^2}+\frac{\partial^2\varphi}{\partial r^2}=0. \tag{15.129}$$

Here, z, r are dimensionless longitudinal and cross-sectional coordinates, respectively. As the linear length scale, a half of the width of the gap between the plates is accepted (for the sake of convenience, let us use the notation R_0 for it). The velocity potential for a flat problem looks like

$$\varphi=-\frac{1}{\pi}\ln\left(z^2+r^2\right)-\frac{2}{\pi}\int\limits_0^\infty\frac{\exp(-y)}{y\sinh y}\cosh yr\cos yz\,dy-\varepsilon z. \tag{15.130}$$

The axial and radial velocity components of the flow are defined by the relations

$$U=\frac{2}{\pi}\frac{z}{z^2+r^2}+\frac{2}{\pi}\int\limits_0^\infty\frac{\exp(-y)}{\sinh y}\cosh yr\sin yz\,dy-\varepsilon, \tag{15.131}$$

$$V=\frac{2}{\pi}\frac{r}{z^2+r^2}-\frac{2}{\pi}\int\limits_0^\infty\frac{\exp(-y)}{\sinh y}\sinh yr\cos yz\,dy. \tag{15.132}$$

The parameter characterizing the relative intensity of the source is equal to

$$\varepsilon\equiv\frac{U_\infty}{U_0}=\coth\frac{\pi}{2}x. \tag{15.133}$$

Finally, the dependence of the Froude number (Eq. (15.87)) on the parameter x is

$$\mathrm{Fr}=\frac{1-\exp(-\pi x)}{\sqrt{3\pi}}. \tag{15.134}$$

Equation (15.134) demonstrates a smooth monotonous character of the increase in the Froude number at the increase of the parameter x (Fig. 15.8).

Thus, our assumptions made at the consideration of the axisymmetric problem are confirmed. In the limit at $x\to\infty$, it can be obtained from Eq. (15.134) that

$$\mathrm{Fr}=1/\sqrt{3\pi}\approx 0.326. \tag{15.135}$$

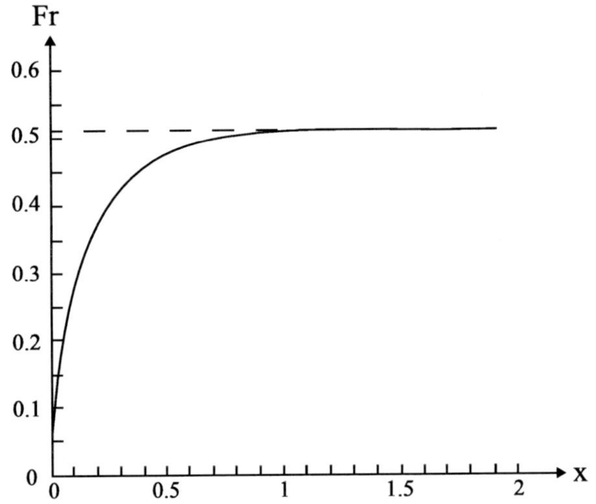

Fig. 15.8 The Froude number versus the parameter x

This value by less than 6% differs from the solution [55] obtained by the methods of the theory of functions of a complex variable. Thus, the analysis of the two-dimensional problem is qualitatively identical to the axisymmetric case, being favorably different from the latter due to the radical simplification of the mathematical calculations.

15.2.7 Non-uniqueness of the Solution

In the conclusion, we will return once again to the property of non-uniqueness of the solution derived above in the course of the approximate approach. Namely, this property served as the reason to search for the additional boundary conditions on the free surface (the method of collocations, asymptotical solution). It is interesting to point out that precisely the same property was also revealed while deriving the "exact solution" of this problem by the methods of the theory of functions of a complex variable [54, 56]. Thus, these additional conditions used within the frames of the approximate solution also allowed achieving the uniqueness of the problem. It should be noticed that such an approach does not function in combination with the exact solution.

Therefore, a paradoxical situation arises here, where the approximate solution is "cleverer" as the exact one! Muskhelishvili [56] pointed out that the domain for the application of the theory of functions of a complex variable is limited. He showed that in order to establish a correspondence between an axisymmetric and a flat problem, it is necessary to prove a special property of the Neumann boundary value problem for the Laplace equation ("property of ellipticity"). In our approach, the "test solution" obtained just as the additional one for the flat problem plays the role of this strict property. In doing so, the same approximate method is applied, as that used at the solution of the initial axisymmetric problem.

Thus, in the present chapter, a correct approximate solution of the problem of the rise of the Taylor bubble in a round tube is presented. The author hopes that he managed to present an evident illustration of the beauty and complexity of the problems dealing with the flows of an ideal fluid with a free surface. At the same time, it is surprising that such a refined mathematical methodology was required "only" to calculate the values of the numerical constants in the "obvious" relations (15.107), (15.128), and (15.135). The main results described in the present chapter were published by the author in the works [57, 58].

15.3 Conclusions

We consider the problem of gas bubble rise in a liquid at rest under the action of the gravity force. A survey is given of the available experimental and theoretical studies of the problem for five conditional regions, which differ from each other by the form of the bubble and particularities of the flow past a bubble. We present a model of bubble rise for the region free from the effect of viscosity (moderately large and large bubbles). In the model based on the "base underpressure" concept, a bubble is assumed to be composed of two parts. The frontal surface is in a stream of an inviscid liquid with separation in the midsection. The aft phase interface is an hydrostatic equilibrium form generated under the joint action of the gravity forces and the surface tension forces. Analysis of the "gravitational" and "capillary" asymptotics of the dependence of the rise velocity on the equivalent radius is given. A general analytical solution of the problem is put forward, which holds for the "inviscid range" of the bubble rise velocity.

The problem of the rise of the Taylor bubble in a round tube is considered. The available solutions are shown to be ill-justified due to the divergence of some involved infinite series. An analytical solution of the problem is obtained based on the collocation method and asymptotical analysis of the solution to the Laplace equation. The method employed was validated on an example of the solution of the corresponding flat problem. As a result, a correct approximate solution of the problem of the rise of the Taylor bubble in a round tube is presented.

References

1. MacCurdy E (1938) The notebooks of Leonardo da Vinci. Reynal & Hitchcock, New York
2. Roberts J (1981) The Codex Hammer of Leonardo da Vinci. Giunti Barbera Florence
3. Stokes GG (1851) On the effect of internal friction of fluids on the motion of pendulums. Trans Cambridge Philosoph Soc 9(ii):8–106
4. Kirby BJ (2010) Micro- and nanoscale fluid mechanics: transport in microfluidic devices. Cambridge University Press

5. Hadamard JS (1911) Motion of liquid drops (viscous). Comp Rend Acad Sci Paris 154:1735–1755
6. Rybczynski W (1911) Über die fortschreitende Bewegung einer flüssigen Kugel in einem zähen Medium. Bull Acad Sci Cracovie, A 40–46.3
7. Allen HS (1900) L. The motion of a sphere in a viscous fluid. London, Edinburgh Dublin Philosoph Mag J Sci 50(306):519–534
8. Miyagi O (1925) The motion of an air bubble rising in water. Phil Mag 50:112–140
9. Davies RM, Taylor GI (1950) The mechanics of large bubbles rising through extended liquids and through liquids in tubes. Proc Roy Soc London A200:375–390
10. Peebles FN, Garber HJ (1953) Studies on the motion of gas bubbles in liquids. Chem Eng Prog 49:88–97
11. Haberman WL, Morton RK (1953) An experimental investigation of the drag and shape of air bubbles rising in various liquids. David Taylor Model Basin Report 802, US Department of the Navy, Washington, DC
12. Hnat JG, Buckmaster JD (1976) Spherical cap bubbles and skirt formation. Phys Fluids 19:182–194
13. Liu L, Yan H, Zhao G, Zhuang J (2016) Experimental studies on the terminal velocity of air bubbles in water and glycerol aqueous solution. Exp Therm Fluid Sci 78:254–265
14. Liu N, Yang Y, Wang J et al (2020) Experimental investigations of single bubble rising in static Newtonian fluids as a function of temperature using a modified drag coefficient. Nat Resour Res 29:2209–2226
15. Saffman PG (1956) On the rise of small air bubbles in water. J Fluid Mech 1:249–275
16. Hartunian RA, Sears WR (1957) On the instability of small gas bubbles moving uniformly in various liquids. J Fluid Mech 3:27–47
17. Aybers NM, Tapucu A (1969) The motion of gas bubbles rising through stagnant liquid. Wärme-Und Stoffübertragung 2:118–128
18. Duineveld PC (1995) The rise velocity and shape of bubbles in pure water at high Reynolds number. J Fluid Mech 292:325–332
19. Ellingsen K, Risso F (2001) On the rise of an ellipsoidal bubble in water: oscillatory paths and liquid induced velocity. J Fluid Mech 440:235–268
20. Tomiyama A, Celata GP, Hosokawa S, Yoshida S (2002) Terminal velocity of single bubbles in surface tension force dominant regime. Int J Multiph Flow 28:1497–1519
21. Wu M, Gharib M (2002) Experimental studies on the shape and path of small air bubbles rising in clean water. Phys Fluid 14(7):L49–L52
22. De Vries AWG, Biesheuvel A, van Wijngaarden L (2002) Notes on the path and wake of a gas bubble rising in pure water. Phys Fluids 28(11):1823–1835
23. Talaia MAR (2007) Terminal velocity of a bubble rise in a liquid column. Int J Phys Math Sci 1(4):220–224
24. Harper JF (1972) The motion of bubbles and drops through liquids. Adv Appl Mech 12:59–129
25. Kulkarni AA, Joshi JB (2005) Bubble formation and bubble rise velocity in gas-liquid systems: a review. Ind Eng Chem Res 44:5873–5931
26. Clift R, Grace JR, Weber ME (1978) Bubbles, drops, and particles. Academic Press
27. Sadhal SS, Ayyaswamy PS, Chung JN (1997) Transport phenomena with drops and bubbles. Springer, New York
28. Birkhoff G, Zarantonello EH (1957) Jets wakes and cavities. Academic Press INC, New-York
29. Batchelor GK (1967) An introduction to fluid dynamics. Cambridge University Press, Cambridge
30. Hua J, Stene J, Lin P (2008) Numerical simulation of 3D bubbles rising in viscous liquids using a front tracking method. J Comp Phys 227(6):3358–3382
31. Feng J (2007) A spherical-cap bubble moving at terminal velocity in a viscous liquid. J Fluid Mech 579:347–371
32. Kang C, Liu H, Mao N, Zhang Y (2019) Motion of bubble. In: Methods for solving complex problems in fluids engineering. Springer, Singapore

33. Weber A, Bart H-J, Klar A (2017) Simulating spiraling bubble movement in the EL approach. Open J Fluid Dyn 07(03):288–309
34. Lamb H (1994) Hydrodynamics, 6th edn. Trinity College, Dover Books on Physics
35. Levich VG (1962) Physicochemical hydrodynamics. Prentice Hall, Englewood Clift
36. Moore DW (1963) The boundary layer on a spherical gas bubble. J Fluid Mech 16:161–176
37. Joseph DD, Wang J (2004) The dissipation approximation and viscous potential flow. J Fluid Mech 505:365–377
38. Schlichting H, Gersten K (2006) Grenzschichttheorie. Springer, Berlin Heidelberg
39. Eckhardt B, Grossmann S, Lohse D (2004) Hundert Jahre Grenzschichtphysik Physik J Nr 10:31–37
40. Mendelson HD (1967) The prediction of bubble terminal velocities from wave theory. AIChE J 13:250–253
41. Baz-Rodríguez S, Aguilar-Corona A, Soria A (2012) Rising velocity for single bubbles in pure liquids. Revista Mexicana De Ingenieria Quimica (Mexico) 11(2):269–278
42. Zudin YB (1975) About the rising of moderately large bubbles in a large volume. Works Moscow Power Eng Inst 268:79–86 (in Russian)
43. Labuntsov DA, Zudin YB (1975) Speed of gravitational rising and form of large bubbles. Works Moscow Power Eng Inst 268:72–79 (in Russian)
44. Zudin YB (1995) Calculation of the rise velocity of large gas bubbles. J Eng Phys Thermophys 68:10–15
45. Myshkis AD, Babskii VG, Kopachevskii ND, Slobozhanin LA, Tyuptsov AD (1987) Low-gravity fluid mechanics. Mathematical theory of capillary phenomena. Springer, Berlin etc
46. Bashforth F, Adams J (1883) An attempt to test the theories of capillary action: by comparing the theoretical and measured forms of drops of fluid. Cambridge University Press, Cambridge
47. Van Hout R, Eisma J, Elsinga GE, Westerweel J (2018) Experimental study of the flow in the wake of a stationary sphere immersed in a turbulent boundary layer. Phys Rev Fluids 3(2):024601
48. Funada T, Joseph DD, Maehara T, Yamashita S (2004) Ellipsoidal model of the rise of a Taylor bubble in a round tube. Int J Multiphase Flow 31:473–491
49. Dumitrescu DT (1943) Strömung an einer Luftbluse im senkrechten Rohr. Z. Angew Math Mech 23:139–149
50. Abramowitz M, Stegun IA (1964) Handbook of mathematical functions: with formulas, graphs, and mathematical tables. National Bureau of Standards, Washington
51. Labuntsov DA, Zudin YB (1976) About emerging of a Taylor bubble in a round tube. Works Moscow Power Eng Inst 310:107–115 (in Russian)
52. Stein EM, Shakarchi R (2003) Fourier analysis: an introduction. Princeton University Press, Princeton
53. Bellomo N, Lods B, Revelli R, Ridolfi L (2007) Generalized collocation methods—solution to nonlinear problems. Birkhäuser, Boston
54. Birkhoff G, Carter D (1957) Rising plane bubbles. J Math Phys 6:769–779
55. Daripa PA (2000) Computational study of rising plane Taylor bubbles. J Comput Phys 157(1):120–142. Driscoll TA, Trefethen LN (2002) Schwarz-Christoffel mapping. Cambridge University Press
56. Muskhelishvili NI (1968) Singular integral equations. Nauka Publishers, Moscow (in Russian)
57. Zudin YB (2013) The velocity of gas bubble rise in a tube. Thermophys Aeromech 20(1):29–38
58. Zudin YB (2013) Analytical solution of the problem of the rise of a Taylor bubble. Phys Fluids 5(5). Paper 053302-053302-16

Chapter 16
Bubbles Dynamics in Liquid

Abbreviations

HN	Homogeneous nucleation
MSBD	Maximum stable bubble diameter

Symbols

d_b	Bubble diameter
l	Length
p	Pressure
Δp	Pressure drop
r	Radial coordinate
Re	Reynolds number
R	Bubble radius
T	Temperature
t	Time
U, u	Velocity
We	Weber number
z	Axial coordinate

Greek Letter Symbols

υ	Kolmogorov velocity microscale
ε	Dissipation
ξ	Darcy friction factor
μ	Dynamic viscosity
ν	Kinematic viscosity
η	Kolmogorov length microscale
ρ	Density
σ	Surface tension

Subscripts

∞	State at infinity

Y. B. Zudin, *Non-equilibrium Evaporation and Condensation Processes*,
Mathematical Engineering, https://doi.org/10.1007/978-3-030-67553-0_16

* Critical

16.1 Bubble Dynamics in a Tube

16.1.1 The Generalized Rayleigh Equation

The derivation of the generalized Rayleigh equation that describes the dynamics of a spherical gas bubble in a tube filled with an ideal liquid is given. Its solution has spherical (the classical Rayleigh equation) and cylindrical (the case of a long tube) asymptotics. An exact analytical solution of the problem on vapor bubble collapse in a long tube was obtained. The classical Rayleigh equation is the basic equation used for studying the dynamics of a spherical gas bubble in an infinite volume of an ideal liquid [1–3]. This equation relates the law of bubble radius variation in time $R(t)$ to the pressure drop $\Delta p(t) = p_{\mathrm{v}} - p_\infty$

$$\frac{\Delta p}{\rho} = \frac{3}{2}\dot{R}^2 + R\ddot{R}, \tag{16.1}$$

where $\dot{R} \equiv dR/dt$, $\ddot{R} \equiv d^2R/dt^2$ are the corresponding derivatives of the bubble radius with respect to time.

The paper [4] gives a brief description of the derivation of an analog to the Rayleigh equation for the case of bubble dynamics in a cylindrical tube filled with an ideal liquid. The present section is devoted to the further development of the indicated equation. A detailed description of the derivation of the generalized Rayleigh equation is given. In the limiting cases, this equation is transformed into the spherical (classical) Rayleigh equation and into its cylindrical analogue. Let a spherical gas (vapor) bubble of radius R_t be located at the center of symmetry of a cylindrical tube of radius R_t and length l (Fig. 16.1a). At the initial moment, the ideal liquid filling the tube is at rest and at $z = \mp l$ the pressure $p_\infty < p_{\mathrm{v}}$. Under such conditions, the bubble grows under the action of a constant pressure difference $\Delta p > 0$, and the liquid flows out from the open sections of the tube with a velocity u_l. The opposite case where $p_\infty > p_{\mathrm{v}}$ at $z = \mp l$ can also be considered. The vapor bubble in this case collapses under the action of a constant pressure difference, and the liquid flows out from the open sections of the tube with a velocity U.

We first consider the limiting case of a long tube under the following assumptions the initial radius of the bubble is much smaller than the tube radius: $R_0 \ll R_t$. The length of the tube is much greater than its radius: $l \gg R_t$. The momentum conservation equation for the liquid is then transformed into the equation of motion of a liquid cylinder

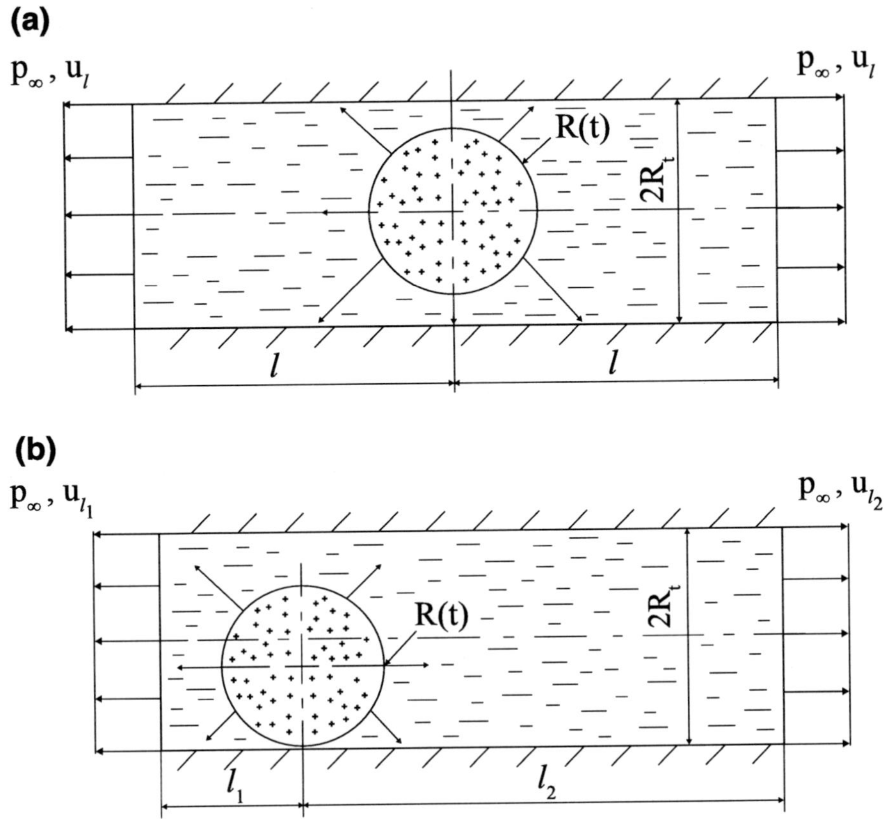

Fig. 16.1 Growth of a vapor bubble in a long tube. **a** Symmetrical case, **b** asymmetrical case

$$-\frac{1}{\rho}\frac{\partial p}{\partial z} = \frac{dU}{dt}. \tag{16.2}$$

In Eq. (16.2), the integral that connects the cylinder acceleration with the pressure gradient has the form

$$\frac{\Delta p}{\rho} = l\frac{dU}{dt}. \tag{16.3}$$

The velocity of cylinder motion U is connected with the radial velocity of bubble expansion $\dot{R}$ by the condition of flow mass constancy

$$R_t^2 U = 2R^2\dot{R}. \tag{16.4}$$

Determining the velocity U from Eq. (16.4) and substituting it into Eq. (16.3), we obtain

$$\frac{\Delta p}{\rho} = 2\frac{lR}{R_t^2}\left(2\dot{R}^2 + R\ddot{R}\right). \tag{16.5}$$

The latter equation represents a "cylindrical analogue" of the Rayleigh equation for the limiting case of a long tube [4]. Let now the center of the spherical bubble be located at different distances from the outlets of the long tube $l_1 \neq l_2$ (Fig. 16.1b).

Our argument will be similar to that employed in the derivation of the cylindrical asymptotics for the symmetric case. In an asymmetric case, a bubble, when it grows, "pushes out" two liquid cylinders of different lengths from the corresponding tube outlets. The momentum conservation Eq. (16.2) for each of the two liquid cylinders can be written as

$$-\frac{1}{\rho}\frac{\partial p}{\partial z} = \frac{dU_1}{dt} = \frac{dU_2}{dt}. \tag{16.6}$$

The integrals of Eq. (16.6) have the form

$$\frac{\Delta p}{\rho} = l_1\frac{dU_1}{dt} = l_2\frac{dU_2}{dt}. \tag{16.7}$$

The velocities of each of the liquid cylinders are described by the relations following from the balances of mass flows

$$R_t^2 U_1 = R_t^2 U_2 = 2R^2\dot{R}. \tag{16.8}$$

Determining the velocities U_1, U_2 from (16.8) and substituting these values into Eq. (16.7), we obtain

$$\frac{\Delta p}{\rho} = 2\frac{l_* R}{R_t^2}\left(2\dot{R}^2 + R\ddot{R}\right). \tag{16.9}$$

Here, the "generalized length" l_* is defined as a combination of the distances l_1, l_2

$$l_* = 2\frac{l_1 l_2}{l_1 + l_2}. \tag{16.10}$$

In combination with (16.10), Eq. (16.9) represents the generalized Rayleigh equation for the asymmetric case of a long tube. From Eq. (16.9), we obtain the above-considered symmetric case ($l_* = l$) described by Eq. (16.5).

Let us now consider the case of a bubble at the center of symmetry of a tube of arbitrary length $0 < l/R_t < \infty$. Here we will retain the constraint imposed on the bubble radius, $R_0 \ll R_t$. To derive the equation of the indicated bubble dynamics, we will use the mathematical apparatus that was used earlier to analyze the bubble rise velocity [5, 6], as well as the bubble drift velocity during bubbly flow in a vertical tube. We consider an ideal liquid flow produced by a point source located at the center of the tube (Fig. 16.1b). The axial u and radial v liquid velocities are defined by the expressions

$$\left.\begin{aligned}\tilde{u} &= \frac{1}{2}\frac{\tilde{z}}{\left(\tilde{z}^2+\tilde{r}^2\right)^{3/2}} + \frac{1}{\pi}\int\limits_0^\infty \frac{\mathrm{K}_1(y)y}{\mathrm{I}_1(y)}\mathrm{I}_0(y\tilde{r})\sin y\tilde{z}dy,\\ \tilde{\mathrm{v}} &= \frac{1}{2}\frac{\tilde{r}}{\left(\tilde{z}^2+\tilde{r}^2\right)^{3/2}} - \frac{1}{\pi}\int\limits_0^\infty \frac{\mathrm{K}_1(y)y}{\mathrm{I}_1(y)}\mathrm{I}_1(y\tilde{r})\cos y\tilde{z}dy.\end{aligned}\right\} \tag{16.11}$$

Here, $\tilde{z} = z/R_t, \tilde{r} = r/R_t$ are the dimensionless coordinates, $\tilde{u} = u/u_\infty, \tilde{\mathrm{v}} = \mathrm{v}/u_\infty$, are the dimensionless velocities, u_∞, is the homogeneous flow velocity for $\tilde{z} \to \mp\infty$ and $\mathrm{I}_1(y)$, $\mathrm{K}_1(y)$ are the modified Bessel functions of the first and second kinds, respectively [7].

For $\tilde{z} \ll 1, \tilde{r} \ll 1$ the system of Eq. (16.11) describes a spherical flow from the source in an infinite volume of liquid

$$\left.\begin{aligned}\tilde{u} &= \frac{\tilde{z}}{2\left(\tilde{z}^2+\tilde{r}^2\right)^{3/2}},\\ \tilde{\mathrm{v}} &= \frac{\tilde{r}}{2\left(\tilde{z}^2+\tilde{r}^2\right)^{3/2}}.\end{aligned}\right\} \tag{16.12}$$

As $\tilde{z} \to \mp\infty, 0 < \tilde{r} \le 1$ the flow becomes homogeneous and cylindrical

$$\left.\begin{aligned}\tilde{u} &= 1,\\ \tilde{\mathrm{v}} &= 0.\end{aligned}\right\} \tag{16.13}$$

It can be easily shown that for the case of a long tube ($l \gg R_t$) the asymptotic equality $u_\infty \approx U$ is valid. Making use of the problem symmetry, we will consider, for the sake of definiteness, a liquid flow along the axis z within the interval $(0 < z \le l)$. We will determine the liquid energy balance in the course of spherical bubble expansion in a tube on the assumption that the total kinetic energy of the liquid has spherical and cylindrical components. In turn, they can be calculated from relations (16.11) and (16.12) (Fig. 16.2) the spherical component reads as

$$0 \le r \le R_t, 0 \le z \le R_t, E_1 = 2\pi\rho R^3\dot{R}^2\left(1 - \frac{R}{R_t}\right), \tag{16.14}$$

the cylindrical component is as follows

$$r = R_t, R_t \le z \le l, E_2 = 4\pi\frac{\rho l R^4\dot{R}^2}{R_t^2}(1 - \xi). \tag{16.15}$$

Here, $\xi = R_t/l$ is the parameter that determines the relative length of the tube. For the case of a long tube, $\xi \ll 1$.

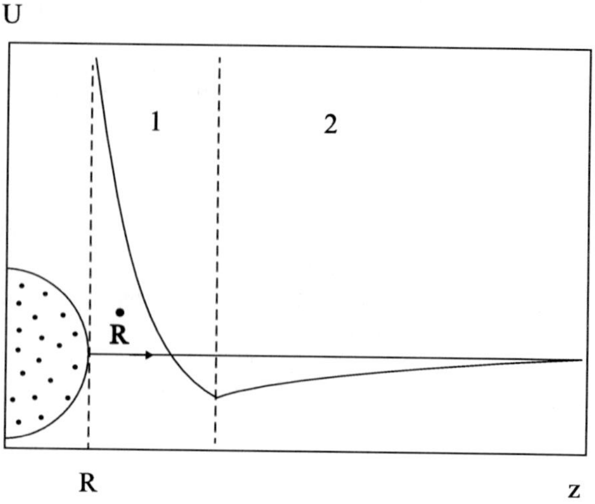

Fig. 16.2 Superposition of the spherical (1) and cylindrical (2) components of the axial liquid velocity

16.1.2 The General Case

For the general case of a tube of arbitrary length, with account for the assumption about the summation of the spherical and cylindrical components of kinetic energies, we obtain

$$E \approx E_1 + E_2 = 2\pi\rho R^3 \dot{R}^2 \left[1 + 2\frac{lR}{R_t^2}\left(1 - \frac{3}{2}\xi\right)\right]. \tag{16.16}$$

From Eq. (16.16), we derive a differential equation that relates the pressure difference Δp in a tube to the characteristics of a bubble in it

$$\frac{\Delta p}{\rho} = \left(\frac{3}{2}\dot{R}^2 + R\ddot{R}\right) + \left(2\frac{lR}{R_t^2} - \frac{R}{R_t}\right)(2\dot{R}^2 + R\ddot{R}). \tag{16.17}$$

In the case $R \ll R_t$ understudy, Eq. (16.17) can be rewritten with sufficient accuracy as

$$\frac{\Delta p}{\rho} = \left(\frac{3}{2}\dot{R}^2 + R\ddot{R}\right) + 2\frac{lR}{R_t^2}(1 - \xi)(2\dot{R}^2 + R\ddot{R}). \tag{16.18}$$

Equation (16.18) represents the desired generalization of the Rayleigh equation. It holds exactly in the following two limiting cases. In the spherical asymptotics $\left(lR/R_t^2 \to 0\right)$, Eq. (16.18) goes over asymptotically into the classical Rayleigh Eq. (16.1). In the cylindrical asymptotics $\left(lR/R_t^2 \to \infty\right)$, Eq. (16.18) describes the case of a long tube (16.5). In the intermediate region $0 < lR/R_t^2 < \infty$, the generalized Eq. (16.18) describes cases that are intermediate between the indicated limiting cases. The passage from the symmetric case (a bubble at the tube center) to the general asymmetric case (a bubble at different distances from the tube outlets) is provided by the simple replacement of l in Eq. (16.18) by the generalized length l_* defined by Eq. (16.10).

It is interesting to note that the principle of summation of two asymptotical components of the pressure difference was also used earlier in investigating the bubble dynamics in a liquid near an infinite planar surface. For this case, the generalized Rayleigh equation of the following form was obtained in [8]:

$$\frac{\Delta p}{\rho} = \left(\frac{3}{2}\dot{R}^2 + R\ddot{R}\right) + \frac{R}{z_0}\left(\dot{R}^2 + \frac{1}{2}R\ddot{R}\right), \tag{16.19}$$

where z_0 is the distance from the center of a spherical bubble to a solid surface. As $z_0 \to 0$, $R/z_0 \to \infty$, Eq. (16.19) becomes the asymptotic equation

$$\frac{\Delta p}{\rho} = \frac{R}{z_0}\left(\dot{R}^2 + \frac{1}{2}R\ddot{R}\right). \tag{16.20}$$

For $z_0 = 2R_t^2/l$, Eq. (16.19) is identical to Eq. (16.18) if it is assumed in the latter that $\xi = 0$. Thus, the two entirely different cases of bubble dynamics (in a tube and near a solid wall) can be described approximately by a single equation. The difference is in the value of the geometrical factor ahead of the parenthetical last term on the right-hand sides of Eqs. (16.18) and (16.19). This analogy confirms the validity of the principle of summation of kinetic energies applied above.

Let us estimate the validity of the assumptions made in the derivation of the generalized Rayleigh Eq. (16.18). For this, it is convenient to use expressions for the corresponding projections of velocities onto their own coordinate axes

$$\tilde{u}_z \equiv \tilde{u}(z,0) = \frac{1}{2\tilde{z}^2} + \frac{1}{\pi}\int_0^\infty \frac{\mathrm{K}_1(y)y}{\mathrm{I}_1(y)} \sin y\tilde{z} dy, \tag{16.21}$$

$$\tilde{\mathrm{v}}_r \equiv \tilde{\mathrm{v}}(0,r) = \frac{1}{2\tilde{r}^2} - \frac{1}{\pi}\int_0^\infty \frac{K_1(y)y}{\mathrm{I}_1(y)}\mathrm{I}_1(y\tilde{r}) dy. \tag{16.22}$$

Formula (16.21) describes the axial projection of the axial velocity, and formula (16.22), the radial projections of the radial velocity. These relations show that each of the two expressions of (16.11) for the velocity components has the form of the superposition of the spherical and cylindrical components. Expanding the integral parts of Eqs. (16.21) and (16.22) into a Taylor series, we obtain

$$\tilde{u}_z = \frac{1}{2\tilde{z}^2} + 0.8\tilde{z} - 0.2\tilde{z}^3 + 0.0621\tilde{z}^5 - 0.019\tilde{z}^7 + \cdots \approx \frac{1}{2\tilde{z}^2} + 0.8\tilde{z}, \tag{16.23}$$

$$\tilde{\mathrm{v}}_r = \frac{1}{2\tilde{r}^2} - 0.4\tilde{r} + 0.075\tilde{r}^3 - 0.0194\tilde{r}^5 + 0.00521\tilde{r}^7 + \cdots \approx \frac{1}{2\tilde{r}^2} - 0.4\tilde{r}. \tag{16.24}$$

Expressions (16.23) and (16.24) in the entire range of variation of the coordinates $\tilde{z}, \tilde{r}$ can be approximated well by the formulas

$$0 < \tilde{z} < \infty : \tilde{u}_z \approx \frac{1}{2\tilde{z}^2} + \tanh 0.8\tilde{z}, \tag{16.25}$$

$$0 < \tilde{r} \leq 1 : \tilde{v}_r \approx \frac{1}{2\tilde{r}^2} - 0.4\tilde{r} - 0.1\tilde{r}^3. \tag{16.26}$$

We will next proceed with the quantitative estimations of the assumptions made In the region with $R_0 \leq z \leq R_t$, $R_0 \leq r \leq R_t$ the flow is spherical, whereas in the region with $R_t \leq z \leq l$, $R_0 \leq r \leq R_t$ it is cylindrical. From Eqs. (16.25) and (16.26), we find the boundary value of the axial coordinate that corresponds to the equality of the indicated components. As a result, we obtain the value $\tilde{z}_* \approx 0.909$, which is very close to the initially adopted one, $\tilde{z}_* \approx 0.909$.

It follows from Eq. (16.11) that the flow is strictly spherical only in the vicinity of the coordinate origin, where $\tilde{z} \to 0, \tilde{r} \to 0$. As $\tilde{z}, \tilde{r}$ increase, the flow becomes progressively deformed under the action of the cylindrical component. Let $R_0 = R_t/4$. Then the cylindrical component will be equal to +2.5% for the axial velocity and to –1.25% for the radial one. Thus, within the range $0 \leq R_0/R_t \leq 1/4$ the bubble boundary practically does not depart from the spherical form. The above estimates support the validity of the assumptions used as a basis for the derivation of the generalized Rayleigh Eq. (16.18).

16.1.3 Collapse of a Bubble in a Tube

As an example, we will consider the problem of the collapse of a vapor bubble in a long tube. Equation (16.9) does not contain the time t in an explicit form. This allows us to go over in this equation from the variables $R(t)$ to the variables $\dot{R}(R)$

$$R^2 \frac{d\dot{R}^2}{dR} + 2R\dot{R}^2 = \frac{\Delta p}{\rho} \frac{R_t^2}{l_*}. \tag{16.27}$$

The solution of Eq. (16.27) has the form

$$\dot{R} \equiv \frac{dR}{dt} = \frac{R_t}{R^2} \left[\frac{2}{3} \frac{\Delta p}{l_* \rho} \left(R^3 - R_0^3 \right) \right]^{1/2}. \tag{16.28}$$

The equation with separable variables (16.28) has an exact analytical solution

$$\frac{R}{R_0} = \left(1 + \frac{2}{3} \frac{R_t^2}{R_0^3 l_*} \frac{\Delta p}{\rho} \right)^{1/3} t^{2/3}. \tag{16.29}$$

When the bubble collapses, we have $p_v < p_\infty$, $\Delta p < 0$, $R/R_0 < 1$. From Eq. (16.29), we find the full time of bubble collapse for the cylindrical problem

$$t_0^{\text{cyl}} = \frac{2}{\sqrt{3}}\left(\frac{R_0^3 l_*}{R_t^2}\frac{\rho}{\Delta p}\right)^{1/2} \approx 1.155\left(\frac{R_0^3 l_*}{R_t^2}\frac{\rho}{\Delta p}\right)^{1/2}. \tag{16.30}$$

As is known, the solution of the classical problem of bubble collapse in an infinite volume of liquid was represented in the form of a quadrature in the classical work of Rayleigh [1]. The expression for the full time of collapse for the spherical problem has the form

$$t_0^{\text{sp}} = \frac{1}{\sqrt{6}}\frac{\Gamma(1/2)\Gamma(5/6)}{\Gamma(4/3)}R_0\left(\frac{\rho}{\Delta p}\right)^{1/2} \approx 0.9147 R_0\left(\frac{\rho}{\Delta p}\right)^{1/2}, \tag{16.31}$$

where $\Gamma(x)$ is the gamma function. Drawing up the ratio of the times of collapse for the two limiting problems, we obtain

$$\frac{t_0^{\text{cyl}}}{t_0^{\text{sp}}} \approx A, \tag{16.32}$$

where $A = R_0 l_*/R_t^2$ is the cylindricity parameter.

The generalized Rayleigh equation describing the dynamics of a spherical gas (vapor) bubble in an ideal liquid flow produced in a tube by a point mass source located at the tube center has been derived. It is shown that in the limit of a spherical flow, the above solution goes over into the classical Rayleigh equation, and its cylindrical asymptotics describes the case of a long tube. Estimates of the validity of the assumptions made in the derivation of the indicated equation have been obtained. An exact analytical solution of the problem on the vapor bubble collapse in a long tube is given. The generalized Rayleigh equation can be used for calculating the processes of growth, collapse, and periodical pulsations of a gas bubble in a tube of arbitrary length. The results given in the present section were published by the author in [9].

16.2 Homogeneous Nucleation

16.2.1 Introduction

In the present section, a rather exotic physical example of phase transitions with a periodic internal structure in the area of nanoscopic scales is treated. According to [10, 11], the emergence of a steam phase in the form of a microscopic inclusion ("vapor cluster") in an infinite volume of a liquid phase ("parent phase") demonstrates itself as a kind of fluctuation. This process is known in the literature as the

"Homogeneous Nucleation" (HN). It should be mentioned that, contrary to this, the process of nucleate boiling on a heated wall considered in Chap. 7 is called the "heterogeneous nucleation". The basic characteristic parameter of the HN process is the frequency J of formation of the primary vapor clusters per unit of a liquid volume

$$J = J_0 \exp\left(-\frac{W}{k_B T}\right). \tag{16.33}$$

Here, W the energy of formation (the "energy barrier") of the vapor cluster, T is the absolute temperature of the liquid, $k_B = 1.38 * 10^{-23}$ J/K is the Boltzmann constant, J_0 is the frequency of HN in the limiting case of zero energy barrier ($W = 0$).

As is known from phenomenological thermodynamics [12], any liquid overheated above the saturation temperature at a given pressure ("metastable liquid") can basically always pass to a vaporous state. However, if "weak places" (i.e., nucleation sites) lowering the energy barrier are absent in the parent phase, it can theoretically continue to exist in the liquid form up to the temperature of the maximum thermodynamic overheating (named in other words a "spinodal temperature"). Thus, the overheated liquid can occupy the entire metastable area located in the pressure–temperature chart between the curves of saturation and spinodal [10, 11]. As was already pointed out by Gibbs (see [10, 11]), the existence of the liquid overheated above the spinodal temperature is thermodynamically impossible, and therefore, under this condition, the explosive formation of vapor occurs.

16.2.2 Classical Theory

An overall objective of the modern "classical" theory [12] is to predict the limiting frequency HN J_0. For this purpose, a variety of theoretical models was offered starting from the first pioneer works of Volmer, Becker, and Doering [10, 11] and finishing with fundamental books of Scripov [10] and Debenedetti [11]. A rather elegant result obtained in the frames of the "hydrodynamic variant" of the HN theories is as follows

$$J_0 \approx \frac{\sigma^{3/2} N_0}{\mu (k_B T)^{1/2}}, \tag{16.34}$$

where σ is the surface tension coefficient, μ is dynamic viscosity of the liquid, N_0 is the number of molecules per unit of the liquid volume.

An extremely important remark here is worth making here. According to the author of the present book, the following basic physical contradiction lies in the bedrock of the classical HN theory. From the point of view of phenomenological thermodynamics, the parent phase should not admit formation inside itself of a microscopic inclusion of a new (vapor) phase lying in the area of nanoscopic scales [13]. Therefore, to explain the phenomenon of HN, it is considered [11, 12] that its origination is caused by a certain fluctuation. On the one hand, such a fluctuation leads to a negative splash in entropy and consequently breaks directly the second law of thermodynamics. But on the other hand, the fluctuation is supposed to be slow enough, so that it becomes possible to apply the first law of thermodynamics to it. Within the frames of such a scheme (with some differences, which are insignificant to be mentioned here), the isobaric-isothermal process of the growth of a hypothetical vapor cluster, which originally contains just a few molecules, is described. The fundamental contradiction specified above can be formulated in more detail as follows:

- On the one hand, a vapor cluster arising in the volume of a parent phase should have a size larger than some critical size: $l > l_*$. Only in this case, it is sustainable in the thermodynamic sense [10, 11] and will continue growing at the expense of evaporation of the liquid mass next to it. In the case the size of the vapor cluster is less than critical ($l < l_*$), its sustainable existence in the liquid is impossible, and it will consequently condense
- On the other hand, as it was already mentioned, within the limits of phenomenological thermodynamics, it is in principle impossible to explain the spontaneous emergence of a vapor cluster of finite size in a parent phase $l = l_*$ [13]
- To reconcile the outlined contradiction, the theory of HN assumes an initial growth of a vapor cluster from a minimum (actually molecular) size up to the achievement of a critical radius ($l = l_*$). Namely the distinction in the considered models of the vapor cluster growth leads to the different variants of the general theory of homogeneous nucleation.

As a result, one is confronted with a physical paradox: a vapor cluster, which from the point of view of thermodynamics should condense, nevertheless steadily grows according to the theory of HN. Thereupon rather promising and interesting are the works that appeared recently on direct numerical modeling of the HN process by the Monte Carlo method [14, 15] and molecular dynamics methodology [16]. However, unfortunately, these approaches are in the initial stage of development and cannot yet give an exhaustive answer to the questions raised above. Before passing to the statement of a new model, it is expedient to summarize the characteristic features of this exotic and intriguing phenomenon. The process of HN: (a) is a physical reality confirmed experimentally (see a description of the investigations by Volmer, Becker, and Doering performed in 1930s and documented in the books [10, 11]), (b) is a subject of the "classical" theoretical description, (c) breaks the second law of thermodynamics ("negative splash in entropy"), (d) is subject to the first law of thermodynamics ("slow fluctuation"), (e) falls into the area of nanoscopic scales (initial vapor cluster contains few molecules).

16.2.3 Quantum Mechanical Model

A possible way to eliminate the specified contradictions in frames of the quantum mechanical model of HN offered by the author of this book is given below. The basic assumption is that a spherical vapor cluster with a critical radius l_* and the surface area of $4\pi l_*^2$ already exists in the uniformly heated liquid. The energy to be passed to forms a vapor cluster in a parent phase (i.e., in the liquid volume) is

$$W = 4\pi l_*^2 \sigma. \tag{16.35}$$

To roughly estimate the cluster radius, it can be set equal to the minimum radius of a molecule $l_* \sim 10^{-10}$ m. Then at the value of $\sigma \sim 10^{-2}$ N/m, one can obtain an extremely small value of the "energy quantum" equal to $W \sim 10^{-21}$ J. To estimate the scale of the energy quantum, one can use the dispersion law known in quantum mechanics [17] and describing the energy spectrum of a "quasiparticle". So, for the superfluid helium [18], the quasiparticles have characteristic energy of the order of $W \sim 10^{-22}$ J. This means that this value has the order of magnitude identical with the energy of formation of the vapor cluster. Thus, if one conventionally considers the vapor cluster as a quasiparticle, its surface energy will approximately correspond to the energy quantum characteristic for the law of dispersion [19]. These estimates allow in principle to consider the formation of a critical vapor cluster from the positions of quantum mechanics.

According to [10, 11], the critical vapor cluster exists in the state of unstable balance: any deviation of its radius from the critical value results either in its growth, ($l > l_*$) or in its collapse ($l < l_*$). We will assume that these two alternative irreversible processes are divided by a certain hypothetical steady state. Strictly speaking, this is a state of absolutely unstable balance. Therefore, it is necessary to impose perturbations with some frequency ω onto this "steady state". While the size of this vapor cluster is in the range of nanoscopic scales, it can be identified with a quantum oscillator [19]. The energy W of the quantum harmonic oscillator is described by the relation

$$W = \hbar\omega\left(\frac{1}{2} + n\right), \tag{16.36}$$

where $\hbar = 6.626 \times 10^{-34}$ J*s is the Planck constant, n is the quantum number (energy level). In agreement with Eq. (16.36), the energy to be passed to enables the formation of a sustainable vapor cluster in a parent phase has a discrete spectrum.

Let us now make the following assumption. We assume that the energy level n of the quantum oscillator is approximately equal to the number of molecules in the vapor cluster. From Eqs. (16.35), (16.36), the following expression for the quantum oscillator frequency can be drawn

$$\omega = 4\pi \frac{l_*^2 \sigma}{h(1/2+n)}. \tag{16.37}$$

Based on the physical reasoning, it is natural to assume that no direction or a point should be given priority in the parent phase. This means that the process of HN is homogeneous and isotropic so that a dense cubic packing of the vapor clusters will emerge in the liquid. As a result, one can obtain from Eq. (16.37)

$$J_0 = \frac{3\pi}{4} \frac{\omega}{l_*^3}. \tag{16.38}$$

Substituting the expression for the quantum oscillator frequency from Eq. (16.37) into Eq. (16.38) one can obtain

$$J_0 = 3 \frac{\sigma}{l_* h(1/2+n)}. \tag{16.39}$$

As is known, there exists the so-called Laplace pressure drop through the spherical surface of the vapor cluster

$$\Delta p_* = 2\frac{\sigma}{l_*}. \tag{16.40}$$

Taking into account Eq. (16.40), we rewrite Eq. (16.39) in the following form

$$J_0 = \frac{3}{2} \frac{\Delta p_*}{h(1/2+n)}. \tag{16.41}$$

As assumed, in the boiling liquid, a vapor cluster arises with a critical radius. The vapor inside the cluster is in the state of thermodynamic equilibrium. It is easy to show that the Laplace pressure drop through the cluster can be expressed from the known Fürth's equation for the spinodal [10, 11]

$$\Delta p_* = 1.32 \frac{\sigma^{3/2}}{(k_B T)^{1/2}}. \tag{16.42}$$

From Eqs. (16.39)–(16.42), one can obtain the required expression for the limiting frequency of formation of a vapor cluster per unit volume of a liquid

$$J_0 \approx \frac{\sigma^{3/2}}{h(1/2+n)(k_B T)^{1/2}}. \tag{16.43}$$

Comparing Eqs. (16.43) and (16.35), one can obtain the following estimate for the energy level of the quantum oscillator

$$n = \frac{\mu}{hN_0}. \tag{16.44}$$

Estimating the number of molecules per unit volume of a liquid as $N_0 \approx 10^{28}$ m^{-3}, one obtains from Eq. (16.44) that $n \approx 10^2$, which corresponds approximately to the number of molecules in the vapor cluster [10, 11]. This circumstance is an indirect confirmation of the plausibility of the physical estimations performed above. Thus, under the condition that equality (16.44) holds, the quantum-mechanical model stated above reduces to Eq. (16.35) derived within the frames of the classical theory of HN (exactly saying, its "hydrodynamic model"). It is necessary to point out that the methods of quantum mechanics were applied earlier in [20] for the analysis of the HN in a vapor parent phase, i.e., as applied to the condensation process. As far as the author is aware, the quantum mechanical model of HN stated above as applied to the boiling process was for the first time derived by the author in [21, 22]. The author would consider his objective attained if on the above exotic example he was able to show the efficiency of simulation of heat processes with the help of the analysis of their periodic inner structure. Such a modeling, as a red thread passing through the entire book, began on the usual "macroscopic level" and finally came to its application in the area of nanoscopic scales.

16.3 Bubble Size in a Turbulent Fluid Flow

16.3.1 Structure of Turbulent Vortices

As is known [23], turbulent flows give an example of a nonlinear mechanical open system with a very large number of degrees of freedom and have the following properties:

- The initial small uncertainty of flow characteristics, as introduced by an external disturbance in the turbulent flow, will increase in time. Therefore, it is impossible to predict the result of perturbation evolution
- Turbulent flow must satisfy the intensive mixing properties
- The pulsation component of a turbulent flow is characterized by a wide range of wavelengths and frequencies.

According to Kelvin's circulation theorem [24], if the nonviscous fluid was potential at the initial time, then it must remain potential at all subsequent times. However, the presence of viscosity in the flow changes the situation drastically. The presence of boundaries imposes the no-slip condition on the flow, which generates vortices in the flow due to the effect of viscosity. Under the influence of various instability mechanisms (for example, stretching of line vortices), the production of the vorticity increases, which results in a turbulence development.

In the region of macroscopic dimensions much larger than molecular sizes, the flow of a liquid satisfies the Navier–Stokes equations. The rigorous mathematical description of turbulence is based on the postulate of the Newtonian deterministic system: if at time t_0 (as well as under given boundary conditions and certain external perturbations) the initial positions of the liquid particles and their velocities are known, then at any time $t > t_0$, the system has a unique state. Physically, this postulate means that the presence of a molecular viscosity smoothes the solutions of the Navier–Stokes equations in order to exclude the appearance of singularities and bifurcations. From these equivalent definitions, it follows that in theory, the turbulence is a deterministic phenomenon, despite its quite involved nonlinear nature. Such a rigorous and precise description of the deterministic development of a turbulent flow for an arbitrary time moment seemed almost until very recently. However, due to rapid progress in modern computer systems, numerical modeling of certain classes of turbulent flows based on the solution of Navier–Stokes equations has now become available.

A twofold role of the viscosity property in the turbulence phenomenon should be pointed out. On the one hand, viscosity contributes to turbulence generation, while on the other hand, it results in the dissipation of the mechanical energy and its transition to thermal energy. The nonlinearity property of the Navier–Stokes equations implies the existence of the main parameter (the Reynolds number) of a turbulent flow: $\mathrm{Re} = Lu/\nu$. The Reynolds number is responsible for the range of flow scales for which the dissipative properties of the system are displayed. The mechanism of turbulence and the internal structure of a turbulent flow can be schematically described as follows:

The motion of large vortex structures of scale L becomes unstable in time and is destroyed due to growing disturbances. Since the perturbation amplitude is bounded from above by scale L, only vortices of small size will be excited, $l < L$. The latter, in turn, will also be unstable for large Reynolds numbers. According to this physical scheme, an extended hierarchy of unstable vortex perturbations is formed, which completely determines the nature of the turbulent flow. When a certain minimum vorticity scale is reached ($l = n$), the continuous process of generating smaller-scale perturbations fades out. Due to their essentially dissipative nature, microscale vortices will already be stable. Their energy, which is spent on overcoming the viscous forces, dissipates and turns into the heat energy. The described "cascade process" of the sequential breakup of all nondissipative perturbations of an ever smaller scale creates a continuous flow of energy "down" the scale.

Note that such a scenario of perturbation development, in which vortices of different scales interact, is possible only in a nonlinear system.

This suggests the following consequences:

- Constant energy flow along the entire spectrum. The rate of energy entering the cascade, transferring it down the cascade, and dissipating on dissipative scales occurs with a constant "dissipation rate" ε.
- Locality of the interaction. Vortices of scale l_n interact only with vortices from neighboring scale links l_{n+1} and l_{n-1}. Interaction of vortices of very different

scales is reduced to a transfer of microvortices between them. This transfer occurs without any energy transfer and is regulated by the velocity field of macrovortices
- Stochastic behavior of energy transport. On each chain of the cascade process, a partial "erasure of information" occurs. Moreover, the smaller the scale of perturbations, the lesser their statistical regime is affected by large-scale effects (anisotropy, unsteadiness, etc.). Here an analogy of the transport of turbulent disturbances with the Markov process is worth mentioning. As is known [25], a Markov process is a stochastic process whose evolution after a given time t does not depend on the evolution before t, given that the value of the process at t is fixed.

According to [26], in addition to the randomness property, the small-scale turbulence is also characterized by the self-organization property. Thus, in the transition from a laminar to a turbulent flow, a part of the energy of thermal chaos (associated with arbitrary fluctuations occurring at the molecular level) passes into the macroscopical motion of ordered dissipative structures. Self-organization increases the internal order of the system compared to the molecular chaos. From these positions, a cascade process of vortices breakup can be looked upon as an ordered sequence of processes of self-organization. Moreover, the set of space-time scales on which this process is played out corresponds to the coherent behavior of vortices, which is expressed in various forms of their motion (collective, coordinated, interdependent, rotational motion). It can be assumed [26] that the cascade process forms such an interaction between the order and chaos of a pulsation flow, which via self-organization forms macroscopic dissipative space-time structures. Based on such representations, one can apply non-equilibrium thermodynamics methods to the description of the cascade process of transport of turbulent energy by vortices of different sizes. In this case, these vortices can be interpreted as excited macroscopic degrees of freedom. In the formalism of non-equilibrium thermodynamics, this corresponds to Prigogine's "order through fluctuation" concept [27].

16.3.2 Richardson–Kolmogorov Cascade

Transfer of energy from large motion scales to small motion scales is referred to as a "direct energy cascade" in the mechanics of solid media [24]. In relation to the phenomenological flow pattern of a turbulent flow, the idea of energy transport downstream the vortices spectrum was first expressed in 1922, by Richardson [28]. He described the physical picture of the transport of turbulent energy on the qualitative level and made up a rhyme: "Big whirls have little whirls that feed on their velocity, and little whirls have lesser whirls and so on to viscosity". Note, however, that Richardson expressed these general considerations in a qualitative form and did not draw any conclusions in the rigorous mathematical language.

The final understanding of the role of the small-scale turbulence in the processes of turbulent transport came after Taylor's paper [29], who introduced the concept of "homogeneous isotropic turbulence". According to Taylor, a turbulence is called homogeneous if the field of pulsation velocities does not change with arbitrary parallel translations. Accordingly, a turbulence is called isotropic when it is invariant with respect to rotations and reflections. Mathematically, this means that the invariance of the turbulence structure with respect to arbitrary orthogonal transformations. It should be noted, however, that the homogeneity and isotropy properties assume that the flow has no boundaries and has strict constant average velocity. Therefore, Taylor's model, strictly speaking is unsuitable for describing real turbulent flows.

Kolmogorov [30], the creator of the universal equilibrium theory of local isotropic turbulence, obtained the most important and fundamental results for small-scale turbulence. The key characteristic of the Kolmogorov's theory is the dissipation rate of turbulent energy ε averaged over the ensemble of possible realizations of the medium flow. The value of ε characterizes the rate of transfer of the kinetic energy of the pulsation motion along the hierarchy of vortices in the cascade process. Below, for brevity, ε will be simply called the "dissipation".

Kolmogorov significantly extended the concept of a cascade process of energy transfer. He assumed a monotonous weakening of the predominant influence of the average flow with each transition to smaller structures. This has led to the first Kolmogorov background hypothesis: the statistical mode of small-scale turbulence is universal and is determined only by two-dimensional parameters: the dissipation ε and the kinematic viscosity ν. According to Kolmogorov's hypothesis, the average flow affects the statistical mode of small-scale turbulence through the amount of energy flow transmitted from the largest structures (with a linear scale L) through their entire hierarchy, up to the smallest (with a linear scale of η). This means that, for sufficiently large Reynolds numbers, the anisotropy and nonstationarity of the averaged flow have no effect on the microstructure of turbulence.

In the framework of Kolmogorov hypothesis, the mathematical machinery of isotropic turbulence, as created by Taylor [26], proved to be very potent for describing the properties of small-scale components of real turbulent flows. Now the statistical specification of these components is naturally considered as homogeneous and isotropic. In other words, any developed turbulence with a sufficiently large Reynolds number can be considered locally isotropic, which radically simplifies its mathematical description. Thus, in the limit of very large Reynolds numbers in a turbulent flow, there is a microscale fluctuation level at which "viscous attrition" of small hydrodynamically stable vortices takes place.

The second Kolmogorov background hypothesis [30] postulates the existence of an "inertial interval" of vortex scales intermediate between the macro- and microvortices zones. In this interval, which is free from generation and dissipation of turbulent energy, statistic conditions of turbulence depend on the single parameter ε. So, for large Reynolds numbers, the following physical dualism holds: "large-scale nonisotropic inhomogeneities are isotropic statistical small-scale perturbations". It is natural to expect that the period of pulsations of various orders will be proportional to their spatial scale, and in the case of pulsations of minimum scale, it will be much smaller than the time for a noticeable change in the characteristics of the average flow. Therefore, the mode of small-scale oscillations will be quasi-stationary, i.e., practically stationary for time intervals that contain a large number of periods.

In view of the above, the Richardson–Kolmogorov cascade mechanism can be schematically represented as a step-by-step process.

(1) If the fluid flow is turbulent, a vortex structure is formed in the pipe with the scale L and the mean square pulsation velocity υ. The turbulence intensity of such energy-containing vortex is determined by the turbulent Reynolds number: $\mathrm{Re}_t = \upsilon L/\nu$.

(2) Due to instability, the energy-containing vortex splits into several smaller vortices of the same size. At the same time, their total of energy is identically equal to the energy of the initial large vortex.

(3) During the further vortex decay process, the inertial interval of energy spectrum is reached. Here the size of each vortex l_i is determined by the pulsation velocity υ and the dissipation ε

$$l_i \approx \frac{\upsilon^3}{\varepsilon}. \tag{16.45}$$

Such vortices are still quite large and are not affected by the viscosity. Thus, the length and velocity scales in the inertial interval are interrelated and cannot be determined separately.

(4) Within the limits of the inertial interval, the vortices continue to split until they become stable due to the influence of the viscosity force. The minimum size of the vortex now depends on two parameters: the dissipation ε and the kinematic viscosity ν

$$\eta \approx \left(\frac{\nu^3}{\varepsilon}\right)^{1/4}. \tag{16.46}$$

The same parameters control the minimum velocity scale

$$\upsilon \approx (\nu\varepsilon)^{1/4}. \tag{16.47}$$

The length scale η and the velocity scale υ are known as "Kolmogorov microscales" Thus, at the last stage of breakup, the vortex structure becomes self-organized and fixed scales (determined up to a numerical constant) are set.

Note the fundamental role played by the dissipation ε in the Richardson–Kolmogorov cascade—it is responsible for the following functions:

- energy supplied to fluid per unit mass and time
- energy cascading from scale to scale, per unit mass and time
- energy dissipated by viscosity, per unit mass and time.

16.3.3 Models of Turbulence

16.3.3.1 Prandtl's Mixing Length Theory

The modern semi-empirical theory of turbulence was started by Prandtl [31], we put forward an extremely fruitful hypothesis to the effect that the local variation of the averaged velocity is governed by its first derivative in the transverse coordinate $\partial u/\partial y$. On this basis by dimensional arguments, one introduces the scale of turbulence (the "mixing length") $l = \kappa y$, where $\kappa = 0.41$ is the Karman constant. It follows the well-known Prandtl formula for a turbulent flow

$$\tau = \pi \kappa^2 \rho y^2 \left(\frac{\partial u}{\partial y}\right)^2. \tag{16.48}$$

which is known under the name "turbulent mixing length hypothesis". Here ρ is the density of liquid, y is the transverse coordinate, u is the averaged axial velocity. Prandtl's formula (16.48) laid the basis for the first historical model of turbulent viscosity in near-wall flows, which is used till now in practical calculations. One of the most important results of Prandtl's theory is the derivation of the universal log law for wall-bounded turbulent flows.

16.3.3.2 Second Generation Turbulence Models

Prandtl's theory is chiefly based on the "locality hypothesis": turbulent shear stresses are completely governed by the local structure of the mean flow. This locality hypothesis was found efficient in describing equilibrium flows. However, the locality hypothesis becomes less justified in the context of essentially non-equilibrium flows (in which the structure of the mean flow does not correspond to the inner structure of the turbulence). It proved, in particular, that the equilibrium between the turbulence and the mean flow is established much slower than that of the internal structure of the turbulence. In order to find the relations between the components of the tensor of turbulent shear stresses and the local parameters of turbulence, new models of turbulence were proposed based on the use of equations for the "second moments" (the kinetic energy and Reynolds stresses). To close these equations, it is required to express the unknown terms in terms of the defining parameters, the number of equations should agree with the number of parameters.

16.3.3.3 Algebraic Models

Models of turbulence involving the equations for second moments are based on the fundamental hypothesis formulated in the 1940s, independently by Kolmogorov, Prandtl, and Wieghardt [31]. This hypothesis relates the coefficient of turbulent viscosity ν_t to the kinetic energy of turbulence K

$$\nu_t = c\frac{L}{\sqrt{K}}. \tag{16.49}$$

Here, L is the integral scale of turbulence, $c \approx 0.09$ is the empirical constant. Kolmogorov–Prandtl hypothesis yields the relation between the rate of dissipation ε, the kinetic energy K and the integral scale L

$$\varepsilon \approx \frac{K^{3/2}}{L}. \tag{16.50}$$

Relation (16.50) is obtained by making Re $\to \infty$. Physically this means that for large Reynolds numbers the rate of dissipation is governed by the "cascade" energy transfer: from large eddies to small ones. The last in this chain is the process of dissipation itself (transition of the kinetic energy of small-scale eddies into heat). Excluding the integral scale L from Eq. (16.49), we obtain from Eq. (16.50) an equation relating the coefficient of turbulent viscosity mt, the kinetic energy of turbulence k and the rate of dissipation ε

$$\nu_t \approx \frac{K^2}{\varepsilon}. \tag{16.51}$$

The free parameter in Eq. (16.50) is the integral scale of turbulence L: In simple ("algebraic") models this quantity is taken to be proportional to the Prandtl mixing length: $L \approx l$. In more involved ("differential") models, the scale of turbulence is determined from the differential equation.

16.3.3.4 Differential Models

A number of practical advantages come from the use, as an additional second parameter of the model, of the dissipation ε, which is related to the turbulence scale by (16.50). Expressing e from the Navier–Stokes equation by means of the averaging procedure, one may arrive at the "$K - \varepsilon$ model" of turbulence. Different variants of this widely useful two-parameter model, as augmented by the algebraic expression (16.51), are chiefly useful for the description of turbulent flows immune to viscosity and near-wall effects. In the context of a turbulent boundary layer, the $K - \varepsilon$ model is used for the description of the flow region lying outside the viscous sublayer. A transition of the boundary conditions from the surface to points outside the range of influence of viscosity is carried out using various "near-wall functions".

For this aim, in the simplest case at some point lying in the region of the logarithmic velocity profile, one sets the conjugation parameters: the averaged flow velocity, the kinetic energy of turbulence, and the dissipation. According to numerous researches, the standard "high-Reynolds version" of the $K - \varepsilon$ model does not describe the effect of a number of factors due to the near-wall effects: large longitudinal pressure differences, small and transient Reynolds numbers, 3D-flows, and so on. It should be noted that the use of various modifications of the near-wall functions was found to be ineffective. This has led to the appearance of the "low-Reynolds versions" of the $K - \varepsilon$-models, which, however, also employ various "damping factors", which take into account the effect of viscosity on the turbulent characteristics near the wall.

The transition from the algebraic models to the differential ones residing on the equations for second moments has enabled us to account for a number of new factors: the convective and diffuse transfer of turbulence, nucleation of turbulence due to shear strains of the mean flow, energy dissipation due to viscosity forces. However, in spite of a considerable advantage in comparison with the Prandtl model, some fundamental "genetic" drawbacks of the $K - \varepsilon$ models were identified: (a) inconsistency in the calculation of turbulent near-wall flows, (b) unphysical behavior of the dissipation when transiting from the flow core in the near-wall region, (c) required "tuning" of the equations in e for a specific character of the velocity profile in the near-wall region.

16.3.3.5 New Models

Critical analysis of differential models of turbulence casts serious doubts on their principal point: an expected larger exploitability in comparison with one one-parameter models. This has led to the appearance as an alternative of universal one-parameter models based on the equations for turbulent viscosity [32]. A special emphasis should be given to promising models of turbulence residing on the equations for the components of the Reynolds stress tensor. Here, however, the problem of closure of these equations issues a matter for the future. The process of modeling separate terms of the equations is chiefly based on physical and dimension considerations, which involves principal difficulties in their practical implementation. As a result, such a model is found not to be aimed at the whole spectrum of scales turbulence, but rather at some or other interval thereof. One of the most representative modern differential models is the two-parameter two-zonal model [33] of "hybrid" nature. This model is obtained by conjugating to regions: in the inner (near-wall) region one works with a version of small-scale turbulence, in the outer region (the flow core) one allies the version of the $K - \varepsilon$ aimed at the description of large-scale coherent structures.

16.3.4 Bubbles Breakup

Based on his universal equilibrium theory of local isotropic turbulence, Kolmogorov [34] proposed a model of breakup of dispersed inclusions (drops and bubbles) in a medium (gas or liquid). Later, Hinze [35] specified the general Kolmogorov model in the specific case of a bubble in a liquid. According to [34, 35], violation of force balance is a criterion of bubble breakup. The stabilizing force is the surface tension at the interface boundary $\sim \sigma/d_b$, which ensures the sphericity of the bubble. The inertial force ρv^2 of the action on the bubble of a turbulent vortex coming from a continuous medium serves as a destabilizing force.

Large vortices will drag the bubble as a whole when they move. Small vortices when they collide with a bubble will not have a noticeable effect on its surface. Therefore, the vortices belonging to the inertial interval corresponding to the Kolmogorov spectrum for energy will be responsible for bubbles breakup: $l \approx l_i \approx d_{b*}$. This interval occupies an intermediate position between the intervals of large and small vortices. Hence, the stability condition of the bubble surface will be read as: $\sigma/d_b \geq \rho v^2$. This gives the following expression for the Maximum Stable Bubble Diameter (MSBD): $d_{b*} \approx \sigma/\rho v^2$. This value corresponds to the critical Weber number beyond which the bubble breakup occurs

$$\mathrm{We}_* = \frac{\rho v^2 d_{b*}}{\sigma}.$$

Setting $d_{b*} \approx l_i$ and using formula (16.45) for the inertial interval of energy spectrum, from (16.48), we get the well-known Hinze formula [35]

$$d_{b*} = c_1 \left(\frac{\sigma}{\rho}\right)^{3/5} \frac{1}{\varepsilon^{2/5}}, \tag{16.52}$$

where $c_1 = 1/\mathrm{We}_*^{3/5}$ is a numerical constant. Batchelor [36] based on the theoretical analysis of isotropic turbulence refined the basic ratio (16.45) for the inertial interval (written by Kolmogorov up to a number constant)

$$v^2 = 2(\varepsilon l_i)^{2/3}.$$

Hence, using (16.45), we get $\mathrm{We}_* = 2$, $c_1 = 0.66$.

In [37, 38], a simulation of bubbles breakup in the absence of gravity was performed using the lattice Boltzmann method. Isotropic turbulence was generated by stochastic mixing of liquid on a three-dimensional periodic grid. The Navier–Stokes equation for a liquid was solved, as well as the kinetic equation for the probability distribution of bubble interaction. The advantage of the numerical method used in [37, 38] is that the computing time is practically independent of the number of bubbles. It was found that the bubble breakup was preceded by stochastic pulsations of its shape. With increasing mixing intensity, a monotonous increase in the linear size of the bubble was observed in the direction of the main axis of deformation. It is shown that the process of bubble breakup was explosive. In [37], we obtained the value of the critical Weber number $\mathrm{We}_* = 3$, which gives the value of the constant $c_1 = 0.517$ in (16.49).

16.3.5 Local Isotropic Turbulence

To perform calculations via (16.52) it is necessary to associate the dissipation included in it with the macroscopic parameters of the turbulent flow in the channel. To this end, the Kolmogorov's theory of local isotropic turbulence should be employed. We will proceed from the energy balance for a section of pipe of the hydraulic diameter d_h and length L

$$\text{Drag Friction}\left(\frac{\xi}{8}\rho u^2\right) * \text{Flow velocity}\,(u) * \text{Surface Area}\,(\pi d_h L)$$
$$= \text{Density}\,(\rho) * \text{Dissipation}\,(\varepsilon) * \text{Volume}\left(\frac{\pi d_h^2}{4}L\right)$$

This gives the expression for the average dissipation over the cross-section

$$\varepsilon = \frac{\xi}{2}\frac{u^3}{d_h}. \tag{16.53}$$

Here ξ is Darcy friction factor, which depends on the correlation given by Filonenko for the turbulent flow within a smooth tube

$$\xi = (0.79\ln\mathrm{Re} - 1.64)^{-2}.$$

The less accurate, but more convenient power law

$$\xi = \frac{0.179}{\mathrm{Re}^{1/5}} \tag{16.54}$$

is also used in the literature. From (16.53)–(16.54), we get the resulting relation for the MSBD averaged over the section

$$d_{b*} = c_2\left(\frac{\sigma^6 d_h^5}{\nu\rho^6 U^{11}}\right)^{1/10}, \tag{16.55}$$

where $c_2 = 1.38$. From (16.53)–(16.55), we get the dependence of the MSBD on the velocity: $d_{b*} \sim u^{-6/5}$.

We briefly list the contents of the Kolmogorov-Hinze model.

- In the neighborhood of a bubble, there are turbulent vortices of the continuous medium that cover the entire spectrum of linear scales.
- Large turbulent vortices transport the bubble in the space without deforming its surface.
- Vortices that are small relative to the bubble do not have the necessary energy to breakup it.
- Vortices of a size comparable to a bubble in their chaotic motion collide with the surface of the bubble, causing its deformation, instability, and subsequent breakup.
- Such bubbles represent the majority of those formed during the breakup process. Vortices of smaller and smaller sizes "chip" smaller and smaller "child bubbles" from the bubble of the maximum size.
- As the size of the vortex decreases, the number of child bubbles decreases monotonically. Finally, small-scale vortices no longer have any noticeable effect on the bubbles. This results in the formation of some distribution of bubbles in sizes.

16.3.6 Empirical Formulas and Experimental Data

At present, there are several semi-empirical relations for MSBD: Levich's formula [39]

$$d_{b*} \approx \frac{\sigma^{0.6}}{\rho^{0.4}\rho_v^{0.2}\varepsilon^{0.4}}, \tag{16.56}$$

Baldyga and Bourne's formula [40]

$$d_{b*} \approx \left(\frac{\sigma}{\rho\varepsilon^{2/3}l^{5/3}}\right)^{0.926}, \tag{16.57}$$

where l is the scale of vortices, Shinnar's formula [41]

$$d_{b*} \approx \frac{1+\mu_v/\mu}{1+1.19\mu_v/\mu}\frac{\sigma\nu^{1/2}}{\mu\varepsilon^{1/2}}. \tag{16.58}$$

The above empirical formulas are given up to numerical constants. In this case, formula (16.57) contains another unknown value l, which must be associated with the flow parameters.

To date, a significant number of experimental studies have been performed on measuring the diameters of bubbles in liquid flows [35, 43–47] for various methods of organization of a turbulent two-phase flow (Table 16.1). In these experiments, We_* was determined indirectly. For this purpose, the dissipation was calculated from the velocity profile and the geometric parameters of the liquid flow measured in the experiments. The value of ε thus determined was then substituted into (16.52).

Table 16.1 Experimental evaluation of the critical Weber number

Source	We_*	Conditions
Hinze [35]	1.17	Turbulent flow between coaxial rotating cylinders, immiscible liquids with $\rho_d \approx \rho_c$
Hesketh et al. [42]	0.9–1.6	Vertical tube, immiscible liquids, gas bubbles in liquid
Sevik, Park [43]	1.26	Turbulent bubble jet in quiescent liquid
Risso, Fabre [44]	2.7–7.8	Gas bubbles in quiescent liquid, isotropic turbulence
Karabelas [45]	1.5	Horizontal tube, immiscible liquids
Evans et al. [46]	1.2	Turbulent bubble flow, isotropic turbulence
Andreussi et al. [47]	1.05	Turbulent bubble flow, horizontal tube
Qian et al. [37]	3	Numerical simulation, isotropic turbulence

16.3.7 Calculation of Turbulence Energy Dissipation

The value of dissipation averaged over the channel cross-section is used when changing from (16.52) to (16.55): $\varepsilon \to \varepsilon$. At the same time, according to various models of turbulence [31–33], the distribution ε along the transverse coordinate is strongly inhomogeneous. Therefore, the MSBD determined by the standard method can be arbitrarily far from the actual value.

Figure 16.3 shows the results of the evaluation of the local dissipation performed using the high-Reynolds version of the $K - \varepsilon$ model.

The figure shows that the distribution of the dissipation has a pronounced maximum at the boundary of the "buffer layer" of the near-wall turbulence: $y^+ \equiv$

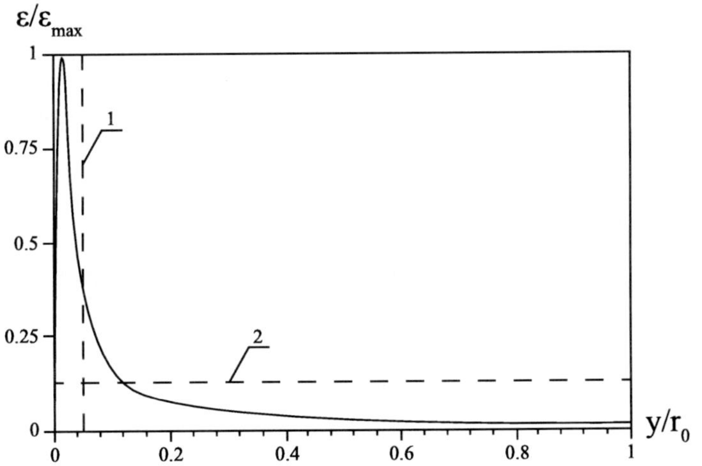

Fig. 16.3 Distribution of the local turbulence energy dissipation along the transverse coordinate of a circular pipe. 1 Buffer layer boundary, 2 dissipation averaged over the cross-section

$u_* y/\nu = 30$. According to the theory of the turbulent boundary layer [48], for smooth pipes, the inertial interval of turbulent scales is concentrated mainly in the region of the "logarithmic wall law". The lower boundary of this area is estimated as $y^+ = 30$–35. Below is the area of Kolmogorov microscale, where the dissipation of vortices occurs. The upper boundary of the inertial interval depends on the Reynolds number. From the dependence $d_{b*} \sim \varepsilon^{-2/5}$, which follows from (16.52), one should expect a noticeable increase in the mean size of the bubble as the central region of the channel is approached (where ε drops significantly). However, experimental observations [43–47] show a very weak variation in d_{b*} in the transverse coordinate. Based on this, the following assumptions can be made:

- Bubble breakup has a nonlocal character and is determined by the dissipation averaged over a certain range of the transverse coordinate
- Bubble breakup does not occur in the entire cross-section of the channel, but only in a certain zone, in which an inertial interval of pulsations is realized comparable to the size of the bubble. The boundaries of this zone should depend both on the turbulent characteristics in the flow core and on the initial bubble sizes.

16.3.8 Modified Kolmogorov–Hinze Model

As already mentioned, the Kolmogorov–Hinze model is derived from the balance of the capillary (stabilizing) and inertial (destabilizing) forces. We note that the dependence $d_{b*} \sim u^{-6/5}$, which follows from (16.55), assumes that the MSBD unboundedly increases with decreasing velocity of the turbulent flow. This anomalous trend suggests the existence of a physical mechanism of bubbles breakup as $u \to 0$. Below we give a simple approximate modified Kolmogorov–Hinze model, in which this paradox is eliminated.

We formulate the following hypothesis: the pulsation energy in the vicinity of the bubble consists of two sources: the motion of turbulent vortices in the liquid $\left(\upsilon_K^2\right)$ and the perturbations caused by rising bubble $\rho\upsilon_b^2$. We write down the total pulsation energy $\rho\upsilon_\Sigma^2$ as

$$\rho\upsilon_\Sigma^2 = k_1\rho\upsilon_K^2 + k_2\rho\upsilon_b^2, \tag{16.59}$$

where k_1, k_2 are numerical constants. We shall assume that υ_b is proportional to the velocity of the bubble rising in the liquid at rest

$$\upsilon_b \sim \frac{gd_b^2}{\nu}. \tag{16.60}$$

Relation (16.60) physically means that the gas bubble due to its gravitational pop-up turbulates the liquid in its vicinity. The condition for the stability of the bubble surface reads as $\sigma/d_b \geq \rho\upsilon_\Sigma^2$, and the critical Weber number, which determines the MSBD, assumes the form

$$\mathrm{We}_* = \frac{\rho v_\Sigma^2 d_{b*}}{\sigma}.$$

Hence, in view of (16.59), we get the modified Kolmogorov–Hinze model

$$d_{b*}\left[k_1(d_{b*}\varepsilon)^{2/3} + \beta\left(\frac{g d_{b*}^2}{\nu}\right)^2\right] = 1. \quad (16.61)$$

From (16.61), we get the equation for MSBD

$$A d_{b*}^{5/3} + B d_{b*}^5 = 1, \quad (16.62)$$

where $A = \alpha \frac{\rho\varepsilon^{2/3}}{\sigma}$, $B = \beta \frac{\rho}{\sigma}\left(\frac{g}{\nu}\right)^2$.

Putting $x = d_{b*}^{5/3}$, we get from (16.62) the following cubic equation

$$Ax + Bx^3 = 1. \quad (16.63)$$

The numerical constants α and k_2 were selected from the results of experiments [48]: $\alpha = 0,532$, $\beta = 7,48 \cdot 10^{-6}$. Making $u \to 0$, we formally get the MSBD

$$d_{b*}|_{u=0} = c_3\left(\frac{\nu^2\sigma^5}{g^2\rho^5}\right)^{1/5}, \quad (16.64)$$

where $c_3 = 10.6$.

Figure 16.4 shows the results of calculation of the MSBD from the standard and modified model in the case of water flow at atmospheric pressure in a pipe of diameter 10 mm at pressures 0.1 and 12.7 MPa. The figure shows that in the range $u \le 2$ m/s, the results predicted by the standard model demonstrate a nonphysical increase. Note that the Kolmogorov–Hinze model is aimed at turbulent flows, and hence is formally inapplicable for $\mathrm{Re} \le 2 \times 10^3$. Hence, its extrapolation to the zero velocity is conventional. At the same time, the modified model is originally free from this constraint. It would be interesting to match the asymptotic expression (16.64) with the value of the capillary constant. For the conditions from Fig. 16.4, we have

$$\frac{d_{b*}|_{u=0}}{b} \approx 0.6 - 0.67.$$

Thus, despite the pressure change by more than two orders of magnitude, the asymptotic value of the MSBD is practically unchanged and is approximately equal to the value of the capillary constant. As is known, the latter appears in the analysis of capillary-gravity waves on a free surface. From this, we can conclude that in the zone of low velocities, the scenario of bubble breakup can be controlled by the Kelvin–Helmholtz instability. To conclude, we note that in the survey paper [50] on bubble flows, in which the bibliography comprises 606 titles, only a very small

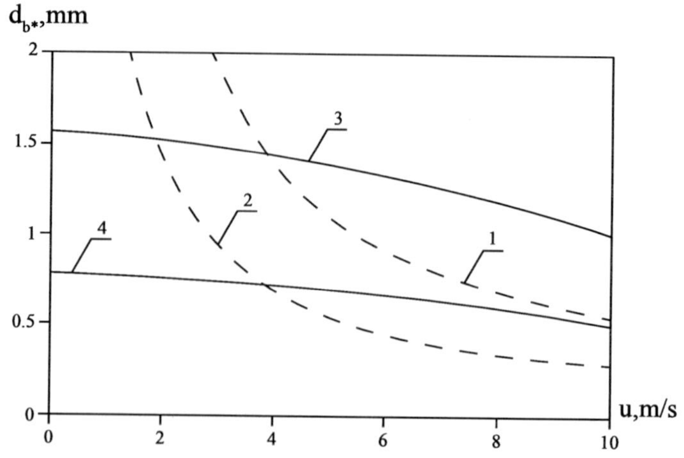

Fig. 16.4 The maximum stable bubble diameter vs. the velocity. Water flow in a circular tube of diameter 10 mm. 1 Standard model, pressure 0.1 and 12.7 MPa, 2 standard model, pressure 12.7 MPa, 3 modified model, pressure 0.1 MPa, 4 modified model, pressure 12.7 MPa

part is devoted to the dependence of the characteristics of bubbles on the liquid flow velocity. The author of [50] even concludes that the motion of fluid has almost no effect on the MSBD. This, once again indicates that the problem of bubbles breakup in the area of low flow rates remains practically unexplored to date.

16.4 Conclusions

The problem of the dynamics of gas bubbles in a liquid under various conditions is considered. Based on the analysis of the Laplace equation for the velocity potential in an ideal fluid, a generalized Rayleigh equation is proposed for the dynamics of a bubble in a circular pipe. The spherical and cylindrical asymptotics are analyzed. An exact analytical solution of the bubble collapse problem in a tube is obtained. A critical analysis of the problem of homogeneous nucleation of vapor bubbles in an unlimited volume of liquid is given. A quantum mechanical model of homogeneous nucleation is proposed. The problem of bubble size in a turbulent fluid flow is considered. A description of the Kolmogorov–Hinze model of bubbles breakup, which is based on the balance of capillary and inertial forces, is given. It is shown that the model assumes an unlimited increase of the maximum stable bubble diameter with a decrease in the velocity of the turbulent flow. The modified Kolmogorov–Hinze model eliminating the above paradox is presented. The main idea of the modified model is the consideration of the balance of the forces acting on the bubble of liquid pulsations caused by the gravitational bubble rise. An expression for the maximum stable bubble diameter in the zero velocity asymptotics of the fluid flow is obtained.

References

1. Rayleigh L (1917) On the pressure developed in a liquid during the collapse of a spherical cavity. Philos Mag 34:94–98
2. Plesset MS, Prosperetti A (1977) Bubble dynamics and cavitation. Ann Rev Fluid Mech 9:145–185
3. d'Agostino L, Salvetti MV (2008) Fluid dynamics of cavitation and cavitating turbopumps. Springer, Vien, New York
4. Zudin YB (1992) Analog of the Rayleigh equation for the bubble dynamics in a tube. Inzh-Fiz Zh 63(1):28–31
5. Zudin YB (1995) Calculation of the rise velocity of large gas bubbles. Inzh-Fiz Zh 68(1):13–17
6. Zudin YB (1998) Calculation of the drift velocity in bubbly flow in a vertical tube. Inzh-Fiz Zh 71(6):996–999
7. Freeden W, Gutting M (2013) Special functions of mathematical (geo-)physics. Appl Numer Harmonic Anal. Springer, Basel
8. Klaseboer E, Khoo BC (2006) A modified Rayleigh-Plesset model for a nonspherically symmetric oscillating bubble with applications to boundary integral methods. Eng Anal Bound Elem 30(1):59–71
9. Zudin YB, Isakov NS, Zenin VV (2014) Generalized Rayleigh equation for the bubble dynamics in a tube. J Eng Phys Thermophys 87(6):1487–1493
10. Scripov VP (1974) Metastable liquids. John Wiley & Sons, New York
11. Debenedetti PG (1996) Metastable liquids: concepts and principles. Princeton University Press, Princeton, New York
12. Perrot P (1998) A to Z of thermodynamics. Oxford University Press
13. Kashchiev D (2000) Nucleation: basic theory with applications. Butterworth-Heinemann, Oxford
14. Horst JH, Kashchiev D (2008) Rate of two-dimensional nucleation: verifying classical and atomistic theories by Monte Carlo simulation. J Phys Chem B 112(29):8614–8618
15. Sekine M, Yasuoka K, Kinjo T, Matsumoto M (2008) Liquid–vapor nucleation simulation of Lennard-Jones fluid by molecular dynamics method. Fluid Dyn Res 40:597–605
16. Chao L, Xiaobo W, Hualing Z (2010) Molecular dynamics simulation of bubble nucleation in superheated liquid. In: Proceedings of the 14th international heat transfer conference IHTC14, August 7–13, Washington. IHTC14-22129
17. Griffiths DJ (2005) Introduction to quantum mechanics, 2nd edn. Prentice Hall International
18. Guénault AM (2003) Basic superfluids. Taylor & Francis, London
19. Cumberbatch E, Uno S, Abebe H (2006) Nano-scale MOSFET device modelling with quantum mechanical effects. Eur J Appl Math 465–489 http://journals.cambridge.org/action/displayJournal?jid=EJM17
20. Keith AC, Lazzati D (2011) Thermal fluctuations and nanoscale effects in the nucleation of carbonaceous dust grains. Mon Not R Astron Soc 410(1):685–693
21. Zudin YB (1998) Calculation of the surface density of nucleation sites in nucleate boiling of a liquid. J Eng Phys Thermophys 71:178–183
22. Zudin YB (1998) The Distance between nucleate boiling sites. High Temp 36:662–663
23. Lesieur M (1997) Turbulence in fluids. Publ, Kluwer Acad
24. Truesdell C (2018) The kinematics of vorticity. Courier Dover Publications
25. Bharucha-Reid AT (1960) Elements of the theory of Markov processes and their applications. McGraw-Hill, New York
26. Prigogine I, Stengers I (1984) Order out of Chaos. Bantam Books, University of Michigan
27. Prigogine I (1961) Introduction to thermodynamics of irreversible processes, 2nd edn. Interscience, New York
28. Richardson LF (1922) Weather prediction by numerical process. Cambridge University Press
29. Taylor G I (1935) Statistical theory of turbulence. In: Proceedings of the royal society of London. Series A, mathematical and physical sciences, vol 151, no 873, pp 421–444

30. Kolmogorov AN (1941) The local structure of turbulence in incompressible viscous fluid for very large Reynolds numbers. In: Proceedings of the USSR academy of sciences (in Russian), vol 30, pp 299–303
31. Moulden TH (1977) Handbook of turbulence. fundamental and applications. In: Frost W, Moulden TH (eds) Plenum Press, New York
32. Spalart PR, Allmaras SR (1992) A one–equation turbulence model for aerodynamic flows. AIAA Paper 92–0439, Jan 1992
33. Menter FR (1993) Zonal two-equation k-x turbulence models for aerodynamic flows. AIAA Paper 93–2306, Jun 1993
34. Kolmogorov AN (1949) On the disintegration of drops in turbulent flow. Dokl Akad Nauk 66:825–828
35. Hinze JO (1955) Fundamentals of the hydrodynamic mechanism of splitting in dispersion processes. AIChE J 1(3):289–295
36. Batchelor GK (1951) Pressure fluctuations in isotropic turbulence. Proc Cambridge Phil Soc 47:359–374
37. Qian D, McLaughlin JB, Sankaranarayanan K, Sundaresan S, Kontomaris K (2006) Simulation of bubble breakup dynamics in homogeneous turbulence. Chem Eng Commun 193:1038–1063
38. Sankaranarayanan K, Shan X, Kevrekidis IG, Sundaresan S (1999) Bubble flow simulations with the lattice Boltzmann method. Chem Eng Sci 54:4817–4823
39. Levich VG (1962) Physicochemical hydrodynamics. Prentice-Hall
40. Baldyga J, Bourne JR (1995) Interpretation of turbulent mixing using fractals and multifractals. Chem Eng Sci 50:381–400
41. Shinnar R (1961) On the behaviour of liquid dispersions in mixing vessels. J Fluid Mech 10:259–275
42. Hesketh RP, Russell TWF, Etchells AW (1987) Bubble size in horizontal pipelines. AIChE J 33(4):663–667
43. Sevik M, Park SH (1973) The splitting of drops and bubbles by turbulent fluid flow. J Fluids Eng 95:53–60
44. Risso F, Fabre J (1998) Oscillations and breakup of a bubble immersed in a turbulent field. J Fluid Mech 372:323–355
45. Karabelas AJ (1978) Droplet size spectra generated in turbulent pipe flow of dilute liquid/liquid dispersions. A.I.C.E.J 24(2):170–180
46. Evans GM, Jameson GJ, Atkinson BW (1992) Prediction of the bubble size generated by a plunging liquid jet. Chem Eng Sci 47(13–14):3265–3272
47. Andreussi P, Paglianti A, Silva FS (1999) Dispersed bubble flow in horizontal pipes. Chem Eng Sci 54(8):1101–1107
48. Schlichting H, Gersten K (1997) Grenzschicht-Theorie. Springer, Berlin Heidelberg, New York
49. Hibiki T, Ishii M, Xiao Z (2001) Axial interfacial area transport of vertical bubbly flows. Int J Heat Mass Trans 44:1869–1888
50. Abdulmouti H (2014) Bubbly two-phase flow: Part I—characteristics, structures, behaviors and flow patterns. Am J Fluid Dyn 4(4):194–240

Chapter 17
Heat Transfer to a Disperse Two-Phase Flow

Abbreviations

MSDD	Maximum stable droplet diameter

Symbols

a	Droplet radius
d	Particle diameter
E	Energy
k	Thermal conductivity
T	Temperature
Pr	Prandtl number
Re	Reynolds number
r_0	Tube radius
St	Stanton number
t	Time
T_*	Temperature scale
u_*	Friction velocity
U	Velocity
x	Vapor quality
y	Transverse coordinate

Greek Letter Symbols

α	Perturbations amplitude
ε	Dissipation
ξ	Darcy friction factor
μ	Dynamic viscosity
ν	Kinematic viscosity
η	Kolmogorov length microscale
ψ	Relative law of heat transfer
ϕ	Angular coordinate
σ	Surface tension

Y. B. Zudin, *Non-equilibrium Evaporation and Condensation Processes*,
Mathematical Engineering, https://doi.org/10.1007/978-3-030-67553-0_17

τ	Kolmogorov time microscale
θ	Dimensionless temperature
ϑ	Kolmogorov velocity microscale
δ	Thickness of viscous sublayer
ω	Frequency
ρ	Density

Subscripts

d	Dispersed medium
eff	Effective
f	Fluid

Superscripts

+	Dimensionless

17.1 Droplet Size in Dispersed Two-Phase Flow

17.1.1 Free Oscillations of Droplet

According to [1], the distribution of spherical liquid droplets dispersed in a vapor flow significantly affects the characteristics of heat transport in a dispersed two-phase flow. In particular, the energy equation for the vapor phase includes mass sources and heat sinks due to the evaporation of the droplet in the superheated vapor. These quantities are determined by the parameters of the dispersed phase: the droplet distribution function, as well as the Maximum Stable Droplet Diameter (MSDD), which generally depends on the radial coordinate. In the analysis of the process of droplet breakup in a turbulent flow of a continuous medium, various semiempirical and empirical representations are used due to the lack of a rigorous theory.

Consider the general case of the breakup of a particle (a droplet or a bubble) in an unbounded medium (gas or liquid). To this end, we make use of the classical problem of oscillations of a droplet of liquid about a spherical form, which was first experimentally studied by Plateau in 1873 [2]. In 1879, Rayleigh [3] carried out a linear analysis of the differential equation of balance of the kinetic and potential energy of an oscillating dispersed particle in an unbounded medium. For the modern account of Rayleigh's analysis, see, for example, [4]. An analytical solution was found for small amplitudes of deviation of the surface of the particle from the spherical one. Rayleigh determined the normal modes of oscillations of a liquid drop about the spherical form and their natural frequencies. Oscillations of a spherical surface result in the periodical appearance of reciprocal transfers of the kinetic energy of the continuous medium to the potential energy of the particle and vice versa.

The classical paper [3] is the basis for the calculation of the oscillations of dispersed particles up to the present day. Below we briefly summarize the main points of the Rayleigh analysis in relation to the problem of droplet oscillations in an unbounded gas medium.

Both media (dispersed and continuous) were considered inviscid and incompressible, the surface perturbations were considered symmetric about the vertical axis. It was shown that only the shape of the droplet surface is subject to perturbations, while its volume is preserved. This means that the "expansion–compression" pulsations of the droplet volume are excluded from consideration.

We shall use the standard method for solving problems of inviscid fluid hydrodynamics. Consider the velocity potential, which after being plugged into the continuity equation gives the Laplace equation. We represent the droplet parameters (the radius a, the surface area S, and the volume V) as sums of the undisturbed values (with the subscript "0") and the small perturbations. The solution for perturbations of the droplet radius is sought as an infinite series in the Legendre polynomials P_n

$$a(\phi, t) = a_0\left(1 + \sum_n \alpha_n(t)\mathrm{P}_n(\cos\phi)\right).$$

Here ϕ is the angular coordinate measured from the vertex of the unperturbed sphere, t is the time. The expressions for the surface and volume of a droplet can be written as

$$S = 4\pi a_0^2 + 2\pi \sum_n \frac{(n-1)(n+2)}{2n+1}\alpha_n^2(t),$$

$$V = \frac{4}{3}\pi a_0^3\left(1 + \frac{3}{\alpha_0^2}\sum_n \frac{1}{2n+1}\alpha_n^2(t)\right).$$

The surface energy of the droplet reads as

$$E_s = 2\pi\sigma \sum_n \frac{(n-1)(n+2)}{2n+1}\alpha_n^2(t).$$

From the point of view of mechanics, the surface energy is the potential energy that periodically accumulates and is lost during the full period of oscillations of the droplet. The kinetic energy of the droplet is given by

$$E_k = 2\pi\rho_f a_0^3 \sum_n \frac{1}{n(2n+1)}\left(\frac{d\alpha_n(t)}{dt}\right)^2.$$

Since we consider the oscillations of a droplet in a gas whose density is negligible compared to that of the liquid $(\rho_f \ll \rho_d)$, the dynamic interaction of a droplet with the gas can be ignored. Then the sum of the potential and kinetic energy should remain unchanged as the ideal fluid moves inside the droplet

$$\frac{d}{dt}(E_k + E_s) = 0.$$

As a result, we obtain the ratio for the eigenfrequency of oscillations of the droplet form

$$\omega_{R,n}^2 = n(n-1)(n+2)\frac{\sigma}{\rho_f a_0^3}. \tag{17.1}$$

It is interesting to note that mathematically the problem of deformation oscillations of a particle is identical to the well-known problem of harmonic oscillations of an elastic string. The generalized Rayleigh solution for oscillations of a spherical particle in fluid is more involved and is not considered here.

17.1.2 *Analysis of the Solution*

Consider is succession the expansion coefficients ("modes") of solution (17.1) with different numbers *n*. The mode n = 0 is responsible for oscillations of the droplet volume of the "expansion–compression" type and falls out of consideration due to the assumption of incompressibility of the liquid. Hence, $\omega_{R,0}^2 = 0$. For $n = 1$ we get the nonphysical cases: the potential energy is absent for the final kinetic energy value. Hence $\omega_{R,1}^2 = 0$. The mode $n = 2$ is the first for which a nonzero eigenfrequency of oscillations occurs

$$\omega_{R,2}^2 = 8\frac{\sigma}{\rho_f a_0^3}.$$

With the frequency $\omega_{R,2}$, the droplet surface oscillates to deform from a flattened to an elongated spheroid. The mode $n = 3$ corresponds to a higher oscillation frequency and leads to a more complex deformation of the droplet surface

$$\omega_{R,3}^2 = 30\frac{\sigma}{\rho_f a_0^3}.$$

Formula (17.1) shows that the eigenfrequency increases with increasing mode number. As the fundamental frequency, we take $\omega_{R,2}$

$$\omega_0 = \omega_{R,2} = \sqrt{8\frac{\sigma}{\rho_f a_0^3}}. \tag{17.2}$$

Table 17.1 shows the values of the circular frequency $f = \frac{\omega_0}{2\pi}$ for water and various hydrocarbons. Calculations were performed for various values of the radius

Table 17.1 Eigenfrequency of a droplet in various liquids

← Droplet radius (mm) →						
Liquid	0.5	0.6	0.7	0.8	0.9	1
Water	348	265	210	172	144	123
Methanol	221	168	134	109	91	78
Ethanol	220	167	133	108	91	77
Propanol	225	171	135	111	93	79
Heptane	224	170	135	110	92	79
Octane	239	182	144	118	99	74
Decane	230	175	139	113	95	81

of an undisturbed droplet in dry air at atmospheric pressure and temperature 20° 17. It can be seen that for hydrocarbons, the eigenfrequencies are quite close to each other, while the eigenfrequency of a water droplet is significantly higher due to higher values of the surface tension and density.

Periodic oscillations are expanded in a Fourier series in harmonics labeled by natural numbers: $n_F = 1, 2, 3, 4, \ldots$ Now let us find out how the eigenfrequencies relate to each other as the mode number is increasing. For this purpose, we consider the ratio of the eigenfrequencies of an arbitrary mode with number $n \geq 2$ and the main mode

$$n_R = \frac{\omega_{R,i}}{\omega_0}, \tag{17.3}$$

$$i = n - 1, n = 2, 3, 4, 5, \ldots$$

Table 17.2 shows that when transiting from the main mode $n = 2$ to the next mode $n = 3$ the eigenfrequency increases by 1.936 times. At the same time, expansion of the oscillations into a Fourier series gives a simple doubling of the frequency. With a further increase in the number of the oscillation mode, the frequency first "catches" the Fourier frequency ($n_R = n_F = 3$), and then "overtakes" it more and more.

Table 17.2 Frequencies of the Fourier series n_F and the frequencies of Rayleigh oscillations of a droplet n_R

n_F	1	2	3	4	5	6	7	8	9	10
n_R	1	1.94	3	4.18	5.48	6.87	8.37	9.95	11.6	13.4

17.1.3 Theory of Locally Isotropic Turbulence

The study of turbulence began more than 500 years ago by Leonardo da Vinci. He first noted turbulence as a special behavior of a fluid and called such movements "turbulence" [5]. A rigorous mathematical treatment of turbulence began only in the first half of the twentieth century. An important milestone was the work of Taylor [6], performed in 1935. When studying turbulence, Taylor [6] was the first to use statistical methods: correlations, Fourier transforms, power spectra. Taylor introduced the concept of "isotropic and homogeneous turbulence" into the scientific literature.

Isotropic turbulence means that there is no mean shear, rotation, or buoyancy effects in the flow as this can lead to anisotropy. Homogeneous turbulence means that there are no mean flow gradients. In another way, isotropy deals with invariance in rotation and homogeneity deals with invariance in translation. Mathematically, the uniformity means the invariance with respect to parallel transfers—the motion of all points in the space in the same direction by the same distance. In turn, the isotropy means the invariance with respect to all rotations around the origin in three-dimensional Euclidean space. The homogeneity and isotropy properties assume that the flow has no boundaries and that the velocity is constant. Hence Taylor's representations are, strictly speaking, unsuitable for describing real turbulent flows that are essentially anisotropic. An important step was the Kolmogorov's theory of locally isotropic turbulence [7] (see also [8]), which is based on two important hypotheses.

Kolmogorov's first similarity hypothesis. Kolmogorov suggested that the directional biases of the large scales are lost in the chaotic scale-reduction process as energy is transferred to successively smaller eddies. Kolmogorov's hypothesis of local isotropy states that at sufficiently high Reynolds numbers, the small-scale turbulent motions are statistically isotropic. Here, the term local isotropy means isotropy at small scales, moreover, large-scale turbulence is essentially anisotropic. Kolmogorov microscales follow from Kolmogorov's first similarity hypothesis. On the basis of two parameters—kinematic viscosity ν and dissipation ε—the following universal microscales are formed:

length

$$\eta = \left(\frac{\nu^3}{\varepsilon}\right)^{1/4}, \tag{17.4}$$

velocity

$$\upsilon = (\nu\varepsilon)^{1/4}, \tag{17.5}$$

and frequency

$$\omega_\eta = \left(\frac{\varepsilon}{\nu}\right)^{1/2}. \tag{17.6}$$

These scales are indicative of the smallest eddies present in the flow, the scale at which the energy is dissipated. Thus, the Reynolds number constructed on Kolmogorov microscales is identically equal to one: $\mathrm{Re} = \eta v/\nu \equiv 1$.

Kolmogorov's second similarity hypothesis. Because the Reynolds number of the intermediate scales l_i is relatively large, they will not be affected by the viscosity ν. Based on that, Kolmogorov's second similarity hypothesis states that in every turbulent flow at sufficiently high Reynolds number, the statistics of the motions of scale l_i in the range $\eta \ll l_i \ll L$ have a universal form that is uniquely determined by e independent of ν.

Thus, the Kolmogorov theory introduces a hierarchy of scales of locally isotropic turbulence. This length scale splits the universal equilibrium range into two subranges. The inertial interval ($\eta < l < L$) where motions are determined by inertial effects and viscous effects are negligible. The dissipation range ($l < \eta$) where motions experience viscous effects. According to the huge bank of experimental data accumulated to date, Kolmogorov hypotheses correctly reflect the real characteristics of the local structure of developed turbulence.

17.1.4 Resonance Model of Droplets Breakup

We proceed from the fact that the mean diameter d of a particle (a droplet or a bubble) corresponds to the inertial interval: $\eta < d < L$. The Rayleigh general solution for the fundamental frequency of a particle in an unbounded medium reads as [3, 4]

$$\omega_0^2 = \frac{48}{\pi^2} \frac{1}{2\rho_f + 3\rho_d} \frac{\sigma}{d^3}. \tag{17.7}$$

Here ρ_f, ρ_d is the density of the solid and dispersed phase, respectively. Formula (17.7) implies the dependence of the eigenfrequency of a particle on its diameter.: $\omega_0 \sim d^{-3/2}$. Now let us estimate the frequency of turbulent pulsations ω_t. We shall use the Batchelor formula [9]

$$u_i^2 = 2(\varepsilon d)^{2/3}, \tag{17.8}$$

where ε is the turbulent energy dissipation. Hence, writing the frequency in the form $\omega_t = u_i/d$, we get

$$\omega_t^2 = 2\frac{\varepsilon^{2/3}}{d^{4/3}}. \tag{17.9}$$

From (17.9), we get an estimate for the frequency of pulsations of a turbulent vortex having dimensions of the order of the particle diameter

Fig. 17.1 The resonant breaking-drop model

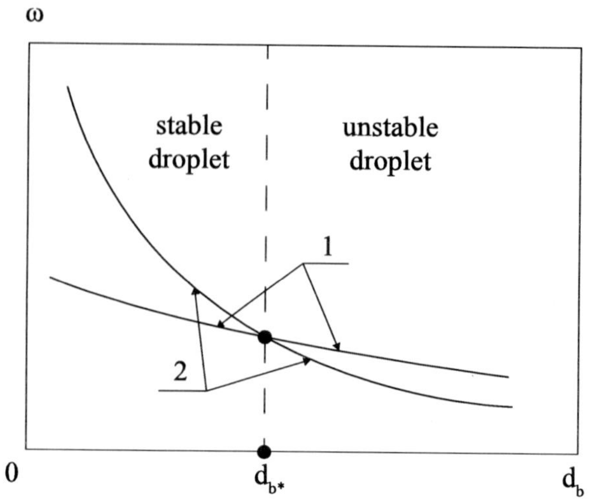

$$\omega_t \sim d^{-2/3}.$$

It is seen that the dependence $\omega_0(d)$ is steeper than the dependence $\omega_t(d)$, and so, for a certain value of the diameter, both curves must intersect (Fig. 17.1). The coincidence of the frequencies should lead to the phenomenon of resonance: a sharp increase in the energy of pulsations of a fluid particle. For physical reasons, it can be assumed that the dispersed particle "does not notice" low-frequency pulsations of the continuous medium, hence the region $\omega_0 > \omega_t$ will correspond to the stability of the droplet (the bubble). For $\omega_0 < \omega_t$, high-frequency external pulsations will "rock" the particle, creating "ripples" on its surface, therefore, this region will correspond to the instability of the droplet (the bubble). As a condition of particle breaking, we take the equality of the frequencies at which the resonance of vibrations should occur

$$\omega_0 = \omega_t. \tag{17.10}$$

The above physical picture will be called as the "resonant model" of droplet break. Let us evaluate the lower boundary of the model applicability. Assume that the particle diameter does not fit into the inertial scale interval and falls into the viscous region: $d \sim \eta$. Then the frequency of turbulent pulsations will no longer depend on the linear size, but will be determined by the ratio: $\omega_t \sim 1/\tau = (\varepsilon/\nu)^{1/2}$. If we now mentally "remove" the particle from the instability region by reducing its diameter, then the curve $\omega_0(d)$ at a certain moment will cross the line ω_t = const and fall into the stable region. It follows that in the case when a particle "falls" in the viscous scale interval, the former condition of "resonant breakup" $\omega_0 = \omega_t$, will be fulfilled, but now not in the inertial, but in the viscous range of turbulence scales. Using the expressions for the corresponding frequencies (17.7) and (C9) in (17.10), we obtain the following general expression for the MSDD

$$d_{p*} = 1.12\left(\frac{\sigma}{\rho_{\text{eff}}}\right)^{0.6}\frac{1}{\varepsilon^{0.4}}. \tag{17.11}$$

Here, $\rho_{\text{eff}} = 2\rho_f + 3\rho_d$ is the effective density value.

Consider the limit variants of the resonance breakup model.

- A bubble in the liquid medium. Here, $\rho_f \gg \rho_d$, $\rho_{\text{eff}} \approx 2\rho_f$. In this case, we have

$$d_{b*} = 1.56\left(\frac{\sigma}{\rho_f}\right)^{3/5}\frac{1}{\varepsilon^{2/5}}. \tag{17.12}$$

- A droplet in a gas medium. Here, $\rho_f \ll \rho_d$, $\rho_{\text{eff}} \approx 3\rho_d$. Using this value in (17.11) and the numerical constant obtained above, we get

$$d_{p*} = 1.22\left(\frac{\sigma}{\rho_d}\right)^{3/5}\frac{1}{\varepsilon^{2/5}}. \tag{17.13}$$

So, the resonance breakup model is capable of evaluating the maximal size of a dispersed particle in the general case (a bubble in liquid, a droplet in gas, a droplet in a different liquid).

17.2 Effect of Droplets on Heat Transfer to a Disperse Two-Phase Flow

17.2.1 Analytical Solutions

As known, the majority of problems of hydrodynamics and heat transfer are described by partial differential equations. So, Navier–Stokes and energy equations represent quasi-linear partial differential equations whose solution in most cases can only be obtained with the help of numerical methods. This can lead to a "natural" conclusion about an absolute priority of numerical solutions in the specified area of research. However, analytical solutions of the fluid flow and heat transfer problems play a significant role even in the current computer age.

They possess the following decisive advantages in comparison with numerical methods.

- The value of the analytical approach consists of an opportunity of the closed qualitative description of the process, revealing the full list of dimensionless characteristic parameters and their hierarchical classification basing on the criteria of their importance.
- Analytical solutions possess a necessary generality, so that a variation of boundary and inlet conditions allows carrying out parametrical investigations.

- In order to validate numerical solutions of the full differential equations, it is necessary to have basic (often rather simple) analytical solutions of the equations for some obviously simplified cases (after an estimation and omission of negligible terms).
- In a global aspect, an analytical solution can be used for direct validation of the correctness in the statement of numerical investigations applicable to a particular problem.

17.2.2 Energy Equation for Dispersed Two-Phase Flow

We accept the following simplifying assumptions of the mathematical description of the problem.

1. Droplets have no effect on the velocity field in the vapor phase. This means that the dispersed phase behaves like a passive admixture in relation to the continuous phase.
2. There is no droplet slip relative to the vapor flow. The velocity of the center of a spherical drop placed at a given point of the vapor phase is equal to that of the vapor before the droplets are introduced into it.
3. The effect of the dispersed phase on the heat transfer is reduced to volumetric heat sinks caused by the evaporation of the droplet in superheated vapor.
4. There are no droplets inside the viscous sublayer of the turbulent vapor flow.
5. The Prandtl number for vapor is equal to 1.

When analyzing the flow in the vapor phase, we will proceed from the law wall (also known as the logarithmic wall law) states that the speed of a turbulent flow at a given point is proportional to the logarithm of the distance from that point to the wall [10]. Using the wall law, it is possible to establish a relationship between the shear stress and the energy flux transferred to the boundary surface at different speeds and temperatures. This connection leads to extended Reynolds analogy, as expressed by Petukhov and Kirillov's relation [11]

$$\mathrm{St}_0 = \frac{\xi/8}{1 + 11.7\sqrt{\xi/8}\left(\mathrm{Pr}^{2/3} - 1\right)}. \tag{17.14}$$

Here,

$$\mathrm{St}_0 = \frac{h}{\rho_f c_f U} \tag{17.15}$$

is the Stanton number for the "reference case" of a single-phase vapor flow, Pr is Prandtl number, ξ is the Darcy friction factor, h is the heat transfer coefficient, U is the vapor flow velocity.

In view of assumption 5, expression (17.14) transforms to the classical Reynolds analogy [10]

$$\mathrm{St}_0 = \frac{\xi}{8}. \tag{17.16}$$

The expression for dissipation for the logarithmic wall law has the form [12]

$$\varepsilon = \frac{u_*^3}{\kappa y}. \tag{17.17}$$

Here y is the transverse coordinate, u_* is the friction velocity, and κ is the von Karman constant. Moreover, in accordance with assumption 4, we have $\delta \le y \le r_0$, where r_0 is the tube radius.

We assume that the mean diameter of the droplet corresponds to the viscous subrange: $l < \eta$. Using (17.17) in (17.6), we get the following expression for the MSDD

$$d_{p*} = \beta_1 \frac{1}{u_*}\left(\frac{\kappa \nu \sigma y}{\rho_p}\right)^{1/3}. \tag{17.18}$$

Thus, according to the resonance breakup model, the MSDD depends on the transverse coordinate by the law: $d_{p*} \sim y^{1/3}$. We introduce the dimensionless variables

$$d_p^+ = \frac{u_* d_p}{\nu},\ y^+ = \frac{u_* y}{\nu},$$

and rewrite (17.18) in the form

$$d_p^+ = \beta_1 \left(\frac{\kappa \sigma y^+}{\rho_p \nu u_*}\right)^{1/3}. \tag{17.19}$$

In the framework of the resonance breakup model, it is assumed that the droplets of size $d_p < d_{p*}$ merge upon impact and reach the maximum stable diameter as a result of some cascade process. In turn, the droplets of size $d_p > d_{p*}$ become unstable and are subject to cascade breakup process up to the size d_{p*}. In accordance with assumption 3, the expression for the heat sinks distributed over the volume of the vapor phase (due to the evaporation of the droplet in the superheated vapor) is written as

$$q_\mathrm{v} = 12\frac{1-\varphi}{\varphi}\frac{k(T - T_s)}{d_p^2}. \tag{17.20}$$

Here k is the vapor thermal conductivity, T is the temperature of vapor with the current value of the transverse coordinate. Assuming that the droplet slip relative to the vapor flow is absent, we write the relation between the void fraction φ and the vapor quality x as

$$\frac{1-\varphi}{\varphi}=\frac{1-x}{x}\frac{\rho}{\rho_p}. \tag{17.21}$$

We approximate dependence (17.19) by

$$d_p^+=\beta_2\sqrt{\frac{\kappa\sigma y^+}{\rho_p \nu u_*}}. \tag{17.22}$$

In fact, relation (17.22) is the additional (sixth) simplifying assumption. In view of (17.19)–(17.22), the energy equation for a turbulent flow can be written as

$$y^+\frac{d}{dy^+}\left(y^+\frac{d\theta}{dy^+}\right)=m^2\theta. \tag{17.23}$$

Here $\theta=\frac{T-T_s}{T_*}$ is the dimensionless temperature difference, T_* is the temperature scale defined by

$$T_*=\frac{\mathrm{St}}{\sqrt{\xi/8}}(T_w-T_0),$$

where T_w, T_0 are the vapor temperature on the wall (with $y=0$) and the flow core (with $y=R_0$), respectively. Equation (17.23) shows that the degree of influence of the droplets on the heat transfer toward the dispersed two-phase flow is determined by the "heat sinks parameter"

$$m=\beta_3\sqrt{\frac{1-x}{x}\frac{\mu u_*}{\sigma}}. \tag{17.24}$$

17.2.3 Analytical Solution

Equation (17.17) has the exact solution

$$\theta=\frac{\left(r_0^+\right)^{2m}-\left(y^+\right)^{2m}}{\kappa m (y^+)^m\left[\left(r_0^+\right)^{2m}-(y^+)^{2m}\right]}, \tag{17.25}$$

satisfying the boundary condition $\theta = 0$ with $y^+ = r_0^+$. Here, $\kappa = 0.41$ is the Karman constant and $r_0^+ = \frac{u_* r_0}{\nu}$ is the dimensionless tube radius.

In the limit $m \to 0$ the influence of drops degenerates, and the solution changes to the classical universal logarithmic profile [10]

$$\theta = \frac{1}{\kappa} \ln \frac{r_0}{y}. \tag{17.26}$$

Using the well-known procedure for crosslinking the temperature profile (17.25) with the linear temperature profile in the thermal sublayer we get the following relation

$$\frac{1}{8} \frac{\xi}{\mathrm{St}} = \delta^+ + \theta_\delta - \langle \theta \rangle. \tag{17.27}$$

Here St is the Stanton number for a dispersed two-phase flow, $\delta^+ = 11.5$ is the dimensionless thickness of the viscous sublayer, $\langle \theta \rangle$ is the dimensionless temperature difference averaged over the tube cross-section, $\theta_\delta = \frac{T_\delta - T_s}{T_*}$ is the dimensionless temperature difference for the turbulent flow core, as defined from (17.21) with $y^+ = \delta^+$.

Integrating (17.25) over the cross-sectional area of tube, we get the averaged temperature difference

$$\langle \theta \rangle = \frac{2A}{\kappa m \left(r_0^+\right)^2 \left[\left(r_0^+\right)^{2m} + 1\right]}. \tag{17.28}$$

Here A is the function given by

$$A = \frac{\left(r_0^+\right)^{2m+1}}{1-m} \left[\left(r_0^+\right)^{1-m} - \left(\delta^+\right)^{1-m}\right] - \frac{r_0^+}{1+m} \left[\left(r_0^+\right)^{1+m} - \left(\delta^+\right)^{1+m}\right] - \frac{\left(r_0^+\right)^{2m}}{2-m} \left[\left(r_0^+\right)^{2-m} - \left(\delta^+\right)^{2-m}\right] + \frac{\left(r_0^+\right)^{2m}}{2+m} \left[\left(r_0^+\right)^{2+m} - \left(\delta^+\right)^{2+m}\right].$$

17.2.4 Relative Law of Heat Transfer

We introduce the relative law of heat transfer

$$\psi \equiv \frac{\mathrm{St}}{\mathrm{St}_0}. \tag{17.29}$$

In the limit $m \to 0$, from (17.24) we get the classical logarithmic friction law [10]

$$\frac{1}{\sqrt{\xi}} = 2\ln\left(\mathrm{Re}\sqrt{\xi}\right) - 0.8. \tag{17.30}$$

In view of the above condition Pr = 1, formula (17.30) leads to the Reynolds analogy (17.16). Hence the influence of the drops on the heat transfer degenerates, and we will have the asymptotics

$$\mathrm{St} \to \mathrm{St}_0 = \frac{\xi}{8}. \tag{17.31}$$

Moreover, the relative law of heat transfer is 1

$$\psi = 1.$$

In the other limit of infinite intensity of heat sinks ($m \to \infty$), we have

$$\mathrm{St} \to \frac{\sqrt{\xi/8}}{\delta^+}. \tag{17.32}$$

This corresponds to the asymptotics of maximum improvement of the heat transfer due to droplet evaporation

$$\psi \to \psi_{\max} = \left(\delta^+\sqrt{\frac{\xi}{8}}\right)^{-1}. \tag{17.33}$$

Figure 17.2 shows the results of evaluation of the relative law of heat transfer depending on the parameter m. It is seen that the maximum intensification of the

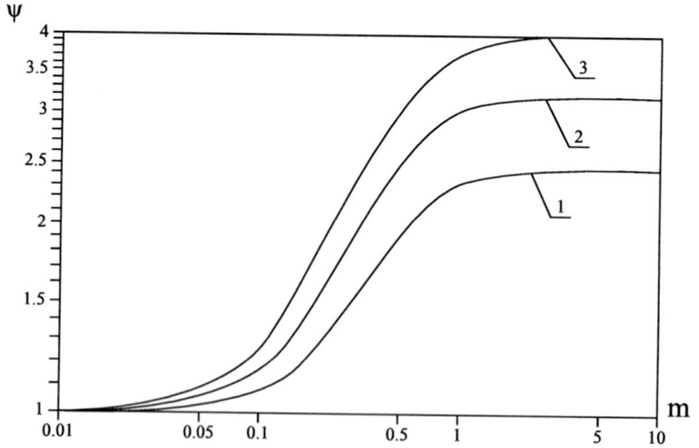

Fig. 17.2 Relative law of heat transfer versus the parameter m

heat transfer occurs already with $m \approx 1$. Thus, the effect of droplets dispersed in the vapor flow on the heat transfer is responsible for a stepwise increase in the heat transfer intensity in comparison with the reference case of a single-phase vapor flow (with the same values of the total temperature difference $T_w - T_0$) and the Reynolds number $\mathrm{Re} = \frac{U d_0}{\nu}$.

This effect has a clear physical explanation. The introduction of thermal effluents into the vapor flow due to the evaporation of droplet leads to a decrease in the thermal resistance of the turbulent flow core. When a certain level of intensity of heat sinks is reached, this thermal resistance becomes negligible. In this case, the heat transfer is carried out by the thermal conduction mechanism through the thermal sublayer

$$q_{\max} = \frac{k(T_w - T_0)}{\delta}. \tag{17.34}$$

The analytical solution (17.25) and the relative law of heat transfer (17.29), which is calculated from (17.25), allowed us to obtain a clear physical picture of the effect of liquid droplets dispersed in the current on the heat transfer. This solution can be used as a closing relation in modern numerical codes for the study of disperse two-phase flows [12].

The above analytical solution was first published by Zudin in [13].

17.3 Conclusions

The problem of the effect of droplets dispersed in a turbulent vapor flow in a channel on the heat transfer characteristics is considered. The process of droplet breakup is investigated on the basis of a semiempirical resonance model. The ratio for the maximum stable droplet diameter is obtained. The classical Prandtl turbulence model was used to study the effect of droplets on heat transfer. An analytical solution of the energy equation with thermal effluents caused by the droplet evaporation is obtained. The relative law of heat transfer is calculated, allowing us to determine the quantitative measure of the effect of the droplets on the heat transfer with a dispersed flow. The asymptotic variants of the analytical solution thus obtained are analyzed.

References

1. Morel C (2015) Mathematical modeling of disperse two-phase flow. Springer
2. Eggers J (1997) Nonlinear dynamics and breakup of free-surface flows. Rev Mod Phys 69(3):865–930
3. Rayleigh L (1879) On the capillary phenomena of jets. Proc R Soc Lond 29:71–97
4. Prosperetti A (1980) Free oscillations of drops and bubbles; the initial value problem. J Fluid Mech 100(2), 333–347
5. Monaghan JJ, Kajtar JB (2014) Leonardo da Vinci's turbulent tank in two dimensions. Eur J Mech B Fluids 44:1–9

6. Taylor GI (1935) Statistical theory of turbulence. Proc R Soc Lond Ser A Math Phys Sci 151(873):421–444
7. Kolmogorov AN (1941) The local structure of turbulence in incompressible viscous fluid for very large Reynolds numbers. Proc USSR Acad Sci (in Russian) 30:299–303
8. Pope SB (2000) Turbulent flows. Cambridge University Press
9. Batchelor GK (1951) Pressure fluctuations in isotropic turbulence. Proc. Cambridge Phil Soc 47:359–374
10. Schlichting H, Gersten K (1997) Grenzschicht-Theorie. Springer, Berlin, Heidelberg, New York
11. Taler D (2017) Simple power-type heat transfer correlations for turbulent pipe flow in tubes. J Therm Sci 26(4):339–348
12. Moulden TH (1977) In: Frost W, Moulden TH (eds) Handbook of turbulence. Fundamentals and applications. Plenum Press, New York
13. Zudin YB (1997) Calculation of the effect of evaporating drops on the relative law of heat transfer with a disperse mist flow. J Eng Phys Thermophys 70:507–510

Chapter 18
Thermal–Hydraulic Stability Analysis of Supercritical Fluid

Abbreviations

BC	Boundary condition
SB	Stability boundary
SCP	Supercritical pressures

Symbols

A	Homochronicity parameter
E	Extension parameter
h	Enthalpy
K	Pressure parameter
p	Pressure
t	Time
u	Velocity
v	Specific volume

Greek Letter Symbols

α	Increment/Decrement
ρ	Density
Ω	Dimensionless frequency
β	True frequency

18.1 Thermal–Hydraulic Instability

Helium at Supercritical Pressures (SCP) is used as a coolant for cryogenic freezing of superconductivity-based objects. Supercritical helium has the following advantages over the boiling working media.

Y. B. Zudin, *Non-equilibrium Evaporation and Condensation Processes*, Mathematical Engineering, https://doi.org/10.1007/978-3-030-67553-0_18

- Low values of the critical temperature (5.2 K) and the critical pressure (0.227 MPa).
- High heat transfer intensity under the turbulent flow regime.
- The absence of phase changes that are deleterious to the cooling conditions and responsible
- For the growth of the hydraulic resistance.

Survey [1] examines the performance of heat transfer correlations adopted at Supercritical Pressures (SCP). It is found in [1] that most of the correlations could predict heat transfer quite well in the low enthalpy region and a few correlations could predict heat transfer in the high enthalpy region near-critical conditions. As a field of application, the works [2, 3] consider the problem of ensuring the effective operation of the thermal control system of superconducting magnets. When a liquid with temperature-dependent density flows in a heated channel, pulsations of parameters can arise caused by two different physical reasons.

(1) It is better to consider pressure pulsations in the Eulerian coordinates. Pressure perturbations will expand upstream and downstream with the speed of sound. This speed can be considered "infinitely large" (i.e., to surpass the flow velocity by several orders of magnitude). Therefore "pressure waves" will move in the liquid almost "instantly". In other words, at any point of the Eulerian coordinates, high-frequency pressure pulsations will take place practically synchronous.

(2) While describing thermal pulsations of density, it is expedient to "track" a chosen liquid particle in the Lagrangian coordinates. Heat supply to a moving liquid will lead to an increase in its temperature. Consequently, the density of the liquid will decrease, and its volume will increase. For fluid flow in a channel of a constant cross-section, it will lead to the linear expansion of a chosen "liquid volume" and therefore to the growth of the wall area "moistened" by this volume. Then, for a preset heat flux per unit of the heating surface, the amount of heat transferred from the wall to the liquid will increase. In turn, this will amplify the density pulsations, etc. Thus, these elementary physical reasoning predicts an avalanche-like growth of the density wave amplitude "drifting" downstream together with the liquid volume.

The main objective is to maintain the stable performance of a cooling systems via preserving the necessary mass flow-rate of helium in them to remove heat from a solid body (the channel wall) so that the limiting permitted level of temperatures will not be exceeded. Usually, cooling systems are calculated under the assumption of a steady-state regime and do not take into account possible pulsations of thermal–hydraulic parameters. Insufficiency of the "steady-state approach" has been convincingly demonstrated in works [2, 3]. It has been experimentally proved that under certain conditions in a cooling system, low-frequency pulsations of the coolant mass flow-rate with noticeable amplitude can emerge. The works [2, 3] demonstrated clearly that pulsations of the "density wave" type took place in this case. Indeed, once emerged, the "pressure waves" would have been "instantly" propagated both downstream and upstream and would have been "damped" on the throttles set at the inlet and the outlet of the heated channel in experiments [2, 3]. And, the most important thing, pressure pulsations would have had a strongly pronounced high-frequency character. Therefore, while analyzing the experimental results of [2, 3], we will further consider exclusively the instability of the "density wave" type.

An extremely dangerous situation can arise with regard to thermostatic systems in superconducting magnets. Here even insignificant temperature pulsations can lead to the phenomenon known as the "destruction of superconductivity" in the special literature [2–4]. This phenomenon is inadmissible from the point of view of practical applications, as it causes irreversible changes in the character of processes in the magnet windings. In other words, the trespassing by the cooling system parameters beyond the stability boundaries is considered as a kind of an "accident". It is clear from here that identifying the boundary values of the regime parameters at which the coolant flow (helium at SCP) loses hydrodynamic stability is of great importance in engineering.

18.2 Small Variations Method

The method of small variations [5] is the classical method of investigating the instability. The sketch of the method is as follows. Pulsation corrections are added to the parameters involved in the equations for conservation of mass, momentum, and energy. Subtracting the original stationary equations from the resulting nonstationary equations and linearizing, we get a system of linear equations for the pulsation quantities. By analyzing them for stability one determines the "stability region", or, in other words, the ranges of the regime parameters in which the perturbations introduced into the system will die out with time. Outside of these ranges (in the "instability region") the pulsation amplitudes exponentially increase with time. These two regions are separated by the "stability boundary" (SB). The main purpose of the present study is to identify and analyze the SB. In general, an SB is a surface or a family of surfaces. For practical applications, one usually uses the boundary stability curves, which reflect the relation between some two of the main parameters, the remaining parameters are "frozen".

The small variations method was used for investigating various forms of instability development, which have become classical areas in mathematical physics.

- The Kelvin–Helmholtz instability of sliding ideal liquids [6].
- The Rayleigh–Taylor instability of a heavy liquid above a light one [7].
- The Tollmien–Schlichting instability of a laminar boundary layer on the onset of turbulence [8].
- The Darrieus–Landau instability of combustion surface in the gravity field with surface tension [6].

The traditional theory of small variations operates with perturbations varying with time along the axial coordinate as a function of the wave running in the flow direction. In the context of coolant flow in a channel, this model describes the "pressure wave"-type instability due to the compressibility of the medium. Here, high-frequency perturbations are propagated in the channel at the velocity of sound, whose wavelength is negligible in comparison with the channel length. Hence the shortwave pressure waves are developed independently of the inlet and outlet conditions. Such a scenario of the development of the instability is typical for acoustic vibrations in high-speed flows of a viscous compressible gas.

For the first time, a theoretical analysis of hydrodynamic stability in a flow of "supercritical helium" by means of the Nyquist plot was performed in [2]. A Nyquist plot is a tool used in automatic control and signal processing to estimate the stability of a system with feedback [9]. It has a form of a graph in polar coordinates, in which the gain-frequency characteristic and phase-frequency characteristic. The plot of these phase quantities exhibits the phase as the angle, while the magnitude is plotted as the distance from the origin. According to [10], the Nyquist plot, as well as the other methods related to it, is a recognized computational tool in the automatic control theory. However, from the point of view of the author of the present book, these methods perform not so effectively for hydrodynamic applications. Indeed, as the theoretical analysis executed in [2, 3] shows, Nyquist plot is capable of conducting only selective calculations of modes with concrete values of parameters. A similar method of "selective numerical shooting", which can though appear useful in practice to simulate certain particular situation, does not however possess a required generality and is not representative of the physical nature of the processes under consideration. In an ideal, an appealing alternative may be provided by the method of analytical prediction (though with a certain inaccuracy) of the SB, i.e., a line or a surface in the space of the parameters limiting the area of stable operating modes of cooling systems.

An important step in this direction was made in the pioneering works of Labuntsov and Mirzoyan [10, 11]. These authors [10, 11], while analyzing the boundaries of flow stability of helium at SCP in heated channels, proposed to branch off the methods of the theory of automatic control. Moreover, they have for the first time applied the approach of the classical linear theory of hydrodynamic stability to the solution of the considered problem.

18.3 Density Wave Instability

18.3.1 Physical Analysis

Let us consider the scenario of the development of thermal–hydraulic instability for a flow in a channel of supercritical helium. The velocity pulsations at the inlet are developed downstream and are responsible for perturbations of the enthalpy of the medium. In turn, such perturbations promote density oscillations. A density wave is carried far downstream, reaches the outlet section, reflects from it, and moves further in the opposite direction. The characteristic lifetime of such oscillations is commensurable with the time during which a fluid particle is over the channel length. A density wave can be schematically represented as low-frequency oscillations of an elastic thread fastened between the inlet and the outlet. So, the wavelength of such oscillations is comparable with the channel length.

A theoretical analysis of thermal–hydraulic instability of helium flow at SCP seems to have been first carried out [2] by numerical methods. In [2], the Laplace transformed equations of perturbed motion were represented in a finite difference form along the axial coordinate. The character of the flow (stable or unstable) was determined using the Nyquist criterion from the automatic control theory. So, the method of [2] is based on a series of variant numerical calculations for each given regime. The SB was determined by drawing a smooth curve along a set of points "depicted" in a chosen frame based on the results of calculations. It seems that this method is reduced in the actual fact to a trial-and-error procedure.

In [10, 11], a method of analysis was developed not requiring artificial recourse to the automatic control theory and based on the real thermal–hydraulic process pattern. The theoretical approach of [10, 11] is based on the following physically justified assumptions.

- The flow of the coolant is described by one-dimensional nonstationary differential equations for conservation of mass, momentum, and energy.
- The laws of heat transfer and hydraulic resistance in a channel obey the quasi-stationary laws.
- At each point along the stream, the coolant is in a local thermodynamic equilibrium.
- Due to low pulsation frequency, the heat response delay of the channel was is not taken into account.
- The heat diffusion in the axial direction is negligible.

Let us consider the method of [10, 11] in accordance with the process setup depicted in Fig. 18.1. From the reservoir (in which the pressure and the temperature

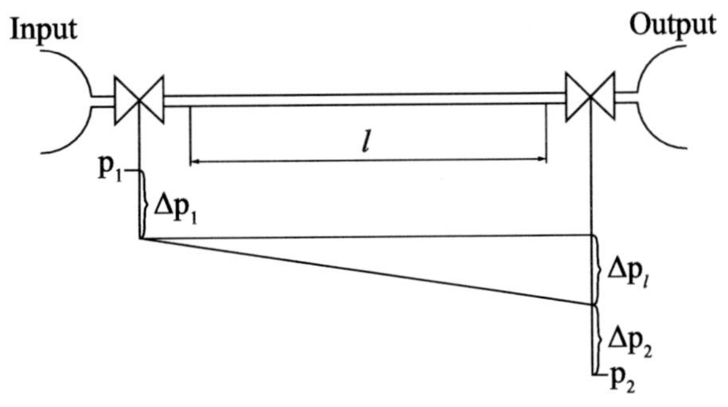

Fig. 18.1 Process setup

(and hence, and the enthalpy) are maintained constant) the coolant is supplied from the inlet throttle to the heated channel. Next, the liquid is fed through the outlet throttle into the second reservoir in which the pressure is also constant. The density of heat sources is maintained constant over the channel length[1]: $q_v = \text{const}$. The pressure is lost along the stream: Δp_1 at the inlet throttle, Δp_l over the channel length, Δp_2 at the outlet throttle.

18.3.2 Mathematical Description

The flow of SCP liquid in a channel is described by one-dimensional nonstationary continuity equation

$$\frac{\partial \rho}{\partial t} + \frac{\partial (\rho u)}{\partial x} = 0, \tag{18.1}$$

as well as by the motion equation

$$\rho \frac{\partial u}{\partial t} + \rho u \frac{\partial u}{\partial x} + \frac{1}{2} \frac{\xi}{d_h} \rho u^2 = -\frac{\partial p}{\partial x}, \tag{18.2}$$

and by the energy equation

$$\rho \frac{\partial h}{\partial t} + \rho u \frac{\partial h}{\partial x} = q_v. \tag{18.3}$$

Here, t is the time, x is the coordinate in the direction of the flow, ρ is the liquid density, u is the mass-average velocity, p is the pressure, h is the enthalpy, ξ is the Darcy friction factor, d_h is the hydraulic diameter of the channel, q_v is heating source density.

According to the physical scheme of the development of a thermal–hydraulic instability, it is generated by velocity pulsations in the inlet section. The flow enthalpy at the channel inlet remains unperturbed, whence we get the first Boundary Condition (BC) for the system of Eqs. (18.1)–(18.3)

$$h_1 = \text{const}. \tag{18.4}$$

The BC (18.4) will be called the "starting" BC. The perturbations of the pressure propagate over the channel at the velocity of sound, that is, practically instantaneously in comparison with the flow velocity. Hence the total pressure differential over the channel length should be constant in time. This is secured by the second BC

[1] Within the one-dimensional problem, this condition is also equivalent to the condition of constant heat flow density over the channel length.

$$\Delta p = \Delta p_1 + \Delta p_2 + \Delta p_l = \text{const.} \tag{18.5}$$

The BC (18.5), which establishes a link between the pressure differentials over the channel length, will be called the "closing" BC. By analyzing the thermodynamic state diagrams for an SCP helium one can approximate each real isobar $\mathrm{v}(h)$ by the linear dependence

$$\mathrm{v} = a + bh.$$

Here, the coefficients a, b are constant for each isobar, but they change their values from one isobar to the next. Correspondingly, the relation between the specific volume and the enthalpy can be written as the state equation for the supercritical helium

$$b = \frac{\mathrm{v}_2 - \mathrm{v}_1}{h_2 - h_1}.$$

This allows one to introduce the physical frequency scale of oscillations of density waves

$$\omega_0 = bq_{\mathrm{v}}. \tag{18.6}$$

Relation (18.6) singles out the instability in the supercritical region in a separate class of problems. Moreover, using (18.6) the mathematical description can be radically simplified. Excluding from (18.1)–(18.3) the density ρ and the enthalpy h, we get, after some algebra, the equations for conservation of mass, momentum, and energy

$$\frac{\partial u}{\partial x} = \omega_0, \tag{18.7}$$

$$\frac{\partial u}{\partial t} + u\frac{\partial u}{\partial x} + \frac{\xi}{2d_h}u^2 = -\mathrm{v}\frac{\partial p}{\partial x}, \tag{18.8}$$

$$\frac{\partial \mathrm{v}}{\partial t} + u\frac{\partial \mathrm{v}}{\partial x} = \omega_0 \mathrm{v}. \tag{18.9}$$

As a result, the starting BC (18.4) for the enthalpy is replaced by the corresponding BC for the specific volume

$$\mathrm{v}_1 = \text{const.} \tag{18.10}$$

We represent the velocity u, the specific volume v, and the pressure p of the real flow as a superposition of the stationary and pulsation terms

$$u = \bar{u} + u', \mathrm{v} = \bar{\mathrm{v}} + \mathrm{v}', p = \bar{p} + p'.$$

Introducing the generalized coordinate

$$z = 1 + \frac{\omega_0 x}{\bar{u}_1},$$

we write down the time averaged Eqs. (18.7)–(18.9)

$$\frac{d\bar{u}}{dz} = \bar{u}_1, \tag{18.11}$$

$$\bar{u}\frac{d\bar{u}}{dz} + \frac{1}{2}\frac{\xi\bar{u}_1}{d_h\omega_0}\bar{u}^2 = -\bar{v}\frac{d\bar{p}}{dz}, \tag{18.12}$$

$$\bar{u}\frac{d\bar{v}}{dz} = \bar{u}_1\bar{v}. \tag{18.13}$$

The system of Eqs. (18.11)–(18.13) has the trivial solution

$$\left.\begin{aligned} \frac{\bar{u}}{\bar{u}_1} = \frac{\bar{v}}{\bar{v}_1} = z, \\ -\frac{d\bar{p}}{dz} = \frac{\bar{u}_1^2}{\bar{v}_1}(1 + Az). \end{aligned}\right\} \tag{18.14}$$

Here, $A = \frac{1}{2}\frac{\xi\bar{u}_1}{d_h\omega_0}$ is the dimensionless parameter, whose physical sense will be discussed below.

The linearization procedure can be applied in the case when the pulsation amplitudes of the parameters are small in comparison with their averaged values. In Eqs. (18.7)–(18.9), we retain only the terms with linear departures of the quantities from their stationary values and subtract the averaged values from the original equations. This gives us the system of conservation equations for the pulsation quantities of the velocity u', the specific volume v', and the pressure p'

$$\frac{\partial u'}{\partial z} = 0, \tag{18.15}$$

$$\frac{\bar{u}_1}{\omega_0}\frac{\partial u'}{\partial t} + \bar{u}\frac{\partial u'}{\partial z} + u'\frac{d\bar{u}}{dz} + \frac{\xi}{d_h}\bar{u}u' = -v'\frac{d\bar{p}}{dz} - \bar{v}\frac{\partial p'}{\partial z}, \tag{18.16}$$

$$\frac{1}{\omega_0}\frac{\partial v'}{\partial t} + \frac{\bar{u}}{\bar{u}_1}\frac{\partial v'}{\partial z} + \frac{u'}{\bar{u}_1}\frac{d\bar{v}}{dz} = v'. \tag{18.17}$$

The BC for the perturbed Eqs. (18.15)-(18.17) follow from the original BC (18.5), (18.10)

$$\mathrm{v}_1' = 0, \tag{18.18}$$

$$\Delta p' = \Delta p_1' + \Delta p_2' + \Delta p_l' = 0. \tag{18.19}$$

We introduce the dimensionless complex pulsation frequency[2]

$$\Omega \equiv \frac{\omega}{\omega_0} = \alpha + i\beta,$$

where ω is dimensional complex frequency.

The quantity β is the true perturbation frequency. The quantity α characterizes the evolution rate of the pulsation amplitude: in the unstable region it is the growth increment ($\alpha > 0$), and in the stable region it is the damping decrement ($\alpha < 0$). On the BS, we have $\alpha = 0$. The solutions of Eqs. (18.15)–(18.17) with the BC (18.18), (18.19) will be sought in the form

$$\frac{u'}{\bar{u}} F_u \exp(\Omega t) = F_u \exp(\alpha t)(\cos\beta t + i\sin\beta t),$$

$$\frac{\mathrm{v}'}{\bar{\mathrm{v}}} F_{\mathrm{v}} \exp(\Omega t) = F_{\mathrm{v}} \exp(\alpha t)(\cos\beta t + i\sin\beta t),$$

$$\frac{\bar{\mathrm{v}}_1 p'}{\bar{u}_1^2} = F_p \exp(\Omega t) = F_p \exp(\alpha t)(\cos\beta t + i\sin\beta t).$$

Here, $F_u(z)$, $F_{\mathrm{v}}(z)$, $F_p(z)$ are pulsation amplitudes, which in general are complex numbers.

From the consideration of the averaged (18.11)–(18.13) and the perturbed (18.15)–(18.17) conservation equations, one can single out three dimensionless parameters describing the above problem.

(1) The extension parameter $E = \bar{\mathrm{v}}_2/\bar{\mathrm{v}}_1$, which characterizes the growth rate of the flow specific volume between the inlet and outlet of the channel.

(2) The pressure parameter $K = \Delta\bar{p}_1/\Delta\bar{p}_2$, which is the ratio of the pressure losses at the inlet and outlet throttles.

(3) The homochronicity parameter $A = \frac{1}{2}\frac{\xi\bar{u}_1}{d_h\omega_0}$ involves two factors, which control the instability development.

[2]The standard complex notation is used to make the calculations more compact. In reality, all the quantities are real.

First, the hydraulic resistance coefficient ξ characterizes the pressure losses along the channel length. Second, the drift time of a liquid particle along the channel is $t_0 \sim \omega_0^{-1}$. So, the quantity $\overline{u}_1 t_0/d_h \sim \overline{u}_1/d_h\omega_0$ characterizes the degree of nonstationarity of the flow.

For the physical analysis of the scenario of the development of the thermal–hydraulic instability, it is expedient to change from the above three-parameter problem to the corresponding two-parameter variants. Here the principal role is played the extension parameter, which relates the channel length l, the frequency scale of oscillations ω_0, and the averaged velocity at the inlet section $\overline{u}_1$

$$E = 1 + \frac{\omega_0 l}{\overline{u}_1}. \tag{18.20}$$

Attaching to the parameter E in succession the parameters K, A (which are responsible for different components of general pressure losses in the channel), we get two limit two-parameter problems.

18.4 Problem 1. Pressure Losses in Throttles

18.4.1 The Process of Solution

Let us consider the case of fluid pressure losses in the inlet and outlet throttles (resistances): $A = 0$ (Fig. 18.1). Now E, K are the governing parameters, the motion Eqs. (18.12), (18.16) falls out from the mathematical description, and the solution for the pressure pulsation amplitude F_p becomes superfluous. The closing BC (18.19) assumes the form

$$\Delta p_1^{'} + \Delta p_2^{'} = 0. \tag{18.21}$$

We have the following equations for the velocity pulsation amplitudes F_u and the specific volume F_{v}

$$\frac{dF_u}{dz} = 0, \tag{18.22}$$

$$\frac{dF_{\mathrm{v}}}{dz} + \frac{\Omega - 1}{z} F_{\mathrm{v}} + \frac{F_u}{z} = 0. \tag{18.23}$$

The BC for the amplitude Eqs. (18.22), (18.23) assume the form

$$F_{\mathrm{v}1} = 0, \tag{18.24}$$

$$2EKF_{u1} + 2F_{u2} - F_{v2} = 0. \qquad (18.25)$$

The solutions of Eqs. (18.24), (18.25) with the BC (18.18), (18.21) can be written as

$$F_u = D, \qquad (18.26)$$

$$F_v = \frac{D}{\Omega - 1}\left(z^{\Omega-1} - 1\right). \qquad (18.27)$$

Having found the values of the amplitudes F_u, F_v from (18.26), (18.27) at the channel end-points, substituting them into the BC (18.24), (18.25), and excluding the constant D, we get the principal equation for the stability analysis (the dispersion relation)

$$2(EK + 1)(\Omega - 1) + 1 - E^{1-\Omega} = 0. \qquad (18.28)$$

For further analysis, it is necessary to separate the real and imaginary parts of the oscillation frequency: $\Omega = \alpha + i\beta$. As a result, we get the system of two transcendental equations relating to K, E, α, β

$$2(\alpha - 1)(EK + 1) + 1 - CE^{1-\alpha} = 0, \qquad (18.29)$$

$$2\beta(\alpha - 1)(EK + 1) + SE^{1-\alpha} + 0. \qquad (18.30)$$

Here, $C = \cos g$, $S = \sin g$, $g = \beta \ln E$ is the generalized parameter involving the oscillation frequency and the extension parameter.

The parameters E, K are determined by the external characteristics of the flow under consideration and should be considered as known. If E, K are fixed, then from the system of Eqs. (18.29), (18.30) one can find the corresponding values of the amplitude α and frequency β, and hence, and determine whether the flow form is stable or unstable. Moreover, from the presence of periodic functions in (18.29), (18.30) it follows that to each specific combination of the parameters E, K there corresponds an infinite discrete spectrum of the quantities α_n, β_n $(n = 1, 2, 3, \ldots)$. Let us consider three possible variants of perturbation evolution.

(1) $\alpha < 0$, the perturbation introduced in the flow are damped out in time: the flow is stable.
(2) $\alpha > 0$, the pulsation amplitude exponentially increases: the flow is unstable.
(3) $\alpha = 0$, the perturbations are neither damped out or increased: the flow is on the SB.

As was already pointed out, the knowledge of the SB is central for practical applications of the results of the analysis. Moreover, the perturbation with the smallest

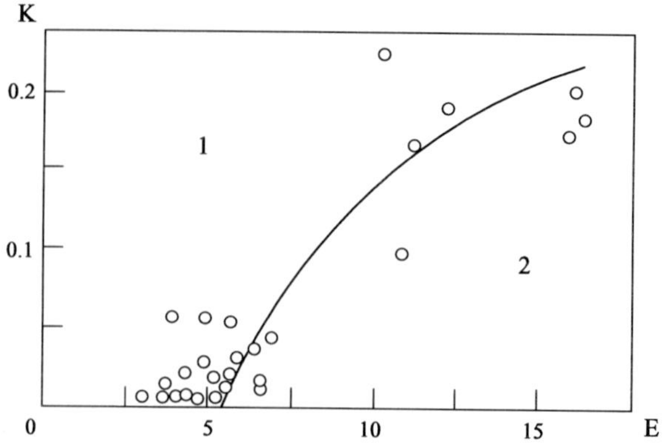

Fig. 18.2 Comparison of the numerical calculations of [10] with experimental findings from [12]. 1 The stability region, 2 the instability region, 3 the stability boundary

frequency ($\beta_1 = 2.673$) will be the most rapidly growing (that is, the most "hazardous"). In [10], a numerical solution of the system of Eqs. (18.29), (18.30) was obtained for the range $5 < E < 15.0 < K < 0.25$ of the regime parameters and compared with that obtained in the experimental study [12].

According to the estimates of [2, 3], for the scheme of thermal control of the superconducting magnets realized in experiments [12], the fluid pressure losses over the channel length will be practically always smaller than those in inlet and outlet resistances (throttles). Figure 18.2 shows that due to the large scatter of experimental data one can speak here only about the qualitative agreement with the numerical results. In this connection, the necessity of a general analytical solution, from which the problem of thermal–hydraulic instability can be parametrically analyzed, becomes still more acute.

Analytical solutions have the following advantages.

- Closed qualitative description of the problem and identification of the complete list of the governing parameters.
- Parametric investigation of the problem by varying the boundary-value and initial conditions.
- The results of the analytical solution can be used for validation of numerical calculations.

An analytical solution of the system of Eqs. (18.29), (18.30) for the SB was obtained using the Maple computer algebra software system [13]. As a result, the SB was calculated (Fig. 18.3), and the asymptotic formulas for the problem with respect to the extension parameter were obtained

$$\left.\begin{array}{c} E = E_0 = 5.439 : K = 0, \\ E \to \infty : K \to \frac{1}{2}. \end{array}\right\} \tag{18.31}$$

It was shown that in the interval $E < E_0$ the system is absolutely stable for each value of the pressure parameter $K = \Delta\bar{p}_1/\Delta\bar{p}_2$. The instability is first manifested

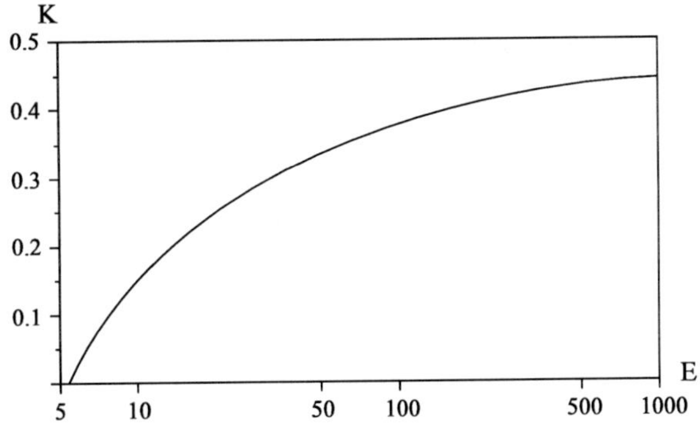

Fig. 18.3 Stability boundary for Problem 1. The pressure parameter versus the extension parameter. Evaluation by formula (18.32)

at the bifurcation point $E = E_0$. The instability region monotonically increases with a further increase of the extension parameter, but it is still confined in the interval $0 < K < 1/2$. So, the pressure parameter plays the role of a stabilizing factor: with increasing throttling at the inlet and (or) decreasing it at the outlet, we can dampen instabilities of the "density wave" type. Physically, this means that inlet throttling stabilizes the system, while outlet throttling promotes its oscillation.

From the solution, we also get the following limit variants of the above problem.

- For $\Delta\bar{p}_1 > 2\Delta\bar{p}_2$, the system becomes absolutely stable.
- With no inlet throttle $\left(\Delta\bar{p}_1 = 0\right)$, we get $K = 0$ for any arbitrarily small value of $\Delta\bar{p}_2$. This means that the system will be absolutely unstable starting from some value of the extension parameter $(E > E_0)$.
- The absence of outlet throttling $\left(\Delta\bar{p}_2 = 0\right)$ means that $K \to \infty$ with arbitrary $\Delta\bar{p}_1 > 0$, and the system will always be absolutely stable.

18.4.2 Approximation of the Solution

An exact analytical solution for all ranges of the parameters is quite bulky and can hardly be written down. Hence, approximations of the analytical solution are required. The following relations for the SB (Fig. 18.3)

$$K = \frac{\sqrt{1 + \left(1 + \beta^2\right)\left(E^2 - 1\right)} - 1 - 2\beta^2}{2E\left(1 + \beta^2\right)}, \tag{18.32}$$

and for the pulsation frequency (Fig. 18.4)

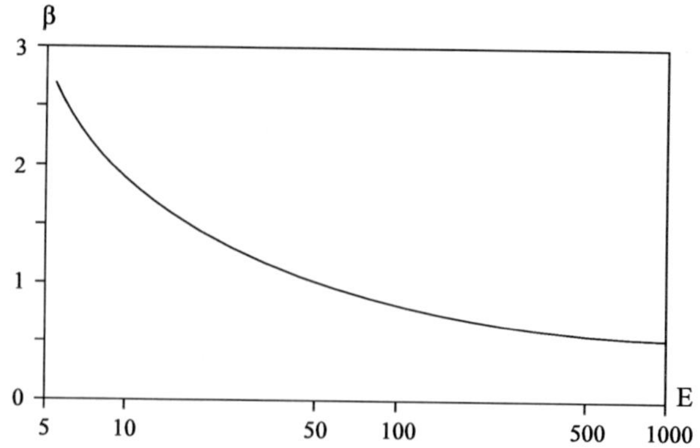

Fig. 18.4 Problem 1. The perturbation frequency versus the extension parameter on the stability boundary. Evaluation by formula (18.33)

$$\beta = \frac{\pi + (4.527 - \pi)(E_0/E)^{0.425}}{\ln E} \tag{18.33}$$

were obtained.

Figure 18.3 shows that with increasing extension parameter the pressure parameter is monotonically increased and tends as $E \to \infty$ to the limit value $K = 1/2$.

Figure 18.4 shows that with increasing E the main perturbation frequency decreases monotonically and tends to zero as $E \to \infty$. This means that, for very large values of the extension parameter, the leading perturbation harmonic is "frozen" and ceases to jiggle the system. At the same time, with increasing E the system crosses in succession new stability boundaries. Here new high-frequency harmonics of more and more increasing frequency that destabilize the system come into play. As a result, in the conjectural case $E \to \infty$, the system will feature an infinite discrete spectrum of exponentially increasing perturbations.

The above results can be applied as follows. Assume that we know the value of the extension parameter ($E > E_0$). Then, in order to achieve a stable state, one needs to select the ratio of the pressure losses at the inlet and outlet throttles that the pressure parameter would exceed the value given by formula (18.32) (see Fig. 18.3). The pulsation frequency of the parameters near the SB should be evaluated by formula (33) (see Fig. 18.4).

18.5 Problem 2. Pressure Losses Over the Channel Length

18.5.1 Construction of the General Solution

Let us now consider the case when the losses at the inlet and outlet throttles are negligible against those over the channel length: $K = 0$ (Fig. 18.1). Here, E, A are the governing parameters. The closing BC (18.19) assumes the form

$$\Delta p_l' = 0. \tag{18.34}$$

The system of equations for (18.29), (18.30) should be augmented with the motion equation

$$\frac{dF_p}{dz} + \frac{dF_u}{dz} + \left(\frac{\Omega + 1}{z} + 2A\right)F_u - \left(\frac{1}{z} + A\right)F_v = 0. \tag{18.35}$$

The BC for Eqs. (18.29), (18.30), (18.35) follow from the BC (18.18), (18.34)

$$F_{v1} = 0, \tag{18.36}$$

$$F_{p2} - F_{p1} = 0. \tag{18.37}$$

The amplitudes of pulsation of the velocity F_u and the specific volume F_v are, as before, given by the above relations (18.26), (18.27). Substituting these relations into Eq. (18.35) and integrating with the use of the BC (18.36), (18.37), we get the following expression for the pressure pulsation amplitude

$$F_p = F_{p1} + D\left[\frac{1}{\Omega - 1}\left(\frac{z^{1-\Omega} - 1}{1 - \Omega} + A\frac{z^{2-\Omega} - 1}{2 - \Omega}\right) + A(1 - 2\Omega)(z - 1)\right]. \tag{18.38}$$

Determining from (18.38) the pressure pulsation amplitude F_{p2} at the channel outlet and substituting it into the BC (18.37), we get the dispersion relation for Problem 2

$$A\left[(E - 1)(1 - 2\Omega) + \frac{E^{2-\Omega} - 1}{2 - \Omega}\right] = \Omega^2 \ln\Omega + \frac{E^{1-\Omega} - 1}{1 - \Omega}. \tag{18.39}$$

As in the above Problem 1, here each specific combination of the external parameters A, E represents some ensemble of real flows, for which the discrete spectrum of $\alpha_n, \beta_n (n = 1, 2, 3, \ldots)$ is infinite.

We have not succeeded in obtaining an exact analytical solution to the two-parameter Eq. (18.39). A possible way for its numerical investigation is as follows. Putting $\alpha = 0$ in (18.39) and separating the real and imaginary parts, we get the system of two equations, which describes the main SB. By giving concrete value to the pair of external parameters A, E, we find the frequency β on the SB. Next, fixing A and specifying a different value of E, we get the successive value of β. We repeat these steps for the successive value of the parameter E, and so on. As a result, we get the SB in the coordinates $E(\beta)$. Next, the same steps are again repeated for the new pair of external parameters A, E, and so on. As a result of this (labor-intensive) exhaustive search of variants, one can obtain the main SB in the coordinates $A(E)$ (in case it exists).

Numerical variant calculations have shown that, for sufficiently large values of the homochronicity parameter ($A > 10$), the SB behaves like a vertical line in the coordinates $A(E)$, which means that the form of the flow (stable or unstable) depends only on the extension parameter E of the flow over the channel length.

18.5.2 Analytical Solution

Let us analyze Eq. (18.39) as $A \to \infty$. In order that its right-hand side would remain finite, it is necessary that the expression in the square brackets on the left be zero

$$A\left[(E-1)(1-2\Omega)+\frac{E^{2-\Omega}-1}{2-\Omega}\right]=0.$$

Using the complex notation $\Omega = \alpha + i\beta$, we get the following pair of equations controlling the SB

$$\left[2\left(\beta^2-\alpha^2\right)+5\alpha-2\right](E-1)+1-CE^{2-\alpha}=0, \qquad (18.40)$$

$$\beta(2\alpha-5)(E-1)-SE^{2-\alpha}=0. \qquad (18.41)$$

Making $\alpha = 0$ in (18.40), (18.41), we get the system of equations describing the SB

$$2\left(\beta^2-1\right)(E-1)+1-CE^2=0, \qquad (18.42)$$

$$5\beta(E-1)+SE^2=0. \qquad (18.43)$$

By solving Eqs. (18.42), (18.43), we get the discrete spectrum of pairs of values E_n, β_n ($n = 1, 2, 3, \ldots$) The first three pairs of these values are given in Table 18.1. The table shows that with increasing extension parameters the perturbation harmonics of more and more increasing frequencies are engaged, which drive the system out of its stability state. Moreover, as each successive boundary is attained, the amplitudes of oscillation due to intersections of the previous boundaries can build up in the system.

Let us write down the approximation of the frequency of the leading harmonic suitable for the entire range of variation of the extension parameter ($1 \leq E < \infty$,

Table 18.1 Calculated values of E_n, β_n for $n = 1, 2, 3$

n	1	2	3
E_n	11.88	28.38	70.57
β_n	2.145	3.562	5.806

Fig. 18.5)

$$1 < E < 10^3 : g = 3.42 + 0.367\arctan(1.72G) + 1.34\mathrm{erf}(1.108G), \quad (18.44)$$

$$10^3 < E < 10^8 : g = 5.22E^m. \quad (18.45)$$

The growth increment of perturbations for the region of instability ($E_1 \leq E < \infty$) is described by the relation (Fig. 18.6)

$$\alpha = \sin\frac{\pi}{2}\left(1 - \frac{\ln E}{\ln E_1}\right). \quad (18.46)$$

Here, $G = (E - 1)^{0.65}$, $m = 3.91*10^{-3}$.

18.5.3 Instability Region

The right-hand branch in Fig. 18.5 shows that with increasing extension parameter the frequency of the leading harmonic β gradually decreases and tends to zero as $E \to \infty$. From (18.46), it follows that the growth increment increases monotonically (Fig. 18.6). As $E \to \infty$, we have $\alpha \to 1$. Moreover, the leading harmonic crosses in succession new and new stability boundaries E_n and meets on these boundaries the perturbations of more and more high frequency β_n which are generated on these boundaries. This trend is illustrated in Table 18.2. Such an evolution of the principal perturbation with increased penetration in the instability region is analogues to the scenario of the development of the perturbation described in Problem 1.

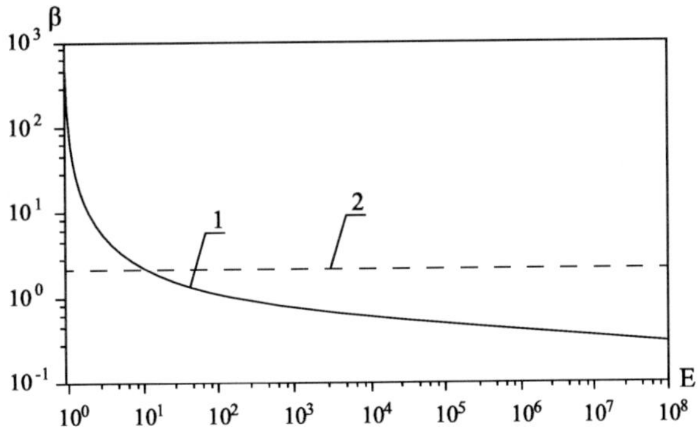

Fig. 18.5 Problem 2. ——— Perturbation frequency versus the extension parameter on the stability boundary, – – – – – stability boundary. Evaluation by formulas (18.44)–(18.45)

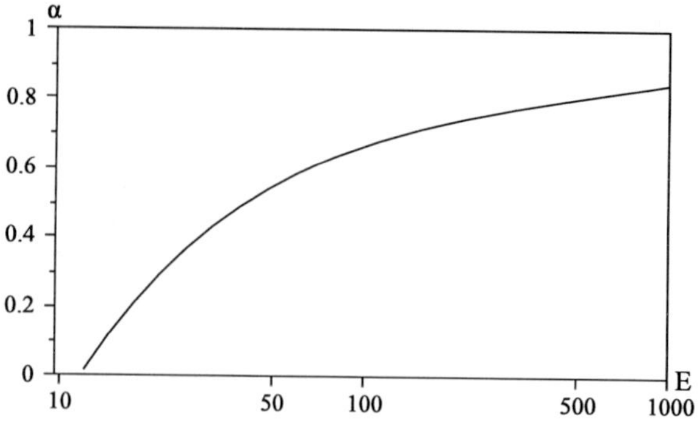

Fig. 18.6 Problem 2. The growth increment of perturbations versus the extension parameter (the stability region). Evaluation by formula (18.46)

Table 18.2 Evolutions of the principal perturbation. E_n, β_n are parameters on the stability boundaries ($n = 1, 2, 3$); α_1, β_1 are parameters of the most hazardous perturbation

n		1	2	3
E_n		11.88	28.38	70.57
β_n		2.145	3.562	5.806
β_1		2.145	1.587	1.247
α_1		0	0.3974	0.6111

18.5.4 Stability Region

When moving back inside the region of absolute stability, β increases more and more and tends to infinity as $E \to 1$ (the left branch in Fig. 18.5). Here, α changes its sign to negative and the growth increment becomes damping decrement of pulsations (Fig. 18.7). The left branch in Fig. 6 shows that the absolute value of the damping decrement increases more and more rapidly. Making $E \to 1$ we have $\alpha \to -\infty$. For comparison, Fig. 18.7 also shows the right (flat) branch of the dependence $\alpha(E)$ for the stability region.

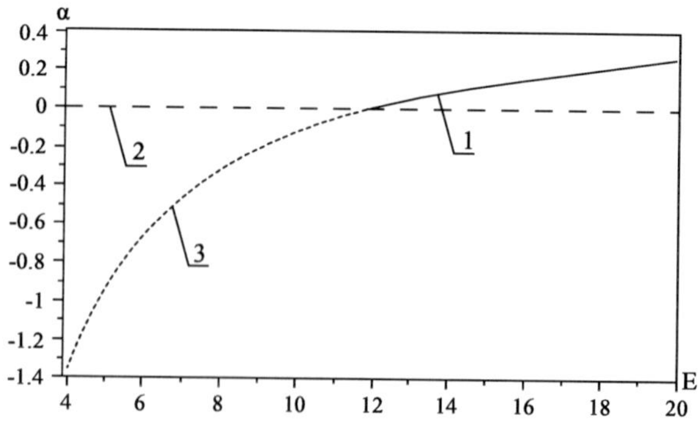

Fig. 18.7 Problem 2. Dependence of α on the extension parameter. ——— The growth increment (the instability region), - - - - - the stability boundary, ● ● ● ● ● ● —the damping decrement (the stability region). Evaluation by formula (18.46)

Thus, the above analysis reveals a quite involved multiparameter mechanism of development of a thermal–hydraulic of the "density wave" type. Relations (18.44)–(18.46), as well as the data from Tables 18.1, 18.2 can be used for evaluating the parameters of a thermal–hydraulic instability in the limit case of small pressure losses at throttles in the interval $A \geq 10$. According to Table 18.1, instability occurs for $E > E_1 = 11.88$, the most hazardous (lowest) frequency of perturbations being $\beta_1 = 2.145$.

Formulas (18.44), (18.45) (the oscillation frequency), and formula (18.46) (the growth increment and the damping decrement) describe the evolution of the first harmonic as the extension parameter moves inside the stable and unstable regions. Table 18.2 gives the values of these parameters for the first three stability boundaries ($n = 1, 2, 3$). Having known the growth increment of the amplitude of perturbations in the instability region for $E > E_1$, one can estimate the characteristic development time of the nonlinear stage of the instability.

The above technique can also be used in the problem of development of a thermal–hydraulic instability for two-phase flows in channels [14]. Note that in [15–17] the above method was applied for the analysis of the conjugate "Fluid–Structure Interaction" problem, where it was shown that consideration of the pulsations of temperatures and heat flows in the channel wall results in the stabilization of the flow. For a thorough survey in this field, we refer to the book [18].

18.6 Conclusions

Analysis of a thermal–hydraulic "density wave"-type instability occurring in flows of supercritical helium in channels is given. It is shown that this instability is described by three dimensionless quantities: the extension parameter, the pressure parameter, and the homochronicity parameter. We consider two problems described by two parameters and distinguished by the type of pressure losses over the channel length. Maple computer algebra software system was used to derive exact analytical solutions determining the stability boundary, the frequency of developed perturbations, and the characteristics of growth increment and damping decrement. It is shown that the system stabilizes with increased throttling at the channel inlet, whereas the final differential at the outlet throttle promotes its oscillation. The asymptotical behavior of the analytical solution is studied and its approximations are constructed. It is shown that with increasing extension parameter the system crosses in succession new stability boundaries, as a result, high-frequency exponentially increasing perturbations of more and more increasing frequency come into play. The main achievement of the above analysis is the construction of the analytical solutions (18.32), (18.33), (18.44)–(18.46), and also Tables 18.1, 18.2, which can be used for evaluation of the parameters of thermal–hydraulic instability of supercritical coolants in long channels.

References

1. Pioro IL, Duffey RB (2007) Heat transfer and hydraulic resistance at supercritical pressures in power engineering applications. ASME, New York
2. Daney DE (1979) An experimental study of thermal-induced flow oscillations in supercritical helium. ASME J Heat Transf 101:9–14
3. Wang Q, Kim K, Park H et al (2004) Heating surge and temperature oscillation in KSTAR PF and TF coils for plasma disruption under continuous plasma discharging conditions. IEEE Trans Appl Supercond 14:1451–1454
4. Sung SW, Lee J, Lee I-B (2009). Process identification and PID control. Wiley-IEEE Press
5. Nayfeh AH (2000) Perturbation methods. Wiley
6. Landau LD, Lifshitz EM (1987) Fluid mechanics. 2nd edn, Pergamon Press
7. Kull H-J (1991) Theory of the Rayleigh-Taylor instability. North-Holland
8. Schlichting H (1974) Boundary layer theory. McGraw-Hill, New York
9. Ackermann J (2010) Robuste regelung. Springer, Berlin
10. Labuntsov DA, Mirzoyan PA (1983) Analysis of boundaries of stability of motion of helium at supercritical parameters in heated channel. Therm Eng 30(3):121–123
11. Labuntsov DA, Mirzoyan PA (1986) Stability of flow of helium at supercritical pressure with non-uniform distribution of heat flux over the length of a channel. Therm Eng 33(4):208–211
12. Jones MC, Peterson RG (1975) A Study of Flow Stability in Helium Cooling Systems. J Heat Transf 97(4):521–527
13. Heck A (2003) Introduction to maple. Springer, New York
14. Delhaye JM, Giot M, Riethmuller ML (1981) Thermohydraulics of two-phase systems for industrial design and nuclear engineering. McGraw-Hill Book Company
15. Zudin YB (1998) Calculation of the thermal effect of the wall on the thermohydraulic stability of a flow of liquid of supercritical parameters. High Temp 36:239–243
16. Zudin YB (1998) The stability of a flow of liquid of supercritical parameters with respect to density-waves. High Temp 36:975–978
17. Zudin YB (2000) A possible scenario for the development of thermohydraulic instability. High Temp 38:156–157
18. Zudin YB (2016) Theory of periodic conjugate heat transfer. 3rd edn. Springer

Chapter 19
Heat Transfer in a Pebble Bed

Abbreviations

BC	Boundary condition
HTC	Heat transfer coefficient
PB	Pebble bed
TTC	Turbulent thermal conductivity

Symbols

d	Pebble diameter
h	Heat transfer coefficient (HTC)
k_t	Turbulent thermal conductivity

Greek Letter Symbols

η	Similarity variable
ϑ	Temperature difference

Subscripts

W	State at wall
∞	State at flow core
δ	State at boundary

19.1 Introduction

Numerous studies were devoted to the investigation of hydrodynamics and heat transfer in a close-packed fixed layer of pebbles (pebble bed), the results of such studies were generalized, in particular, in the monographs [1–5]. In analyzing the flow in the space between pebbles, which exhibits a complex three-dimensional pattern, it is as a rule assumed that the heat transfer occurs owing to the mixing of

Y. B. Zudin, *Non-equilibrium Evaporation and Condensation Processes*,
Mathematical Engineering, https://doi.org/10.1007/978-3-030-67553-0_19

differently directed jets of liquid (similarly to the mechanism of turbulent transfer in jet flows). Note, however, that the term "turbulent", which is generally employed in describing flows of the class under consideration, must not be understood in a literal sense. Under conditions of filtration of a medium through a pebble bed (PB), the values of the Reynolds number constructed by the pebble diameter d do not as a rule exceed 1000, therefore, the flow almost always remains laminar.

The main objective of studying heat transfer in a PB is that of determining the turbulent thermal conductivity (TTC) k_t which is used in the Fourier law, $q = -k_t \partial T/\partial x$. Experimental investigations were performed for two Boundary Conditions (BC): T_w = const [6, 7] and q_w = const [8]. From the physical standpoint, it is clear that the manner of heat delivery must not affect the thermohydrodynamics of flow in the bed located in a channel [9]. However, it is often difficult in the experiments to provide for the constant temperature conditions throughout the length of the experimental section [10]. In view of the significant intensity of heat transfer at high velocities of flow, the radial temperature profiles in the case of round pipe flow turn out to be very "flat", which makes difficult the processing of experimental data. By locating a PB in the annular gap between two cylinders [11], one can provide for an adequate "steepness" of temperature profile and, thereby, reduce the error of processing of measurement results.

Previous investigations were largely performed for the case of flow of air in a PB. In so doing, the marked difference between the thermal conductivities of the moving phase and of the "skeleton" of the PB resulted in significant difficulties in the processing of experimental results [12]. Free of such disadvantages is the study of Dekhtyar et al. [13], who investigated the flow of water in a bed of glass pebbles (with a thermal conductivity close to that of water) for the BC q_w = const. The use of two types of geometry of the working section (round pipe and annular channel) made it possible to compare the obtained dependences of TTC on the process parameters.

19.2 Experimental Facility

Previous studies [14, 15] involved experimental investigations of hydrodynamic drag under conditions of flow of water and steam–water mixture in a PB for wide ranges of variation of the process parameters, namely, pressure from 0.9 to 15.6 MPa, mass velocity from 107 to 770 kg/(m^{2}s), and steam quality from zero to 0.49. Polished pebbles of stainless steel with an average diameter of 2.12 mm were used as the PB. In [16], the theoretical model of flow of a two-phase mixture in a PB, which was used for generalizing the experimental data of [14, 15], is described.

The present chapter, which is a further development of [14, 15], gives the results of experimental investigation of heat transfer under conditions of longitudinal flow of water and steam–water mixture in a PB (calibrated glass pebbles 2 mm in diameter) past a flat heated wall. The experiments involved measurements of the temperature of the heated wall (in four cross-sections throughout the bed height), as well as the temperature distribution over the cross section of the channel at the outlet from the PB.

The experimentally obtained temperature profiles were processed using the mathematical model of the process by numerical optimization techniques. The processing was performed in view of the "two-layer" structure of the process, namely, the wall region (with a width of the order of pebble diameter d) with linear temperature profile was mated with the central part (core) of the bed (this central part was characterized by a constant rate of filtration). As a result, the values of the TTC in the PB were obtained as a function of the rate of filtration and heat flux. Note that the previous experiments were performed, as a rule, for the conditions of temperature stabilization of flow. Therefore, it is of interest to investigate the flow in the thermal initial segment of the channel. And, finally, the heat transfer was hardly studied for the case of the flow of liquid in a PB under conditions of wall boiling.

The experimental setup is shown schematically in Fig. 19.1. The working section was a rectangular 40-by-64 mm channel of height 370 mm. The channel wall made of stainless steel was 1.5 mm thick. The close-packed bed of glass pebbles of diameter

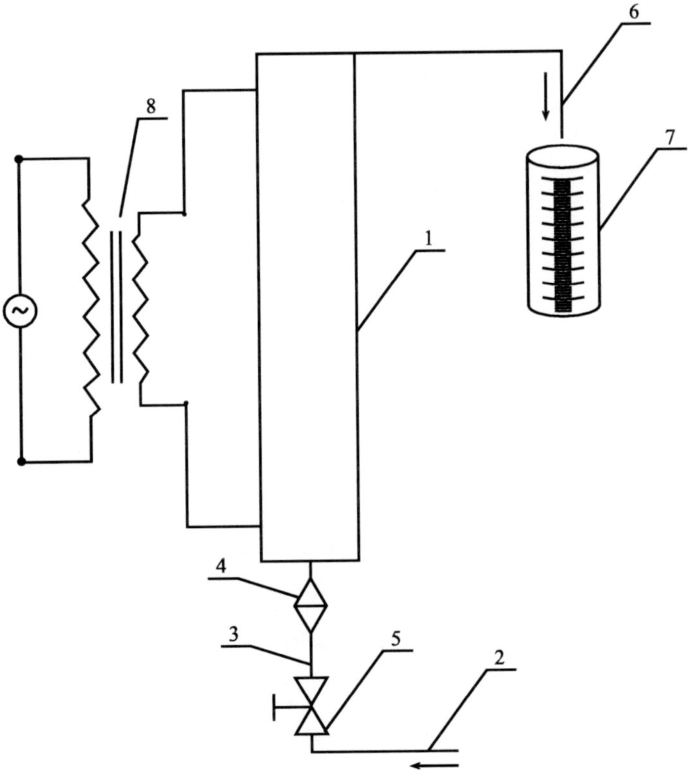

Fig. 19.1 Experimental setup. 1. Working section, 2. water delivery line, 3. collector, 4. filter, 5. control valve, 6. water removal line, 7. measuring tank, 8. transformer with variable output voltage

$d = 2$ mm was fixed by two grates (top and bottom) of steel gauze with a cell size of 1 mm by 1 mm.

The BC q_w = const was maintained by electric heating. The heater was provided by a stainless steel ribbon 0.3 mm thick, 30 mm long, and with a heated height of 303 mm, attached to the external surface of one of the walls 40 mm long.

The outer insulation of the working section was provided by a layer of kaolin wool 30 mm thick. The temperature distribution over the height of the heated wall was measured using five Chromel Copel cable thermocouples located at distances of 55, 115, 165, 215, and 265 mm from the beginning of the section being heated.

The hot junctions of the thermocouples were welded to the wall at the vertexes of horizontal triangular recesses at a distance of 1 mm from the inner surface of the wall. For ensuring a uniform distribution of the flow of water over the channel cross section, the heating of the wall was begun after water passed the 40 mm long region of hydrodynamic stabilization.

The water flow rate through the PB was determined from the measurements of the volume of water leaving the working section during the preassigned time. A mercury thermometer was used for measuring the temperature of water at the inlet and outlet of the working section.

The hot junctions of eight cable thermocouples were introduced via top grate into the bed to a depth of 5 mm, these thermocouples were installed along the longitudinal axis of the working section at distances of 2, 4.5, 8, 10.5, 15, 20, 30, and 39 mm from the surface being heated.

The cross-sectional average porosity of the PB was determined by the volumetric method in individual experiments and was $m = 0.375$.

19.3 Measurement Results

The basic results of the present chapter are the experimental data on the distribution of temperature of the heated wall T_w throughout the height of the bed, as well as of the water temperature over the depth of the PB. These data were obtained for different values of the velocity u and of heat fluxes q from the heated wall.

Figure 19.2 gives the values of the heated wall–flow core temperature differences ($\vartheta_w = T_w - T_\infty$) measured in four cross-sections over the bed height, where y is the longitudinal coordinate. In so doing, the inlet temperature of the liquid was taken to be T_∞. One can see in Fig. 19.2a, b that, for a fixed heat flux, the wall temperature in each cross section decreases with increasing velocity of liquid. This result is physically obvious: the intensity of convective heat transfer must increase with velocity. Less clear at first sight is the reason for the very gently sloping dependence of the wall temperature on the longitudinal coordinate. For the maximal value of velocity $u =$ 52.1 mm/s, the first three values of temperature difference lie on the horizontal shelf. It follows from Fig. 19.2c, d that, for fixed velocity, the wall temperature in each cross section decreases with the heat flux, and this is physically obvious. However,

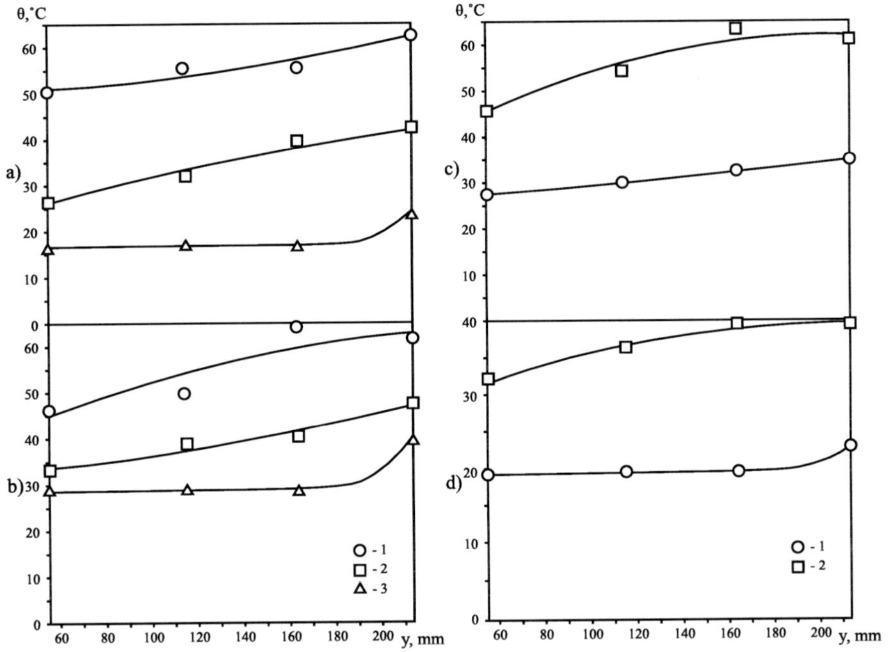

Fig. 19.2 Wall–flow core temperature difference versus the bed height. **a** $q = 45$ kW/m^2, 1. $u = 6.31$ mm/s, 2. $u = 19.5$ mm/s, 3. $u = 52.1$ mm/s. **b** $q = 86$ kW/m^2, 1. $u = 12.9$ mm/s, 2. $u = 30.7$ mm/s, 3. $u = 52.1$ mm/s. **c** $u = 12.8$ mm/s, 1. $q = 45$ kW/m^2, 2. $q = 86$ kW/m^2. **d** $u = 52$ mm/s, 1. $q = 45$ kW/m^2, 2. $q = 86$ kW/m^2

the shelf distribution of temperature difference is observed for the region of low heat fluxes and high velocities as well (Fig. 19.2d).

Figure 19.3 gives the distributions of the heat transfer coefficient over the heated wall height, obtained for the same heat fluxes and different velocities of flow.

Here and below, u is the velocity of filtration of flow (volumetric flow rate of liquid related to the cross section of the channel without the bed). One can see in the figures that the HTC increases with velocity for each given heat flux, this confirms the convective pattern of heat transfer. In so doing, the intensity of heat transfer along the wall subjected to flow must decrease because of the increasing thickness of the thermal boundary layer. This is demonstrated by curves 1 and 2 in Fig. 19.3a and by curve 1 in Fig. 19.319. However, some experimentally obtained dependences $h(y)$ exhibit a weakly defined tendency for an increase in the HTC over the height with its subsequent abrupt decrease (curve 3 in Fig. 19.3a and curves 2 and 3 in Fig. 19.3b).

One possible reason for the foregoing effects is the presence in the bed of a thin wall layer of significant thermal resistance which is approximately constant throughout the heated height. At the same time, a thermal boundary layer develops on the outer boundary of the wall layer, i.e., in the core of the bed, with the thermal resistance of this layer monotonically increasing from zero (the beginning of heating) to some

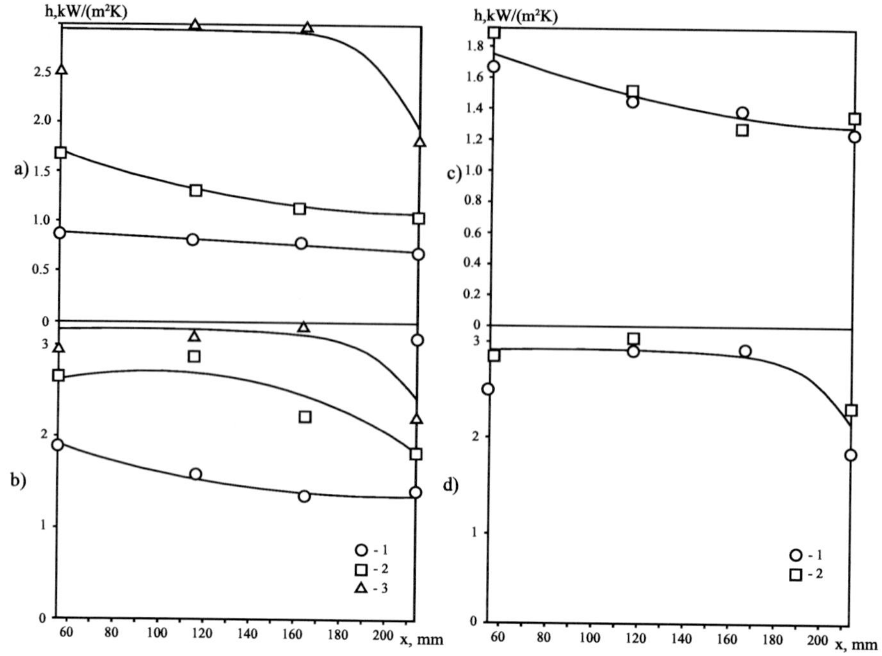

Fig. 19.3 Heat transfer coefficient versus the bed height. **a** $q = 45$ kW/m^2, 1. $u = 6.31$ mm/s, 2. $u = 19.5$ mm/s, 3. $u = 52.1$ mm/s. **b** $q = 86$ kW/m^2, 1. $u = 12.9$ mm/s, 2. $u = 30.7$ mm/s, 3. $u = 52.1$ mm/s. **c** $u = 12.8$ mm/s, 1. $q = 45$ kW/m^2, 2. $q = 86$ kW/m^2. **d** $u = 52$ mm/s, 1. $q = 45$ kW/m^2, 2. $q = 86$ kW/m^2 °C

fixed value at the outlet from the bed. With this two-layer pattern of heat transfer, the contribution of the wall zone to the overall thermal resistance will be prevailing throughout the significant height of the heated wall. This, in turn, must cause a gently sloping distribution of $h(y)$.

On the other hand, the thickening of the thermal boundary layer in height must result in that the importance of the flow core in overall heat transfer will become ever more pronounced on approaching the outlet from the bed. This would explain the drop of the HTC in the top part of the bed, which was mentioned above. As to the tendency for some increase in the HTC in height observed in some modes (though within the experimental error), this tendency may be caused in particular by the temperature dependence of viscosity of water. Note that the foregoing reasoning further gives a qualitative explanation of the very gently sloping distributions of temperature difference observed in Fig. 19.2. Figure 19.3c, d give the results of measurements of HTC, performed for the same velocities and for sharply differing heat fluxes. Analysis of the figures reveals the actual independence of intensity of heat transfer of heat flux, which is physically obvious.

Figure 19.4 gives the distributions of temperature differences over the bed depth in its outlet cross section (at a distance of 215 mm from the beginning of heating).

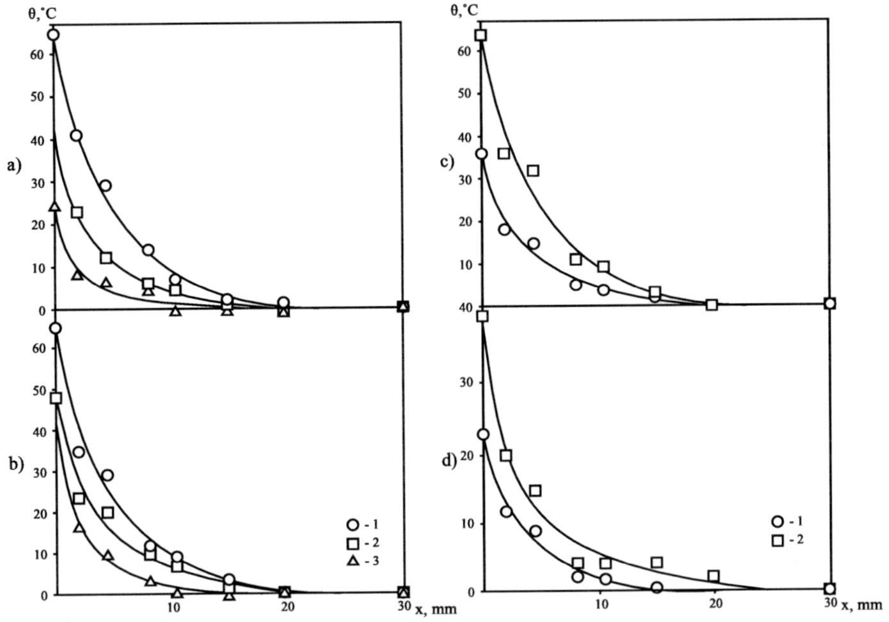

Fig. 19.4 Transverse distributions of the temperature difference in a pebble bed. **a** $q = 45$ kW/m^2, 1. $u = 6.31$ mm/s, 2. $u = 19.5$ mm/s, 3. $u = 52.1$ mm/s. **b** $q = 86$ kW/m^2, 1. $u = 12.9$ mm/s, 2. $u = 30.7$ mm/s, 3. $u = 52.1$ mm/s. **c** $u = 12.8$ mm/s, 1. $q = 45$ kW/m^2, 2. $q = 86$ kW/m^2. **d** $u = 52$ mm/s, 1. $q = 45$ kW/m^2, 2. $q = 86$ kW/m^2

Plotted on the ordinate is the temperature difference $\vartheta = T - T_\infty$. One can see that all $\vartheta(x)$ curves arrive at zero level in the case of high values of transverse coordinate. This supports the estimate made prior to the experiment that the of the boundary layer at the height at which the temperature profile is measured does not reach the distance of 40 mm where the last (most removed from the wall) thermocouple is located.

Experiments involving the flow of water in a PB for the case of wall boiling were performed in the following range of process parameters: $u = (2–50)$ mm/s, $q = (27–86)$ kW/m^2, and $T_\infty = (14–68)$ °C. In so doing, the temperature of the heat transfer surface was $T_w = (103–120)$ °C. The measured temperature distributions over the bed depth are given in Fig. 19.5. With the heat flux $q = 52$ kW/m^2, the temperature profiles for different velocities have the form typical of a single-phase medium (Fig. 19.5a). The same pattern is observed at high values of heat flux and velocity (Fig. 19.5b). However, a clearly defined transition to boiling at a distance of up to 5 mm from the heat transfer surface occurs when the velocity is reduced to $u = 3.3$ mm/s: the constancy of temperature $T = 100$ °C is observed up to the third thermocouple. It follows from Fig. 19.5c that, at low velocities and heat fluxes, the temperature is not registered even by the first (from the wall) thermocouple, so that the temperature profiles exhibit a typical "single-phase" shape. However, when the heat flux increases, a gently sloping segment of distribution of temperature $T =$

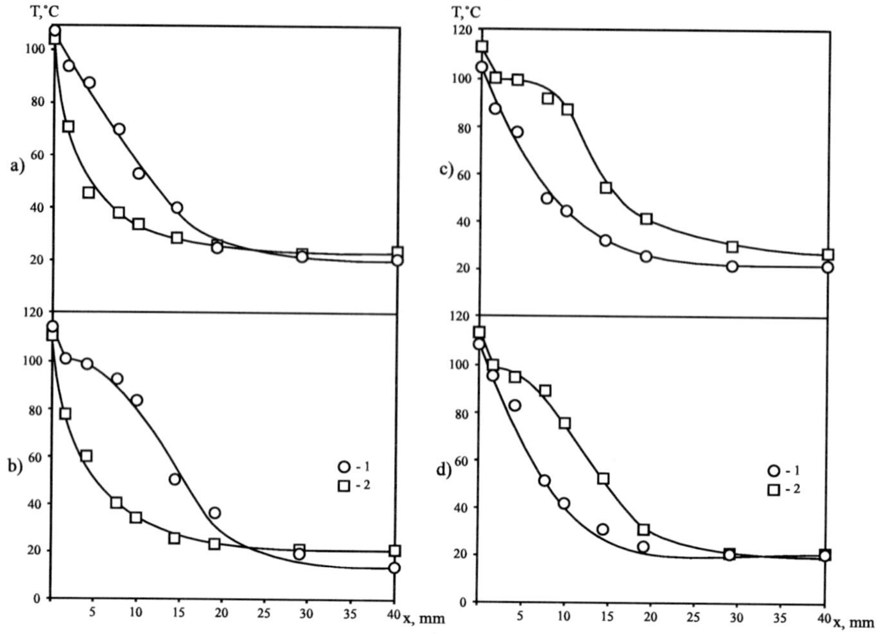

Fig. 19.5 Transverse distributions of temperature in a pebble bed in boiling. **a** $q = 52$ kW/m^2, 1. $u = 3.3$ mm/s, 2. $u = 7.8$ mm/s. **b** $q = 86$ kW/m^2, 1. $u = 3.3$ mm/s, 2. $u = 12.8$ mm/s. **c** $u = 2$ mm/s, 1. $q = 27$ kW/m^2, 2. $q = 53$ kW/m^2. **d** $u = 4.8$ mm/s, 1. $q = 53$ kW/m^2, 2. $q = 83.1$ kW/m^2

100 °C reappears, which is indicative of boiling. A similar tendency is observed for other values of process parameters (higher values of u and q, see Fig. 19.5d).

19.4 Processing of the Results

The data obtained for temperature distributions may be used for solving a number of problems in thermohydrodynamics of single-phase and two-phase flows in a PB such as the calculation of the TTC for single-phase flow, the calculation of the HTC to the wall for single-phase and two-phase flows, and the construction of models of wall boiling. Within the framework of the present chapter, we restrict ourselves to the first problem. It follows from geometric considerations [1–3] that, on approaching a flat wall, the value of PB porosity must abruptly increase on a scale of the order of d. The distribution $m(y)$ maybe described by the empirical formula [17]

$$m = m_{\infty}\left[1 + 1.36\exp\left(-5\frac{x}{R}\right)\right]. \tag{19.1}$$

Here, R is the pipe radius, x is the transverse coordinate reckoned from the wall.

According to Eq. (19.1), the value of m on the wall exceeds the respective value in the flow core (uniform cellular structure) by a factor of 2.36. The abrupt increase in porosity must cause a significant transformation of the thermohydrodynamic pattern on approaching the wall. Therefore, in performing theoretical analysis, one usually proceeds from the two-layer pattern of PB [18–20], i.e., the presence of uniform transverse distributions of porosity (and, as a consequence, of flow velocity) in the central part of the flow and the presence of peaks of velocity and temperature gradients in the wall region. It is the objective of the present chapter to determine the TTC k_t which is used in the Fourier law

$$q = -k_t \frac{\partial T}{\partial x}.$$

The analogy with free turbulent flow, which was mentioned above, leads one to assume the following tendencies of the dependence of k_t: $k_t \sim u$, $k_t \sim d$. Hence follows, in view of dimensional considerations, the equation for TTC

$$k_t = b\rho c u d. \tag{19.2}$$

Here, c and ρ denote the specific heat capacity and density of the medium, respectively, b is a numerical constant.

Note that, in experiments in air filtration, the molecular component reflective of the contribution by the thermal conductivity of the bed skeleton must be taken into account in addition to the turbulent component of thermal conductivity [10]. In the previous studies, this molecular component was separately determined in the experiments, based on the results of measurements in a gas-filled packed bed in the absence of filtration, and then used in the form of an addition to k_t in formula (19.1). It is clear that the error of determination of k_t in this case could be very significant. In view of this, the combination of continuous (water) and disperse (glass pebbles) media with close values of thermal conductivity, which was employed in our experiments, appears to be optimal from the standpoint of attaining the thermal uniformity of the bed.

In analyzing the flow in the core of PB, the real flow (three-dimensional jet flow in the space between the pebbles) is replaced by a homogeneous medium with a fictitious (related to the total cross section of the bed) velocity of filtration u [4–6] (Fig. 19.6). Given the validity of conditions $u = \text{const}$ and $k_t = \text{const}$, the energy equation for flow in the bed core is written in the form of unsteady-state heat equation

$$c\rho \frac{\partial \vartheta}{\partial t} = k_t \frac{\partial^2 \vartheta}{\partial x^2}. \tag{19.3}$$

In accordance with the results of the analysis made by the authors of [18–20], we will assume that the heat flux is transferred via thermally thin wall layer without distortions: $q_w = q_\delta = \text{const}$. As a result, we have the problem for heat Eq. (19.3)

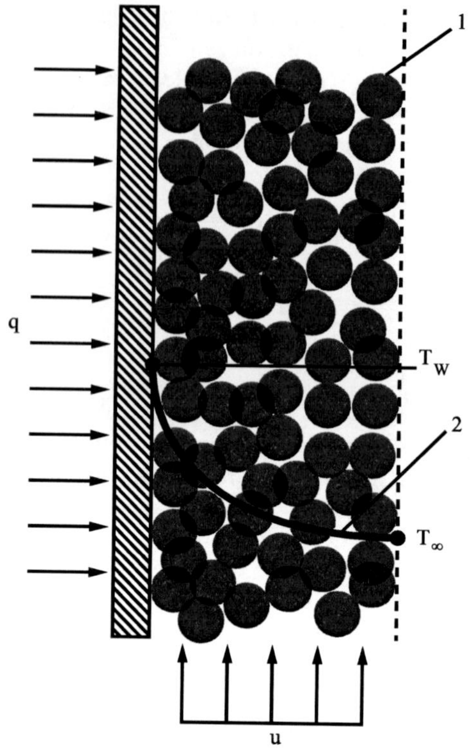

Fig. 19.6 Scheme of the process. 1. Pebble bed, 2. temperature profile

with the BC $q = \text{const}$ at $x = \delta$ the method of whose solution is well known [21]. We introduce the similarity variable

$$\eta = \frac{1}{2}(x - \delta)\sqrt{\frac{c\rho}{k_t t}},$$

where δ is the boundary of the wall region of thickness of the order of pebble diameter [4, 5]. The solution of Eq. (19.3) is sought in the form

$$\vartheta = \vartheta_\delta(t)\theta(\eta).$$

The temperature difference $\vartheta_\delta(t)$ on the boundary $x = \delta$ is defined by the formula

$$\vartheta_\delta = \frac{2}{\sqrt{\pi}} q_\delta \sqrt{\frac{t}{k_t c\rho}} = \frac{2}{\sqrt{\pi}} \frac{q_\delta}{c\rho u} \sqrt{\frac{y}{d}}. \tag{19.4}$$

Here, $\vartheta = T - T_\infty$, $t = y/u$ is the time of motion of liquid particle from the inlet to the bed to the assigned value of longitudinal coordinate y. The function $\theta(\eta)$

satisfies the equation following from Eq. (19.3)

$$2\left(\theta - \eta\frac{d\theta}{d\eta}\right) = \frac{d^2\theta}{d\eta^2}. \tag{19.5}$$

The solution of Eq. (19.5) will be

$$\theta = \exp(-\eta^2) - \sqrt{\pi}\eta\,\mathrm{erfc}(\eta), \tag{19.6}$$

where erfc(η) is complementary error function [21].

A noteworthy feature of the case under consideration of heat transfer in the thermal initial segment is the possibility of calculating the transverse distribution of heat flux by the measured temperature profile. For this purpose, we rewrite Eq. (19.5) in the form

$$2\left(\theta - \eta\frac{d\theta}{d\eta}\right) = -\frac{df}{d\eta}, \tag{19.7}$$

where $f = -d\theta/d\eta$. We integrate Eq. (19.7) with respect to η with BC $\eta \to \infty$: $\theta = f = 0$, to derive

$$f = 2\theta\eta + 4\int_{\eta}^{\infty}\theta d\eta.$$

Or, in the dimensions form

$$q = \frac{\rho c u}{H}\left(\frac{1}{2}\vartheta x + \int_{x}^{\infty}\vartheta dx\right). \tag{19.8}$$

Here, $H = 215$ mm is the height of the segment being heated. Strictly speaking, relation (19.8) is valid only for the bed core ($x > \delta$). However, with $q_w = q_\delta$, this relation may be approximately extended to be wall region as well. Then, at $x = 0$, formula (19.8) transforms to the equation of heat balance over the height of the working section

$$q_w = \frac{\rho c u}{H}\int_{0}^{\infty}\vartheta dx. \tag{19.9}$$

Figure 19.7 gives the transverse profiles of heat flux, calculated by relation (19.9) and by the measured temperature profiles. The results are given in dimensionless form, $\tilde{q} = q/q_w$. Note that, in accordance with Fig. 19.7, the distribution of heat flux in the neighborhood of the wall does not assume a horizontal pattern, as would have followed from the assumption that $q_w = q_\delta$. For example, at distance x $= d/2 =$

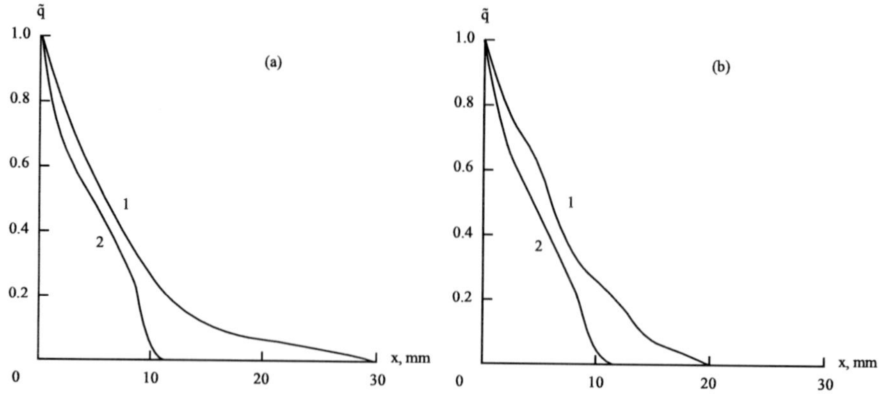

Fig. 19.7 Transverse distribution of the heat flux in a pebble bed. **a** $q = 45$ kW/m^2, 1. $u =$ 6.31 mm/s, 2. $u = 52.1$ mm/s. **b** $q = 86$ kW/m^2, 1. $u = 12.9$ mm/s, 2. $u = 52.1$ mm/s

1 mm from the wall, we have $\tilde{q} \approx 0.9$ for $u = (6.3-12.9)$ mm/s and $\tilde{q} \approx 0.84$ for $u = 52.1$ mm/s. Therefore, the pattern of variation of parameters in the wall region is apparently more complex than it was assumed in the models of [18–20].

We will now turn back to the experimental data on the distribution of temperature differences over the height of the working section. It follows from Eq. (19.4) and from the assumption of linearity of temperature profile in the wall zone that $\vartheta_w \sim \vartheta_\delta \sim \sqrt{y}$. However, one can see in Fig. 19.2 that the dependence $\vartheta_w(y)$ is much weaker and that, for the maximal value of velocity $u = 52.1$ mm/s, the first three values of temperature difference (Fig. 19.2a, b, d) actually lie on the horizontal. The qualitative explanation of these tendencies on the basis of "two-layer" pattern of flow in the bed was given above in analyzing the measurement results.

The experimentally obtained distributions $\theta(\eta)$ were approximated by dependences (19.4) and (19.6) with two free parameters (the temperature on the boundary of the flow core and the TTC) and then processed by numerical optimization methods. The sought value of TTC for each experiment was obtained as a result of the minimization of mean-square deviations between the experimental and calculation data. A certain indeterminacy of the employed calculation procedure consisted of preassigning the concrete thickness of the wall zone. As a result of multivariate calculations, it was found that an increase in the wall zone thickness in the range $\delta = (1/3-2/3)d$ causes a decrease in the TTC by $\approx$8%. In the final version, it was assumed that $\delta = d/2$

$$\frac{k_t}{k} = b\text{Pe}. \tag{19.10}$$

The obtained results are described by the formula (Fig. 19.8). Here, $b = 0.0818$

$$\text{Pe} = \frac{ud}{\alpha}$$

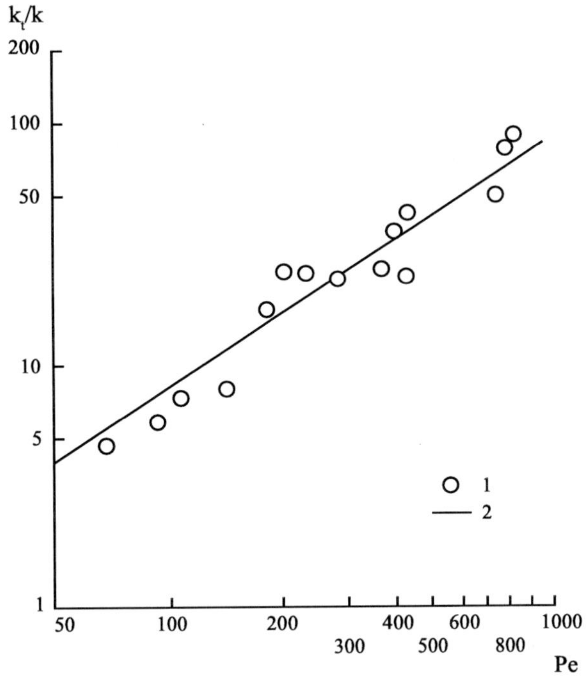

Fig. 19.8 Dimensionless pseudo-turbulent thermal conductivity as a function of the Peclet number. 1. Our experimental data, 2. calculated from (19.10)

is the Peclet number, and k is the thermal conductivity of liquid. It is interesting to note that dependence (19.10) almost coincides with that suggested by Dekhtyar et al. [13] ($b = 0.083$).

We will now consider the important question of the possible effect of the viscosity of liquid on the TTC. With the porosity $m = 0.375$ and velocities of filtration $u = (6.31-52.1)$ mm/s, the values of the true velocity of flow of liquid in the space between the pebbles will be $U = u/m = (16.8-140)$ mm/s. The Reynolds number, which is constructed on the pebble diameter with the viscosity of water in the bed core ($T_\infty \approx 20\,°C$) equal to $\nu \approx 1$ mm²/s, varies in the range Re $= Ud/\nu \approx 33.6$–280, this pointing to the purely laminar pattern of flow.

The characteristic time of viscous relaxation of velocity disturbances may be estimated at $t_\nu \approx d^2/\nu \approx 4$s. The characteristic time of inertial transport of disturbances is $t_U \approx d/U \approx (0.0144$–$0.12)$ s. Then the ratio between these times will be $t_\nu/t_U \approx$ (30–300). Therefore, the smoothing of disturbances owing to the effect of viscosity will proceed at a rate that is one or two orders of magnitude slower than their inertial transport. In view of this, in spite of the clearly laminar pattern of flow, the effect of viscosity on convective heat transfer in the investigated range of parameters must be negligibly small, as is confirmed by the turbulent structure of formula (19.2).

19.5 Conclusions

An experimental investigation was performed of turbulent heat transfer under conditions of the flow of water in a pebble bed of glass located in a channel of rectangular cross section and consisting of glass pebbles 2 mm in diameter. The experiments involved measurements of the temperature of the heated wall, as well as of the temperature distribution over the channel cross section at the outlet from the pebble bed. Use was made of a method of processing of experimental data, which enables one to determine the turbulent thermal conductivity without differentiation of the experimentally obtained temperature profile. The solution of unsteady-state heat equation, obtained for the conditions of thermal initial segment, was used for this purpose. The experimental data for single-phase flow were described using the mathematical model of the process with two free parameters (the temperature on the boundary of the flow core and the turbulent thermal conductivity) and then processed by numerical optimization methods. Temperature profiles were obtained for the case of boiling on the pebble bed wall, and qualitative analysis of these profiles was performed. The material from this Appendix was published in [22].

References

1. Aerov MA, Todes OM (1968) The hydraulic and thermal principles of operation of apparatuses with stationary and fluidized packed bed. Khimiya, Leningrad (in Russian)
2. Goldshtik MA (1984) Transfer processes in a packed bed. Izd. SO AN SSSR Siberian Div., USSR Acad. Sci, Novosibirsk (in Russian)
3. Bogoyavlenskii RG (1978) Hydrodynamics and heat transfer in high-temperature nuclear reactors with spherical fuel elements. Atomizdat, Moscow (in Russian)
4. Tsotsas E (1990) Über die Wärme- und Stoffübertragung in durchströmten Festbetten, VDI-Fortschrittsberichte. Reihe 3/223. VDI-Verlag, Düsseldorf
5. Bey O, Eigenberger G (1998) Strömungsverteilung und Wärmetransport in Schüttungen. VDI-Fortschrittsberichte. Reihe 3/570. VDI-Verlag, Düsseldorf
6. Ziolkowski D, Legawiec B (1987) Remarks upon thermokinetic parameter. Chem Eng Process 21:64–76
7. Freiwald MG, Paterson WR (1992) Accuracy of model predictions and reliability of experimental data for heat transfer in packed beds. Chem Eng Sci 47:1545–1560
8. Nilles M (1991) Wärmeubertragung an der Wand durchströmter Schüttungsrohre. VDI-Fortschrittsberichte Reihe 3/264. VDI-Verlag, Düsseldorf
9. Martin H, Nilles M (1993) Radiale Wärmeleitung in durchströmten Schüttungsrohren. Chem Ing Tech 65:1468–1477
10. Bauer R, Schlünder EU (1977) Die effektive Wärmeleitfahigkeit gasdurchströmter Schüttungen. Verfahrenstechnik 11:605–614
11. Dixon AG, Melanson MM (1985) Solid conduction in low dt/dp beds of spheres, pellets and rings. Int J Heat Mass Transf 28:383–394
12. Bauer M (2001) Theoretische und experimentelle Untersuchungen zum Wärmetransport in gasdurchströmten Festbettrohrreaktoren. Dissertation, Universität Halle-Wittenberg
13. Dekhtyar RA, Sikovsky DP, Gorine AV, Mukhin VA (2002) Heat transfer in a packed bed at moderate values of the Reynolds number. High Temp 40(5):693–700
14. Avdeev AA, Balunov BF, Zudin YB, Rybin RA, Soziev RI (2006) Hydrodynamic drag of a flow of steam-water mixture in a pebble bed. High Temp 44(2):259–267

15. Avdeev AA, Balunov BF, Rybin RA, Soziev RI, Zudin YB (2007) Characteristics of the hydrodynamic coefficient for flow of a steam-water mixture in a pebble bed. ASME J Heat Transf 129:1291–1294
16. Avdeev AA, Soziev RI (2008) Hydrodynamic drag of a flow of steam-water mixture in a pebble bed. High Temp 46(2):223–228
17. Vortmeyer D, Haidegger E (1991) Discrimination of three approaches to evaluate heat fluxes for wall-cooled fixed bed chemical reactors. Chem Eng Sci 46:2651–2660
18. Schlünder EU, Tsotsas E (1988) Wärmeubergang in Festbetten, durchmischten Schüttungen und Wirbelschichten. Georg Thieme Verlag, Stuttgart-New York
19. (1997) VDI – Wärmeatlas, Abschnitt Mh. Wärmeleitung und Dispersion in durchströmten Schüttungen. Springer, Berlin, Heidelberg
20. Dixon AG (1988) Wall and particle-shape effects on heat transfer in packed beds. Chem Eng Commun 71:217–237
21. Carslaw HS, Jaeger JC (1988) Conduction of heat in solids, 2nd edn. Clarendon Press, Oxford
22. Avdeev AA, Balunov BF, Zudin YB, Rybin RA (2009) An experimental investigation of heat transfer in a pebble bed. High Temp 47:692–700

Appendix A
Film Boiling Heat Transfer

Abbreviations

BC Boundary condition

Symbols

c_p Isobaric heat capacity
k Heat conductivity
L Heat of phase transition
q Heat flux
S Stefan number
u Longitudinal velocity
v Transverse velocity
x Longitudinal coordinate
y Transverse coordinate

Greek letter symbols

δ Film thickness
ρ Density
ϑ Temperature difference

Subscripts

$*$ Dimensionless variable
eff Effective value
s Saturation state
w State at infinity

The heat energy during film boiling is known to be transferred from a hot hard surface to the saturated liquid through the wall-adjacent vapor film. Under the conventional approach to the transfer calculation, it is assumed that the heat transport across the laminar vapor film is effected by the mechanism of heat conduction

Y. B. Zudin, *Non-equilibrium Evaporation and Condensation Processes*, Mathematical Engineering, https://doi.org/10.1007/978-3-030-67553-0

$$h = \frac{k}{\delta}. \tag{A.1}$$

The costs of heat energy for vapor superheating in the film are taken into account by introducing the effective heats of phase transition

$$L_{ef} = L + \frac{1}{2} c_p \vartheta_w. \tag{A.2}$$

Here, k, c_p are, respectively, the heat conductivity and the specific heat capacity at a constant pressure of vapor, L is the heat of phase transition, δ is the vapor film thickness, $\vartheta_w = T_w - T_s$, T_w is the surface temperature, and T_s is the saturation temperature.

It is worth pointing out that such an approach has no rigorous substantiation and is in essence semi-empirical. Below, we shall propose an approximate physical model of heat transfer during film boiling, which is capable of calculating the effective thermophysical properties of vapor. Evaporation of liquid results in the formation of a vapor flow, which is injected in the film through the interphase surface and then spreads over the hard hot surface. We shall assume that the interphase obeys the laws of laminar boundary layer. The differential equation of energy balance reads as

$$c_p \rho \frac{\partial}{\partial x}(u\vartheta) + c_p \rho \frac{\partial}{\partial y}(\mathrm{v}\vartheta) = -\frac{\partial q}{\partial y}. \tag{A.3}$$

Here, x, y are the longitudinal and transverse coordinates, u, v are the longitudinal and transverse velocities, ρ is the vapor density, $\vartheta = T - T_s$ is the temperature difference. Averaging both sides of Eq. (A.3) in the film thickness, we obtain

$$c_p \rho \int_0^\delta \frac{\partial}{\partial x}(u\vartheta) dy + c_p \rho \, (u\vartheta)|_0^\delta = q_w - q_s. \tag{A.4}$$

Here, q_w, q_s are, respectively, the heat flux on the hot surface and on the interphase surface. In the approximation of the boundary layer, one may write the equality

$$c_p \rho \int_0^\delta \frac{\partial}{\partial x}(u\vartheta) dy = c_p \rho \frac{d}{dx} \int_0^\delta u\vartheta dy - c_p \rho \, (u\vartheta)|_0^\delta \frac{d\delta}{dx}. \tag{A.5}$$

The Boundary Conditions (BC) for the energy Eq. (A.3) are set as follows:

$$\left. \begin{array}{l} y = 0 : u = 0, \\ y = \delta : u = 0. \end{array} \right\} \tag{A.6}$$

It implies that the last term on the right of Eq. (A.5) is zero. In view of Eqs. (A.5), (A.6), the integral equation (A.4) assumes the form

$$c_p\rho\frac{d}{dx}\int\limits_0^\delta u\vartheta dy = q_w - q_s. \tag{A.7}$$

Equation (A.7) has the transparent physical sense: the heat flux q_w, which is transferred from the hot surface, is spent on the evaporation of liquid (q_s) and on the superheat of vapor which is supplied to the film ($q_w - q_s$). As a result, we have $q_w > q_s$. It is worth noting that the heat transport through the vapor film is affected not only by the heat conductivity. The flow of the vapor injected in the film through the interphase surface results in the formation of the additional (convective) component of the heat transfer.

Labuntsov [1] was the first to consider the effect of the convection on the heat transfer during film boiling. Below we shall briefly outline the model of [1] with some modification. The energy balance relation on the interphase surface is as follows

$$q_s = L\rho\frac{d}{dx}\int\limits_0^\delta u dy. \tag{A.8}$$

Using Eq. (A.8) in Eq. (A.7), we find the specific form of the energy conservation law for the process of film boiling

$$q_w = L\rho\frac{d}{dx}\int\limits_0^\delta u\left(1+\frac{c_p\vartheta}{L}\right)dy. \tag{A.9}$$

Let us find the distributions of the temperature and heat flux across the vapor film. For this purpose, we shall use the following BC

$$\left.\begin{aligned} y = 0 &: q = q_w, \\ y = 0 &: \tfrac{\partial q}{\partial y} = 0, \\ y = \delta &: q = q_s. \end{aligned}\right\} \tag{A.10}$$

Taking into account the equation of continuity, the differential equation of energy balance (A.3) can be put in the form

$$c_p\rho u\frac{\partial\vartheta}{\partial x} = c_p\rho \mathrm{v}\frac{\partial\vartheta}{\partial y} = -\frac{\partial q}{\partial y}. \tag{A.11}$$

Another BC for $\partial q/\partial y$ is obtained from Eq. (A.11) with $y = \delta$

$$c_p \rho u_\delta \left(\frac{\partial \vartheta}{\partial x}\right)_\delta + c_p \rho v_\delta \left(\frac{\partial \vartheta}{\partial y}\right)_\delta = -\left(\frac{\partial q}{\partial y}\right)_\delta. \tag{A.12}$$

We shall assume that the temperature field is homogeneous in the longitudinal coordinate x. Hence, the total increment of ϑ along the film surface is zero

$$d\vartheta \equiv \left(\frac{\partial \vartheta}{\partial x}\right)_\delta dx + \left(\frac{\partial \vartheta}{\partial y}\right)_\delta d\delta = 0. \tag{A.13}$$

Using Eq. (A.13), this gives

$$\left(\frac{\partial \vartheta}{\partial x}\right)_\delta = -\left(\frac{\partial \vartheta}{\partial y}\right)_\delta \frac{d\delta}{dx}. \tag{A.14}$$

Taking into account the equality

$$-\left(\frac{\partial \vartheta}{\partial y}\right)_\delta = \frac{q_s}{k}, \tag{A.15}$$

the left-hand side of Eq. (A.8) assumes the form

$$-c_p \rho \left.\frac{\partial \vartheta}{\partial y}\right|_\delta \frac{d}{dx} \int\limits_0^\delta u dy. \tag{A.16}$$

From Eq. (A.16), we get the fourth BC for the heat flux

$$y = \delta : \frac{\partial q}{\partial y} = -\frac{q_s^2 c_p}{Lk}. \tag{A.17}$$

The heat flux distribution will be sought along the transverse coordinate in the form of the following polynomial

$$q = a_0 + a_1 y + a_2 y^2 + a_3 y^3. \tag{A.18}$$

The polynomial coefficients (A.18) are determined form the BC (A.10) and are written as follows

$$\left.\begin{aligned} a_0 &= q_w, \\ a_1 &= 0, \\ a_2 &= 3\tfrac{q_w - q_s}{\delta^2} + \tfrac{q_s^2 c_p}{Lk\delta}, \\ a_3 &= 2\tfrac{q_w - q_s}{\delta^3} - \tfrac{q_s^2 c_p}{Lk\delta^2}. \end{aligned}\right\} \tag{A.19}$$

As a result, we get the following heat flux distribution

$$q = q_w - (q_w - q_s)\left(3Y^2 - 2Y^3\right) + \frac{q_s^2 c_p \delta}{Lk}\left(Y^2 - Y^3\right). \tag{A.20}$$

where $Y = y/\delta$ is the dimensionless transverse coordinate.

Moreover, the BC for the energy equation (A.11) are rewritten as

$$\left.\begin{array}{l} Y = 0 : u = 0, \\ Y = 1 : u = 0. \end{array}\right\} \tag{A.21}$$

Integrating Eq. (A.20) and taking into account Fourier's law

$$q = -\frac{k}{\delta}\frac{\partial \vartheta}{\partial Y}, \tag{A.22}$$

we get the following temperature distribution in the vapor film

$$\frac{k}{\delta}(\vartheta_w - \vartheta) = q_w Y - (q_w - q_s)\left(Y^3 - \frac{1}{2}Y^4\right) + \frac{q_s^2 c_p \delta}{Lk}\left(\frac{1}{3}Y^3 - \frac{1}{4}Y^4\right). \tag{A.23}$$

Setting $Y = 1$ in (A.23) and employing the second BC (A.21), this establishes

$$\frac{k}{\delta}\vartheta_w = \frac{1}{2}(q_w + q_s) + \frac{1}{12}\frac{q_s^2 c_p \delta}{Lk}. \tag{A.24}$$

The commonly used relation (A.1) for heat transfer calculation can be rewritten as follows for the film boiling

$$\frac{k\vartheta_w}{\delta} = q_w. \tag{A.25}$$

A comparison of Eqs. (A.24) and (A.25) clearly shows that Eq. (A.25) is a great simplification of the mechanism of heat transfer and is incorrect in the general case. Below, we shall present a refined approach based on the actual physical process pattern. Consideration of the convective film boiling heat transfer can be made on the basis of the following simple relations:

$$q_w = \frac{k_{\text{ef}}\vartheta_w}{\delta}, \tag{A.26}$$

$$q_w = L_{\text{ef}}\rho\frac{d}{dx}\int_0^{\delta} u\,dy. \tag{A.27}$$

Here, k_{ef}, L_{ef} are the effective values of thermal conductivity and heat of phase transition. Let us introduce the dimensionless variables

$$k_* = \frac{k_{\text{ef}}}{k},\ L_* = \frac{L_{\text{ef}}}{L}.$$

They will be sought in the form of universal functions $k_*(\mathrm{S})$, $L_*(\mathrm{S})$. Here,

$$\mathrm{S} = \frac{c_p \vartheta_w}{L} \tag{A.28}$$

is the Stefan number defined as the ratio of the superheating enthalpy of the vapor to the phase transition enthalpy (both quantities are referred to as a unit mass).

The method for dealing with the vapor superheat and the influence of convection on the temperature field in a vapor film depends on Eqs. (A.26) and (A.27) with the use of the universal effective values of thermal conductivity and heat of phase transition. From Eqs. (A.27) and (A.9) it follows that

$$L_* = \frac{\int_0^\delta u\left(1 + c_p \vartheta / L\right) dy}{\int_0^\delta u dy}. \tag{A.29}$$

Equation (A.26) in view of Eqs. (A.9) and (A.27) assumes the form

$$\frac{1}{k_*} = \frac{1}{2}\frac{1 + L_*}{L_*} + \frac{1}{12}\frac{k_*}{L_*^2}\mathrm{S}. \tag{A.30}$$

With the available temperature and velocities of distributions in the vapor film, the system of equations (A.29), (A.30) determines the required dependences $k_*(\mathrm{S})$, $L_*(\mathrm{S})$. The temperature distributions in the film are described by relation (A.14). The longitudinal velocity distribution $u(Y)$ should be specified from the type of a problem under consideration. We first consider the parabolic distribution

$$u = 6uY(1 - Y), \tag{A.31}$$

where $u = \int_0^1 udY$ is the average velocity of the vapor film.

Using Eqs. (A.26), (A.27) in Eq. (A.29), this gives

$$L_* = 1 + \frac{1}{70}\mathrm{S}\left[44 - k_*\left(13 - \frac{4}{L_*}\right)\right]. \tag{A.32}$$

The system of equations (A.30), (A.32) determines the required dependences for problems with parabolic velocity distribution. For the linear velocity distribution

$$u = 2uY, \tag{A.33}$$

instead of (A.32) we have

$$L_* = 1 + \frac{S}{15}\left[6 - k_*\left(2 - \frac{1}{L_*}\right)\right]. \tag{A.34}$$

The system of equations (A.30), (A.34) describes the case of linear velocity distribution. Both systems of equations are reducible to cubic ones in k_*, L_*. However, the analytical solutions of these equations are very bulky and unfit for practical calculations. Hence it seems reasonable to search for appropriate approximations. We first consider the asymptotic behavior of the above solutions. For the parabolic velocity distribution (A.31), we have

$$\left.\begin{aligned} S \to 0 : L_* = 1 + \tfrac{1}{2}S,\, k_* = 1 + \tfrac{1}{6}S, \\ S \to \infty : L_* \to \tfrac{9}{35}S,\, k_* \to 2. \end{aligned}\right\} \tag{A.35}$$

For the linear velocity distribution (A.33), this gives

$$\left.\begin{aligned} S \to 0 : L_* = 1 + \tfrac{1}{3}S,\, k_* = 1 + \tfrac{1}{12}S, \\ S \to \infty : L_* \to \tfrac{2}{15}S,\, k_* \to 2. \end{aligned}\right\} \tag{A.36}$$

From Eqs. (A.35), (A.36) it follows that the dependences of the effective values of thermal conductivity and heat of phase transition of superheated vapor on the Stefan number are of principally different character. So, the value of k_{ef} changes only qualitatively: it doubles as S increases from zero to infinity (for both types of velocity distribution). For its part, the value of L_{ef} with S $\ll$ 1 varies according to the law

$$L_{ef} = L + k_1 c_p (T_w - T_s), \tag{A.37}$$

where $k_1 = 1/2$ for a parabolic distribution and $k_1 = 1/3$ for a linear distribution. However, for $S \gg 1$, the dependence described by Eq. (A.37) is qualitatively different

$$L_{ef} = k_2 c_p \vartheta_w. \tag{A.38}$$

Here, $k_2 = 9/35$ for a parabolic distribution and $k_2 = 2/15$ for a linear distribution. The asymptotic formula (A.38) has an interesting physical interpretation: for $c_p\vartheta_w \gg L$ the effective heat of phase transition ceases to depend on the reference value of L, as taken for the saturation temperature, but rather depends on the superheating enthalpy of the vapor

$$L_{ef} \sim c_p \vartheta_w.$$

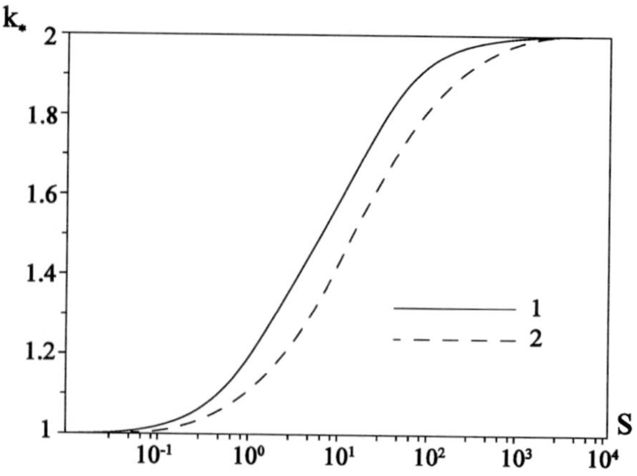

Fig. A.1 Efficient heat conductivity versus the Stefan number. 1 The parabolic profile, 2 the linear profile

The solutions thus obtained are approximated to error up to 1% by the following relations

for the parabolic velocity distribution

$$L_* = \left(1 + 0.765\mathrm{S} + 9.66 * 10^{-2}\mathrm{S}^2 + 1.54 * 10^{-3}\mathrm{S}^3\right)\left(1 + 0.266\mathrm{S} + 6 * 10^{-3}\mathrm{S}^2\right)^{-1},$$

$$k_* = \left(1 + 0.576\mathrm{S} + 3.4 * 10^{-2}\mathrm{S}^2\right)\left(1 + 0.409\mathrm{S} + 1.7 * 10^{-2}\mathrm{S}^2\right)^{-1}.$$

fFor the linear velocity distribution

$$L_* = \left(1 + 0.55\mathrm{S} + 4.9 * 10^{-2}\mathrm{S}^2 + 3.4 * 10^{-4}\mathrm{S}^3\right)^{-1},$$

$$k_* = \left(1 + 0.247\mathrm{S} + 3.6 * 10^{-3}\mathrm{S}^2\right)\left(1 + 0.164\mathrm{S} + 1.8 * 10^{-3}\mathrm{S}^2\right)^{-1}.$$

These equations culminate the above analysis. They are capable of taking into account the effects of the vapor superheat in a film and the influence of convection on the effective values of thermal conductivity and the heat of phase transition of superheated vapor. The corresponding calculated curves are shown in Figs. A.1, A.2.

It is worth pointing out that in the problems of film boiling the case $\mathrm{S} \gg 1$ corresponds to anomalous superheats of the vapor film and may not be realized in practice. As a rule, the Stefan number varies in the range $0 < \mathrm{S} \leq 1$. To conclude, we note that from the known effective values of thermal conductivity and heat of phase transition one may calculate the temperature distribution across the film. Using Eq. (A.23), we obtain

$$\frac{\vartheta}{\vartheta_w} = \frac{T - T_s}{T_w - T_s} = \left(1 - 4Y^3 + 3Y^4\right) - k_*\left(Y - 3Y^3 + 2Y^4\right) + \frac{k_*}{L_*}\left(Y^3 - Y^4\right).$$

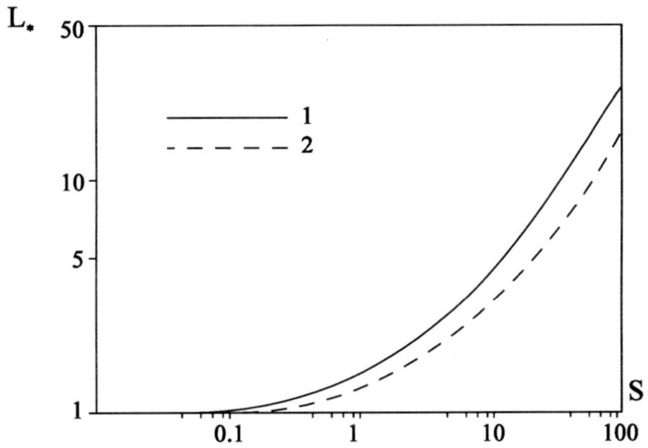

Fig. A.2 Efficient heat of phase transition versus the Stefan number. 1 The parabolic profile, 2 the linear profile

Appendix B
Critical Heat Flux and Landau Instability

Abbreviations

CHF	Critical heat flux

Symbols

g_*	Accelerations of the body forces
H	Pitch of twisted tape
j	Mass flow of evaporation
q	Heat flux
t	Time

Greek Letter Symbols

ξ	Amplitude of perturbations
ω	Increment of perturbations
Ω	Vorticity
$\Re e$	Real part
ρ	Density
σ	Surface tension

Subscripts

1	Liquid
2	Vapor
c	Critical
max	Maximal
W	State at wall

Y. B. Zudin, *Non-equilibrium Evaporation and Condensation Processes*, Mathematical Engineering, https://doi.org/10.1007/978-3-030-67553-0

Landau Instability

Problems of Hydrodynamic Instability

In this section, we consider the physical relation of the model of pool boiling crisis and the development of instability for evaporation from the interphase boundary. In 1944, Landau [2] formulated and solved the following three problems of hydrodynamic instability of the rupture surface in a perfect liquid.

- The tangential discontinuity. It is required to determine the stability of the interface between two liquid layers of different densities sliding relative to each other (Helmholtz solution 1868). The Helmholtz solution is shown to be absolutely unstable.
- The combustion front. The source gas flows over the interface surface and burns in an adjacent infinitely thin layer, becoming the combustion product of lesser density due to energy release. Additional volume of gas generated by thermal expansion pushes the combustion products normal to each local point of the surface, causing its curvature. It is required to determine the possible recovery of a plane view of combustion front after the introduction of random perturbations in the gas. That the combustion front is shown to be absolutely unstable with respect to spatial distortions of any size.
- The evaporation front. The previous problem on the combustion front is considered, but combustion occurs at the surface of the evaporating liquid. It is required to find conditions for the stability with due account for the stabilizing effect of gravity and surface tension. The maximum value of the mass evaporation velocity at which the combustion mode is stable is obtained. A detailed description of Landau problems listed above and methods of their solution can be found, in particular, in [3–5].

Problem Statement

Following the analysis of Landau [2], we study the problem of hydrodynamic instability on the boundary between the "liquid–vapor" saturated semi-infinite arrays. Assume that as the mass flow of evaporation j passes through the initially planar interphase boundary (evaporation front) the latter begins to experience fluctuations with amplitude ξ_0.

We choose the coordinate system so that the unperturbed evaporation front is at rest, superimpose the plane surface with the coordinate plane (y, z) and direct the x—axis along the normal to it (to the area filled with vapor), $x = 0$ corresponds to the surface of the unperturbed front. In the laboratory coordinate system, the evaporation front propagates over the resting vapor. In the coordinate system associated with the interphase, consider a moving liquid that flows to the surface with velocity V_1 equal to the normal velocity of evaporation front propagation, the vapor flows out through the surface with velocity V_2. The system of Euler equations in both contacting phases is written as follows:

$$\left.\begin{aligned} \frac{\partial \mathrm{v}}{\partial x} + \frac{\partial u}{\partial y} &= 0, \\ \frac{\partial \mathrm{v}}{\partial t} + V\frac{\partial \mathrm{v}}{\partial x} &= -\frac{1}{\rho}\frac{\partial p}{\partial x}, \\ \frac{\partial u}{\partial t} + V\frac{\partial u}{\partial x} &= -\frac{1}{\rho}\frac{\partial p}{\partial y}. \end{aligned}\right\} \tag{B.1}$$

The pulsation flow results from random perturbations in contacting media near the evaporation front. Perturbations of the interphase ξ are set as a progressive wave propagating along the y coordinate along the surface

$$\xi = \xi_0 \exp(iky + \omega t). \tag{B.2}$$

We seek solutions for perturbations in the form

$$\mathrm{v} = f_{\mathrm{v}} \exp(iky + \omega t), u = f_u \exp(iky + \omega t), p = f_p \exp(iky + \omega t). \tag{B.3}$$

Here, t is the time, k is the wavenumber, ω is the increment of perturbations (a complex value in the general case), v, u are pulsations of normal (on x axis) and tangential (on y axis) velocity components, p are the pressure pulsations, and $f_{\mathrm{v}}(x)$, $f_u(x)$, $f_p(x)$ are the pulsation amplitudes. In the framework of the two-dimensional problem, expressions (B.3) describe the perturbations that are homogeneous relative to the transverse coordinate z.

The qualitative nature of the flow is determined by the sign of the real part of the increment of perturbations.

- At $\Re\mathrm{e}(\omega) > 0$ perturbations introduced in the system exponentially increase in time (the instability region).
- At $\Re\mathrm{e}(\omega) < 0$, fluctuations of parameters attenuate (the stability region).
- The case of $\Re\mathrm{e}(\omega) = 0$ describes the stability boundary: the frequency of pulsations becomes zero, on the rupture surface there is a range of periodic spatial inhomogeneities determined by the relevant wave numbers.

Systems of equations (B.1) considering (B.2) and (B.3) are solved in the standard way. Substituting values included in (B.1) as a sum of stationary and pulsation components, we average the so-obtained equations for the period of pulsations and linearize them. Then we subtract these equations from the original equations and as a result obtain the system of linear equations relative to pulsations. The solution of these equations for region 1 takes the form

$$\left.\begin{aligned} f_{1\mathrm{v}} &= A \exp(kx), \\ f_{1u} &= iA \exp(kx), \\ f_{1p} &= -A\rho_1\left(\tfrac{\omega}{k} + V_1\right) \exp(kx). \end{aligned}\right\} \tag{B.4}$$

According to Kelvin's theorem [6], the source potential flow in the liquid phase will continue until the interphase boundary. At crossing a curved evaporation front due to jumps of the parameters in it, vortex formation occurs in the flow. Potential component for a semiclosed array of vapor has the same form as in [2]

$$\left.\begin{aligned} f_{2v} &= B\exp(-kx), \\ f_{2u} &= -iB\exp(-kx), \\ f_{2p} &= B\rho_2\left(\tfrac{\omega}{k} - V_2\right)\exp(-kx). \end{aligned}\right\} \tag{B.5}$$

The vortex component of the flow for this area is described by the system of equations (B.1), assuming that $\partial p/\partial x = \partial p/\partial y = 0$,

$$\left.\begin{aligned} f_{2v} &= C\exp\left(-\tfrac{\omega x}{V_2}\right), \\ f_{2u} &= -\tfrac{iC\omega}{kV_2}\exp\left(-\tfrac{\omega x}{V_2}\right), \\ f_{2p} &= 0. \end{aligned}\right\} \tag{B.6}$$

Formulas (B.4)–(B.6) involve the unknown constants A, B, C.Using (B.6), we find the vorticity in area 2

$$\Omega \equiv \frac{\partial v}{\partial y} - \frac{\partial u}{\partial x} = \left[iC\left(k - \frac{\omega^2}{kV_2^2}\right)\right]\exp\left(iky + \omega t - \frac{\omega x}{V_2}\right).$$

It is easy to show that the derivative of vorticity in the vapor phase along the streamline is equal to zero

$$\frac{d\Omega}{dt} \equiv \frac{\partial \Omega}{\partial t} + V_2\frac{\partial \Omega}{\partial x} = 0.$$

Hence, we can conclude that in the Lagrangian coordinate system associated with a separated individual vapor particle, the vorticity that has arisen on the discontinuity surface remains along the trajectory of motion—$\Omega = \text{const}$.

Consistency Conditions

On the surface of discontinuity, the following consistency conditions must be met: the equality of tangential velocity components

$$u_1 + V_1\frac{\partial \xi}{\partial y} = u_2 + V_2\frac{\partial \xi}{\partial y}, \tag{B.7}$$

the equality of normal velocity components

$$v_1 - \frac{\partial \xi}{\partial t} = v_2 - \frac{\partial \xi}{\partial t} = 0, \tag{B.8}$$

the continuity of the normal component of the tensor of the momentum flow

$$p_1 - p_2 = -\sigma \frac{\partial^2 \xi}{\partial y^2} + (\rho_1 - \rho_2) g \xi. \tag{B.9}$$

We now briefly repeat the main provisions of the theory of Landau instability. In 1868, Helmholtz found a class of solutions for the Euler equations describing the mutual sliding of two liquid layers. In 1944, Landau showed that discontinuous solutions obtained by Helmholtz were absolutely unstable. To prevent tangential discontinuities in the analysis of the stability of the flame front, Landau formulated the additional BC (B.7). Since the theory of ideal fluid does not allow restrictions of tangential velocity, the resulting mathematical description is unclosed. For its closing, the authors of [2] obtained an additional vortex solution (B.6) for the area of combustion products. The physical reason of vortex formation is an abrupt increase of the normal velocity component on a curved rupture surface while maintaining its tangential component. The superposition of potential (B.5) and vortex (B.6) particular solutions leads to the general solution of the third problem of the Landau instability.

Analysis of Stability

Material balance implies the condition of conservation of the mass flow of matter in both phases

$$j = \rho_1 V_1 = \rho_2 V_2 = \text{const.}$$

Below we restrict ourselves to the practically important case of a strong inequality of phase densities

$$\rho_2/\rho_1 \ll 1. \tag{B.10}$$

It implies the condition: $V_1/V_2 \ll 1$. Substituting the expressions for the perturbation of parameters in the consistency conditions (B.7)–(B.9), we get a homogeneous system of four linear equations for four unknowns: ξ_0, A, B, and C. By solving this system, we find the characteristic equation for the increment of perturbations ω

$$a_2\omega^2 + a_1\omega + a_0 = 0. \tag{B.11}$$

The coefficients of the quadratic equation (B.11) may be written as

$$\left.\begin{aligned} a_2 &= V_2, \\ a_1 &= 2kV_1V_2, \\ a_0 &= \left[-k^2V_2 + \tfrac{gk\rho_1+\sigma k^3}{j}\right]V_1V_2. \end{aligned}\right\} \tag{B.12}$$

We change the amplitude of perturbations in time at each point of the interface occurs according to the law $\sim \exp(\Re e(\omega)t)$. Therefore, the stability of the evaporation front is determined by the sign of the free term a_0 of the characteristic equation (B.11). The following three situations are possible:

- $a_0 < 0 : \Re e(\omega_1)\langle 0, \Re e(\omega_2)\rangle 0$ (the instability region). The first solution exponentially attenuates and the second one exponentially grows.
- $a_0 = 0 : \Re e(\omega_1) < 0, \Re e(\omega_2) = 0$ (the boundary of stability). The first solution exponentially attenuates, and the second one neither attenuates nor grows.
- $a_0 > 0 : \Re e(\omega_1) < 0, \Re e(\omega_2) < 0$ (stability region), the first and second solutions exponentially attenuate.

Substituting $a_0 = 0$ in the last equation (B.12), we obtain the expression for the stability boundary

$$j^2 = \frac{\rho_2\left(\sigma k^2 + \rho_1 g\right)}{k}, \tag{B.13}$$

where the minimum values are as follows: $k_{\mathrm{min}} = (\rho_1 g/\sigma)^{1/2}$. Substituting this value into (B.13), we get the maximum mass flow through the interphase boundary, for which the "liquid–vapor" system remains on the stability boundary

$$j\mathrm{max} = \sqrt{2}\rho_2^{1/2}(\rho_1\sigma g)^{1/4}. \tag{B.14}$$

The heat flux supplied to the system is related to the mass flow via the heat balance condition

$$q = h_{fg}j, \tag{B.15}$$

where h_{fg} is heat of phase transition. From (B.14), (B.15) we get the following expression for the maximum heat flux:

$$q\mathrm{max} = \sqrt{2}h_{fg}\rho_2^{1/2}(\rho_1\sigma g)^{1/4}. \tag{B.16}$$

Critical Heat Flux

Kutateladze–Zuber Model

Cooling of technological equipment using a boiling liquid should ensure trouble-free operation of the equipment, and thus it assumes the absence of heat transfer crisis. The boiling crisis phenomenon is probably the most important problem of boiling heat transfer especially in view of practical applications. Therefore, the calculation of the Critical Heat Flux (CHF) q_{cr} is necessary to ensure boiling safety in a controlled heat flux pool. A huge number of works have been published on this topic, so we will limit ourselves to a reference to the well-known book [7]. The "hydrodynamical model" of boiling crisis was initially developed in the seminal Kutateladze's paper [8] on the basis of dimensional analysis. Kutateladze obtained the following dependence for CHF for boiling of saturated liquid[1]:

$$q_c = k\sqrt{2}h_{fg}\rho_2^{1/2}(\rho_1\sigma g)^{1/4}. \tag{B.17}$$

Subsequently, Zuber [9] established the "hydrodynamic instability model" developing the Kuateladze approach. He applied instability theories to the CHF phenomena: the Rayleigh–Taylor instability and the critical Kelvin–Helmholtz wavelength. Where constant k according to Zuber,s theory is found to be in the range from 0.119 to 0.157. The average value of k recommended by Kutateladze on the basis of experimental data is 0.14.

It is interesting to note that Kutateladze's formula (B.17) is identical to (B.16), but differs from it by a numerical coefficient (which is precisely one order smaller). This allows us to hypothesis that there is an internal relationship between two instability processes: evaporation from the surface of the liquid mass and bubble boiling on the heated wall. As for such a large difference in the values of the numerical multiplier before the formula, we can draw an analogy with another type of instability.

As is known, if there are tangents relative to the discontinuity surface, the latter may experience a jump called the "tangential discontinuity". This results in the Kelvin–Helmholtz instability [2] of inertial nature. Numerous experiments showed that the theoretical value of the velocity jump leading to the development of the Kelvin–Helmholtz instability was overestimated by approximately five times. This quantitative discrepancy can be explained by the lack of consideration of the viscosity of the liquid and the turbulent pulsations in the gas phase. At present, it is assumed that the theoretical critical velocity from the Kelvin–Helmholtz theory corresponds to the transition of the instability to the nonlinear stage development, rather than to its emergence.

[1]Formula (B17) is derived in view of condition (B10).

Boiling in a Tube with Twisted Tape

The Kutateladze–Zuber model [3, 4] was successfully applied for the generalization of a large bank of experimental data accumulated to date for a wide variety of liquids. On this basis, we can assume that formula (B.17) should describe CHF not only under normal gravity conditions, but also for the general case of mass forces of different physical nature.

One of the methods for heat exchange intensification for boiling of a subcooled liquid under conditions of a turbulent flow in a channel is the use of twisted-tape swirl promoters [10]. A unique experiment in this area was conducted by Gambill et al. [11], in which burnout heat fluxes at swirl flow boiling of water were investigated and the advantages of a swirl flow for CHF increase was showed. Tubes of copper, aluminum, and nickel with twisted- tape inserts were tested. Tube diameters ranged from 3.45 to 10.2 mm IC. The inlet axial velocities ranged from 4.57 to 47.55 m/s. The test sections were heated with AC electrical power. The experiments of [11] demonstrated the independence of the swirl flow burnout heat flux on either the pressure level or the degree of liquid subcooling.

This fact suggests an idea that at high centrifugal accelerations of the body forces (g_*) entirely determine the hydrodynamical conditions near the heated wall under boiling. This assumption means that the boiling crisis can be calculated on the basis of the correlations for pool boiling CHF if g_* is used instead of the standard gravitational one g. Gambill et al have presented 39 experimental values of CHF for water swirl flow at centrifugal acceleration from 100 g to about 57,000 g.

When calculating the hydrodynamic characteristics of a swirl flow, a simplified model for adding two independent flows has given good results: the rectilinear flow in a circular pipe with velocity u and the rigid-body rotation coming from the twisted tape [12]. Under this approach, the tangential velocity is calculated as

$$w = 2\pi \frac{ru}{H}. \tag{B.18}$$

Here H is the pitch of twisted tape (the length of full rotation of the tape), r is the radial coordinate. The centrifugal acceleration is given by

$$g_* = \frac{w^2}{r}. \tag{B.19}$$

The process of boiling of a subcooled liquid is always realized in a thin near-wall layer superheated with respect to the saturation temperature [12]. Hence when evaluating the CHF one should plug the centrifugal acceleration with $r = r_0 = d/2$ into (B.17). Using (B.18), (B.19), we get

$$g_{*w} \equiv g_*|_{r=r_0}. \tag{B.20}$$

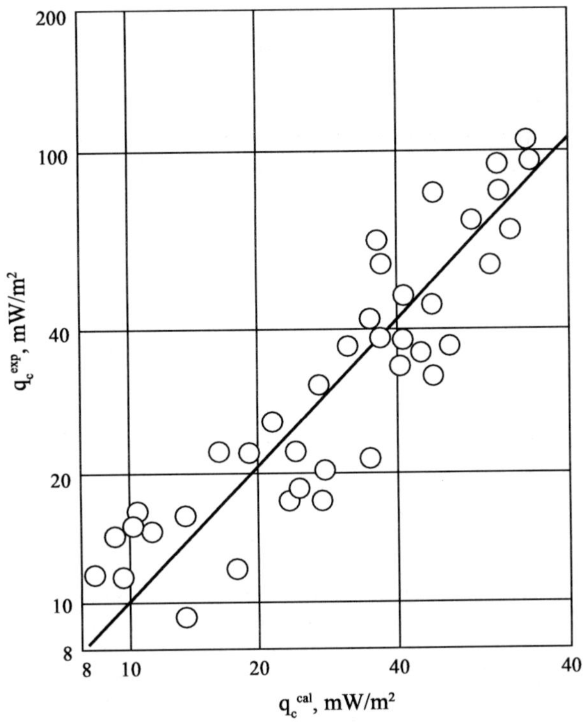

Fig. B.1 Results of evaluation of CHF (————) at swirl flow boiling of water versus the experimental results of [11] (○). q_c^{calc}, the theoretical CHF, q_c^{exp}, the experimental CHF

Substituting g_{*w} into (B.17) in place of the gravitational acceleration, we get the required formula for evaluation of the CHF for a boiling liquid in a pipe with twisted tape. The comparison of these CHF results with calculated ones (Fig. B.1) demonstrated quite satisfactory correspondence with the experimental data [11]. The difference between the calculated $\left(q_c^{calc}\right)$ and experimental $\left(q_c^{exp}\right)$ results are at most 30%, which is quite satisfactory for problems on boiling systems.

The above results serve as a testimony to the universality of the "hydrodynamical model" of boiling crisis, which describes the CHF in the field of body forces independently of their physical nature. These results were first published in [13]. It should be noted that the semi-empirical method [12], which is based on the summation of two main components of fluid motion in a twisted-tape swirl promoter—the axial flow and the solid-state rotation—has proved itself well in the calculation of g_c.

Conclusions

Analysis of the physical relationship between the hydrodynamic model of pool boiling crisis and the development of the instability during evaporation from the interfacial surface (the Landau instability) is given. It is shown that in both cases the resulting dependences for the limit heat flow are identical in both cases and differ only by a numerical coefficient. An analogy with the development of Kelvin–Helmholtz instability is given. Calculations of the critical heat flow at liquid boiling in a pipe under conditions of a turbulent flow in a channel are performed. It is found that the use of the centrifugal acceleration in Kutateladze's formula provides a satisfactory agreement with the experimental data.

References

1. Labuntsov DA (2000) Physical foundations of power engineering. Selected works, Moscow Power Energetic Univ. (Publ.), Moscow (in Russian)
2. Landau LD, Lifshitz EM (1987) Fluid mechanics, 2nd edition. Pergamon Press
3. Zeldovich YB, Barenblatt GI, Librovich VB, Makhviladze GM (1985) The mathematical theory of combustion and explosions. Plenum Publishing Co., New York
4. Matalon M (2007) Intrinsic flame instabilities in premixed and nonpremixed combustion. Ann Rev Fluid Mech 39:163–191
5. Bychkov V, Petchenko A, Akkerman V (2007) On the theory of turbulent flame velocity. Combust Sci Technol 179:137–151
6. Schobeiri MT (2014) Fluid mechanics for engineers. Springer, Berlin
7. Carey VP (2020) Liquid-vapor phase-change phenomena: an introduction to the thermophysics of vaporization and condensation processes in heat transfer equipment, third edn. CRC Press
8. Kutateladze SS (1950) Hydromechanical model of the crisis of boiling under conditions of free convection. J Tech Phys USSR 20(11):1389–1392
9. Zuber N (1959) Hydrodynamic aspects of boiling heat transfer. AEC Report, AECU–4459, Los Angeles
10. Liang G, Mudawar I (2020) Review of channel flow boiling enhancement by surface modification, and instability suppression schemes Int J Heat Mass Transf 146:118864
11. Gambill WR, Bundy RD, Wansbrough RW (1961) Heat transfer, burnout and pressure drop for water in swirl flow through tubes with internal twisted tapes. Chem Eng Progr Symp Ser 57(32):127–137
12. Hasanpour A, Farhadi M, Sedighi K (2014) A review study on twisted tape inserts on turbulent flow heat exchangers: the overall enhancement ratio criteria. Int Comm Heat Mass Transf 55:53–62
13. Yagov VV, Zudin YB (1994) Mechanistic model for nucleate boiling crisis at different gravity fields. In: Proceedings of 10th international heat transfer conference vol 5. Pool boiling, Brighton, 14–18 August 1994, pp 189–194